Lecture Notes in Mechanical Engineering

Lecture Notes in Mechanical Engineering (LNME) publishes the latest developments in Mechanical Engineering—quickly, informally and with high quality. Original research or contributions reported in proceedings and post-proceedings represents the core of LNME. Volumes published in LNME embrace all aspects, subfields and new challenges of mechanical engineering.

To submit a proposal or request further information, please contact the Springer Editor of your location:

Europe, USA, Africa: Leontina Di Cecco at Leontina.dicecco@springer.com
China: Ella Zhang at ella.zhang@cn.springernature.com
India, Rest of Asia, Australia, New Zealand: Swati Meherishi at swati.meherishi@springer.com

Topics in the series include:

- Engineering Design
- Machinery and Machine Elements
- Mechanical Structures and Stress Analysis
- Automotive Engineering
- Engine Technology
- Aerospace Technology and Astronautics
- Nanotechnology and Microengineering
- Control, Robotics, Mechatronics
- MEMS
- Theoretical and Applied Mechanics
- Dynamical Systems, Control
- Fluid Mechanics
- Engineering Thermodynamics, Heat and Mass Transfer
- Manufacturing Engineering and Smart Manufacturing
- Precision Engineering, Instrumentation, Measurement
- Materials Engineering
- Tribology and Surface Technology

Indexed by SCOPUS, EI Compendex, and INSPEC.

All books published in the series are evaluated by Web of Science for the Conference Proceedings Citation Index (CPCI).

To submit a proposal for a monograph, please check our Springer Tracts in Mechanical Engineering at https://link.springer.com/bookseries/11693.

Numan M. Durakbasa · Kemal Güven Gülen
Editors

Sustainable Green Conversion

Selected Papers from ISPR2024, October 10–12, 2024 Budva-Montenegro, Volume 1

Springer

Editors
Numan M. Durakbasa
Institute of Production Engineering
and Photonic Technologies
TU Wien
Vienna, Austria

Kemal Güven Gülen
Department of Industrial Engineering
Namık Kemal University
Tekirdağ, Türkiye

ISSN 2195-4356 ISSN 2195-4364 (electronic)
Lecture Notes in Mechanical Engineering
ISBN 978-3-031-83582-7 ISBN 978-3-031-83583-4 (eBook)
https://doi.org/10.1007/978-3-031-83583-4

This Springer imprint is published by the registered company Springer Nature Switzerland AG
The registered company address is: Gewerbestrasse 11, 6330 Cham, Switzerland

This book is dedicated to the memory
of Prof. Dr. Güneş Gençyılmaz,
with respect and gratitude…

Organization

ISPR2024 was organized by TU Wien, Austria, and the Society for Production Research, Istanbul, Türkiye. The symposium took place in Budva, Montenegro.

Editors

Numan M. Durakbasa
Kemal Güven Gülen

Co-editors

Selim Zaim
Jorge Martin Bauer
Serpil Erol
Semra Birgün
Alptekin Erkollar
Jan Torgersen
Nükhet Tunçbilek
Ismail Böğrekçi
Pınar Demircioğlu

Honorary Chairs

Kurt Matyas, TU Wien, Austria
M. Güneş Gençyılmaz, Society for Production Research, Istanbul, Türkiye

Symposium Chairs

Numan M. Durakbasa, TU Wien, Austria
Kemal Güven Gülen Namık Kemal University, Tekirdağ, Türkiye

International Honorary Committee

Prof. Dr. Albert Weckenmann, Friedrich Alexander Universität Erlangen, Germany
Prof. Dr. Alice E. Smith, Auburn University, USA
Prof. Dr. Jerzy A. Sładek, Faculty of Mechanical Engineering, Cracow University of Technology, Poland
Prof. Dr. Andrew Kusiak, The University of Iowa, USA
Prof. Dr. Daniela Popescu, Technical University of Cluj Napoca, Romania
Prof. Dr. Detlef Gerhard, Ruhr-University Bochum, Germany
Prof. Dr. Duc Truong Pham, The University of Birmingham, UK
Prof. Dr. Friedrich Bleicher, TU Wien, Austria
Prof. Dr. Jozef Peterka, Slovak University of Technology, Slovakia
Prof. Dr. Lubomir Šooš, Slovak University of Technology Slovakia
Prof. Dr. Sorin Popescu, Technical University of Cluj-Napoca, Romania
Prof. Dr. Stanislaw Adamczak, Kielce University of Technology, Poland
Prof. Dr. Vidosav Majstorovic, University of Belgrade, Serbia

Organizing Committee

Selim Zaim, Türkiye
Erol Güçlü, Austria
Semra Birgün, Türkiye
Serpil Erol, Türkiye
Berna Dengiz, Türkiye
Kemal Güven Gülen, Türkiye
Gül E. Kremer, USA
Alptekin Erkollar, Türkiye
Eva Maria Walcher, Austria
Ferhan Çebi, Türkiye
Dursun Delen, USA
Günther Poszvek, Austria
Şakir Esnaf, Türkiye
Şükrü Alp Baray, Türkiye
Lukas Kräuter, Austria,
Nükhet Tunçbilek, Canada

Tufan Tunçbilek, Canada
David Riepl, Austria
Hatice Camgöz Akdağ, Türkiye
Haluk Soyuer, Türkiye
Hür Bersam Sidal, Türkiye
Ece Soyuer, Austria
Jorge Bauer, Austria
Tuğçe Apaydin, Türkiye
Aziz Kemal Konyalioğlu, The UK
Lianc Höllcr, Austria
Gökçen Baş, Austria
Gamze Uğur Tuncer, Austria
Osman Bodur, Austria

This Book is Prepared for Publishing By

Tuğçe Apaydin, Istanbul-Türkiye
Aziz Kemal Konyalioğlu, Glasgow-The UK

Scientific Committee

S. Adamczak, Kielce University of Technology, Poland
H. Camgöz Akdağ, Istanbul Technical University, Türkiye
A. Akdoğan, Yıldız Technical University, Türkiye
K. Altinel, Bosphorus University, Türkiye
İ. Ar, Yıldırım Beyazıt University, Türkiye
B. Baki, Karadeniz Technical University, Türkiye
A. Baray, Istanbul University, Türkiye
E. Balaban, Istanbul University, Türkiye
G. Baş, CypherX, Cyprus
N. Başoğlu, İzmir Institute of Technology, Türkiye
J. Bauer, National Technological University Buenos Aires, Argentina
A. Baykasoğlu, Dokuz Eylül University, Türkiye
E. Bayraktar, The American University of the Middle East, Kuwait
F. Berto, Norwegian University of Science and Technology, Norway
S. Birgün, Doğuş University, Türkiye
P. Blecha, Brno University of Technology, Czechia
F. Bleicher, TU Wien, Austria
F. Bozbura, Bahçeşehir University, Türkiye
İ. Böğrekci, Adnan Menderes University, Türkiye
I. Budak, Faculty of Technical Sciences in Novi Sad, Serbia

A. Kusiak, University of Iowa, USA
V. Majstorovic, University of Belgrade, Serbia
I. Mankova, Technical University of Kosice, Slovakia
A. Markopoulos, National Technical University of Athens, Greece
K. Matyas, Vienna University of Technology, Austria
M. Novák, J. E. Purkyně University in Ústí nad Labem, Czech Republic
B. Oberer, ETCOP Institute for Interdisciplinary Research, Austria
Y. Omurtag, Robert Morris University, USA
N. Özçakar, Istanbul University, Türkiye
D. Özdemir, Bilgi Univcrsity, Türkiye
A. Özok, Okan University, Türkiye
E. Öztemel, Marmara University, Türkiye
J. Peterka, Slovak University of Technology, Slovakia
D. Pham, University of Birmingham, UK
M. Pokusová, Slovak University of Technology in Bratislava, Slovakia
S. Popescu, Technical University of Cluj Napoca, Romania
D. Popescu, Technical University of Cluj Napoca, Romania
E. Pucher, TU Wien, Austria
K. Rezaie, University of Tehran, Iran
A. Sağbaş, Namık Kemal University, Türkiye
B. Sağbaş, Yıldız Technical University, Türkiye
J. Sanz Juan, Polytechnic University of Valencia, Spain
A. Sharif, University of Bradford, UK
H. B. Sidal, Istanbul Technical University, Türkiye
A. Smith, University of Auburn, USA
Ľ. Šooš, Slovakia University of Technology in Bratislava, Slovakia
A. Sorguç, Middle East Technical University, Türkiye
H. Soyuer, Ege University, Türkiye
K. Stepien, Kielce University of Technology, Poland
P. Tamas, Budapest University of Technology and Economics
B. Tan, Koç University, Türkiye
M. Tanyaş, Maltepe University, Türkiye
H. Taşkin, Sakarya University, Türkiye
J. Torgersen, Technical University of Munich, Germany
O. Torkul, Sakarya University, Türkiye
B. Tunaboylu, Marmara University, Türkiye
G. Ulusoy, Sabancı University, Türkiye
G. Varga, University of Miskolc, Hungary
K. Velíšek, Slovak University of Technology, Slovakia
J. Wang, National University of Singapore
A. Weckenmann, University of Erlangen Nürnberg, Germany
M. Yalçintaş, Istanbul Commerce University, Türkiye
M. Yenisey, Istanbul University, Türkiye
M. E. Yurci, Yıldız Technical University, Türkiye

H. Yüksel, Dokuz Eylül University, Türkiye
S. Zaim, Ibn Haldun University, Türkiye
W. Zebala, Cracow University of Technology, Poland
M. Zerenler, Selçuk University, Türkiye

Reviewers

A. Akdogan
Ş. Baray
G. Bas
J. Bauer
S. Birgün
I. Böğrekçi
I. Budak
A. Bulutsuz
H. Camgöz Akdağ
L. Crisan
F. Çebi
P. Demircioglu
B. Dengiz
M. Dragomir
A. Dregelyi-Kiss
N. Durakbasa
A. Erkollar
S. Erol
Ş. Esnaf
Z. Gergin
S. Grozav
K. Gülen
I. Islamov
G. Kremer
I. Mankova
A. Öner
M. Pokusova
K. Rezaie
B. Sagbas
A. Sorguc
H. Soyuer

K. Stepien
J. Torgersen
B. Tunaboylu
G. Varga
S. Zaim

Foreword

Dear Colleagues and Friends,

It is a great pleasure to welcome you to the "**International Symposium for Production Research—ISPR 2024**", **Budva/Montenegro**, from 10 to 12 October 2024, with the overall theme of "**Sustainable Green Conversion**". As in all previous years, we continue to cherish and benefit from our successful collaboration with the Society for Production Research, Istanbul, Turkey, in the organization of this event.

The aim of the symposium is to bring together researchers, scientists, and leading experts from universities, companies, institutions, communities, associations, and societies from around the world to share ideas, experiences, research results, and visions on the topic of production and operations management and technology. This organization will also provide a forum for experts and professionals to discuss the challenges and opportunities of this year's symposium and drive innovation.

We hope that this year's gathering, once again hybrid, will provide an opportunity to share ideas and discuss topics in person and virtually with attendees and speakers.

We wish you a successful and productive symposium.

October 2024

Prof. Dr. Kurt Matyas
TU Wien
Vienna, Austria

Preface

This book consists of the selected papers presented at the 24th International Symposium for Production Research (ISPR2024) held from 10 to 12 October 2024 in hybrid form, face-to-face and virtually with the participants and speakers who came to Budva/Montenegro on the first day, and virtually in the presentation sessions on the subsequent days, covering the overall theme of "Sustainable Green Conversion".

This symposium was organized by the Society for Production Research, İstanbul, Türkiye, and TU Wien, Austria, for the seventh year in a row.

The generic theme of "Industry 4.0" was first adopted in the symposium held in 2016 and maintained in the following seven symposiums in 2017, 2018, 2019, 2020, 2021, 2022, and 2023 but each time with an updated emphasis on the relevant developments and progress on the various aspects of this theme.

In the Industry 4.0 era, Industrial Engineering plays a central role as a fundamental discipline with a broad impact. It is now widely accepted that Industry 5.0 complements the current Industry 4.0 paradigm by shifting research and innovation towards a sustainable, human-centered, and resilient industry. It leverages the power of advanced technologies, data analytics, and automation to orchestrate a symphony of operational excellence. Predictive maintenance systems are meticulously built to minimize downtime, supply chains are optimized with data-driven insights, and product quality is improved with automated quality control processes. It was inevitable that these developments in industry would affect education, especially higher education. Indeed, it has been observed that with the spread of Industry 4.0, the curricula of some engineering fields have changed and some courses have been added to the curricula to include the fundamentals of Industry 4.0 and 5.0. For this reason, the general theme of last year's symposium was determined as "Industrial Engineering in the Industry 4.0 Era".

This year, we gather under the motto "Sustainable Green Conversion", addressing the essential transition towards environmentally responsible and energy-efficient production. In the concept of Industry 5.0, new technologies will be incorporated in order to achieve environmental targets. It is necessary to achieve the production processes with regard to the environmental impact. As global industries navigate the shift to greener processes, our focus is on pioneering research, advanced technologies, and scalable strategies that support resource efficiency, reduced emissions, and circular economy practices.

Given the importance of this theme, ISPR2024 hosted numerous distinguished speakers from both the academia and the industry to share their views on the applications of Industry 4.0 and 5.0 on the various components of production systems and the influence of Industry 4.0 and 5.0 on Sustainable Green Conversion.

A total of over 223 participants attended this year's symposium—academics, practitioners, and scientists from 17 countries, who contributed 13 invited talks and 73 papers on the plenary, special, workshop, and ordinary sessions. The symposium program included keynote addresses (opening/closing session), breakout sessions, and workshop discussions.

This book contains 73 refereed selected papers in 12 categories shown in the contents of the book.

We would like to express our gratitude to Prof. Kurt Matyas, Honorary Chairman of the Scientific Committee of this symposium, for his leadership and generous support. Our thanks also go to Prof. Friedrich Bleicher, Head of the Institute for Production Engineering and Photonic Technologies for his interest and support for this symposium.

We would like to thank all the keynote and invited speakers whose contributions enhanced the success of the symposium. In organizing this event, our colleagues in Vienna and Istanbul contributed endless hours of hard work, energy, and wisdom to make this event a success. On the Vienna side, our sincere thanks go to the staff of the Research Group Production Metrology and Adaptronic Systems of the Institute for Production Engineering and Photonic Technologies, and Hasan Teoman and Dipl. Ing. Erol Güçlü for his invaluable leadership. On the Istanbul side, we are grateful to Prof. Dr. Hatice Camgöz Akdağ from Istanbul Technical University and Asst. Prof. Dr. Zeynep Gergin from Istanbul Kültür University, for their invaluable efforts in presenting the keynote speakers and managing the first day activities with great competence. Our special thanks go to Asst. Prof. Dr. Tuğçe Apaydın at Istanbul Kültür University and Dr. Aziz Kemal Konyalıoğlu at Istanbul Technical University and University of Strathclyde, Glasgow who prepared this book with great care and Comp. Eng. Tufan Tunçbilek, Amazon Company Canada who managed the virtual sessions, and Biset Toprak, Mehmet Güdelek from Istanbul Sabahattin Zaim University, Paşa Çiçeklidağ University of Western Australia, and Muhammet Ali Erdem who spent great efforts in managing virtual rooms under the guiding of Mr. Tunçbilek. We also thank Dr. Gökçen Baş and to research assistants at TU Wien, Dr. Günther Poszvek, Dr. Osman Bodur, and Dipl. Ing. Eva Walcher for their hard and dedicated work and finally to Prof. Dr. Gül E. Kremer Dean of Engineering Faculty, University of Dayton USA, Prof. Dr. Berna Dengiz, Industrial Engineering Department Başkent University Türkiye, Prof. Dr. Bahadır Tunaboylu, Faculty of Engineering Marmara University, Prof. Dr. Hatice Camgöz Akdağ, Faculty of Management Istanbul Technical University Türkiye, Prof. Dr. Pınar Demircioğlu, Aydın Adnan Menderes University, Prof. Dr. İsmail Böğrekci, Aydın Adnan Menderes University, Prof. Dr. Anil Akdogan, Assoc. Prof. Dr. Binnur Sagbas and Assoc. Prof. Dr. Asli Günay Bulutsuz,

Yıldız Technical University, and Dr. Ömer Şahin Ganiyusufoğlu, QIAP China for their valuable role and efforts in the Organizing Committee.

We would like to express our gratitude to the board members of the Society for Production Research in Istanbul, Türkiye, and in other countries for their strong support and involvement in the successful organization of the symposium, Prof. Dr. Selim Zaim, Prof. Dr. Semra Birgün, Prof. Dr. Dursun Delen (USA), Prof. Serpil Erol, Prof. Dr. Alptekin Erkollar (Austria), and Asst. Prof. Dr. Nükhet Tunçbilek (Canada).

Our very special thanks go to our colleagues and the participants of this symposium. Undoubtedly, they were the core component of this organization.

We would like to recognize and thank our dear colleagues who graciously accepted to join the honorary and scientific committees or who served as peers in this event.

Finally, no such event is possible without the generous support of patrons and sponsors. In this regard, we would like to thank Dr. Michael Ludwig, the Mayor of Vienna, for his continued generous support of this symposium series and to all the corporations and individuals who provided invaluable financial and intellectual contributions.

Vienna, Austria
Tekirdağ, Türkiye

Prof. Dr. Numan M. Durakbasa
Prof. Dr. Kemal Güven Gülen

Sustainable Green Conversion

Sustainable Green Conversion focuses on issues and recent developments in sustainability in production economies through processes that minimize negative environmental impacts. These are based on a range of new technologies and management approaches relevant to the industry, such as sophisticated methods for the development of smart, autonomous, self-organizing, durable, sustainable, and high-quality production systems, and involve the employment of innovative workers with high qualifications and creativity to increase productivity. The production of the future will be characterized by a fully integrated and highly flexible production processes with integrated sophisticated systems in line with international standards. In the context of Industry 4.0 and Industry 5.0, this concept aligns the Sustainable Green Conversion of the environmental needs of industries with industrial demands, focusing on reducing energy consumption, minimizing resource use, diminishing waste, lowering carbon emissions, and increasing production efficiency. This approach not only involves adopting eco-friendly methods but also aims to create a sustainability-driven approach across all stages, such as design, production, consumption, and End of Life (EoL).

In mass production industries like automotive, a focus on circular economy principles helps reduce waste and promotes the reuse of materials. For instance, many automotive companies are finding innovative ways to recycle old batteries, reducing the need for new materials and lowering the energy used in producing new ones. This approach not only conserves resources but also contributes to the creation of new jobs and industries.

Energy-efficient smart factories provide another example of the critical role digitalization plays in sustainable Green Conversion. In modern industries, factories equipped with IoT (Internet of Things) sensors and data analytics continuously monitor energy use, helping businesses save energy and reduce emissions in real-time. This proactive approach benefits the environment, enhances efficiency, and reduces costs for businesses.

Sustainable green conversion supports eco-friendly production which avoids harming the natural environment, produces no waste, and uses the existing waste as raw material. This model aims for efficient use of natural resources without depletion and prevents the formation of environmentally harmful components like greenhouse gases. Clean production based on production processes aims to reduce waste and pollution during production. Clean production based on products minimizes the environmental impact of products throughout their entire lifecycle. Clean production based on services reduces the environmental impact of services provided throughout a product's lifespan. This holistic approach promotes sustainable and environmentally conscious production.

Sustainable green conversion spans a wide range of areas, including renewable energy sources, energy storage and efficiency, waste management, clean water and air technologies, agricultural and food technologies, and transportation technologies. This transformation is applied across various sectors to protect the environment and ensure sustainable resource use. Sustainable green conversion aims to minimize environmental issues, promote the efficient use of natural resources, and enhance quality of life. Focusing on areas such as energy efficiency and clean energy production not only protects the environment, but also provides economic benefits and prevents resource depletion. Ultimately, investments in sustainable green conversion and the adoption of eco-friendly practices play a crucial role in building a sustainable future and contribute to the goal of leaving a clean world for future generations.

Prof. Dr. Numan M. Durakbasa
Prof. Dr. Kemal Güven Gülen

Contents

Invited Articles

A Framework for Aggregating Timely and Potential Risks to Create Robust Supply Chains

Elif Elçin Günay, Abdulkadir Günay, Wei-Chih Chern, Ahmad E. Elhabashy, Jaemun Sim, Karl R. Haapala, Kyoung-Yun Kim, and Gül E. Okudan Kremer

Abstract Timely evaluation of risk factors is essential for building resilient supply chains and requires continuous monitoring and updating. This paper proposes a conceptual model for supply chain risk management that uses large language models (LLMs) to detect a disruption trigger event from news data, determines its impact, and develops a proper risk mitigation strategy. Even though the conceptual model can accommodate many kinds of supply chain risks, this study focuses on natural hazard risks affecting global supply chains. Because natural hazards are related to

E. E. Günay (✉) · A. Günay · W.-C. Chern · G. E. O. Kremer
School of Engineering, University of Dayton, Dayton, USA
e-mail: egunay1@udayton.edu

A. Günay
e-mail: agunay1@udayton.edu

W.-C. Chern
e-mail: chernw1@udayton.edu

G. E. O. Kremer
e-mail: gkremer2@udayton.edu

E. E. Günay
Industrial Engineering Department, Sakarya University, Sakarya, Turkey

A. E. Elhabashy · K. R. Haapala
School of Mechanical, Industrial, and Manufacturing Engineering, Oregon State University, Corvallis, USA
e-mail: elhabasa@oregonstate.edu

K. R. Haapala
e-mail: karl.haapala@oregonstate.edu

A. E. Elhabashy
Production Engineering Department, Alexandria University, Alexandria, Egypt

J. Sim · K.-Y. Kim
Department of Industrial and Systems Engineering, Wayne State University, Detroit, USA
e-mail: jaemun.sim@wayne.edu

K.-Y. Kim
e-mail: kykim@wayne.edu

N. M. Durakbasa and K. G. Gülen (eds.), *Sustainable Green Conversion*, Lecture Notes in Mechanical Engineering, https://doi.org/10.1007/978-3-031-83583-4_1

the locations of the suppliers, the model uses country level natural hazard risk scores to evaluate suppliers. Here, a country's risk score determined by the Index for Risk Management (INFORM) is used as a baseline risk score for each country, which are then, updated by the disruption trigger event's features captured by the LLM. Lastly, a supply chain optimization model is used to determine alternative suppliers to mitigate the impact of risk on disruption. The conceptual model is demonstrated through the case study of a bicycle company. Preliminary results indicate that the proposed model can be used to improve the resilience of supply chains against natural hazards.

Keywords Supply chain risk management · Large language models (LLMs) · Natural hazard risk · Supply chain optimization · Supply chain resilience

1 Introduction

To reduce costs and increase manufacturing flexibility, companies have increasingly outsourced a significant portion of their manufacturing needs from global suppliers. While global outsourcing may offer a low-cost advantage, it brings substantial risks for manufacturing companies [1]. For instance, the COVID-19 pandemic led to significant disruptions in global supply chains because of border closures, facility lockdowns, and labor shortages [2]. Similarly, the Suez Canal incident in 2021 halted marine transportation and severely hampered the flow of goods [3]. The ongoing Russia-Ukraine war, which started in 2022, resulted in a shortage of natural gas, oil, and crucial metals, causing a notable increase in prices and production delays across multiple industries [4]. These brief examples underscore the importance of effective supply chain risk management to increase manufacturing resilience [5]. Effective supply chain risk management includes risk identification, assessment, and mitigation planning [6]. Risk in a supply chain context can be defined as the likelihood of occurrence of an adverse event that significantly jeopardizes the flow of material and business operations and leads to a supply chain disruption. Supply chain disruption triggers can be categorized into natural (earthquakes, tropical cyclones, etc.) or man-made (war, accidents, etc.) triggers. Here, the trigger refers to the reason for risk to occur [7].

This study focuses on the identification and management of disruption triggers and proposes a conceptual model for supply chain risk management that reduces the impact of disruptions. The model uses different risk assessment procedures depending on the occurrence of a trigger event. Before the disruption trigger (pre-trigger phase), all suppliers are assigned a baseline risk score for any potential disruption. This baseline risk could be thought of as a static component, as it remains relatively constant and offers a foundational reference point for decision-making. In contrast, after the triggering event (post-trigger phase), the risk scores are adjusted in response to the disruption. For instance, the annual probability of a natural hazard, such as an earthquake or hurricane, occurring in a particular region can serve as the baseline risk score in the pre-trigger phase risk assessment. However, once such a natural hazard occurs, the situation shifts from predictive to reactive. At this point, another risk assessment procedure must be applied to address the immediate impact of the event and adjust the risk evaluation accordingly.

To capture the disruption triggers promptly, continuous monitoring is needed. Timely sources, such as news articles, are essential to detect disruptive triggers and strengthen supply chain resilience. The advancement in large language models (LLMs) makes the analysis of timely resources feasible, impactful, and easier compared to applying conventional machine learning approaches. For instance, LLM applications, such as Bidirectional Encoder Representations from Transformers (BERT) and Generative Pre-training Transformer (GPT), have shown noteworthy performance in understanding and summarizing huge amounts of text data [8]. LLMs exhibit great potential to replace the traditional supervised machine-learning algorithms for text analysis. LLMs are powerful, especially GPT-4, in assisting humans in performing complex knowledge-intensive tasks [9]. Moreover, most of LLM applications provide open application program interfaces (APIs), and thus anyone with an internet connection can access them immediately without any need for larger infrastructures [10]. These capabilities enable LLMs to assist supply chain managers in obtaining the most recent information needed for risk assessment by analyzing news data to capture critical details about supply chain disruption triggers. For instance, Zhao et al. [11] employed GPT-3.5 to identify and classify supply chain risks from news data to aid supply chain managers. Similarly, Shahsavari et al. [12] used GPT 3.5 for proactive risk management to predict the risk event considering its dependency on other triggering events.

This study introduces a framework to implement the conceptual model for supply chain risk management by using LLMs. First, news data is analyzed to capture the triggering event and extract its features. Next, if a triggering event is not detected, the baseline risk scores of the suppliers for the disruption are used in the risk assessment. However, if a triggering event is detected, the features extracted by LLMs are used to assess the severity and impact of the event and update suppliers' risk scores. Finally, as a mitigation approach, specific suppliers are selected to minimize the overall supply chain risk based on their current risk scores. The organization of the paper is as follows. Section 2 presents the relevant studies from the literature. Section 3 introduces the proposed framework. Section 4 demonstrates the application of the proposed framework through a case study. Section 5 concludes the paper while discussing the limitations and possible future research avenues.

2 Literature Review

Xu et al. [13] discussed how LLMs can be used in supply chain management. Accordingly, one of the potential applications for LLMs is demand forecasting and inventory management. Considering their capability to integrate external sources with historical data, LLMs can be an alternative approach to traditional forecasting methods.

For instance, Skórnóg and Kmiecik [14] evaluated the demand forecasting capability of GPT- 3.5 by comparing it with ARIMA models. The results showed GPT-3.5 provides reasonably accurate predictions compared to the ARIMA prediction model, and GPT-3.5 has the potential to be used by manufacturers in the demand forecasting. Another application of LLMs in supply chain management mentioned by Xu et al. [13] is route planning for logistics, wherein real-time traffic and weather conditions can be analyzed, and then the route can be dynamically optimized for the transportation of goods. To help supply chain professionals who might lack an optimization background, Li et al. [15] developed the OptiGuide framework, which leverages an LLM with optimization to evaluate different what-if scenarios for supply chain operations. The framework converts user queries into optimization tasks and allows users to understand the impact of what-if scenarios through natural language. LLMs can also be used in risk monitoring using external data to capture supply chain disruptions. Zhao et al. [11] proposed a risk monitoring system for companies that leverages GPT-3.5 to extract risk features from news data and then categorizes risks for supply chain risk managers. The proposed model is demonstrated through a case study, where Apple Inc. is the focal company monitoring the potential risks of its suppliers. In another relevant research, Sun et al. [16] analyzed news data to predict major supply chain risks and assess their impact using LLMs. Shahsavari et al. [12] developed a Lightweight Unsupervised Event Identification (LUEI) framework that is used to identify risk-contributing events and predict potential supply chain risks using news data. While these studies focused on predicting major supply chain disruption risks, our research focuses on assessing the severity of risks by extracting risk features to support SC managers' decision-making.

LLMs recently have been also used in risk monitoring of natural disasters. Colverd et al. [17] developed a tool to generate flood disaster impact reports called FloodBrain. In traditional practice, human experts prepare disaster impact reports, which is time-intensive and affects the delivery time of humanitarian relief. The performance of different LLMs (i.e., GPT-4, GPT-3.5, and PaLM-Text-Bison) is compared with human-written reports. It was found that reports prepared by GPT-4 had the highest overlap with the human-written reports. In a separate study, Hostetter et al. [18] utilized LLMs (GPT and Bard) to predict forest fires and assess evacuation scenarios. The models were used to verify information and perform calculations, with GPT achieving an 88% accuracy rate, outperforming Bard's 80%. They also observed that the accuracy of responses improved when more detailed questions were asked. Similarly, Ziaullah et al. [19] used LLMs, specifically Llama-2 13B and Mistral 7B v1.0, to monitor critical infrastructure during natural disasters by analyzing social media data. The study evaluated the disaster's impact and severity, as well as the operational status of relevant facilities. Their results showed that LLMs achieved reasonable prediction outputs for the disaster's impact and severity; however, they exhibited lower performance in identifying facility operational status.

3 Methodology

We propose a supply chain risk management framework, as outlined in Fig. 1. Accordingly, supply chain risk management starts with the disruption trigger detection stage. This stage leverages LLMs to analyze news data to identify supply chain disruption triggers. Next, the risk assessment stage operates in one of two phases: the pre-trigger phase or the post-trigger phase. If there is no disruption event detected, risk assessment proceeds in the pre- trigger phase, where baseline risk scores of the suppliers are used to evaluate suppliers. However, once a disruption trigger is detected, risk assessment shifts to the post-trigger phase, where suppliers' risk scores are recalculated based on the impact of the event. Finally, the risk scores of the suppliers whether derived from the pre- or post- trigger phase are fed into the supply chain design stage, where suppliers are decided considering their up-to-date risk scores. The following sub-sections discuss each stage presented in Fig. 1.

A. Detection Stage

This sub-section presents the details of the first stage which detects disruption trigger events for the supply chain. Since news data offers timely information relevant to the risk criteria, our approach uses natural language processing to extract information about a disruption event. The process begins with identifying "seed" keywords tied to specific risk factors. For instance, the natural risk might involve keywords like "earthquake," "tropical cyclones," or "flood." For global supply chain risk categories and the associated risk terms, we benefit from our earlier work that provided a global supply chain risk management framework to identify global supply chain risks [1]. The keywords considered within the risk terms guide our collection of relevant news articles. Then, LLMs are used to extract features related to the identified risks. For global supply chains, the key features such as affected countries, duration, and severity of events are crucial for updating the risk scores of the involved countries.

To explain the general framework with a tangible example, typhoons, a common location-specific natural hazard that occurs in East Asia, are selected. Considering

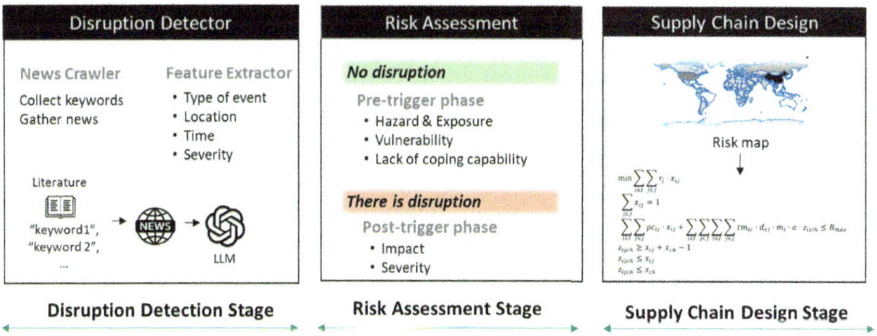

Fig. 1 General framework for supply chain risk management

the dependencies of many industries on the suppliers located in Asia, typhoon risk should be assesed in timely manner. According to the literature, Category 3 typhoons can cause power outages lasting from several days to weeks, and typhoons classified as Category 4 and above result in catastrophic damage and can lead to power outages lasting between 7 to 60 days [20]. Moreover, typhoons not only affect the supplier production times, due to power outages, but also impact international maritime shipping, due to port closures.

To extract the information from the news data, user content (prompt) is a critical element to guide LLMs in accurately understanding and responding to the provided input. Herein, the prompt is generated using template-based prompts [21–23] where users insert target weather event(s) for responses. This prompt is fed into GPT-4o mini and the results are collected through calling OpenAI's API. To interact with the model, system content provides context to the model, then user content is the user's question that the model generates responses to. In this work, the system and user contents that are required by OpenAI's API are as follows:

1) "Role": "system", "content": "You are a supply chain professional in risk analysis. Answer the question given the news articles, and note that not all articles are relevant: [NEWS_ARTICLES]"
2) "Role": "user", "content": "Summarize the information of [WEATHER_NAME] according to the following rules. Answer the question as precisely as possible according to the news article with the following information. Provide the information of the [WEATHER_NAME] location, time, duration, and magnitude. Additionally, further detailed information of evacuation, fatalities, power outages, flooding, port closures, and shipping delays that impact business risk to supply chain companies should also be provided. If the answers could not be found, answer N/A."

B. Risk Assessment Stage

In the pre-trigger phase, the framework assesses the baseline risk scores of suppliers against potential natural and human-made hazards. There are various indexes for assessing natural or human hazard risks. In this study, the Index for Risk Management (INFORM) is selected. INFORM provides a composite indicator to evaluate the risk of humanitarian crises and disasters at the country level. It presents a multi-dimensional risk index measured across three factors: hazards & exposure, vulnerability, and lack of coping capability. Hazards & exposure refers to the probability of occurrence of natural hazards, such as earthquakes, tsunamis, floods, tropical cyclones, and droughts, and human hazards, such as civil wars, crime, and terrorism. Vulnerability refers to the susceptibility of the exposed population to adverse effects of the hazards. It reflects the economic, political, and social aspects of the community disrupted by a hazard. Lack of coping capability is used to measure the country's ability to manage disaster through formal governance structures to mitigate its severe effects [24]. By integrating these three dimensions, natural and human hazard risk is evaluated not only by its probability but also by the affected population's vulnerability and the government's coping capability.

When a disruption trigger is detected, the risk assessment continues with the post-trigger phase to update the risk scores of the affected countries. Here, expert opinions and literature are used to adjust the baseline risk scores. Returning to the typhoon example, the severity of the typhoon is assessed based on its category in the Saffir- Simpson scale [20]; and Category 3 and above typhoons cause devastating or catastrophic consequences. The features extracted by the LLM, such as evacuations, fatalites, port closures, and shipping delays, are used to understand its impact on the supply chain. We reached five supply chain experts with more than 10 years of experience to assist in prioritizing risk features regarding their impact on supply chain disruption. Based on their feedback, the operational state of the main transportation lines was rated as the most important feature. If the operational state is "closed," its impact on supply chain disruption is considered to be "high". Therefore, this feature is separately evaluated from other features. The other features, i.e., evacuation, fatality, power outages, and flooding are considered to evaluate the overall impact of the hazard. The overall impact of the hazard is considered in three levels: severe, moderate, and minor. For instance, high rates of power outages, fatalities (10 + per 10,000 million [25]), and large-scale evacuations (1000 + people [26]) are evaluated as "severe" overall impact. Considering experts' feedback, the matrix in Table 1 is developed to evaluate the impact of a hazard on supply chain disruptions. It takes into account the operational state of main transportation lines and the overall impact of the hazard. For instance, when the overall impact of the hazard is severe and the operational state of main transportation lines is closed, the hazard's impact on supply chain disruption is considered to be "high". It is important to note that, the objective of this paper is not to define hazard- specific features and assess their impact on the supply chain; but rather to present the general risk management framework. For each disruption, the features and the decision criteria to assess its impact should be decided by supply chain risk experts.

C. Supply Chain Design

This stage focuses on developing a mitigation strategy for the risk of supply chain disruption that can be triggered by the discussed hazard impacts. To develop an adequate mitigation strategy, a Mixed-Integer-Programming (MIP) model is proposed to decide on the manufacturers that minimize the overall risk score using the most recent risk scores. The MIP decides the optimal manufacturers to supply a modular product (assembly) in the context of collaborative manufacturing through a manufacturing-as-a-service (MaaS) system, where various third-party manufacturers

Table 1 Hazard impact on supply chain disruption

Overall hazard impact	Operational state of main transportation lines		
	Open	Partially open	Closed
Severe	Medium	High	High
Moderate	Low	Medium	High
Minor	Low	Low	High

are involved to fulfill the production needs [27]. It is important to mention that, in the context of MaaS, manufacturers are referred to as suppliers, as they offer production capabilities to the companies that need products. We assume that the product involves multiple items, which are referred to as the parts and modules needed to manufacture the product. The supply chain for the items is structured using a transition matrix, which defines the parent–child relationship between items in the product structure. For instance, if items A and B are assembled to manufacture item AB, then A and B are considered as the children of item AB. For more details on the transition matrix, we direct readers to an earlier work [28]. The mathematical model presented here is built based on the earlier optimization model to incorporate the risk scores of the suppliers. Below, the sets, indices, parameters, and decision variables associated with the mathematical model are presented.

Sets		*Parameters*	
I:	Set of items, indexed by i	pc_{ij}:	Unit process cost of item i at manufacturer j
J:	Set of manufacturers, indexed by j	d_{jh}:	Distance between manufacturers j and h, $h \in J$
Decision variables		$m_{i:}$	Mass of item i
x_{ij}:	Binary variable, 1 if item i is produced by manufacturer j; 0 otherwise	r_j:	Risk score of manufacturer j
z_{ijch}:	Binary variable, 1 if item i is produced by manufacturer j and its parent item c is manufactured by h, $c \in I$, $h \in J$	tm_{ic}:	Transition matrix value, 1 if item i is child of item c; 0 otherwise
–		B_{max}:	Upper bound on overall cost

$$\min \sum_{i \in I} \sum_{j \in J} r_j \cdot x_{ij} \tag{1}$$

$$\sum_{j \in J} x_{ij} = 1 \ \forall i \in I \tag{2}$$

$$\sum_{i \in I} \sum_{j \in J} pc_{ij} \cdot x_{ij} + \sum_{i \in I} \sum_{j \in J} \sum_{c \in I} \sum_{h \in J} tm_{ic} \cdot d_{jh} \cdot m_i \cdot \alpha \cdot z_{ijch} \leq B_{\max} \tag{3}$$

$$z_{ijch} \geq x_{ij} + x_{ch} - 1 \ \forall i \in I, j \in J, c \in I, h \in J \tag{4}$$

$$z_{ijch} \leq x_{ij} \ \forall i \in I, j \in J, c \in I, h \in J \tag{5}$$

$$z_{ijch} \leq x_{ch} \ \forall i \in I, j \in J, c \in I, h \in J \tag{6}$$

The objective function of the mathematical model presented in Eq. (1), is to minimize the overall risk score of the suppliers. Equation (2) ensures that each item is manufactured by a single supplier. Equation (3) limits the total manufacturing and transportation costs for the product. This is needed to have a compromised solution between risk and total cost. The first portion of Eq. (3) calculates the production cost, whereas the second portion calculates the transportation cost. For calculating the transportation cost, the path between parent and child items is needed, a decision variable z_{ijch} is included to represent whether the item i is produced by the manufacturer j and its parent item c is manufactured by h. Using this variable, the transition matrix, the distance between suppliers, and the mass of the transported items, the transportation cost between manufacturers is calculated. The constant α shows the cost of transporting per kg per km. Equations (4–6) are needed to define the bounds for z_{ijch}, which verifies that z_{ijch} is 1 only if x_{ij} and x_{ch} are 1. Otherwise, z_{ijch} equals to 0.

4 Case Study

This section demonstrates the proposed framenwork for a previously reported case study [28, 29]. The case study involves a bicycle company with three headquarters located in three different countries: USA, Italy, and Taiwan. The company wants to decide on the product design (product architecture) that minimizes the expected overall total cost, including manufacturing and transportation costs for each headquarters. The optimized product architecture for the headquarters in the USA is taken from Günay et al. [28] as presented in Fig. 2. Accordingly, there are a total of six main items which are the saddle, frame, fork, brake, wheels, and transmission. Based on the optimal product design, first, the frame and fork and then the saddle should be assembled to manufacture module ABC. To manufacture the DEF module, first, the wheels and transmission modules and then the brakes should be assembled. The final product bicycle encompasses the assembly of ABC and DEF modules. Manufacturers for each item are located in various countries, including Taiwan, the USA, Italy, France, and Japan. In this case study, using the optimized product architecture [28], optimal suppliers are selected to minimize the overall risk score of the supply chain.

A. Results of the Disruption Detection

The supply chain disruption risk management starts with the disruption trigger detection stage. In this stage Google News with the keywords of “Taiwan” and “weather,” where the time frame was set from March 30, 2024 to September 19, 2024, is used to explore if the LLM can capture the major natural hazards that had occurred. During the given time frame, there were three severe weather conditions: an earthquake on April 3, 2024, Typhoon Gaemi in late July 2024, and Typhoon Bebinca in mid-September 2024. The intention is to collect as many news articles as

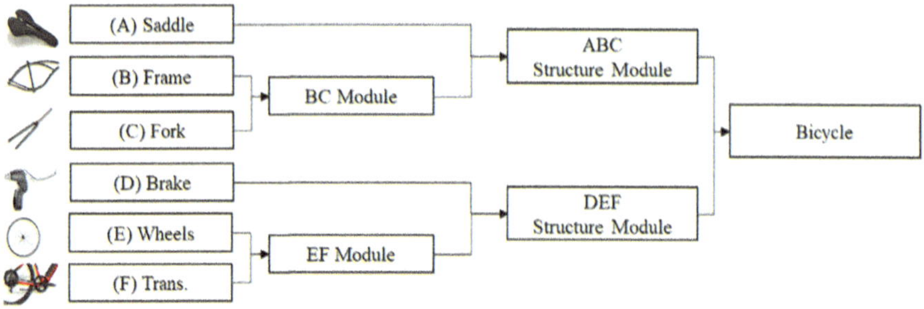

Fig. 2 Optimum product architecture (A-1) for the headquarters in the USA (adapted from [28])

possible with fewer keywords and utilize an LLM to search and provide related information directly. A total of 95 news articles were found, and 85 of them were successfully scraped for the content. The remaining news articles were inaccessible due to automatic scraping restrictions. GPT-4o mini was used to answer questions from users, and the model was accessed by using OpenAI's API. The prompt presented in Section 3-A was given to GPT-4o mini to extract critical risk features. GPT was successful in detecting these three hazards. The output of the LLM for the Typhoon Gaemi is as follows.

Typhoon Gaemi

- Location: Eastern Taiwan, near Hualien City
- Time: Struck on April 3, 2024, around 7:58 a.m. local time
- Duration: Impacted the region from April 3, 2024, with ongoing effects due to aftershocks and subsequent severe weather
- Magnitude: 7.4
- Details:
- Evacuations: Over 8,500 people were evacuated from precarious conditions across Taiwan, particularly in Hualien.
- Fatalities: At least 13 confirmed deaths due to the typhoon, primarily from falling debris and rockslides.
- Power Outage: More than 91,000 households were left without electricity following the storm.
- Flooding: Extensive flooding occurred across Hualien County, blocking roads and damaging many buildings.
- Port Closure: Several port operations were affected, and ferry services were suspended due to the storm.
- Shipping Delays: Significant delays in shipping and logistics occurred due to damage to transportation infrastructure, including highways and tunnels blocking access in Hualien.

Because we do not use an existing benchmark dataset, the accuracy of features predicted by the LLM is evaluated by human experts. Consequently, three experts read all the news data fed into the LLM and independently evaluated the correctness of the LLM output for each feature. The evaluations of the three experts as "correct" or "false" for each feature are listed in Table 2. Next, two different counting metrics are employed to determine the accuracy of the responses. In the first, Metric 1, the LLM output for the feature is counted as true if two-thirds of the experts evaluate it as "correct." In the second, Metric 2, the LLM output for the feature is accepted as true, if all the experts evaluate it as "correct." For instance, the feature location is evaluated as "correct" by Experts 2 and 3, then counted as "TRUE" for Metric 1. However, it is counted as false in Metric 2. For Metric 1, the accuracy is 80% as 8 out of 10 features are correctly predicted. For Metric 2, the accuracy is 70%.

B. Risk Assessment Stage

Due to the detection of a disruption trigger event, Typhoon Gaemi, risk assessment continued with the post- triggering phase. The impact of the event was evaluated based on its category (a Category 3 typhoon) and its associated features, including evacuations, fatalities, power outages, port closures, and shipping delays. According to the hazard impact evaluation matrix in Table 1, Typhoon Gaemi's impact on supply chain disruption is evaluated as a high-risk category due to disruption in transportation lines, including port closure and shipping delays, and the severe overall impact of the hazard. Consequently, the risk categories of the manufacturers located in Taiwan are updated to the high-risk category. Table 3 presents the risk scores and the risk class of the countries for the potential suppliers. The risk scores of the countries except Taiwan are acquired from the INFORM risk index 2024. To have consistency with the INFORM risk index scale, a risk score of 6 is assigned to Taiwan *(updated from a value of 3.0)*.

C. Supply Chain Design Stage

The MIP model is developed in Python, using Pyomo and GLPK libraries. The updated risk scores in Table 3 and parameters from past work [28–30] are used to solve the supply chain optimization model that minimizes the overall risk score for the supply chain. Figure 3 presents the selected manufacturers from each country

Table 2 Expert evaluations for the LLM output

Features	Expert 1	Expert 2	Expert 3	Metric 1 (2/3)	Metric 2 (all)
Location	Wrong	Correct	Correct	TRUE	FALSE
Time	Wrong	Wrong	Wrong	FALSE	FALSE
Duration	Wrong	Wrong	Wrong	FALSE	FALSE
Magnitude/category	Correct	Correct	Correct	TRUE	TRUE
Evacuations	Correct	Correct	Correct	TRUE	TRUE
Fatalities	Correct	Correct	Correct	TRUE	TRUE
Power outage	Correct	Correct	Correct	TRUE	TRUE
Flooding	Correct	Correct	Correct	TRUE	TRUE
Port closure	Correct	Correct	Correct	TRUE	TRUE
Shipping delays	Correct	Correct	correct	TRUE	TRUE

Table 3 The risk scores of the countries where suppliers are located

Bicycle items	USA	Taiwan	France	Japan	Italy
(A) Saddle	3.2	3.0→High (6)	–	–	2.5
(B) Frame	3.2	3.0→High (6)	–	–	–
(C) Fork	3.2	3.0→High (6)	–	–	–
(D) Brake	3.2	–	–	2.3	2.5
(E) Wheels	–	3.0→High (6)	2.4	2.3	–
(F) Trans	3.2	3.0→High (6)	–	2.3	–
BC Module	3.2	3.0→High (6)	–	–	–
EF Module	3.2	–	2.4	2.3	2.5
ABC Structure Module	3.2	3.0→High (6)	–	–	–
DEF Structure Module	3.2	–	–	2.3	2.5

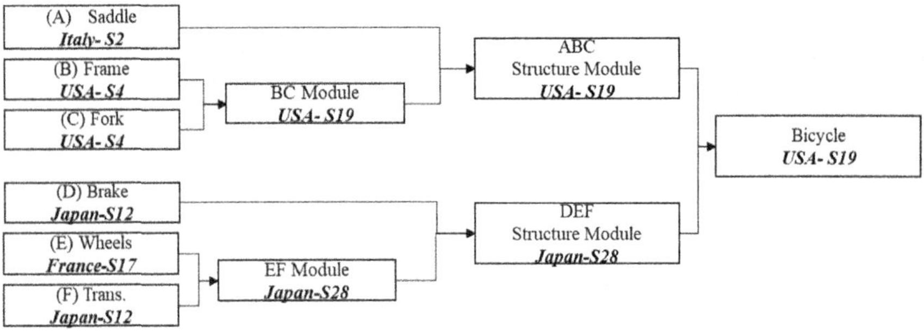

Fig. 3 Manufacturers selected for each item for bicycle production

for each item. Selected manufacturers from each country are presented according to their codes (S1-S28) used in the prior study [28]. For instance, the saddle is supplied by manufacturer S2 located in Italy. Due to the increase in its risk score, no manufacturers were selected from Taiwan. The overall risk score of the supply chain is 30. Using the 0–2.2 very low risk, 2.3–3.2 low risk, 3.3–5.1 medium risk, 5.2–6.9 high risk, and 7 + very high-risk category, a supply chain of 10 suppliers with an overall risk score of 30 falls into the low-risk category.

5 Conclusions

This study proposes a conceptual supply chain risk management framework by leveraging LLMs to detect disruption trigger events from news data, determine its impact, and develop a risk mitigation strategy. These disruptive triggering events can be natural hazards, such as earthquakes and tsunamis, or man-made hazards, such as political tension, war, or a financial crisis. Early detection of disruptive events is critical for developing mitigation strategies and maintaining supply chain resilience. Studies on LLMs show that they are capable of processing, understanding, and summarizing vast amounts of text data. The proposed framework utilizes LLMs to detect the disruption trigger event and extract its features from news data. While the framework is generalizable for different natural or man-made hazards, the application is demonstrated using typhoon natural hazards. Upon detection of the disruptive event, LLM-extracted features were used in assessing the severity and impact of the event, allowing for a more dynamic response to risk. Finally, to mitigate the risks, a MIP model is developed to decide on the alternative manufacturers considering their up-to-date risk scores. The overall framework is demonstrated using a case study. The performance of the LLM in correctly identifying the natural hazard and its features is assessed by three experts. Our preliminary results show that the LLM can be used to adequately capture the disruption trigger events and their features. The risk scores of the suppliers are then updated accordingly to assist supply chain managers in developing mitigation strategies.

However, there are a few limitations that need to be addressed. First, the specific features of each disruption event are context-dependent and need to be defined based on the nature of the risk. For instance, evacuation may be relevant in the context of natural hazards but less so in cases of political tension. Second, expert judgment has a critical role in risk assessment; in order to determine especially the impact in the case of disruption in supply chain, such expert feedback is need. For instance, experts should evaluate what tariffs are likely to do to their business. Third, more experimentation is needed to validate the correctness of the LLM outputs across a wider range of disruption triggers. For example, it is worth examining how effectively LLMs can extract features for political risks. A potential direction for future study is to define various disruption triggers and map them to relevant features that can be extracted from news data. Then, automating the news acquisition with LLM analysis to provide timely alerts to supply chain managers will enable proactive risk management.

Acknowledgements The authors gratefully acknowledge support from the U.S. Department of Commerce and MxD through a project funded by President Joe Biden's American Rescue Plan, awarded by the National Institutes of Standards and Technology (NIST).

References

1. Chu C-Y, Park K, Kremer GE (2020) A global supply chain risk management framework: An application of text-mining to identify region-specific supply chain risks. Adv Eng Inform 45:101053. https://doi.org/10.1016/j.aei.2020.101053
2. Chowdhury P, Paul SK, Kaisar S, Moktadir MdA (2021) COVID-19 pandemic related supply chain studies: a systematic review. Transp Res E Logist Transp Rev 148:102271. https://doi.org/10.1016/j.tre.2021.102271

3. Lee JM, Wong EY (2021) Suez Canal blockage: an analysis of legal impact, risks and liabilities to the global supply chain. MATEC Web Conf 339:01019. https://doi.org/10.1051/matecconf/202133901019
4. Liadze I, Macchiarelli C, Mortimer-Lee P, Sanchez Juanino P (2023) Economic costs of the Russia-Ukraine war. World Econ 46(4):874–886. https://doi.org/10.1111/twec.13336
5. El Baz J, Ruel S (2021) Can supply chain risk management practices mitigate the disruption impacts on supply chains' resilience and robustness? Evidence from an empirical survey in a COVID-19 outbreak era. Int J Prod Econ 233:107972. https://doi.org/10.1016/j.ijpe.2020.107972
6. Heckmann I, Comes T, Nickel S (2015) A critical review on supply chain risk—Definition, measure and modeling. Omega (Westport) 52:119–132. https://doi.org/10.1016/j.omega.2014.10.004
7. Bugert N, Lasch R (2018) Supply chain disruption models: a critical review. Bundesvereinigung Logistik (BVL). https://doi.org/10.23773/2018_5
8. Basyal L, Sanghvi M (2023) Text summarization using large language models: a comparative Study of MPT-7b-instruct, Falcon-7b-instruct, and OpenAI Chat-GPT Models
9. Bubeck S et al (2023) Sparks of artificial general intelligence: early experiments with GPT-4
10. Hendriksen C (2023) Artificial intelligence for supply chain management: Disruptive innovation or innovative disruption? J Supply Chain Manag 59(3):65–76. https://doi.org/10.1111/jscm.12304
11. Zhao M, Hussain O, Zhang Y, Saberi M (2024) Optimizing supply chain risk management: an integrated framework leveraging large language models. In: 2024 IEEE Conference on Artificial Intelligence (CAI). IEEE, pp 1057–1062. https://doi.org/10.1109/CAI59869.2024.00192
12. Shahsavari M, Hussain OK, Saberi M, Sharma P (2024) Event identification for supply chain risk management through news analysis by using large language models. Rev Socionetwork Strat. https://doi.org/10.1007/s12626-024-00169-z
13. Xu Z, Lingfeng G, Shuwen Z, Runze S, Kaiyi N (2024) Enterprise supply chain risk management and decision support driven by large language models. Appl Sci Eng J Adv Researh 3(4):1–7
14. Skórnóg D, Kmiecik M (2023) Supporting the inventory management in the manufacturing company by ChatGPT. Logforum 19(4):535–554. https://doi.org/10.17270/J.LOG.2023.917
15. Li B, Mellou K, Zhang B, Pathuri J, Menache I (2023) Large language models for supply chain optimization
16. Sun J, Wen X, Ping G, Zhang M (2024) Application of news analysis based on large language models in supply chain risk prediction. J Comput Technol Appl Math 1(3):55–65
17. Colverd G, Darm P, Silverberg L, Kasmanoff N (2023) FloodBrain: flood disaster reporting by web-based retrieval augmented generation with an LLM
18. Hostetter H, Naser MZ, Huang X, Gales J (2024) Large language models in fire engineering: an examination of technical questions against domain knowledge
19. Ziaullah AW, Ofli F, Imran M (2024) Monitoring critical infrastructure facilities during disasters using large language models
20. Nguyen H, Sharkey TC, Wheeler S, Mitchell JE, Wallace WA (2021) Towards the development of quantitative resilience indices for Multi-Echelon assembly supply chains. Omega (Westport) 99:102199. https://doi.org/10.1016/j.omega.2020.102199
21. Cui L, Wu Y, Liu J, Yang S, Zhang Y (2021) Template-Based named entity recognition using BART. Available: http://arxiv.org/abs/2106.01760
22. Schick T, Schütze H (2020) Exploiting cloze questions for few shot text classification and natural language inference
23. Gao T, Fisch A, Chen D (2020) Making Pre-trained language models better few-shot learners
24. Marin-Ferrer M, Poljanšek K, Vernaccini L (2017) Index for risk management inform concept and methodology report
25. World Health Organization, The Global Health Observatory-Average death rate due to natural disasters (per 100 000 population). Last accessed: 25 Sep 2024. Available: https://www.who.int/data/gho/indicator-metadata-registry/imr-details/4665

26. Lindell MK (2021) Emergency management: evacuations. In: Encyclopedia of Security and Emergency Management. Springer International Publishing, Cham, pp 272–286. https://doi.org/10.1007/978-3-319-70488-3_106
27. Elhabashy AE, Günay EE, Sim J, Clements MW, Benaggoun H, Kabiri M, Haapala KR, Okudan Kremer GE, Kim KY (2024 Aug 25) Development of a rapid response online product marketplace. In: International design engineering technical conferences and computers and information in engineering conference, vol 88391, p V005T05A002. American Society of Mechanical Engineers
28. Günay EE, Park K, Okudan Kremer GE (2021) Integration of product architecture and supply chain under currency exchange rate fluctuation. Res Eng Des 32(3):331–348. https://doi.org/10.1007/s00163-021-00361-0
29. Chiu M.-C, Okudan G (2011) An integrative methodology for product and supply chain design decisions at the product design stage. J Mech Des 133(2). https://doi.org/10.1115/1.4003289
30. Chiu M-C, Kremer GEO (2014) An investigation on centralized and decentralized supply chain scenarios at the product design stage to increase performance. IEEE Trans Eng Manag 61(1):114–128. https://doi.org/10.1109/TEM.2013.2246569

Beyond Bricks and Mortar: Can Buildings Be Designed and Constructed like a Machine?

Arzu Gönenç Sorguç

Abstract The construction sector faces increasing environmental and technological challenges, with buildings contributing 40% of global energy consumption and significant greenhouse gas emissions. This paper explores the integration of Industry 4.0 technologies, such as Building Information Modelling (BIM) and smart materials, within the Architecture, Engineering, and Construction (AEC) sector. It highlights the shift toward modular design, digitalization, and the circular economy as essential for sustainability. By examining the sector's slow adaptation to these innovations, the paper emphasizes the need for a more holistic, performance-based approach to design and construction through system understanding.

Keywords Construction 4.0 · Circular design · Building Information Modelling (BIM) · Sustainable construction · Modular design · Advanced technologies in AEC

1 Introduction

The construction sector, with its many auxiliary components, plays a significant role in the economies of countries. The sector's prominence is driven by two opposing forces: the growing population and the consequent demand for housing, and the diminishing natural resources coupled with escalating environmental challenges. With buildings accounting for a significant portion of global energy consumption and raw material use, the sector faces immense pressure to evolve. The sector's centrality becomes particularly evident when considering its consumption of natural resources and energy. In 2021, buildings accounted for 40% of global energy consumption and contributed to one-third of greenhouse gas emissions [1]. This percentage did not change in 2024, and an average increase of 5% was even observed [2]. Additionally, buildings account for approximately 50% of the annual consumption of raw materials used during construction [3].

A. G. Sorguç (✉)
METU Design and Prototyping Application and Research Center, Ankara, Turkey
e-mail: arzug@metu.edu.tr

N. M. Durakbasa and K. G. Gülen (eds.), *Sustainable Green Conversion*, Lecture Notes in Mechanical Engineering, https://doi.org/10.1007/978-3-031-83583-4_2

The environmental impact of the construction sector extends beyond energy consumption. According to a report by the World Green Building Council, 32% of landfill waste generated globally comes from construction sites, and 13% of this material is sent directly to landfills without being used [4].

Given the substantial impact of the built environment on the natural world, the economy, and human life, it is imperative to critically examine developments within the construction sector and explore innovative strategies for its future.

In this context, understanding the influence of emerging concepts such as digitalization, green transformation, and the circular economy, which have gained traction in the last decade, is crucial. Additionally, the shift in design and production processes brought about by Industry 4.0 has had profound implications for the construction sector, affecting the way buildings are designed and constructed.

The architecture, engineering and construction sector (AEC), presents a diverse and asymmetrical landscape. Consequently, this study aims to analyse the production and construction processes within these sectors, considering "buildings" as produced/fabricated/manufactured objects.

2 Industry 4.0 and Construction 4.0?

The emergence of Industry 4.0, which relies on heavy digitalization, automation, and the use of big data, has a profound impact on various sectors. Khan et al. characterized Industry 4.0 by interoperability (Cyber-Physical Systems and IoT), virtualization (visualization and simulation), decentralization (autonomous decision-making), real-time data acquisition, service-oriented decision-making, and modularity (quick-action decision-making in production processes) [5]. The reduction of human involvement is a cornerstone of Industry 4.0, leading to the integration of physical and digital realms and the creation of new digital ecosystems across multiple sectors.

The ongoing paradigmatic shift within AEC has garnered significant attention from both researchers and practitioners, prompting the question: What constitutes Construction 4.0? Echoing Khan's framework, Sawhney et al. propose a model for 'Construction 4.0,' characterized by the integration of cyber-physical systems, the Internet of Things, data, and services, alongside Building Information Modeling (BIM) and Common Data Environments (CDE). This framework facilitates the efficient design, construction, and operation of assets by integrating organizations, processes, and information [6].

The transformation from the project phase to the delivery of buildings, as demonstrated in various examples, is set to yield innovative, efficient, and sustainable solutions within the AEC industry. Already, numerous advanced technologies and methodologies are being employed in the construction of buildings. Smart buildings, performance-based design, parametric design, modular buildings, Building Information Models (BIM), digital twins, and robotic construction are all advancements that have emerged with technological progress in the AEC sector. However, the sector still faces significant challenges in fully harnessing the opportunities presented by Industry 4.0. The industry remains highly fragmented, with disjointed processes and asymmetries arising from differences in scale, technology, materials, functions, skilled labour, and logistical requirements. Additionally, the AEC sector has been slower to adopt new technologies and production methods compared to other industries.

Crowley in his early work discusses this reluctance or slowness in adapting new technologies in the AEC drawing a comparison with the automotive industry: "The automotive industry moved from craft-based production to mass production almost a century ago, and yet people still talk of moving the construction industry from a craft-based to a manufacturing industry". He also noted that AEC sector considers "one-of-a-kind, project-based nature of construction" [7] as an asset which ultimately becomes an obstacle to the sector's rapid advancement and improvement. Therefore, the construction industry has much to learn from sectors such as automotive and aerospace. In this context, it is insightful to conceptualize the building process as a manufacturing process and to regard buildings as products at every scale, adopting a lean and integrated systems approach.

3 What Has Happened in AEC in the Last 15 years? Impact of the Technology

To effectively comprehend how technology can be integrated into buildings, it is beneficial to conceptualize buildings as complex dynamic systems. This perspective allows for a more systematic approach to identifying opportunities for technological adaptation and optimization within the built environment. Several models exist for this purpose, and among them, the 'shearing layers model' proposed by Brand is well-accepted and applied for diverse purposes, including life cycle assessment, management, and circularity [8] (Figs. 1 and 2).

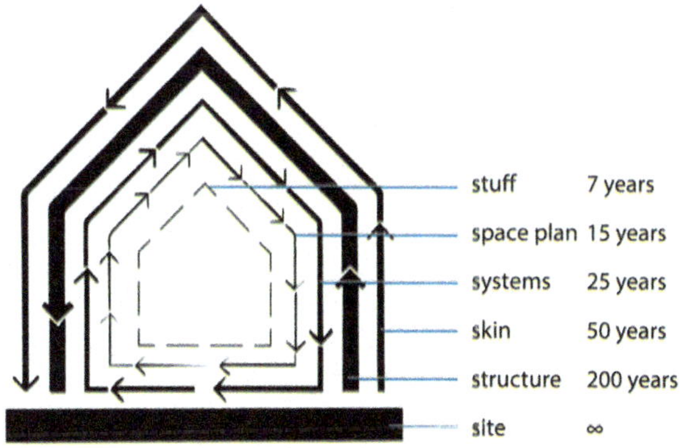

Fig. 1 Shearing layer concept based on brand's framework [9]

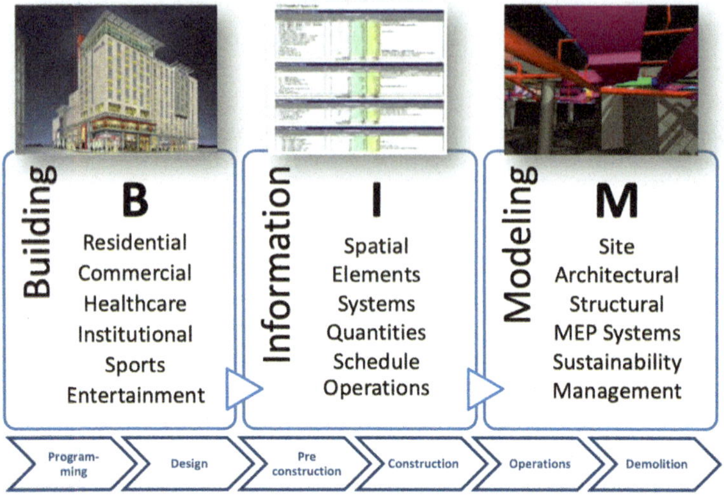

Fig. 2 A typical BIM process [10]

As the model clearly illustrates, although the building is composed of multiple layers with varying levels of complexity, and despite efforts to adhere to a holistic design approach, the qualitative, quantitative, and functional differences among these layers lead to fragmented and disjointed processes during design, construction, and even afterwards. The building is not an assembly of inert construction materials and components; it is a live act embodying human memory, values, and actions.

In this context, the increasingly widespread use of BIM tools and processes is one of the significant steps forward taken by the AEC industry. While the philosophy of BIM dates back to the 1950s, it has been elevated to a different level today with the advent of digitalization and technology. BIM is a software (or a set of software working in alignment with each other) that represents a building as an object-oriented model linked to a database. This enables collaboration among different stakeholders, facilitating communication, decision-making, simulation, visualization, and documentation—all aimed at reducing fragmentation in the construction process.

In the last five years, the use of BIM has been encouraged by many stakeholders like policymakers, funding organizations/institutions, and building rating systems like BREAM, DGNB, LEED etc. Many recent mega structures built all around the world illustrate the use of BIM. Statoil Regional and International Offices (by a-lab 2012), Megatall skyscraper Shanghai Tower (by Gensler 2015), Helsinki Airport Extension (by ALA Architects 2021), Randselva Bridge, Honefoss, Norway (by PNC Norge 2022) are among many examples in which BIM is used in several phases of design and construction. There are also examples of BIM in small-scale buildings and heritage conservation, simulation and visualization, 3D, and 4D parametric (or on-parametric models) digital twins and construction management (Figs. 3, 4, 5 and 6).

In alignment with technological advancements, the construction industry has witnessed a surge in the development and adoption of new and innovative materials. Self-healing concrete [15], use of aerogel in insulation, graphene-based materials, cross-laminated timber, phase-change materials (PCMs), 3D printed materials, recycled and reclaimed materials, transparent aluminium (Aluminum Oxynitride), smart glass, self-cleaning materials, carbon-fibre reinforced polymers (CFRP),

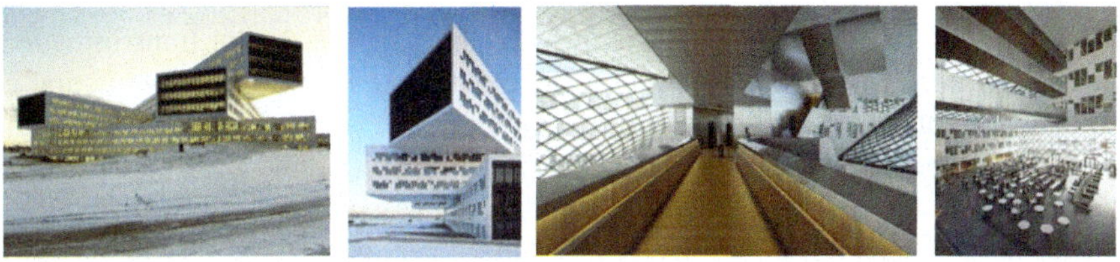

Fig. 3 Statoil regional and international offices by a-lab [11]

Fig. 4 Shanghai tower by Gensler architects [12]

Fig. 5 Helsinki airport extension by ala architects [13]

Fig. 6 Randselva bridge, Honefoss, Norway (by pnc norge 2022) [14]

hydrophobic cement, nano-engineered materials, bamboo are examples used in construction.

These materials are at the forefront of innovation in the construction industry, offering enhanced sustainability, durability, energy efficiency, and cost-effectiveness.

Another noteworthy advancement in design and construction processes is the initial application of 3D printing technology to produce small-scale structures, which has recently expanded to include the production of multi-story buildings. This progression has also facilitated the growing use of robotic construction technologies on-site, marking a shift towards more prevalent and sophisticated construction practices.

Despite the rise of these technologies and the ongoing digital transformation, it remains crucial to address the prevailing perception of building construction as an activity that primarily occurs on-site. This perception is largely shaped by the traditional approach, where the main structure of a building is constructed on-site. However, a building should be understood as a complex system composed of various modules, each assembled with intricate details and an intertwined system of design and construction based on close interaction between the designer and the product. These components are often manufactured using different materials and production techniques, either in factories or on-site by various sources. Within this context, "modularity" in construction processes necessitates not only a modular design but also a modular understanding of the system's components. This approach introduces several uncertainties, influenced by design, materials, construction methods, environmental conditions, and human factors.

Addressing this complexity necessitates a redefined perspective on buildings throughout the entire lifecycle, from design through construction and beyond, accompanied by the implementation of revised regulations and standards. Coupled with the advancements of Industry 4.0 and the transition to a circular economy, the Architecture, Engineering, and Construction (AEC) industry is presented with numerous new opportunities for exploration.

4 Impact of the Economy: Circular Economy to Circular Building

Departing from the linear economy which is characterized by the "take-make-use-dispose" model to a circular economy (CE) requires a fundamental shift in mindset for all stakeholders involved in economic growth and environmental protection. There are many different ways to describe CE. For example, one of the most commonly referenced definitions states "The circular economy is a system where materials never become waste and nature is regenerated" [16]. CE is usually explained by its 7 pillars as 7R's; redesign, reduce, reuse, repair, renovate, recycle and recover.

In CE, as the concept of the 7Rs implies, there is a significant emphasis on the management of materials [17] like the "realization of closed loop material flow in the whole economic system" [18], and "the use of raw materials and energy through multiple phases" [19]. This also implies the need for crucial changes in product design strategies to improve material efficiency. As Allwood and et al. describe, these strategies include "more material services with less production and processing, comprise longer-lasting products, modularization and remanufacturing, component re-use, and designing products with less material" [20]. In parallel with changes in material flow, remanufacturing emerged as a research field in the late 1990s, alongside the restructuring of supply chains (SC). This evolution ultimately gave rise to concepts extending beyond recycling/remanufacturing and life cycle management to include waste management.

Given the complexity, fragmentation, and asymmetries of the sector, this paradigmatic change significantly impacts not only architectural design but also all the actors involved, from design to demolition, from supply chains (SC) to waste management. It requests a systems approach, standardization, and quality control of materials, components, and services. This, in turn, compels the industry to expand its conventional understanding of buildings to include 'circular buildings' (CB).

Various definitions have been proposed for CB, with one of the most cited one being "a building that is designed, planned, built, operated, maintained, and deconstructed in a manner consistent with CE principles" [21] which later expanded by Leising as "a lifecycle approach that optimizes the buildings' useful lifetime, integrating the end-of-life phase in the design and uses new ownership models where materials are only temporarily stored in the building that acts as a material bank" [22].

An example of this concept is the 'Demountable Building' designed by Cepezed in Delft. The building was constructed in six months and is fully demountable, allowing it to be relocated and reassembled at any other site. It can flexibly adapt to new functions assigned to it in the future (Figs. 7 and 8).

Another example of a demountable building is Tridos Bank designed by RAU in the Netherland. It is energy-neutral, hybrid construction and it has a BIM-based material passport providing the records of the materials used in the building for future use.

The number of such buildings is steadily increasing. The driving forces behind this growth are the increasing prominence of the circular economy as a dominant paradigm, coupled with rising environmental challenges, climate crises, and the

Fig. 7 Helsinki airport extension by ala architects [23]

Fig. 8 Tridos bank by RAU [26]

ongoing depletion of raw materials. A closer examination of these examples reveals that the transformation extends beyond materials and construction, and details, encompassing a broader shift in the conceptualization of space itself.

As these examples show, circular buildings (CB) necessitate a holistic reconsideration of 'buildings,' from the earliest stages of design through to completion and beyond. This approach demands a new level of precision, accuracy, and resource management in both design and construction, ultimately reviving the notion of buildings functioning "like a machine".

5 If a Building Would Be a Car or a Plane!

On one hand, the rapid progression of technology and digitalization, and on the other, the influence and pressures exerted by the shift toward a circular economy and a green future—characterized by diverse scales and strategies—are contributing various examples to our built environment within the AEC sector. As this transformation unfolds, the discourse on how processes should be structured when conceptualizing a building as either a product like a car [25] (or not), or a machine [26] can be elevated to a more formal and concrete platform and free AEC to provide more innovative and disruptive solutions.

The analogy between the industrially produced building (and prefabrication) and car manufacturing is not new and despite numerous arguments and studies, it has long remained merely "a dream" for long time [26]. The concept known as "Henry Ford Syndrome," which articulates the building-car analogy while explaining why buildings cannot be like cars. This can be summarized as shown in Fig. 9 based on Aitchison [26] (Table 1).

When examining the table presented above, it is important to remember that, unlike a car or an airplane, which is constructed with variations on a single platform, a building is designed and exists within its unique context, as previously mentioned. However, these constraints or differences should not hinder the transformation of buildings. In this transformation, considering buildings as systems, as well as how

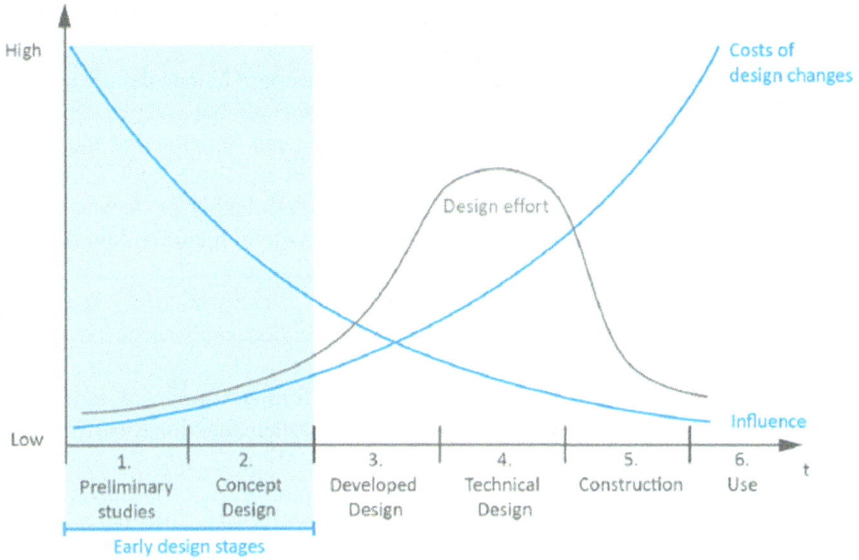

Fig. 9 Paulson curve (revised version of Paulson JR (1976) diagram, in the study of Hollberg (2016) [30]

Table 1 A house is not a car: based on Aitchison [26] adopted by AYÇA AYAZ ERDAĞ [27]

12 Arguments	House	Car
1. Sheer size	Diverse	Diverse
2. Parts and compartmentalization	Big parts	Small parts
3. Site conditions	Stable/specific place	Movable/anywhere
4. Users	Multiuser/family	Single user
5. Climate	Changeable/adaptable by mechanic solutions	Unchangeable for climate designs remains
6. Codes and regulations	Regulations and national standards	Some car can be driven anywhere in the world
7. Material and construction	Houses that look the same may be built with different materials	Almost all cars are made with same materials
8. Tolerances	Different companies have different targets for the proportions of on-/off-site labor	Assembling a series of known and discrete parts
9.Customization	User/functional diversity	All cars' 90 percent is the same
10. Cost relative income	More expensive	Relatively less expensive
11. Supply chain	More diversity	Less diversity
12.Industry setup	?	Diverse

these systems can be decomposed, will play a key role highlighting and advancing the concepts like adaptability and performance.

Adaptability and sustainability are interrelated concepts. Graham defines sustainable building as "a building is not one that must last forever, but one that can easily adapt to change" [28]. Adaptability on the other hand can be defined as the ability to reduce the need for new construction (reduce), to reactivate underutilized or vacant building stock (reuse), and to facilitate the disassembly or deconstruction of components (reuse, recycle), thereby extending the useful life of buildings" [29].

Building performance, on the other hand, is a broad concept that includes criteria such as energy efficiency, the use of environmentally friendly materials, the provision of a healthy and comfortable environment for occupants, acoustic performance, indoor air quality, and more. In line with these considerations, the traditional paradigm of building design has shifted towards a performance-oriented framework, wherein structures are evaluated not solely based on their functional performance but also their performative qualities.

This shift necessitates a broader understanding of performance that encompasses the entire system as well as its individual components. Consequently, the design of details is being reconceptualized and elevated to a more sophisticated level within the design process for which the early design phase decisions which determine the subsequent design process need to be more considered than before as shown in Fig. 9 [30].

The conceptualization of buildings as integrated systems necessitates a re-evaluation of the interrelationship between architectural details and the overall structure. Architectural details not only determine the structural integrity, aesthetic value, and quality of a building but also play a critical role in how the building is maintained, repaired, and upgraded throughout its life cycle. Moreover, these details influence the processes by which the building may be dismantled and its components reused at the end of its operational life. In this context, architectural details become increasingly important for achieving sustainability. The significance of these elements is especially evident when considering the life cycles of individual components and their interrelationships, particularly given their differing service lives.

While it is the standard practice to determine, from the initial design phases, when a vehicle will require servicing and which parts will need replacement, applying the same level of foresight and planning to buildings is not as straightforward, as evidenced by the differences outlined in the table. Despite standards like ISO 15686–1/2/7 and 8 (ISO, 2008, 2011b, 2012, 2017), which define the service life information of building components, addressing the service life of buildings in the same manner as vehicles remains challenging. This is largely due to significant asymmetry in data and standardization across the Architecture, Engineering, and Construction (AEC) sector and its various actors.

6 One Day

Amidst these paradigmatic shifts and transformations, design methodologies have increasingly embraced parametric approaches, performative applications, and intelligent models, all of which have been seamlessly integrated into Building Information Modeling (BIM). The construction industry, drawing on parametric modeling techniques from the automotive and aerospace sectors, has concurrently reinforced integrative design paradigms by incorporating performance-based design frameworks alongside BIM. This evolution has facilitated the sequential application of various modeling strategies, leading to more comprehensive and interconnected design processes. As these processes become more holistic and integrative, the discourse surrounding the interrelationship between individual components and the broader system has gained heightened prominence, as highlighted in this study.

In response, strategies aimed at equipping current and future professionals with the resilience skills necessary to navigate these transformations have become critical focal points for educational institutions and policymakers. These strategies, tailored for the AEC sector, have the potential to serve as a model for other disciplines and industries making buildings as "machine-like".

References

1. Tricoire J.-P (2021) Why buildings are the foundation of an energy-efficient future, World economic forum 22 Feb 2021. Available: https://www.weforum.org/agenda/2021/02/why-the-buildings-of-the-future-are-key-to-an-efficient-energy-ecosystem/. Accessed 10 Sep 2024
2. Bpie U (2022) Global status report for buildings and construction 11 Oct 2022. Available: https://www.unep.org/resources/publication/2022-global-status-report-buildings-and-construction. Accessed 10 Sep 2024
3. Mollaei A, Byers B, Christovan C, Olumo A, Wolf CD, Bachmann C, Haas C (2023) A global perspective on building material recovery incorporating the impact of regional factors. J Clean Prod 429:139525
4. Brady C, Kawamura S, Din A (2024) An integrated approach to a sustainable built environment: the co-benefits of resources & circularity, World green building council. Available: https://worldgbc.org/article/an-integrated-approach-to-a-sustainable-built-environment-the-co-benefits-of-resources-circularity/. Accessed 10 Sep 2024
5. Khan M, Wu X, Xu X, Dou W (2017) Big data challenges and opportunities in the hype of Industry 4.0. In: 2017 IEEE International Conference on Communications ICC, pp 1–69
6. Sawhney A, Riley M, Irizarry J, Riley M, Sawhney A, Riley M, Irizarry J (2020) (eds) Construction 4.0
7. Crowley A (1998) Construction as a manufacturing process: lessons from the automotive industry (1998) 389±400. Comput Struct 67:389–400
8. Brand S (1994) How Buildings Learn. Viking, New York
9. Manifesto OpenBuilding.co (2021). Available: https://www.openbuilding.co/manifesto. Accessed 10 Sep 2024
10. Salman A, Khalfan M, Maqsood T (2012) Building information modeling (BIM): now and beyond. Australas J Constr Econ Build 12(4):15
11. Statoil Regional and International Offices/a-lab. Available: https://www.archdaily.com/359599/statoil-regional-and-international-offices-a-lab. Accessed 10 Sep 2024

12. Shanghai Tower/Gensler (2024). Available: https://www.archdaily.com/783216/shanghai-tower-gensler. Accessed 10 Sep 2024
13. Helsinki Airport Extension/ALA Architects. Available: https://www.archdaily.com/973369/helsinki-airport-extension-ala-architects. Accessed 10 Sep 2024
14. Randselva bridge: closing the gap on drawing-free design. Available: https://www.tekla.com/resources/case-studies/randselva-bridge-closing-the-gap-on-drawing-free-design. Accessed 10 Sep 2024
15. Li VC, Herbert E (2012) Robust self-healing concrete for sustainable infrastructure. J Adv Concr Technol 10:207–218
16. What is a circular economy?. Available: https://www.ellenmacarthurfoundation.org/topics/circular-economy-introduction/overview. Accessed 10 Sep 2024
17. Lieder M, Rashid A (2016) Towards circular economy implementation: a comprehensive review in context of manufacturing industry. J Clean Prod 115
18. Geng Y (2010) Emergy analysis of an industrial park: the case of Dalian China. Sci Total Environ 408(22)
19. Yuan Z, Bi J, Moriguichi Y (2006) The circular economy: a new development strategy in China. J Ind Ecol 10
20. Allwood JM (2011) Material efficiency: a white paper. Resour Conserv Recycl 55(3)
21. Pomponi F, Moncaster A (2017) Circular economy for the built environment: a research framework. J Clean Prod 143:710–718
22. Leising E, Bocken N, Quist J (2018) Circular economy in the building sector: three cases and a collaboration tool. J Clean Prod 176:976–989
23. Building d (Emountable). Available: https://www.cepezed.nl/en/project/building-demountable/28429/. Accessed 10 Sep 2024. Available: https://archello.com/project/triodos-bank. Accessed 10 Sep 2024
25. Corbusier L (1923) Versune architecture
26. Aitchison M (2017) A house is not a car (Yet). J Arch Educ 71(1):10–21
27. Erdağ AA (2024) Architectural intelligent smart coat 'shape or ship out. PhD Dissertation
28. Graham, as cited by Robert Schmidt III and Simon Austin, Adaptable Architecture, Routledge, 2018, 2005
29. Austin R.S.I.A.S (2018) Adaptable architecture, Routledge
30. Poinet P (2020) Enhancing collaborative practices in architecture, engineering and construction (PhD Thesis), Aarhus School of Architecture/Design School Kolding/KADK

Ecologizing Change: Can Humanity Afford NOT Extending the SI to Human, Social, and Natural Capital?

William P. Fisher Jr.

Abstract Increasingly catastrophic instances of human suffering, social discontent, and environmental degradation demand decisive action. The successes of global science and commerce giving rise to the emergence of improved standards of living around the world over the last 200 years have fundamentally altered the planetary environment. The logic and methods by which those successes were won no longer solve problems, they create them. The need for a new type of thinking poses the complex problem of figuring out exactly what it means to think differently. Plainly, to date, individual-level changes in the content of what is thought about and what is emphasized in our thinking have been inadequate to the task of inspiring drastically transformed ways of thinking and acting. The crux of the matter concerns the fact that the human, societal, and environmental problems created by global science and commerce are being addressed by fragmented and disconnected solutions. Until human, social, and natural capital are measured and managed on the same locally adaptable global scale as manufactured capital, little in the way of progress toward the needed solutions can be expected. Cultural progress depends on the developmental level of support provided by the institutional infrastructures of science and commerce. The logic, aesthetics, and ethics informing a participatory culturing of new ecological economies—a process of "ecologizing change"—offers practical methods for more effectively addressing the urgent challenges faced by humanity today.

1 Introduction

Referring to the 2023 United Nations climate change report, Sarah Kaplan, writing in the 20 March 2023 issue of the *Washington Post*, states:

Extending the SI, the International System of Units, is crucial to that process.

W. P. Fisher Jr. (✉)
Living Capital Metrics LLC, Sausalito CA, USA
e-mail: wfisher@berkeley.edu; wpfisherjr@livingcapitalmetrics.com

BEAR Center, Berkeley School of Education, University of California, Berkeley, CA, USA

N. M. Durakbasa and K. G. Gülen (eds.), *Sustainable Green Conversion*, Lecture Notes in Mechanical Engineering, https://doi.org/10.1007/978-3-031-83583-4_3

> "The world is likely to pass a dangerous temperature threshold within the next 10 years, pushing the planet past the point of catastrophic warming—***unless nations drastically transform their economies*** and immediately transition away from fossil fuels, according to one of the most definitive reports ever published about climate change."

As circumstances continue to worsen, it is painfully apparent that we do not have anything in the way of a coherent response to this challenge. Where is the inspiring rallying cry and compelling response to the urgent need to drastically transform our economies? Those calling for the political will to do what needs to be done find themselves trapped between economy-crushing costs and increasingly catastrophic storms, droughts, fires, floods, and heat waves. Despite years of effort, we do not have a galvanizing vision or unifying plan for how this drastic transformation is to be brought about.

New kinds of measurement systems are needed because we seriously need to start imagining and inventing a new future. But what does it mean for nations to "drastically transform their economies"? What are the foremost aspects of the current economy that need to change? Einstein is often quoted as saying something to the effect that "Today's problems cannot be solved using the same kind of thinking that created them." That is well stated, even if these were not exactly his words [1] (p. 292). What he actually said involved the need for a new type of thinking if humanity is to survive in the global environment changed by high technology, especially nuclear weapons. The impetus to hard and clear thinking stimulated by Einstein, however, has not led to solid results [2, 3].

There is broad consensus on the idea that the existing economy's linear destruction of resources is too wasteful to continue unchanged. There is also wide consensus on the idea that a circular cycle of resource regeneration and use would be more sustainable. But several decades of efforts on massive scales undertaken at great expense are not producing the desired results or the needed "drastic transformation." The need to "immediately transition away from fossil fuels" becomes more urgent every day.

So, we might ask just how "drastic" is the linear to circular transformation? Are economic fundamentals actually being addressed? Have we even considered what the economic fundamentals are, and how they might be "drastically transformed" in ways that address the root causes of the impending climate change catastrophes, and that offer realistic possibilities for immediately transitioning on mass scales to alternative energy sources?

Four sets of economic fundamentals need to be addressed in order to effect drastically transformed economies. These concern:

1. The complex interactions of science, the law, markets, and communications.
2. The sources of capital value invested in production.
3. How capital comes to life in self-sustaining ecological economies.
4. New possibilities for theory and practice opened up by advanced technical developments.

Unfortunately, the vast majority of efforts aimed at rethinking economic theory and reinventing capitalism [4–16] do not closely examine the interactions of these four domains and the associated mystery of capital [17] or the soul of capitalism [18]. Most pointedly, they do not address the vitally important fourth point in that list concerning the opportunities opened up by complex adaptive systems perspectives [19–21] on advanced measurement and metrology [22, 23].

Those issues are to be considered here with the additional concern that the utilitarian logic characterizing these economic fundamentals is likely inadequate to the task of drastic transformation. That logic likely needs to be complemented by aesthetic and ethical considerations to be made emotionally accessible and politically compelling. And even then, because of the technical complexities involved in putting these ideas to work, these kinds of arguments focused on persuading and educating individuals are still likely to fall short of inspiring imaginative new ways of thinking and acting on the needed global scale. For that captivation of the human spirit to occur, experiential examples of the collective processes involved in opening up transformed horizons and culturing new kinds of ecological economies will be needed. In what follows, we will take up this sequence of logic, aesthetics, ethics, and examples.

2 Logic

A. *The Four Domains Creating the Prosperity of Modern Life*

Over the last 40 years, a number of Nobel-winning economists [24–29] have expanded in various ways on the general point that:

> "Markets are not spontaneously generated by the exchange activity of buyers and sellers. Rather, skilled actors produce institutional arrangements, the rules, roles and relationships that make market exchange possible. The institutions define the market, rather than the reverse." [30] (p. 710)

Bernstein [31], then, focuses on four fundamental domains contributing to the prosperity of the last 200 years, namely:

- Scientific rationality
- Legal property rights
- Capital markets
- Communications and transportation networks

Bernstein points out that science contextualizes the other three domains by informing an atmosphere valuing innovation and a moral environment in which change agents need not fear retribution.

Crucially, without the science of experimentally validated, theoretically explained, and regulated metrological unit standards, legally defensible property rights and capital markets are inconceivable. As North [26] (pp. 18–19, 36) put it:

> "Without some form of measurement, property rights cannot be established nor exchange take place."

> "One must be able to measure the quantity of a good in order for it to be exclusive property and to have value in exchange. Where measurement costs are very high, the good will be a common property resource. The technology of measurement and the history of weights and measures is a crucial part of economic history since as measurement costs were reduced the cost of transacting was reduced. So long as some characteristic of a good that has economic value is not measured, then there is divergence between private and social cost."

Or, as Benham and Benham [32] (p. 369) say,

> "Economic theory suggests that changes in transaction costs have a first-order impact on the production frontier. Lower transaction costs mean more trade, greater specialization, changes in production costs, and increased output."

By expressing how much of a quantity of a good is available in an easily understood and universally accessible unit, metrologically traceable measurements lower transaction costs. Goods that cannot be measured in quality assured common metrics, such as contemporary quantifications of the intangible assets of human, social, and natural capital, incur transaction costs too high to associate measured amounts with property rights, and are typically treated as common pool resources managed via government-administered infrastructural and welfare economics.

Before expanding further on the key role played by metrological systems of measurements in the economy in facilitating abstract representations of property in courts and banks [17, 33], let us first attend to the second domain of fundamentals.

B. *The Sources of Capital Value*

The traditional definition of the economy as comprised of investments in and profits from land, labor, and manufactured capital has been found incomplete in recent years [8, 9, 34, 35].

- Land is better conceived in terms of natural capital and the huge value provided by nature's regenerative capacities and provision of air, water, sunlight, fisheries, forests, minerals, etc.
- Labor is better conceived in terms of a diverse array of forms of human capital, such as literacy, numeracy, health, and other forms of individuals' resourceful contributions.
- And then there is also social capital, the trust, commitment, loyalty, and relational value without which none of the other forms of capital have any economic value at all.

Social capital is of particular importance for the role it must play in a circular and lean ecological economy. To avoid the excesses and disparities characterizing the current economy, social capital will have to provide checks and balances capping accumulations associated with any kind of negative impacts on individuals, communities, or the environment. For instance, minimum social capitalization requirements might replace taxes to the point that they would be sufficient not only for maintaining investments in physical infrastructure but also for paying reparations and guaranteed minimum incomes.

C. *Bringing Capital to Life*

Table 1 juxtaposes these two domains of economic fundamentals. As suggested by the row of four entries of "Yes," the prosperity of the modern age has been created largely via science and commerce involving manufactured capital, which of course includes real estate and buildings in addition to agriculture, electrical appliances, machinery, and other technologies.

When the four institutional infrastructures of science, law, markets, and communications networks necessary to the creation of self-sustaining economies are considered in relation to all four forms of capital, the question emerges as to whether a "drastic transformation of the economy" can be achieved without bringing natural, human, and social capital to life. The potential for a truly new type of thinking is suggested by the fact that a metrological infrastructure of scientific measurement seems to be literally inconceivable in the contemporary discourse around potential solutions for problems widely recognized as threatening catastrophic consequences.

The nearly complete unmentionability of the idea that an extension of the International System of Units (commonly referred to as the metric system, the SI units, or simply, the SI) is technically feasible [36] conceals the further distinct possibilities that it is urgently needed, intensely beautiful in its mathematics, and profoundly ethical [37]. The unexamined assumption that human, social, and environmental qualities are absolutely unmeasurable is so deeply embedded in the collective consciousness that urgently needed models and methods have been ignored for decades, and viable paths toward global solutions have been completely unimaginable. This circumstance is being altered by a new vision blending technical, aesthetic, and ethical considerations.

D. *New Possibilities Opened up by Advanced Technical Developments*

Largely unnoticed outside of fairly restricted technical circles, key advances in measurement modeling, instrument calibration, and unit quantity definitions have emerged over the last 100 years in psychology and the social sciences. These developments have been recognized by metrologists, physicists, and engineers in the last decade or so as opening new metrological vistas [22, 23, 36, 38–41]. As established in that literature, probabilistic model structures, results replications, and wide range of successful applications point toward methods for making complexity science practical on broad scales. Multiple examples of these kinds of applications are available in the domain of sustainability [2, 42–54].

Table 1 The four capitals in relation to the four key contributors to prosperity

Form of capital	Scientific rationality	Legal property rights	Capital markets	Communications networks
Human	No	No	No	No
Social	No	No	No	No
Natural	No	No	No	No
Manufactured	Yes	Yes	Yes	Yes

A crucial idea introduced in this work shifts attention away from the assumption that social and psychological measurement is primarily a process of statistical data analysis. Instead, attention is focused on the metrological idea that measurement is primarily a matter of reading calibrated instruments distributed to end users who are enabled to align and coordinate their decisions and behaviors at a distance, virtually, without having to negotiate specific details in direct communications. A long history of replicated results of this kind supports an optimistic assessment that new potentials could be realized in the economics of human, social, and natural capital [2, 39, 55–61]. The difference between the data analytic sense of measurement and the metrological sense of measurement as a distributed cognitive system is vitally important. It provides us with a fundamental shift in the way we think, not just a superficial shift in what we think about. Examples of differences in ecological thinking obtained across contexts varying in their situational constraints are provided by Kaiser [48, 50, 51, 62].

Metrologically traceable instruments distributed to end users for on-the-spot interpretation and use have existed in psychology and the social sciences for decades, but they have not been conceived or presented in metrological terms [39, 44]. This is changing as new possibilities emerge from measurement research for empowering individuals and communities with new kinds of knowledge and feedback [21] on the outcomes of their investments in education, health care, employment, social service, and environmental management. By attending to the multilevel nature of irreducibly complex systems, we can see how axiomatic ideals, operational abstractions, and empirical observations are all implicated simultaneously, just as are both particle and wave are in the quantum phenomenon.

That complexity offers otherwise unavailable advantages for adaptive resilience. As Dewey [63] (p. 277) wrote: "Every science and every highly developed technology is a systematic network of connected facts and operations." Whitehead [64] (p. 107), similarly, noted in the wake of the quantum revolution that it was not the case that all of the physicists suddenly became more imaginative at the same time. Rather, they had better instruments enabling them to think together in new ways. Latour [65] (p. 247–257), accordingly then emphasizes the essential role played by metrological networks in coordinating individuals' behaviors and decisions virtually, at a distance, with no need to communicate or negotiate shared understandings. The crucial role of metrology is spelled out by Latour [65] (pp. 248–252) when he writes:

- "Every time you hear about a successful application of a science, look for the progressive extension of a network."
- "The predictable character of technoscience is entirely dependent on its ability to spread networks further."
- "Metrology is the name of this gigantic enterprise to make of the outside a world inside which facts and machines can survive."
- "Scientists build their enlightened networks by giving the outside the same paper form as that of their instruments inside. [They can thereby] travel very far without ever leaving home."

- "That much more effort has to be invested in extending science than in doing it may surprise those who think it is naturally universal."
- "The paramount importance of metrology (like that of development and industrial research) gives us a measure, so to speak, of our ignorance."

Situating measurement in the context of the new institutional economics' documentations of the metrological replications and translations occurring across the scientific, legal, market, and communications domains, we can see the truth and value in Callon's [66] (p. 58), [67] idea that "Translation networks weave a socionature," a shared social reality structured to make objectively measurable entities key partners in science and commerce.

And so, with better instruments for measuring and managing literacy, health, community, and environmental capital, might we similarly catalyze a new paradigm of imaginative possibilities for transforming our economies? In fact, can we afford not to?

3 Aesthetics

Past these purely utilitarian considerations, I would like to also include the more affective, aesthetic, and ethical aspects concerning those capacities for personally rewarding growth and emotionally satisfying feelings of connection and community life. Although experiences of beauty, joy, and meaningfulness are too rare for much of the population even in the best of times, access to such experiences would seem to have greatly increased over the last 200 years in proportion to what came in the centuries before. Today, in contrast, these experiences are again in decline.

I emphasize this point because the complexity of the issues involved can make it difficult to keep things in proportion and maintain a broader perspective. Scientific rationality is not inherently reductionistic or homogenizing. Music gives an excellent example of how high technology, powerful electronics, and precision instrumentation become the media for some of the most subjectively absorbing and spiritually rewarding experiences possible. Capacities for personally rewarding growth and emotionally satisfying feelings of connection and community life are as important as the more utilitarian factors. Many other kinds of similar experiences made possible by scientific rationality are also productively seen in analogous terms.

To say the least, however, much remains to be done. We can begin from Plato's *Symposium*, where Socrates recounts Diotima's story of Eros. Being the son of Poros, an immortal god of expedience and abundance, and Penia, a poor mortal woman, Eros is suspended between full possession and complete dispossession. Eros then teaches us how to understand meaning in language [68], as we never obtain absolute transparency in discourse nor are we completely unaware of it, just as in love we are similarly simultaneously desiring and possessing, blind and visionary, and ignorant and wise. Accordingly, just as the pillars of the Parthenon were designed with

Diotima's aesthetic in mind—symbolizing democracy via their deliberately introduced imperfections—each uniquely curved but harmonized in a united effort [69, 70]—so we should also strive to make our measurements of human, social, and natural capital democracies in which unique individuals are joined in common cause.

Doing so will follow through from Gadamer's [71] (p. 478) point that, "When we understand a text, what is meaningful in it captivates us just as the beautiful captivates us." He continues, suggesting that this captivation with meaning and beauty can be extended in that:

> "The path of love that Diotima teaches leads beyond beautiful bodies to beautiful souls, and from there to beautiful institutions, customs, and laws, and finally to the sciences."

Each step along this path intensifies and complexifies the paradoxical simultaneity of irreducibly and systematically integrated, discontinuous but isomorphic, formal, abstract, and concrete levels of beauty and meaning. Though some uncritically impose a reductionist sequence of Platonic idealizations on movement up the ladder [72, 73], others recognize the value obtained when Eros' characteristic ambiguities are retained [74–76].

Linguistic and musical standards convincingly demonstrate that standardization need not compel totalizing narratives of domination, control, and dehumanization. Musical standards are liberating, joyful, participatory, and playful ways of using instruments tuned to common scales that can be creatively improvised in jazzy, blue, and bent notes. And as further evidence of humanity's place in nature, the same mathematics that structures musical scales, nautilus shells, and galaxies is also found in the scaling algorithms that map dialogical question and answer processes onto equal interval measurements [77].

Absorption into the playful flow of subjective experience is the defining feature of the aesthetic attitude [71] (p. 116). That flow is inherently embodied in a nondualistic unity of subject and object that characterizes human experience in the same way as nature [71] (p. 104). The universality of the concept begins from the undifferentiated commonality emerging from diverse arrays of singular observations [71] (pp. 350–351). We methodically retain a memory and record of that commonality when we "save the phenomena" by devising reproducible representations of the activity of the thing itself experienced in thought [71] (p. 474).

Art and science are both creative in this way and both tap into the ways in which instruments extend embodied experience by mediating unified flows of thing and thought. As Gadamer [71] (p. 479) says,

> "The close connection between the mathematical orders of the beautiful and the order of the heavens means that the cosmos, the model of all visible harmony, is at the same time the supreme example of beauty in the visible sphere. Harmonious proportion, symmetry, is the decisive condition of all beauty.
>
> "As we can see, this kind of definition of the beautiful is a universal ontological one. Here nature and art are not in antithesis to each other. This means, of course, that in regard to beauty the priority of nature is unquestioned."

The mathematical order of the beautiful begins from the erotic or pragmatic idealism of integrated levels of complexity already present in geometry. Here, in the insight foundational to the origin of philosophy [78] (pp. 35, 100–103), [79] (pp. 75–76), [80] (p. 255), the undrawable and unwritable infinity of irrational numbers like pi and the square root of 2 plainly must be distinct from the drawn line or curve and the written number. The aesthetic priority of nature can be extended to human nature in that the stochastically invariant, or deterministically chaotic [20], objectively repeatable and reproducible structures of learning progressions, developmental sequences, and healing trajectories are just as harmoniously proportionate and symmetrical as any physical, cosmological, or biological phenomenon. The paradoxical unity and separation of thing and thought caught up together in the mutually implicating dance of subject and object then lends itself as well to measurements of human and social attributes as it does in the natural sciences [81, 82].

4 Ethics

Ricoeur [83] contends that the fundamental challenge facing the project of a social ethic involves reconciling the unifying value of addressing humanity taken as a whole with the opposed task of maximizing the personalization of relations which tend too easily to become abstract, anonymous, and inhuman in today's society. Ricoeur [83] (p. 166) recognizes then that "a social ethic cannot spring from a system but from a paradox" because "it aims at two opposed things: human totality and human singularity." He nonetheless asserts: "I want both."

We achieve both via the overt and deliberate incorporation of multiple levels of complexity into our communications and metric standards. Star and Griesemer's [84] (p. 392) concept of the boundary object echoes Scott's [85] (p. 357) recommendation that we take language as a model for systems intended to improve the human condition, since language is "a structure of meaning and continuity that is never still and ever open to the improvisations of its speakers." Star and Griesemer [84] (p. 392) similarly say that:

> "Boundary objects are objects which are both plastic enough to adapt to local needs and the constraints of several parties employing them, yet robust enough to maintain a common identity across sites. They are weakly structured in common use, and become strongly structured in individual site use. These objects may be abstract or concrete. They have different meanings in different social worlds but their structure is common enough to more than one world to make them recognizable, a means of translation. The creation and management of boundary objects is a key process in developing and maintaining coherence across intersecting social worlds."

Galison's [86] (p. 49) study of scientific subcultures in microphysics arrives at the same kind of result, as he notes "our general linguistic ability to set broader meanings aside while regularizing different lexical, syntactic, and phonological elements to serve a local communicative function." Developmental psychologists similarly have a long history of recognizing that cognitive skills are simultaneously general and specific, with the complexity afforded by the linguistic infrastructure intimately involved in the skill level exhibited by a person performing a task [87, 88]. Social and psychological measurement's systematic integrations of construct maps, Wright maps, and "kidmaps" [22, 46] correspond with these formal theoretical, abstract instrumental, and concrete data levels of complexity [89, 90].

By more explicitly incorporating theoretical ideals, metrological standards, and local adaptations addressing the interrelated but distinct needs of conceptually-, instrumentally-, and empirically-oriented communities into our economic systems, we tap into Plato's philosophical mathematics [91] and are enabled to more fully embed the balance scale as a symbol of justice in the background assumptions structuring science and commerce [92] (pp. 2, 3, 325). This deepening of our shared sense of fair play follows naturally from the distinction between figure and meaning, or between name and concept, foundational to the origins of philosophy in ancient Greece.

This distinction resonated throughout the political discourse of those times. Plato required students in his Academy to be experienced in geometry [91] because it primed them to the fundamental philosophical insight that drawn figures and written numbers are not the mathematical relationships they represent [78] (pp. 35, 100–103), [79] (pp. 75–76), [80] (p. 255). Geometry's simultaneous involvement of concretely unique line segments and formally axiomatic ideals provides a model for abstract standards mediating relationships, and so also for unique citizens joined in common cause. The Parthenon, for instance, symbolizes democracy in that its columns are each curved in slightly different ways and the floor is not parallel with the roof, but the whole is harmonized in a united effort [69, 70].

The same kind of co-evolving relationships can be seen to have occurred again in the late eighteenth and early nineteenth centuries, with the simultaneity of the second scientific revolution, the French and American political revolutions, and the industrial revolution. It is no mere coincidence that metrology was born at this time [92]. Feudalistic and religious hierarchies were displaced as the spread of the printing press, expanded literacy, the emergence of scientific discoveries, and commercially successful technical innovations distributed a sense of agency and personal potentials to larger and larger populations. Knowledge and truth were no longer reserved to restricted segments of the population. The complex relations of unique individuality participating in collectively constituted larger wholes were evolving in a fractally resonating way throughout all segments of society.

Similar processes are again at work today as global communications intensify and expand yet again, simultaneously redefining individuality in yet more ways. Greider [18] (p. 47) remarks that:

> "The essential challenge of reinventing capitalism requires us to restitch the relationships and alter mechanisms so that people may reclaim their individuality. This does not displace self-interest, but broadens its meaning beyond the narrowness of economics. It can empower individuals to assert themselves and their values, but it also depends upon fashioning new webs of mutual obligation and accountability."

A tangible vision of actionable recommendations as to how such new webs of mutual obligation and accountability might be formed follows from well-established measurement science. There is no principled reason why the key role played by metrology in organizing property rights, capital markets, and communications standards in the economy of manufactured goods could not also be played in the domains of natural, social, and human capital.

How would that enable people to reclaim their individuality, to broaden the meaning of self-interest, to empower the assertion of deeply held values? Consider this: for decades, there have been hundreds of millions of people around the world trying to create more sustainable systems [93]. Their efforts have been for naught because our modern institutional infrastructures are systematically disempowering, built as they are around dysfunctional and discredited conceptions of measurement information wrongly presuming that isolated, disconnected, reductionist, homogenizing, incomparable, and incommensurable numeric representations are the only option available for quantifying human, social, and natural capital.

- But what if entrepreneurs everywhere could profitably pursue their visions, implement their plans, and scale up their operations because those activities were actively supported by the scientific, legal, accounting, and communications standards embedded in the social environment?
- What if individuals globally were empowered to act and to take the initiative instead of having to wait on leaders to show the way?
- What if the need to "immediately transition away from fossil fuels" [94] became a realistic possibility because innovators were able to leverage tools for creating self-sustaining ecological economies of human, social, and natural capital?
- What if drastically transformed infrastructures empowered creative innovators everywhere by aligning authentic human, social, and environmental wealth with finance?
- What if today's invisible and voiceless members of society were able to monetize their measured contributions in the arts, nonprofits, homemaking, volunteering, and caregiving?
- What if philanthropic capital could be invested in consistently productive efforts enabled by comparable service outcome measurements to become globally scalable?
- What if literacy, health care, environmental management, and other efforts could compete in open markets financially rewarding lower costs and higher quality?
- What if employees could more easily identify markets for their skills and scale up the value they are able to offer because their performance metrics are comparable across learning environments and employers?
- What if consistently priced outcomes told customers where to find the best value in the educational, health care, social services and other products they seek?
- What if everyone felt personally invested in ways of representing themselves that really express who they feel they are, that gave them a sense of ownership enabling them to fulfill their dreams?

The existing institutional infrastructure disempowers producers, consumers, suppliers, and investors instead of providing them with rules, roles, and responsibilities [30] supporting their various entrepreneurial initiatives in bringing human, social, and natural capital to market. We need to put the new institutional economics of human, social, and natural capital metrology to work to provide people with comparable, quality assured measurements and legally defensible ownership of their own shares of their personal stocks of human, social, and natural capital. We need to create an investment climate in capital markets where monetary profits simply cannot be had when human, social, or natural capital is being degraded or destroyed. We need investments in intangible assets to be accountable and tracked in ways that distribute universal access to information and opportunities for using it.

What if the maxim of doing well by doing good got some real traction in actual scientific and legal alignments of measured, financial, and social values?

5 Culturing Participatory Social Ecological Economies: Ecologizing Change

I propose that we initiate popular movements creating new scientific, legal, market, and communications infrastructures. I think everyone everywhere ought to be made aware in their own personal experience that numbers are not automatically quantitative, that creating measurements takes skill and technical expertise, that having people answer different questions in a domain does not mean their measurements are inherently incomparable. I think CEOs and teachers, clinicians and entrepreneurs, dishwashers and therapists all ought to have personal experience with improved representations of the value they create, tools that can be had and can be put to good use in applications advancing their personal interests. I think that, instead of assuming that locally created surveys and assessments are inherently restricted in their relevance and applicability, everyone ought to be informed that there are ways in which all the instruments and values could be synched up in a kind of organically spreading social ecology. I think that, instead of waiting on political and corporate leaders to figure out what needs to be done, entrepreneurially minded people everywhere ought to be empowered to follow through on their creative efforts to produce authentically profitable wealth.

I think that, instead of the depressing focus on shouldering the burden of completely unfathomable costs, we have clear opportunities for galvanizing inspired and passionate responses to today's urgent demands for drastic transformations of our economies [2, 37, 39, 44, 55–61]. Making the needed new technologies widely available and inexpensive will likely require AI applications, which opens up another array of issues too complex to take up here [95–97].

But even setting aside those issues for later consideration leaves us engaged with technical matters impossible to address in the typical modern fashion of centrally planned operations methodically following a set of rules to achieve ends reducible to one or another idealist, operationalist, instrumentalist, or empiricist criterion. What is needed instead is an unmodern, ecologizing approach [98, 99] focused on culturing spontaneously self-organized relationships responsive to and mediated by collectively projected structural invariances.

We encounter here the challenge of operationalizing Gadamer's points concerning the definition of beauty in measure, and the role of play in method. How can we make practical use of structures of proportionate invariance that reveal themselves in collective linguistic processes not under the control of any individual? The problem of method in this context is well stated by the Nobel economist, Hayek [100] (p. 88):

> "The problem is precisely how to extend the span of our utilization of resources beyond the span of the control of any one mind; and, therefore, how to dispense with the need of conscious control and how to provide inducements which will make individuals do the desirable things without anyone having to tell them what to do."
>
> "The problem [of the use of knowledge in society] which we meet here is by no means peculiar to economics but arises in connection with nearly all truly social phenomena, with language and with most of our cultural inheritance, and constitutes really the central theoretical problem of all social science. As Alfred Whitehead [101] (p. 61) has said in another connection, 'Civilization advances by extending the number of important operations which we can perform without thinking about them.' This is of profound significance in the social field. We make constant use of formulas, symbols, and rules whose meaning we do not understand and through the use of which we avail ourselves of the assistance of knowledge which individually we do not possess."

This problem, of course, is not entirely new. No one needs to understand thermodynamics to know how to make good use of a thermometer. The problem of the use of knowledge in society has not only been successfully addressed throughout the entire history of manufactured capital economics, it is inherently incorporated in the functions of everyday language, where fluent speakers need not concern themselves about knowing why or how alphabets, phonemes, and grammars came to take the form they have. The unsolved problem we face is one of conceptualizing how to model relationships involving human, social, and natural capital in ways that adapt to their specific requirements and attributes while also facilitating the needed means for addressing the problems of the day.

For an historical example of ecologizing change in the domain of manufactured capital, consider Thomas Edison. We think of Edison as the inventor of the lightbulb, but he did not invent them; they already existed when he came into the field. Instead, he not only improved them, he more importantly created electrical generation and distribution systems that standardized interoperable current and resistance metrics. Edison made electrical metrology FAIR—Findable, Accessible, Interoperable, and Reusable [102–104]—on mass scales.

Edison recognized that the electrical systems of his day were both the problem and the opportunity [105–107]. Edison saw the inefficiency and waste of the existing disconnected electrical systems' needs for skilled engineers to run engines and generators, and to set and maintain unique voltage and power standards. Edison then envisioned and operationalized distributed electrical networks capable of powering low maintenance, quality assured, and interoperable electrical appliances. For this, he had to coordinate the efforts and interests of engineers, lawyers, accountants, managers, customers, investors, and many others.

Similarly, we today can see the inefficiency and waste of disconnected human, social, and natural capital measurement and management systems in education, health care, human resources, social services, environmental management, etc. Skilled psychometricians and statisticians are needed to design and execute centrally planned methods of quantification that result in incommensurable values isolated from others' information systems, rarely tailored to individual needs, and incapable of being scaled up to meet urgent demands for actionable, virally communicable knowledge. Methods exist for altering these conditions. No matter how irreconcilably built into today's institutional infrastructures these conditions may seem, the mathematical equivalence and objective repeatability of measured results across the sciences compels us to recognize the incredible value that stands to be created if we focus on networked organism-environment unities as the source of innovative connectivity. The ecosystem, not the invention, is the innovation.

We would not know Edison's name today if he had not had the crucial insight into the opportunity for altering the environment in which the lightbulb and other electrical appliances could not only take on self-sustaining economic lives of their own but could also evolve [21]. The progressive transformations of electronics from tubes to transistors to integrated circuits and chips became possible as a consequence of the electrical ecosystems created by Edison and his collaborators. Instead of conceiving of the advancement of science solely as a function of brilliant individuals who possess the genius to create insightful innovations out of their own innate capacities, we will make far more rapid progress if we create systems enabling the design of tools structurally integrated with infrastructural supports.

This shift in psychology and the social sciences from incommensurable systems of isolated individuals and groups to interconnected and comparable metrics contextualized by institutional infrastructures may be implicated in the meaning of a quotation often attributed to Edison's competitor, Nicola Tesla: "The day science begins to study non-physical phenomena, it will make more progress in one decade than in all the previous centuries of its existence." By so scaffolding the institutional environment to include intangible assets, we will thereby further enable end users to communicate and advance toward their own self-determined goals and aspirations, FAIRly and meaningfully.

Three approximations involving measurements in education and healthcare are available. Stenner and colleagues devised a measure of reading comprehension that has served as the medium of a literacy capital marketplace since 1995, and which currently serves over 35 million measurements annually in three languages in 80 countries [108, 109]. Morrison and colleagues [21, 110–112] working in the area of STEM education, improved communications across teachers, informal educators, city governments, employers, philanthropists, and other stakeholders by replicating and translating their interests in cultivating empowered participation. In the domain of low vision rehabilitation, Massof and colleagues [113–117], created an Activity Inventory enabling optometry and ophthalmology providers, patients and families, care advocates, equipment suppliers, and community organizations to better coordinate their services and products.

Each of these tools and systems for their application were directed and managed so far as possible in the traditional manner requiring centrally planned and executed organizations and reporting relationships. This was most true in the case of the most commercially successful ecological economy, that involving literacy capital, and was least true in the case of the STEM learning ecosystems, which are philanthropically funded and which also have not enjoyed nearly the scale of uptake seen in the literacy capital domain. But in each case, so far as they were involved in the measurement research or application, participants were recruited and responsible for expertise in a specific area and were under the supervision of a more senior authority capable of judging and guiding contributions. Leadership in the various domains was required to have a clear understanding of the purpose and mission, with the coordination of replications and translations handled by between-group facilitators.

That is to say, there was no intention in any of these three examples of devising large scale publicly available ecological economies. No property rights conferring legally defensible and financially accountable ownership of investments in or profits from unit gains in literacy capital, for instance, were conceived or planned. This means, then, that a significant degree of modern control over the means of production and the value so produced was retained. The viability of a more ubiquitously distributed unmodern multilevel ecological economy requires, of course, the active participation of stakeholders representing the broadest possible and most inclusive societal levels, as is already the case with the authorities and responsibilities possessed by national metrology institutes.

A more distributed and overtly leaderless development process might, then, be most effectively cultivated by bringing together representatives of the metrological, legal, market, and communications domains with the express purpose of engaging in discussions around the viability, feasibility, and desirability of science and commerce in human, social, and natural capital. And beyond mere theory, such a convening ought to be organized to consolidate a shared vision and to formulate the terms in which new directions for addressing the urgent challenges of the day might be imagined and pursued.

Toward these ends, a game and a screenplay, both referred to as "Ecopoly," enacting fictional dramas illustrating new social, political, and economic possibilities are in development. In these contexts, players and audience members actively participate in events as they unfold by means of a smartphone app. The login accounts created and the relationships formed in the course of the game and theatrical play are then retained by participants for later use continuing the process of ecologizing change in real life.

For groups ("ecopolists") interested in a more intensive involvement in developing local ecological economies ("ecopolizing"), the Ecopoly game and the play have been incorporated in a curriculum of training seminars intended to follow through on developing the relationships initiated. Contact the author for more information.

6 Conclusion

Coherent and compelling new directions for sustainability research and practice—and for the future of humanity and the Earth—are indicated by the contrast between the modern Western dualist metaphysics of subjectivity alienated from an independently existing, objective universe that science can only describe, and an unmodern nondualist metaphysics of mutually implicated subjects and objects absorbed together into playfully dancing flows of dialogue. Several decades of popular press on the emergence of a new science of irreducible complexity and unfragmented wholeness [118–126] have not, however, catalyzed a new type of thinking or a transformation of our institutions from systematically disempowering to empowering.

Physical science provides us with both a particularly compelling source of fragmented thinking and an even more powerful source of unified thinking. A consensus perspective has yet to emerge on the paradigmatic differences between the Newtonian mechanical clockwork universe and the relativistic quantum universes described by Einstein and Bohr. But there is nonetheless a plain point of contrast relevant to transforming the way we think. That contrast concerns the metaphysical root metaphor or world hypothesis minimizing human experience and alienating it from an incomprehensibly miniscule and cosmic universe, versus a vision of an alternative universe that includes human being as a fundamental participant in its unfolding.

The practical consequence of this contrast concerns the fragmented versus unified forms of thinking that follow from differences between measurement methods assuming that the description of objective reality is the goal, and those that instead prescribe the existence of an objectively reproducible quantity. Taking the former perspective of subjectivity separated from objects, the assignment of numbers is said to result in interval or ratio quantities when objective reality is involved, but human subjectivity is not objective, and so incommensurable, scale- and sample-dependent ordinal scores are held to be the inevitable, unavoidable, and necessary result.

But in adopting the latter perspective of mutually implicated subjects and objects, the mere assignment of numbers to observations is deemed inadequate in all sciences. History shows in fact that metrological uniformity is a social accomplishment [65, 127, 128], not something pre-existing in nature. Where modernism sees number as existing in nature, as the Pythagoreans did, the unmodern sees both nature and humanity as mathematical, in the manner of Plato.

This transformed way of thinking characterizes the shift from the modern concept of science as describing a mechanical clockwork universe to an unmodern concept of science as playfully participating in a living and complex universe. Multiple physicist philosophers speak of humanity absorbed in a new participatory dialogue with nature [37, 118, 123, 126, 129–136]. Like light through the leaves, puppies and kittens, and waves on the beach, humanity TOO plays and is in fact inescapably absorbed in the play of its language games, as Wittgenstein [137] put it. Gadamer [71] (pp. 101–104) similarly says that humanity's natural enjoyment of play means we, too, are natural. Gadamer [71] (p. 104) emphasizes that this "permits us to draw an important methodological conclusion" that follows from the conceptual roots of method in the ancient Greek meta-odos [71, 81, 82, 91], where desires for meaning, beauty, and existence absorb attention and thought, inducing them to follow along after (meta) things themselves on the paths (odos) they take of their own accord.

This shift in thinking from being separate and alienated from a mechanistic universe to participatory belonging in a living universe marks a drastic transformation in humanity's self image. Here we have begun to flesh out the form and function of new unmodern ways of thinking and living that stand as truly paradigmatically distinct from contemporary modern thinking and living [63, 98, 99]. Reconceiving and operationalizing science in unmodern participatory and playful terms is key to envisioning the drastic transformations of the economy that we need if we are to succeed in addressing the problems we face. This shift has remained a latent possibility not yet made real because of the apparent lack of any means or methods for operationalizing it on mass scales in logically, aesthetically, and ethically satisfying ways. New progress in drastically transforming our economies becomes possible with the advent of validated ideas and methods in support of metrologically viable and personally meaningful measurement modeling and standards in the domains of human, social, and natural capital.

Acknowledgements The author thanks John Baron, Bob Massof, Jan Morrison and Mark Wilson for ongoing conversations, questions, and insights. All errors and omissions are the responsibility of the author. This material was presented in various forms at the inaugural conference of the Society for the Study of Measurement at the University of California, Berkeley, in early August 2024; at the IMEKO World Congress in Hamburg, Germany, in late August 2024; and at the International Symposium for Production Research in Budva, Montenegro (via Zoom), in early October, 2024.

References

1. Calaprice A (ed) (2005) The new quotable Einstein. Princeton University Press
2. Fisher WP Jr (2020a) Contextualizing sustainable development metric standards: Imagining new entrepreneurial possibilities. Sustainability 12(9661):1–22. https://doi.org/10.3390/su12229661
3. Fisher WP Jr (2020) Measuring genuine progress: an example from the UN millennium development goals project. J Appl Meas 21(1):110–133. https://papers.ssrn.com/sol3/papers.cfm?abstract_id=3503715
4. Anielski M (2007) The economics of happiness: building genuine wealth. New Society Publishers
5. Arthur WB (2014) Complexity economics: A different framework for economic thought. In: Arthur WB (ed) Complexity and the economy. Oxford University Press, pp 1–29
6. Beinhocker E, Hanauer N (2014) Capitalism redefined: resolving the tension between a prosperous world and a moral one. Juncture 21(1):12–24. https://doi.org/10.1111/j.2050-5876.2014.00774.x
7. Eisler R (2007) The real wealth of nations: creating a caring economics. Berrett-Koehler Publishers, Inc
8. Ekins P (1992) A four-capital model of wealth creation. In: Ekins P, Max-Neef M (eds) Real-life economics: Understanding wealth creation. Routledge, pp 147–155

9. Ekins P, Dresner S, Dahlstrom K (2008) The four-capital method of sustainable development evaluation. Eur Environ 18(2):63–80
10. Farmer JD (2024) Making sense of chaos. Yale University Press
11. Hanauer N, Beinhocker E (2014) Capitalism redefined. Democracy 31:30–44
12. Stiglitz JE, Fitoussi J-P, Durand M (eds) (2019a) For good measure: an agenda for moving beyond GDP. The New Press
13. Stiglitz JE, Fitoussi J-P, Durand M (2019b) Measuring what counts: the global movement for well-being. The New Press
14. Stiglitz JE, Sen A, Fitoussi J-P (2009) Report by the commission on the measurement of economic performance and social progress, pp 292. https://ec.europa.eu/eurostat/documents/8131721/8131772/Stiglitz-Sen-Fitoussi-Commission-report.pdf
15. Stiglitz JE, Sen A, Fitoussi J-P (2010) Mismeasuring our lives: why GDP doesn't add up. The New Press
16. Wilson J, Tyedmers P (2013) Rethinking what counts. Perspectives on wellbeing and genuine progress indicator metrics from a Canadian viewpoint. Sustainability 5(1):187–202
17. De Soto H (2000) The mystery of capital: why capitalism triumphs in the West and fails everywhere else. Basic Books
18. Greider W (2003) The soul of capitalism: opening paths to a moral economy. Simon & Schuster
19. Fisher WP Jr (2017) A practical approach to modeling complex adaptive flows in psychology and social science. Procedia Comput Sci 114:165–174. https://doi.org/10.1016/j.procs.2017.09.027
20. Fisher WP Jr (2024a) Entropy, deterministic chaos, and new forms of intelligibility: a shared frame of reference for physics and psychology. Int J Adv Prod Res 1(1):46–83. https://dergipark.org.tr/en/pub/ijapr/issue/83619/1453834
21. Fisher WP Jr, Morrison J (2024) Ecologizing STEM education: measuring and managing for stakeholder empowerment. Meas Sens 101311. https://doi.org/10.1016/j.measen.2024.101311
22. Mari L, Wilson M, Maul A (2023) Measurement across the sciences: Developing a shared concept system for measurement, 2nd edn. Springer Series in Measurement Science and Technology, Springer
23. Pendrill LR (2019) Quality assured measurement: unification across social and physical sciences. Springer Series in Measurement Science and Technology, Springer
24. Ménard C, Shirley MM (2008) Handbook of new institutional economics. Springer
25. Ménard C, Shirley MM (2022) Advanced introduction to new institutional economics. Edward Elgar Publishing
26. North DC (1981) Structure and change in economic history. Norton
27. Ostrom E (2015) Governing the commons: the evolution of institutions for collective action. Cambridge University Press (Original work published 1990)
28. Williamson OE (1981) The economics of organization: the transaction cost approach. Am J Sociol 87(3):548–577
29. Williamson OE (1991) Economic institutions: Spontaneous and intentional governance. J Law Econ Organ: Pap Conf New Sci Organ 7:159–187
30. Miller P, O'Leary T (2007) Mediating instruments and making markets: capital budgeting, science and the economy. Account Organ Soc 32(7–8):701–734. https://www.sciencedirect.com/science/article/pii/S0361368207000220
31. Bernstein WJ (2004) The birth of plenty: how the prosperity of the modern world was created. McGraw-Hill
32. Benham A, Benham L (2000) Measuring the costs of exchange. In: Ménard C (ed) Institutions, contracts and organizations: Perspectives from new institutional economics. Edward Elgar, pp 367–375
33. Fisher WP Jr (2002) The mystery of capital and the human sciences. Rasch Meas Trans 15(4):854. http://www.rasch.org/rmt/rmt154j.html

34. Gleeson-White J (2015) Six capitals, or can accountants save the planet? Rethinking capitalism for the 21st century. Norton
35. Hawken P, Lovins A, Lovins HL (1999) Natural capitalism: creating the next industrial revolution. Little, Brown, and Co
36. Fisher WP Jr, Pendrill L (2024) Models, measurement, and metrology extending the SI: trust and quality assured knowledge infrastructures. In: Sommer K-D, Fröhlich T (eds) De Gruyter Series in Measurement Sciences. De Gruyter Oldenbourg. https://www.degruyter.com/document/isbn/9783111036496/html
37. Fisher WP Jr (2024c) Measurement logic, aesthetics, and ethics across the sciences: extending the SI units. In: Fisher WP Jr, Pendrill L (eds) Models, measurement, and metrology extending the SI: Trust and quality assured knowledge infrastructures, pp 193–270. De Gruyter Oldenbourg. https://www.degruyter.com/document/isbn/9783111036496/html
38. Cano S, Pendrill L, Melin J, Fisher WP Jr (2019) Towards consensus measurement standards for patient-centered outcomes. Measurement 141:62–69
39. Fisher WP Jr (2023a) Foreword: Koans, semiotics, and metrology in Stenner's approach to measurement-informed science and commerce. In: Fisher WP Jr, Massengill PJ (eds) Explanatory models, unit standards, and personalized learning in educational measurement: Selected papers by A. Jackson Stenner. Springer. https://link.springer.com/content/pdf/bfm:978-981-19-3747-7/1?pdf=chapter%20toc
40. Mari L, Wilson M (2014) An introduction to the Rasch measurement approach for metrologists. Measurement 51:315–327
41. Pendrill L, Fisher WP Jr (2015) Counting and quantification: comparing psychometric and metrological perspectives on visual perceptions of number. Measurement 71:46–55
42. Alvarez P (2005) Several noncategorical measures define air pollution. In: Bezruczko N (ed) Rasch measurement in health sciences. JAM Press, pp 277–293
43. Baldi L, Trentinaglia MT, Mancuso T, Peri M (2021) Attitude toward environmental protection and toward nature: how do they shape consumer behaviour for a sustainable tomato? Food Qual Prefer 90:104175
44. Fisher WP Jr (2021) Bateson and Wright on number and quantity: how to not separate thinking from its relational context. Symmetry 13(1415). https://doi.org/10.3390/sym13081415
45. Fisher WP Jr, Pendrill L, Lips da Cruz A, Felin A (2019) Why metrology? fair dealing and efficient markets for the United Nations sustainable development goals. J Phys: Conf Ser 1379(012023). [http://iopscience.iop.org/article/https://doi.org/10.1088/1742-6596/1379/1/012023]. https://doi.org/10.1088/1742-6596/1379/1/012023
46. Fisher WP Jr, Wilson M (2019) The BEAR assessment system software as a platform for developing and applying UN SDG metrics. J Phys Conf Ser 1379(012041). https://doi.org/10.1088/1742-6596/1379/1/012041
47. Kaiser FG, Biel A (2000) Assessing general ecological behavior: a cross-cultural comparison between Switzerland and Sweden. Eur J Psychol Assess 16(1):44–52
48. Kaiser FG, Midden C, Cervinka R (2008) Evidence for a data-based environmental policy: induction of a behavior-based decision support system. Appl Psychol 57(1):151–172
49. Kaiser FG, Wilson M (2000) Assessing people's general ecological behavior: a cross-cultural measure. J Appl Soc Psychol 30(5):952–978
50. Kaiser FG, Wilson M (2004) Goal-directed conservation behavior: the specific composition of a general performance. Personality Individ Differ 36(7):1531–1544. https://doi.org/10.1016/j.paid.2003.06.003
51. Kaiser FG, Wilson M (2019) The Campbell paradigm as a behavior-predictive reinterpretation of the classical tripartite model of attitudes. Eur Psychol 24(4):359–374. https://doi.org/10.1027/1016-9040/a000364
52. Lips da Cruz A, Fisher WP Jr, Felin A, Pendrill L (2019) Accelerating the realization of the United Nations Sustainable Development Goals through metrological multi-stakeholder interoperability. J Phys: Conf Ser 1379(012046 [http://iopscience.iop.org/article/https://doi.org/10.1088/1742-6596/1379/1/012046])

53. Moral FJ, Rebollo FJ, Méndez F (2014) Using an objective model to estimate overall ozone levels at different urban locations. Stoch Env Res Risk Assess 28(3):455–465
54. Moral FJ, Rebollo FJ, Paniagua LL, García A, de Salazar EM (2016a) Application of climatic indices to analyse viticultural suitability in Extremadura, south-western Spain. Theoret Appl Climatol 123(1–2):277–289
55. Fisher WP Jr (2011) Bringing human, social, and natural capital to life: Practical consequences and opportunities. In: Brown N, Duckor B, Draney K, Wilson M (eds) Advances in Rasch Measurement, vol 2. JAM Press, pp 1–27
56. Fisher WP Jr (2012a) Measure and manage: Intangible assets metric standards for sustainability. In: Marques J, Dhiman S, Holt S (eds) Business administration education: Changes in management and leadership strategies. Palgrave Macmillan, pp 43–63
57. Fisher WP (2012b) What the world needs now: a bold plan for new standards [Third place, 2011 NIST/SES World Standards Day paper competition]. Standards Engineering 64(3):1, 3–5. http://ssrn.com/abstract=2083975
58. Fisher WP Jr (2022a) Aiming higher in conceptualizing manageable measures in production research. In: Durakbasa N, Gençyilmaz MG (eds) Digitizing production systems: selected papers from ISPR2021, 07–09 Oct 2021. Springer Verlag, Turkey. https://link.springer.com/content/pdf/bfm%3A978-3-030-90421-0%2F1
59. Fisher WP Jr (2022b) Contrasting roles of measurement knowledge systems in confounding or creating sustainable change. Acta IMEKO 11(4):1–7. https://acta.imeko.org/index.php/acta-imeko/article/view/1330
60. Fisher WP Jr (2023c) Measurement systems, brilliant results, and brilliant processes in healthcare: untapped potentials of person-centered outcome metrology for cultivating trust. In: Fisher WP Jr, Cano S (eds) Person-centered outcome metrology: principles and applications for high stakes decision making, pp 357–396. Springer. https://link.springer.com/book/https://doi.org/10.1007/978-3-031-07465-3
61. Fisher WP Jr (2023d) Separation theorems in econometrics and psychometrics: Rasch, Frisch, two Fishers, and implications for measurement. J Interdiscip Econ 35(1):29–60. https://journals.sagepub.com/doi/https://doi.org/10.1177/02601079211033475
62. Kaiser FG, Keller C (2001) Disclosing situational constraints to ecological behavior: a confirmatory application of the mixed Rasch model. Eur J Psychol Assess 17(3):212–221. https://doi.org/10.1027/1015-5759.17.3.212
63. Dewey J (2012) Unmodern philosophy and modern philosophy. In: Deen P (eds) Southern Illinois University Press
64. Whitehead AN (1925) Science and the modern world. Macmillan
65. Latour B (1987) Science in action: how to follow scientists and engineers through society. Harvard University Press
66. Callon M (1995) Four models for the dynamics of science. In: Jasanoff S, Markle GE, Petersen JC, Pinch T (eds) Handbook of science and technology studies. Sage Publications, pp 29–63
67. Arias-Maldonado M (2015) The socionatural entanglement. In: Environment and Society: Socionatural relations in the Anthropocene. Springer, pp 55–74. https://doi.org/10.1007/978-3-319-15952-2_4
68. Gelven M (1984) Eros and projection: Plato and Heidegger. In: Shahan RW, Mohanty JN (eds) Thinking about Being: Aspects of Heidegger's thought. Oklahoma University Press, pp 125–136
69. Blücher H (Bard College, Professor). (1968). XIII. A Dialogue with students. In: Blücher Archive Lecture Transcripts (eds) Senior Symposium. http://www.bard.edu/bluecher/lectures/dialogue/dialogue_pf.htm
70. Cook TA (1914/1979) The curves of life. Dover
71. Gadamer H-G (1989) Truth and method. Crossroad
72. Irigaray L, Kuykendall EH (1988) Sorcerer love: a reading of Plato's *Symposium*, Diotima's speech. Hypatia 3(3):32–44
73. Schott R (1988) Cognition and eros: A critique of the Kantian paradigm. Beacon Press, Boston
74. Nye A (1988) The hidden host: Irigaray and Diotima at Plato's *Symposium*. Hypatia 3(3):45–61

75. Nye A (2015) Socrates and Diotima: Sexuality, religion, and the nature of divinity. Springer
76. Orr D (2006) Diotima, Wittgenstein, and a language for liberation. In: Orr D, McAlister LL, Kahl E, Earle K (eds) Belief, bodies, and being: Feminist reflections on embodiment. Rowman & Littlefield, New York, pp 59–80
77. Stone MH (2002) Musical temperament. Rasch Meas Trans 16(2):873. http://www.rasch.org/rmt/rmt162b.html
78. Gadamer H-G (1980) Dialogue and dialectic: eight hermeneutical studies on Plato. Yale University Press
79. Heidegger M (1967) What is a thing? Regnery/Gateway
80. Heidegger M (1977) Modern science, metaphysics, and mathematics. In: Krell DF (ed) Basic writings [reprinted from Heidegger M (1967). What is a thing? pp 66–108. Regnery/Gateway.]. Harper & Row, pp 243–282
81. Fisher WP Jr (2004) Meaning and method in the social sciences. Human Studies: A J Philos Soc Sci 27(4):429–454. https://rdcu.be/dOzA1
82. Fisher WP Jr (2019) A nondualist social ethic: fusing subject and object horizons in measurement. TMQ Techniques, Methodologies, and Quality, 10(Special Issue on Health Metrology), pp 21–40. https://portal2.ipt.pt/media/manager.php?src=servico&cmd=file&target=m1_MTc2NDQ#page=21
83. Ricoeur P (1974) The project of a social ethic. In: Stewart D, Bien J (eds) (1974) Political and social essays. Ohio University Press, pp 160–175. (Rpt. from Ricoeur P (1966) Le projet d'une morale sociale. Le Christianisme Social 74:285–295)
84. Star SL, Griesemer JR (1989) Institutional ecology, translations and boundary objects: amateurs and professionals in Berkeley's museum of Vertebrate Zoology, 1907–1939. Social studies of science, vol 19(3), pp 387–420
85. Scott JC (1998) Seeing like a state: how certain schemes to improve the human condition have failed. Yale University Press
86. Galison P (1997) Image and logic: a material culture of microphysics. University of Chicago Press
87. Commons ML, Goodheart EA (2008) Cultural progress is the result of developmental level of support. World Futur: J New Parad Res 64(5–7):406–415. https://doi.org/10.1080/02604020802301360
88. Fischer KW, Farrar MJ (1987) Generalizations about generalization: How a theory of skill development explains both generality and specificity. Int J Psychol 22(5–6):643–677
89. Fisher WP Jr, Melin J, Möller C (2021) Metrology for climate-neutral cities (RISE Research Institutes of Sweden AB No. RISE Report 2021:84). Gothenburg, Sweden: RISE, pp 79 http://ri.diva-portal.org/smash/record.jsf?pid=diva2%3A1616048&dswid=-7140
90. Fisher WP Jr, Oon EP-T, Benson S (2021) Rethinking the role of educational assessment in classroom communities: how can design thinking address the problems of coherence and complexity? Educ Des Res 5(1):1–33. https://doi.org/10.15460/eder.5.1.1537
91. Harries K (2010) Let no one ignorant of geometry enter here: ontology and mathematics in the thought of Martin Heidegger. Int J Philos Stud 18(2):269–279
92. Alder K (2002) The measure of all things: the seven-year odyssey and hidden error that transformed the world. The Free Press
93. Hawken P (2007) Blessed unrest: how the largest movement in the world came into being and why no one saw it coming. Viking Penguin
94. Kaplan S (2023) World is on brink of catastrophic warming, U.N. climate change report says. Washington Post (Washington, DC). https://www.washingtonpost.com/climate-environment/2023/03/20/climate-change-ipcc-report-15/
95. Barney M, Barney F (2024) Transdisciplinary measurement through AI: hybrid metrology and psychometrics powered by large language models. In: Fisher W.P, Pendrill L (eds) Models, measurement, and metrology extending the SI. De Gruyter Oldenbourg, pp 103–131. https://www.degruyter.com/document/isbn/9783111036496/html
96. Barney M, Fisher WP Jr (2016) Adaptive measurement and assessment. Annu Rev Organ Psych Organ Behav 3:469–490

97. Barney M, Fisher WP (2017) Avoiding AI armageddon with metrologically-oriented psychometrics. In: 18th International Congress of Metrology 09005, pp 1–6. https://doi.org/10.1051/metrology/201709005
98. Fisher WP Jr, Stenner AJ (2018) Ecologizing vs modernizing in measurement and metrology. J Phys Conf Ser 1044(012025). [http://iopscience.iop.org/article/https://doi.org/10.1088/1742-6596/1044/1/012025]
99. Latour B (1995) Moderniser ou écologiser? A la recherche de la septième Cité. Ecologie Politique, 13, 5–27. In: Castree N, Willems-Braun B (eds) (1998) To modernize or to ecologize? that's the question. In Remaking reality: Nature at the millennium. Routledge, pp 221–242
100. Hayek FA (1945) The use of knowledge in society. American Economic Review 35, 519–530. In: Hayek FA (eds) Individualism and economic order. University of Chicago Press, pp 77–91
101. Whitehead AN (1911) An introduction to mathematics. Henry Holt and Co, New York
102. Krans NA, Ammar A, Nymark P, Willighagen EL, Bakker MI, Quik JT K (2022) FAIR assessment tools: evaluating use and performance. NanoImpact 27(100402)
103. Wilkinson MD, Dumontier M, Aalbersberg IJ, Appleton G, Axton M, Baak A, Mons B (2016) The FAIR guiding principles for scientific data management and stewardship. Sci Data 3:160018. https://doi.org/10.1038/sdata.2016.18
104. Wilkinson MD, Sansone S.-A, Schultes E, Doorn P, da Silva Santos LOB, Michel Dumontier M (2018) A design framework and exemplar metrics for FAIRness. Sci Data 5. https://doi.org/10.1038/sdata.2018.118
105. Akrich M, Callon M, Latour B (2002) The key to success in innovation part II: the art of choosing a good spokesperson. Int J Innov Manag 6(2):207–225
106. Brown T (2008) Design thinking. Harv Bus Rev 86(6):84–92
107. Hargadon A (2003) How breakthroughs happen: the surprising truth about how companies innovate. Harvard Business School Press
108. Stenner AJ (2001) The Lexile framework: a common metric for matching readers and texts. Calif Sch Libr J 25(1):41–42. (Rpt. in Fisher WP Jr, Massengill PJ (eds) (2023) Explanatory models, unit standards, and personalized learning in educational measurement. Springer. pp 251–268. https://link.springer.com/book/https://doi.org/10.1007/978-981-19-3747-7?page=1#toc)
109. Stenner AJ, Fisher WP Jr, Stone MH, Burdick DS (2013) Causal Rasch models. Front Psychol: Quant Psychol Meas 4(536):1–14. https://doi.org/10.3389/fpsyg.2013.00536. (Rpt. in Fisher WP Jr, Massengill PJ (eds) (2023) Explanatory models, unit standards, and personalized learning in educational measurement. Springer, pp 223–250. https://link.springer.com/book/https://doi.org/10.1007/978-981-19-3747-7?page=1#toc)
110. Morrison J, Fisher WP Jr (2018) Connecting learning opportunities in STEM education: ecosystem collaborations across schools, museums, libraries, employers, and communities. J Phys: Conf Ser 1065(022009). https://doi.org/10.1088/1742-6596/1065/2/022009
111. Morrison J, Fisher WP Jr (2019) Measuring for management in science, technology, engineering, and mathematics learning ecosystems. J Phys: Conf Ser 1379(012042). https://doi.org/10.1088/1742-6596/1379/1/012042
112. Morrison J, Fisher WP Jr (2021) Caliper: measuring success in STEM learning ecosystems. Meas: Sens 18:100327. https://doi.org/10.1016/j.measen.2021.100327
113. Brown JC, Goldstein JE, Chan TL, Massof R, Ramulu P, Low Vision Research Network Study Group (2014) Characterizing functional complaints in patients seeking outpatient low-vision services in the United States. Ophthalmology 121(8):1655–1662
114. Chan TL, Perlmutter MS, Andrews M, Sunness JS, Goldstein JE, Massof RW, Low Vision Research Network (LOVRNET) Study Group (2015) Equating visual function scales to facilitate reporting of medicare functional G-code severity/complexity modifiers for low-vision patients. Arch Phys Med Rehabil 96(10):1859–1865
115. Goldstein JE, Chun MW, Fletcher DC, Deremeik JT, Massof RW, Low Vision Research Network Study Group (2014) Visual ability of patients seeking outpatient low vision services in the United States. JAMA Ophthalmol 132:1169–1177

116. Goldstein JE, Fenwick E, Finger RP, Gothwal V, Jackson ML, Lamoureux E, Rhees G, Massof R (2018) Calibrating the Impact of Vision Impairment (IVI): creation of a sample-independent visual function measure for patient-centered outcomes research. Transl Vis Sci Technol 7(6):38
117. Massof RW, Ahmadian L, Grover LL, Deremeik JT, Goldstein JE, Rainey C, Epstein C, Barnett GD (2007) The Activity Inventory: an adaptive visual function questionnaire. Optom vis Sci 84:763–774
118. Bohm D (1980) Wholeness and the implicate order. Routledge & Kegan Paul
119. Farnsworth KD, Nelson J, Gershenson C (2013) Living is information processing: from molecules to global systems. Acta Biotheor 61(2):203–222
120. Gare A (2020) Semiosis and information: meeting the challenge of information science to post-reductionist biosemiotics. Biosemiotics 13:327–346. https://doi.org/10.1007/s12304-020-09393-w
121. Gell-Mann M (1994/2023) The quark and the jaguar: adventures in the simple and the complex. Macmillan
122. Kauffmann S (1996) At home in the universe: the search for laws of self-organization and complexity. Oxford University Press
123. Prigogine I, Stengers I (1984) Order out of chaos: Man's new dialogue with nature. Bantam Books
124. Swimme BT, Tucker ME (2011) Journey of the universe: An epic story of cosmic, earth, and human transformation. Yale University Press
125. Taylor MC (2003) The moment of complexity: emerging network culture. University of Chicago Press
126. Wheeler JA (1994) At home in the universe. American Institute of Physics Press
127. Hunt BJ (1994) The ohm is where the art is: British telegraph engineers and the development of electrical standards. In: van Helden A, Hankins TL (eds) Instruments [Special issue]. Osiris: A Research Journal Devoted to the History of Science and Its Cultural Influences 9:48–63. University of Chicago Press
128. Wise MN (1995) Precision: agent of unity and product of agreement. Part III Today Precision Must Be Commonplace. In: Wise MN (eds) The values of precision, Princeton University Press, pp 352–361
129. Bohm D (1984) Imagination and the implicate order. In: Cazenave M (ed) Science & consciousness: Two views of the universe. Pergamon, pp 375–380
130. Bohm D (1994) Thought as a system. Routledge
131. Bohm D, Peat FD (1987) Science, order, and creativity: a dramatic new look at the creative roots of science and life. Random House
132. Prigogine I (1976) Order through fluctuation: Self-organization and social system. In: Jantsch E, Waddington C (eds) Consciousness and evolution: Human systems in transition. Addison Wesley, pp 93–130
133. Prigogine I (1977) Physics and metaphysics. In: Lawrence JH, Gofman JW, Hayes TL (eds) Advances in biological and medical physics, vol 16. Academic Press, pp 241–263
134. Prigogine I (1986) Science, civilization and democracy: values, systems, structures and affinities. Futures 18(4):493–507
135. Wheeler JA (1974) The universe as a home for man. American Scientist 62:683–691. https://www.jstor.org/stable/27845173
136. Wheeler JA (1981) The participatory universe. Science 81(2):5
137. Wittgenstein L (1958) Philosophical investigations. Anscombe G.E.M Trans 3rd edn. Macmillan

Recent Progress in Disassembly Research at the University of Birmingham

Duc Truong Pham

Abstract Remanufacturing, a process that revitalises end-of-life products, is a key component of a circular economy offering substantial environmental and economic benefits. This paper introduces the concept of smart remanufacturing, which leverages Industry 4.0 technologies such as artificial intelligence (AI), cloud computing, the Internet of Things (IoT), and digital twins to enhance the sustainability and efficiency of remanufacturing processes. Recent research at the University of Birmingham, conducted as part of the Engineering and Physical Sciences Research Council-funded AutoReman project, is reviewed. AutoReman promotes robotic disassembly as a key enabler of smart remanufacturing, focusing on understanding disassembly science, developing effective robotic disassembly strategies, and creating robotic disassembly cells for both fully autonomous and collaborative human–robot disassembly. Overall, this paper underscores the potential of smart remanufacturing to contribute to a more sustainable future.

Keywords Remanufacturing · Smart remanufacturing · Digital remanufacturing · Disassembly · Robotic disassembly · Industry 4.0 · Digital twins

1 Introduction

The manufacturing industry is currently undergoing significant transformation, primarily driven by a growing awareness of environmental issues, the scarcity of resources, and the emergence of Industry 4.0. Remanufacturing, a process that revitalises end-of-life products, is at the forefront of this change, offering substantial environmental and economic advantages.

This paper aims to elucidate the concept of smart remanufacturing, exploring its benefits, the critical role of robotic disassembly, and the fundamental disassembly research undertaken by the author's team. Despite significant advancements in remanufacturing, the challenges in automating disassembly processes remain a critical

D. T. Pham (✉)
University of Birmingham, Birmingham, UK
e-mail: d.t.pham@bham.ac.uk

N. M. Durakbasa and K. G. Gülen (eds.), *Sustainable Green Conversion*, Lecture Notes in Mechanical Engineering, https://doi.org/10.1007/978-3-031-83583-4_4

bottleneck, a gap that our research directly addresses through the development of advanced robotic technology.

The remainder of this paper is organised as follows. Section 2 explains remanufacturing and its benefits. Section 3 discusses leveraging Industry 4.0 technologies to realise smart remanufacturing. Section 4 reviews recent work at our Autonomous Remanufacturing Laboratory. Section 5 takes a glimpse at the future of smart remanufacturing. Section 6 concludes the paper.

2 Remanufacturing: A Cornerstone of the Circular Economy

Remanufacturing, as defined by British Standard BS 8887—Part 2, involves restoring a product to its original performance level, accompanied by a warranty that is equal to or surpasses that given to a newly manufactured product. This distinguishes it from refurbishing, which primarily focuses on aesthetics, and reconditioning, which merely aims to restore basic functionality. Remanufacturing goes beyond these processes, delivering a product that is essentially "as new," and in some cases, even better than the original due to improvements based on failure analysis and technological upgrades.

The remanufacturing process typically involves a series of steps: disassembly of the end-of-life product into its constituent components; cleaning of the components to remove dirt, grime, and contaminants; inspection of each component to assess its condition and identify any signs of damage or wear; repair or replacement of damaged or worn-out components with new ones; reassembly of the product using the restored or new components; and testing of the remanufactured product to ensure it meets the required performance standards.

Remanufacturing plays a key role in a circular economy, offering significant advantages over traditional manufacturing processes that rely on virgin materials. These advantages are evident in two primary areas: environmental and economic.

A. *Environmental Benefits*

Remanufacturing significantly reduces the demand for raw materials, thereby minimising the environmental impact associated with extraction, processing, and transportation. By reusing existing components, remanufacturing requires less energy than making new products, resulting in lower greenhouse gas emissions and a smaller carbon footprint. Studies indicate that remanufacturing can lower energy consumption by up to 86% compared to manufacturing. Moreover, remanufacturing diverts products from landfills, cutting waste and minimising pollution, with the potential to decrease the material sent to landfills by up to 99%. Additionally, remanufacturing processes often consume less water than traditional manufacturing, contributing to water conservation efforts. In certain cases, water consumption can be reduced by up to 93%. These environmental benefits align with several of the United Nations' Sustainable Development Goals (SDGs), including those related to good health and well-being, clean water and sanitation, sustainable cities and communities, and life on land.

B. *Economic Benefits*

Remanufacturing can also be more cost-effective than producing new goods, despite being labour-intensive. The reduced need for raw materials and energy translates into lower production costs, contributing to increased profitability. Remanufactured products can be sold at lower prices than their new counterparts while still generating higher profit margins due to these reduced input costs. Additionally, remanufacturing creates skilled jobs in areas such as disassembly, inspection, repair, and reassembly. Companies engaged in remanufacturing can gain a competitive advantage by offering high-quality products at lower prices. These economic benefits contribute to achieving SDGs such as those related to no poverty, zero hunger, decent work and economic growth, industry innovation and infrastructure, and reduced inequalities.

3 Smart Remanufacturing: Leveraging Industry 4.0 for Enhanced Sustainability

Remanufacturing is not without problems. A significant challenge in remanufacturing has been the uncertainties inherent in the process. Uncertainties in the quality, quantity, and timing of returned products, or 'cores,' coupled with variations in the disassembly, repair, and rebuild routes and times, have traditionally made remanufacturing heavily reliant on human intervention. This dependence on human skill and judgement is a major contributor to the labour-intensive nature of the remanufacturing sector, leading to labour costs that could be double those of manufacturing new products from virgin parts and materials.

To address this, we have proposed the concept of smart remanufacturing [1], a technologically-based approach that aims to optimise the efficiency, flexibility, and sustainability of remanufacturing processes and systems and maximise the aforementioned economic and environmental benefits. Smart remanufacturing leverages technologies like intelligent product digital twins and simulation tools to predict supply timing and minimise uncertainties in the condition of incoming products. In cases where eliminating uncertainties is not economically feasible, process digital twins integrated with virtual reality and augmented reality technologies and supported by human input can enable semi-autonomous remanufacturing systems effectively to manage and navigate residual uncertainties.

Smart remanufacturing enhances remanufacturing by integrating Industry 4.0 technologies, such as cloud computing, the Internet of Things (IoT), and artificial intelligence (AI). This approach enables real-time data analysis, where sensors embedded in products collect data on their condition and performance during use. This data can then be analysed in real time to predict the timing, quality and quantity of returned cores, enabling proactive planning and optimised resource allocation.

Smart Industry 4.0 technologies also facilitate process optimisation, enhancing every stage of the remanufacturing chain, from disassembly and cleaning to inspection and reassembly. This results in increased efficiency, reduced downtime, and improved resource utilisation. Furthermore, AI algorithms can analyse data from sensors to predict when components are likely to fail, enabling proactive maintenance and preventing costly downtime.

Smart technologies can also be used to automate quality control processes, ensuring that remanufactured products meet the highest standards. Finally, a distributed network of remanufacturing facilities can be connected through a cloud-based platform, enabling efficient data sharing, collaboration, and optimised resource allocation. This leads to the idea of cloud remanufacturing [2].

A. *Cloud Remanufacturing: A Distributed and Sustainable Approach*

Cloud remanufacturing (Fig. 1) represents a paradigm shift in the remanufacturing landscape, leveraging cloud computing and other Industry 4.0 technologies to make remanufacturing smarter, i.e., more efficient, flexible, and sustainable. It is built on a foundation of enabling technologies, including cloud computing to store, process and share data, the IoT to connect machines and equipment to the cloud for real-time data collection and analysis, big data analytics to identify trends and optimise processes within large datasets, and artificial intelligence to automate tasks, improve decision-making, and enable predictive maintenance.

Key features of cloud remanufacturing include a distributed network, where resources and facilities can be located at different geographic sites, closer to the

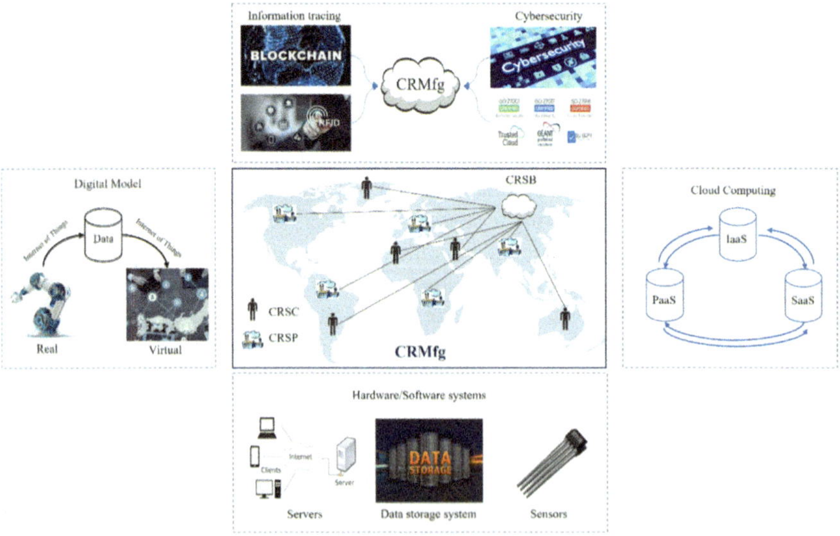

Fig. 1 Cloud remanufacturing [2]

source of used products or customer demand. A centralised, cloud-based platform connects and manages the entire network, facilitating seamless communication and coordination. Real-time data on inventory, demand, capacity, and other key parameters is shared across the network, enabling optimised decision-making. The cloud platform also allows optimised scheduling, logistics, and resource allocation, improving efficiency and reducing costs. As a result, the network can quickly adapt to changes in demand and supply, ensuring responsiveness and resilience.

Cloud remanufacturing provides several benefits, including improved efficiency and productivity through streamlined processes, reduced downtime, and optimised resource utilisation. Lower transportation costs, improved inventory management, and better utilisation of resources contribute to overall cost savings. The sustainability of the remanufacturing process is enhanced through reduced waste, lower energy consumption, and optimised logistics. Improved communication and coordination among stakeholders foster collaboration and innovation. Finally, operations can be easily scaled up or down based on demand, providing flexibility and adaptability.

B. *A Framework for Smart Remanufacturing Research*

To address the challenges in remanufacturing and unlock its full potential, we have developed a framework for smart remanufacturing research that integrates data capture, analysis, and intelligent decision-making throughout the product lifecycle (Fig. 2). The framework illustrates the interconnectedness of various elements throughout the product lifecycle, from Beginning of Life (BoL) to End of Life (EoL). It highlights the importance of data capture and analysis at each stage, again emphasising key enablers like the IoT for real-time monitoring and cloud computing for data storage and accessibility. It underscores the need for intelligent and adaptable remanufacturing facilities capable of handling uncertainties about core quality and quantity. The framework also emphasises the role of AI solutions in optimising processes like disassembly planning and scheduling. Finally, it highlights the importance of collaboration between Original Equipment Manufacturers (OEMs), Product Service System (PSS) providers, and customers, as well as the need for a skilled workforce trained in sustainable manufacturing practices.

4 The AutoReman Project: Enabling Technology for Autonomous Remanufacturing

Disassembly, often the first and most labour-intensive stage of remanufacturing, is in need of automation to reduce labour costs, ensure consistent quality and enable large throughputs. Our research in robotic disassembly focuses on understanding the science behind common disassembly operations to develop efficient robotic strategies. It also involves developing advanced strategies employing techniques such as active compliance, reinforcement learning, and vibration-assisted disassembly to

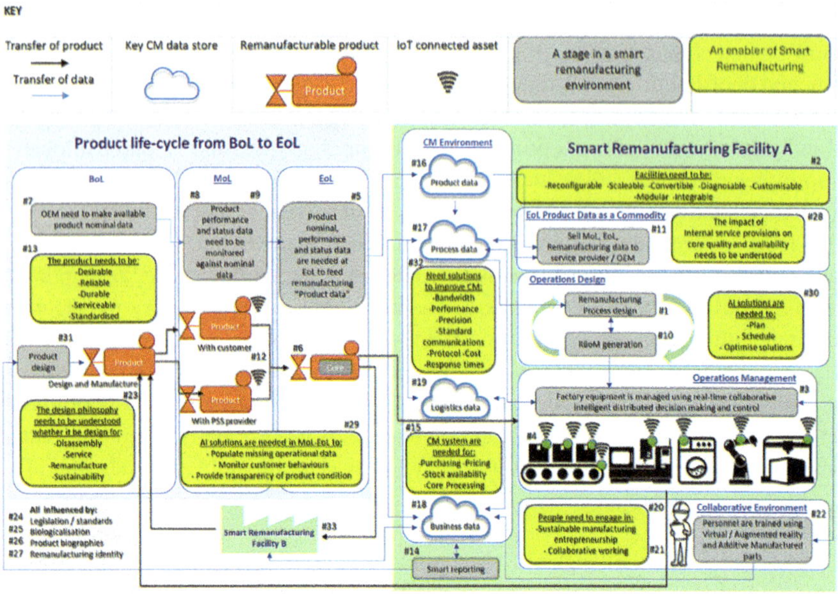

Fig. 2 Smart remanufacturing research framework [3]

enable robots reliably to perform complex tasks. Another important focus is collaborative disassembly, which integrates human expertise with robotic capabilities for tasks that require dexterity and problem-solving skills.

The AutoReman project, a research initiative focused on autonomous remanufacturing led by the author at the University of Birmingham, is developing robotic disassembly technologies to make remanufacturing more efficient and cost-effective. The following summarises the recent robotic disassembly work done by the AutoReman team that will contribute to achieving smart remanufacturing. The research is motivated by the fact that disassembly is labour-intensive, requires skills and dexterity, and is prone to human error. The work includes creating pilot robotic disassembly cells to demonstrate collaborative human–robot disassembly of complex products like water pumps and turbochargers. The AutoReman team is also utilising digital twin technology, creating virtual models of disassembly processes to optimise robot control and improve reliability and efficiency.

A. *Disassembly Science and Strategies*

The AutoReman project has conducted extensive research to understand the science underpinning disassembly operations and develop effective strategies for robotic disassembly. The reason for investigating disassembly science is to gain a better understanding of disassembly. An aim is to develop strategies based on this new understanding to enable robots to perform disassembly tasks reliably and efficiently.

The first task in the research was to determine which disassembly operations to concentrate efforts on. To do that, the AutoReman team studied more than 400

Fig. 3 Robotic bolt loosening tool [5]

products to identify the most common operations. The analysis of over 12,000 operations revealed three common families: unscrewing, pulling, and separation. This research has focused on two main areas: unscrewing and pulling which form 60% of all operations for the products surveyed.

In the area of unscrewing, the project has investigated the phenomenon of self-loosening through vibrations, which can be used to automate the unscrewing process. This led to the development of a robotic bolt-loosening tool that can be mounted on a robot to perform efficient unscrewing operations (Fig. 3).

In the area of pulling, the project has investigated the use of active compliance control or active force accommodation to enable robots to perform pulling operations with varying levels of force and precision [4]. The team looked at the operation of pulling a shaft from a bore while avoiding the shaft being wedged inside the bore. Active force accommodation allows the robot to adapt to the resistance encountered during pulling, which helps prevent damage to the components.

The AutoReman team has also investigated pulling combined with twisting [6]. From common experience and using simple mechanical analysis, it is known that combining twisting and pulling reduces the pulling force required to extract a tightly fitting male component from a female component. We have found in both theory and experiments that this strategy enables a reduction in the pulling force (Fig. 4).

In another study on pulling operations, the task was to slide a pin along a slot before pulling it out. The robots were equipped with force sensors to measure the forces on the pin and programmed to follow the force accommodation strategy. It was found that the operation of the robot was helped by applying small vibrations at the gripper. Vibrations reduced the contact between the pin and the slot and therefore decreased friction, speeding up the process.

Instead of programming the robot to follow a predetermined strategy such as active accommodation, we can allow it to learn the task by itself [7]. We have used reinforcement learning which involved the robot trying to act in such a way as to maximise a reward function. At first, the robot performed poorly and kept failing.

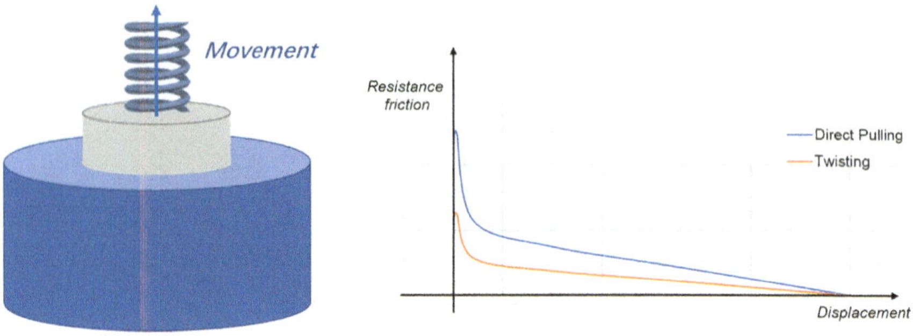

Fig. 4 Combined pulling and twisting of press-fitted components [6]

After many learning episodes, the robot succeeded in sliding the pin to the end of the slot without having been explicitly programmed to do so.

B. *Demonstration case studies*

As human operators will still be needed for tasks that are too costly or difficult to robotise, the research has also involved looking at techniques for human–robot collaboration in disassembly. Two case studies have been completed in which a human operator worked with robots to disassemble two different products, namely, a water pump [8] and a turbocharger [9].

These two case studies have resulted in human–robot collaborative disassembly cells that showcase the potential of collaboration in remanufacturing. These cells feature collaborative robots (cobots) which are designed to work safely alongside humans. This allows for a combination of human expertise and robotic capabilities, enabling efficient and flexible disassembly processes. The cells are also equipped with advanced sensors and vision systems that enable the robots to perceive their environment, identify components, and perform precise manipulations. Additionally, they are designed with flexible tooling that can be easily adapted to different product types and disassembly tasks.

A third case study focused on the scenario where complete automation was necessary because of the nature of the product and the hazards involved in disassembling it. The product was an electric vehicle (EV) battery. A fully robotised system was required for the safe and efficient disassembly of this product which risks causing electric shocks, exploding and emitting noxious fumes. The team's efforts were spent developing a system that can autonomously undertake the complex process of disassembling the EV battery [10]. This is also a critical area of research, as the demand for EV batteries is expected to grow significantly in the coming years, and the safe and sustainable disposal of end-of-life batteries is a major environmental concern.

In the EV battery case study, the use of digital twin technology to optimise robotic disassembly processes was explored (Figs. 5 and 6). A digital twin is a virtual representation of a physical product or process, which can be used for simulation, analysis, and optimisation. In the context of remanufacturing, digital twins can be used to simulate the disassembly process in a virtual environment, identifying potential challenges

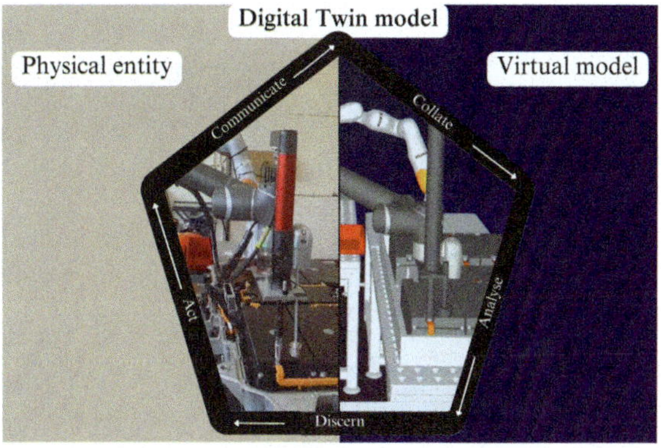

Fig. 5 Digital twin concept [11]

and optimising robot trajectories and control strategies before implementing them in the real world. They can also be used to train robots on disassembly tasks in a safe and controlled environment, reducing the need for physical prototypes and minimising the risk of damage to real components. Additionally, data collected from sensors during the disassembly process can be used to update the digital twin, providing valuable insights into the condition of components and the efficiency of the process.

C. *Disassembly planning*

For smart autonomous remanufacturing, systems need to be able to plan their actions. Plans prepared offline in advance often become invalid because the actual condition of the product at the end of its service life is different from the nominal condition of the product for which the plan was drawn up. The ability of a system to replan online is therefore essential.

Together with colleagues at Wuhan University of Technology [12, 13], Beihang University [14] and the University of Castilla la Mancha [15], our AutoReman research team has explored the area of disassembly planning, a critical aspect of optimising the remanufacturing process. This involves strategising the sequence of disassembly operations to maximise efficiency and minimise costs.

This paper does not have the space to cover the specifics of disassembly planning. Interested readers can refer to Laili et al. [14] and Pham and Hartono [16]. The former focuses on the optimisation of robotic disassembly for remanufacturing and the latter details the Bees Algorithm, a novel optimisation technique developed by the team, inspired by the foraging behaviour of honeybees. This algorithm offers a simple yet effective approach to solving complex optimisation problems in disassembly planning.

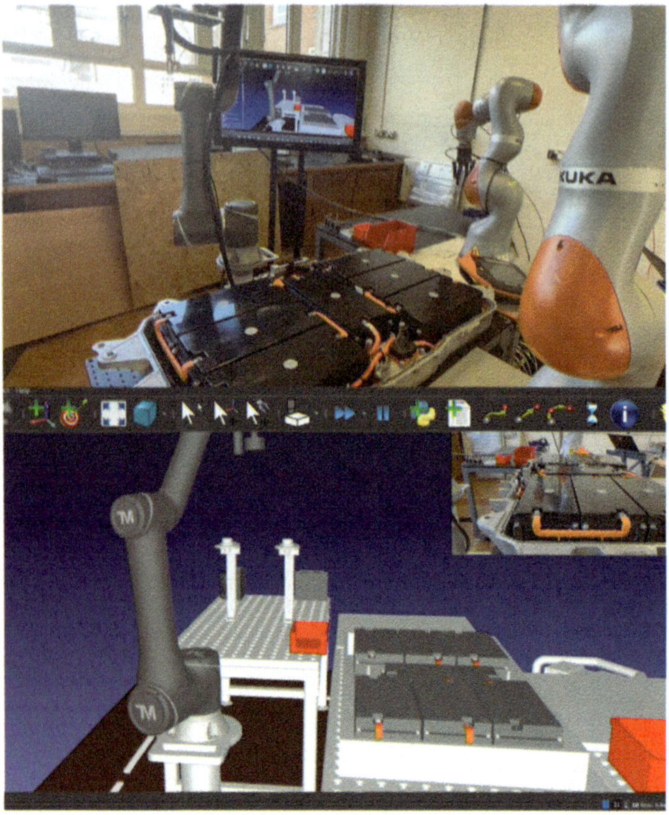

Fig. 6 Digital twin of the unscrewing operation

5 The Future of Smart Remanufacturing

Smart remanufacturing, powered by robotics, AI, and other Industry 4.0 technologies, has the potential to contribute significantly to a sustainable circular economy. By reducing waste, conserving resources, and creating new economic opportunities, smart remanufacturing offers a promising path towards a more sustainable future. Key trends shaping this future include increased automation through robotics and AI, leading to greater efficiency, data-driven optimisation using analytics and machine learning, more widespread cloud-based collaboration and resource sharing, greater integration of remanufacturing into the circular economy, with closer ties between manufacturers, remanufacturers, and consumers, and a continued focus on sustainability as a driver of innovation, with a goal of reducing environmental impact and maximising resource efficiency.

To realise the potential of smart remanufacturing fully, governments and industry must collaborate to create a supportive environment. This includes policy support, such as tax breaks, subsidies, and extended producer responsibility schemes, to incentivise remanufacturing. Continued investment in research and development is crucial to advance the technologies and processes that underpin smart remanufacturing. It is also essential to invest in education and training programs to develop the skills needed for the remanufacturing workforce. Lastly, collaboration among manufacturers, remanufacturers, and other stakeholders is crucial to create a thriving remanufacturing ecosystem.

The AutoReman project actively promotes collaboration and knowledge sharing in autonomous remanufacturing through the annual International Workshop on Autonomous Remanufacturing (IWAR). This event brings together researchers, industry professionals, and policymakers to discuss advancements, challenges, and opportunities in this rapidly evolving field. IWAR aims to foster collaboration among these experts, providing a forum for sharing research findings, best practices, and case studies. It serves as a platform for identifying challenges and opportunities in autonomous remanufacturing and encouraging innovation in the development of new technologies, processes, and business models.

At the time of writing this article, the 8th IWAR held at Wayne State University in Detroit Michigan had just concluded [17]. Preparations are underway for the 9th IWAR organised in 2025 by the University of Castilla la Mancha at its Toledo campus to which participants the world over will be welcome.

6 Conclusion

Underpinned by Industry 4.0 technologies, smart remanufacturing is not merely an incremental improvement but a paradigm shift in the manufacturing landscape. It offers a sustainable and economically viable complement to traditional linear manufacturing models, promoting resource efficiency, waste reduction, and the creation of new economic opportunities. The AutoReman project, with its pioneering research in autonomous remanufacturing and robotic disassembly, exemplifies this transformative approach, demonstrating the potential to revolutionise the way we manage product lifecycles.

The insights and technologies developed through this research have far-reaching implications, extending beyond the realm of remanufacturing to influence broader sustainability efforts and circular economy practices. By fostering collaboration, knowledge sharing, and continuous innovation, initiatives like the AutoReman project and the International Workshop on Autonomous Remanufacturing can be instrumental in accelerating the widespread adoption of smart remanufacturing principles.

As these advancements continue to unfold, we can anticipate a future where the synergy between human ingenuity and technological prowess drives us towards a more sustainable and prosperous world, where resource scarcity and environmental degradation are mitigated through intelligent and responsible manufacturing practices. The journey towards this future is undoubtedly challenging, but the potential rewards, both for our planet and our economies, make it a worthwhile pursuit.

Acknowledgements AutoReman was funded by the Engineering and Physical Sciences Research Council (Grant Number EPN0185241). The work was supported by BorgWarner, Caterpillar, HSSMI, Meritor, MG Motor, MTC and Recoturbo. The author wishes to thank the project funder and partners and the undergraduate and postgraduate students and members of the AutoReman Laboratory at the University of Birmingham, and researchers at Wuhan University of Technology, Beihang University, Universidad de Castilla—La Mancha and Università degli Studi della Campania "Luigi Vanvitelli" who contributed to the success of the project. Thanks are due to Professor Dr. Numan Durakbasa and the ISPR 2024 Organising Committee for the opportunity to present this update on AutoReman at the Symposium. The author first reported on autonomous remanufacturing at ISPR 2017. He owes a debt of gratitude to the late Professor Dr. Güneş Gençyılmaz for the introduction to the ISPR community, and for his continued support and friendship until his untimely passing.

References

1. Pham D (2024) Is smart remanufacturing a tautology? The Future Factory, https://www.thefutfactory.com/blog/6. Last Accessed: 10 Nov 2024
2. Caterino M, Fera M, Macchiaroli M, Pham DT (2022) Cloud remanufacturing: remanufacturing enhanced through cloud technologies. J Manuf Syst 64:133–148. https://doi.org/10.1016/j.jmsy.2022.06.003
3. Kerin M, Pham DT (2020) Smart remanufacturing: a review and research framework. J Manuf Technol Manag 31, ISSN: 1741–038X, 1205–1235. https://doi.org/10.1108/JMTM-06-2019-0205
4. Zhang Y, Lu H, Pham DT, Wang Y, Qu M, Lim J, Su S (2019) Peg–hole disassembly using active compliance. R Soc Open Sci 6(8). https://doi.org/10.1098/rsos.190476
5. Micheli M, Pham DT (2024) A robotic bolt-loosening tool, Technical Report MM24-1. University of Birmingham, UK, AutoReman Laboratory
6. Xu S, Pham DT, Su S (2024) Robotized unplugging of a cylindrical peg press-fitted into a cylindrical hole. R Soc Open Sci 11. https://doi.org/10.1098/rsos.230872
7. Qu M, Wang Y, Pham DT (2023) Robotic disassembly task training and skill transfer using reinforcement learning. IEEE Trans Ind Inform. https://doi.org/10.1109/TII.2023.3242831
8. Huang J, Pham DT, Wang Y, Qu M, Ji C, Su S, Xu W, Liu Q, Zhou Z (2019) A case study in human-robot collaboration in the disassembly of press-fitted components. In: Proceedings of the Institution of Mechanical Engineers, Part B: Journal of Engineering Manufacture. https://doi.org/10.1177/0954405419883060
9. Huang J, Pham DT, Li R, Qu M, Wang Y, Kerin M, Su S, Ji C, Mahomed O, Khalil R, Stockton D, Xu W, Liu Q, Zhou Z (2021) An experimental human-robot collaborative disassembly cell. Comput Ind Eng 143, ISSN: 0360–8352. https://doi.org/10.1016/j.cie.2021.107189
10. Qu M, Pham DT, Altumi F, Gbadebo A, Hartono N, Jiang K, Kerin M, Lan FY, Micheli M, Xu SH, Wang Y (2024) Robotic disassembly platform for disassembly of a plug-in hybrid electric vehicle battery: a case study. Automation 5(2):50–67. https://doi.org/10.3390/automation5020005
11. Gbadebo A, Altumi F, Pham DT (2024) Digital twin of screw unfastening based on model predictive control, Technical Report AG1-24. University of Birmingham, UK, AutoReman Laboratory
12. Fang Y, Xu H, Liu Q, Pham DT (2020) Evolutionary optimization using epsilon method for resource-constrained multi-robotic disassembly line balancing. J Manuf Syst 56:392–413, ISSN: 0278–6125. https://doi.org/10.1016/j.jmsy.2020.06.006
13. Xu W, Tang Q, Liu J, Liu Z, Zhou Z, Pham DT (2020) Disassembly sequence planning using discrete bees algorithm for human-robot collaboration in remanufacturing. Robot Comput-Integr Manuf 62, ISSN: 0736–5845. https://doi.org/10.1016/j.rcim.2019.101860

14. Laili Y, Wang Y, Fan Y, Pham DT (2022) Optimisation of robotic disassembly for remanufacturing. Springer Cham, Switzerland, pp. 183, ISBN: 978–3–030–81798–5
15. Hartono N, Ramírez FJ, Pham DT (2023) Optimisation of product recovery options in end-of-life product disassembly by robots. Automation 4(4):359–377. https://doi.org/10.3390/automation4040021
16. Pham DT, Hartono N (2025) (eds) Intelligent engineering optimisation with the bees algorithm. Springer Cham, Switzerland, pp. 390, ISBN 978-3-031-64935-6
17. Anon: 8th international workshop on autonomous remanufacturing, https://www.iwarconference.org/. Accessed: 10 Nov 2024

The European Green Deal: A Roadmap to a Sustainable Future

Birgit Oberer and Alptekin Erkollar

Abstract The European Green Deal provides a detailed plan to attain climate neutrality in Europe by 2050, tackling essential challenges like climate change, environmental degradation, and socio-economic inequality. This paper examines the Green Deal's strategic priorities, implementation strategies, and alignment with national commitments, including Austria's. It evaluates the potential environmental, economic, and social impacts of the Green Deal, exploring the challenges and opportunities it presents. Additionally, the paper provides insights into pathways for achieving a sustainable and resilient future for Europe.

Keywords European green deal · Climate neutrality · Agenda 2030 · Sustainable development goals · Renewable energy · Green economy

1 Introduction

The European Green Deal, introduced by the European Commission, seeks to make Europe the first continent to achieve climate neutrality by 2050. This bold initiative confronts the urgent problems of climate change and environmental degradation, while at the same time encouraging economic growth and enhancing social well-being.

The European Green Deal is an economic transformation initiative with the objective of establishing a sustainable future for the EU. It is founded upon the principles of increased climate ambition, the guarantee of clean and affordable energy, the promotion of a circular economy, and the enhancement of energy efficiency in buildings. Furthermore, the plan underscores the necessity of curbing pollution, safeguarding biodiversity, advancing sustainable agriculture, accelerating smart mobility,

B. Oberer (✉)
Research and Projects, ETCOP Institute for Interdisciplinary Research, Klagenfurt, Austria
e-mail: oberer@etcop.at

A. Erkollar
International Relations, ETCOP Institute for Interdisciplinary Research, Klagenfurt, Austria
e-mail: erkollar@etcop.at

N. M. Durakbasa and K. G. Gülen (eds.), *Sustainable Green Conversion*, Lecture Notes in Mechanical Engineering, https://doi.org/10.1007/978-3-031-83583-4_5

and bolstering research and innovation, while guaranteeing a just and inclusive transition for all citizens [1]. Central to the Green Deal's approach is the aim to decouple economic growth from resource consumption, enabling Europe to prosper without exhausting natural resources or exacerbating global warming. The emergence of the Green Deal was driven by a growing consensus among scientists, policymakers, and the public on the necessity for immediate and sustained action to avert catastrophic climate consequences. With global temperatures rising, biodiversity declining, and pollution impacting millions, the Green Deal offers a comprehensive strategy that combines environmental sustainability with economic resilience and social inclusion. To realize its ambitious targets, the European Green Deal focuses on several core priorities: the promotion of clean and renewable energy sources, the advancement of a circular economy, the implementation of sustainable agricultural practices through a farm-to-fork approach, and the preservation of biodiversity. Additionally, it highlights the significance of social equity, aiming to make the transition to a green economy just and inclusive, ensuring that the advantages are shared by all regions and communities throughout Europe. The Green Deal's alignment with the United Nations' Agenda 2030 and the Sustainable Development Goals (SDGs) reflects Europe's dedication to maintaining global leadership in sustainability. National initiatives, such as those adopted by Austria, reflect the Green Deal's influence in shaping domestic policies towards a shared vision of a sustainable future [2, 3].

This paper analyzes the strategic priorities outlined in the European Green Deal and the detailed implementation strategies intended to achieve these goals. Additionally, it assesses how these initiatives align with the Sustainable Development Goals (SDGs) of the 2030 Agenda. The analysis also delves into the potential environmental, economic, and social impacts of the Green Deal, highlighting both the challenges and opportunities presented by this transformative strategy. The discussion concludes by identifying potential pathways that could guide Europe towards a sustainable, climate-neutral future by 2050.

2 Strategic Priorities of the European Green Deal

The European Green Deal presents a comprehensive framework designed to transform the European Union into a sustainable, climate-neutral economy by 2050. By addressing the interconnected challenges of climate change, environmental degradation, and socio-economic inequality, the Green Deal aims to create a cohesive strategy for a sustainable future. Central to this initiative is the objective of achieving climate neutrality, primarily through substantial reductions in greenhouse gas emissions. This ambition is legally reinforced by the proposed European Climate Law and includes interim targets, such as a 55% reduction in emissions by 2030 [2].

A fundamental component of the Green Deal is the transition to clean and renewable energy sources, coupled with enhanced energy efficiency and the integration of energy markets to ensure a resilient and reliable energy supply. Achieving these goals requires substantial investment in renewable technologies, including wind and solar power, as well as the modernization of energy infrastructure. This approach aims to develop a decentralized, secure energy system capable of supporting Europe's long-term sustainability. In addition to energy reforms, the Green Deal advocates for the promotion of a circular economy, prioritizing extended product lifecycles, waste reduction, and optimized recycling practices. Key sectors, such as electronics, textiles, and construction, are targeted for improvements to minimize waste and maximize resource efficiency. The Renovation Wave initiative further supports this transition by enhancing the energy efficiency of buildings, which will reduce emissions, lower costs, and generate employment opportunities in construction and related industries [3, 4] (Fig. 1).

The Green Deal places a strong focus on sustainable mobility, promoting the extensive use of electric vehicles, greater investment in public transportation, and the creation of smart mobility networks. These initiatives aim to lower emissions from the transportation sector, enhance the quality of urban life, and support broader environmental sustainability efforts. Complementing these efforts, the Farm to Fork strategy promotes sustainable food systems by reducing the environmental footprint of agriculture, supporting organic farming, enhancing biodiversity, and encouraging healthier diets. Protecting biodiversity is another critical aspect of the Green Deal, involving the safeguarding of natural habitats, restoration of ecosystems, and promotion of sustainable land use practices. These actions aim to enhance ecological resilience and ensure the long-term health of Europe's natural environment. Reaching zero pollution is equally crucial, with focused initiatives to substantially

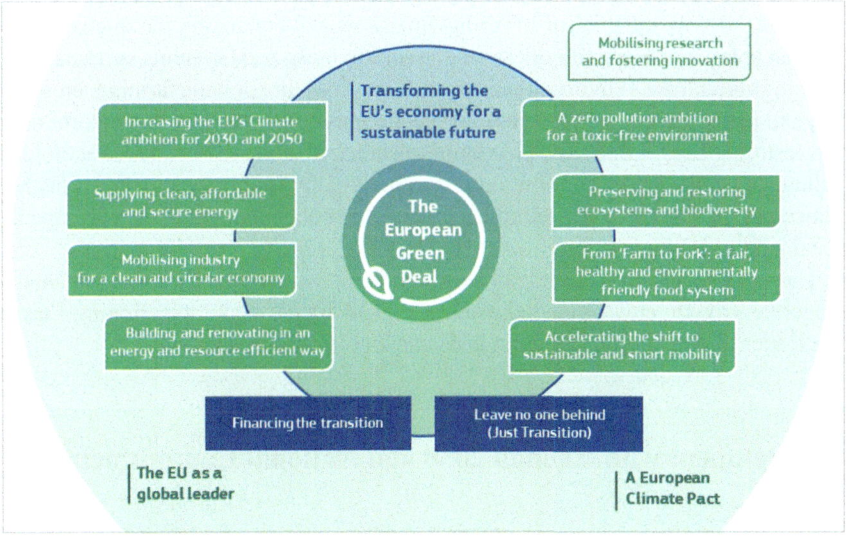

Fig. 1 EU green deals key areas [1]

decrease pollution in the air, water, and soil. These actions are essential for safeguarding public health and maintaining environmental integrity, aligning with the overarching goals of the European Green Deal [3, 4].

3 Implementation Strategies

The European Green Deal presents a strategic vision to transition the European Union into a sustainable, climate-neutral economy by 2050, tackling the intertwined issues of climate change, environmental degradation, and socio-economic inequality within a cohesive framework. The core aim is to reach climate neutrality by substantially cutting greenhouse gas emissions, a goal reinforced by the proposed European Climate Law and interim targets, such as a 55% reduction in emissions by 2030. A key component of the Green Deal is the transition to clean, renewable energy, enhanced energy efficiency, and integrated energy markets to ensure a resilient supply [5]. This shift requires substantial investment in renewable technologies, such as wind and solar, and the modernization of energy infrastructure to support a decentralized and secure energy system. The Green Deal also promotes a circular economy by extending product lifecycles, reducing waste, and optimizing recycling in sectors like electronics, textiles, and construction. The Renovation Wave initiative focuses on improving building energy efficiency, lowering emissions and costs, and creating jobs. Sustainable mobility is advanced by encouraging the use of electric vehicles, enhancing public transportation systems, and creating smart mobility networks, all with the goal of lowering emissions and enhancing the quality of life in urban areas. The Farm to Fork strategy seeks to make food systems sustainable by reducing agriculture's environmental impact, emphasizing organic farming, biodiversity, and healthier diets. Biodiversity protection involves safeguarding natural habitats, restoring ecosystems, and promoting sustainable land use to enhance ecological resilience. Achieving zero pollution is another goal, with measures to significantly reduce air, water, and soil pollution, crucial for public health and environmental sustainability [6, 7].

These strategic priorities establish a comprehensive framework for sustainable economic growth, environmental protection, and social equity, positioning Europe as a leader in global climate action and sustainable development.

4 Alignment with Agenda 2030 and National Commitments

The European Green Deal represents a regional strategy for environmental and economic transformation, while also constituting a vital component of Europe's contribution to global sustainability goals. Aligned with the United Nations' Agenda 2030 and its Sustainable Development Goals (SDGs), the Green Deal offers an integrated framework to tackle the interconnected issues of climate change, environmental degradation, and socio-economic inequality.

A. *Synergies with the Sustainable Development Goals (SDGs)*

The European Green Deal is designed to advance several Sustainable Development Goals (SDGs), with a particular emphasis on environmental sustainability, economic growth, and social inclusion. For SDG 13: Climate Action, the Green Deal targets climate neutrality by 2050, responding to the global call for urgent climate action by setting ambitious goals for greenhouse gas reduction and promoting renewable energy to mitigate climate change and enhance resilience. In support of SDG 7, the Green Deal prioritizes transitioning to a clean, affordable, and secure energy system. This approach includes expanding the use of renewable energy sources, enhancing energy efficiency, and guaranteeing access to sustainable energy for all individuals. These efforts enhance energy security and promote renewable technologies across the EU. The Green Deal aligns with SDG 11: Sustainable Cities and Communities by advocating for sustainable urban development. Initiatives like the Renovation Wave aim to improve building energy efficiency, while sustainable mobility strategies encourage the use of public transport and electric vehicles, contributing to more sustainable, resilient, and inclusive urban environments. Aligned with SDG 12, the Green Deal's Circular Economy Action Plan emphasizes reducing waste, enhancing recycling, and promoting sustainable production practices. By advancing a circular economy, the Green Deal supports SDG 12 goals of sustainable resource management and minimizing environmental impacts. The Green Deal also contributes to SDG 15: Life on Land by committing to biodiversity conservation, protecting natural habitats, and promoting sustainable land use. These efforts help preserve ecosystems and biodiversity, thereby strengthening the health and resilience of the natural environment [8, 9].

B. *Contribution to Advancing the Green Deal and Agenda 2030*

Austria has shown a strong commitment to the European Green Deal and the Sustainable Development Goals (SDGs) of the 2030 Agenda, aiming to achieve climate neutrality by 2040, a full decade ahead of the EU's target. This ambitious goal highlights Austria's leadership in climate action and its alignment with the objectives established by the Green Deal. The national policies of the Austrian government align with the key priorities outlined in the Green Deal, focusing particularly on renewable energy, enhancing energy efficiency, promoting sustainable urban development, and preserving biodiversity. Moreover, Austria's commitment to sustainable agriculture and forestry is consistent with the farm-to-fork strategy and biodiversity conservation objectives set forth in the Green Deal. By incorporating sustainability tenets into its economic and social policies, Austria is actively contributing to the implementation of the European Green Deal and the advancement of its SDG commitments. The national actions taken by Austria serve to exemplify the way domestic strategies can be aligned with and supportive of broader European and global sustainability objectives [10, 11].

C. *Regional and Global Leadership*

The alignment of the European Green Deal with the Agenda 2030 Sustainable Development Goals (SDGs) positions the European Union as a global leader in combating climate change and promoting sustainable development. The Green Deal serves as a model for other regions, demonstrating the effectiveness of comprehensive policy frameworks in addressing environmental, economic, and social challenges in an integrated way. The EU's leadership in international climate diplomacy further strengthens its influence, advocating for stronger global climate commitments and supporting developing countries in their sustainable development efforts. By aligning its policies with the SDGs, the EU boosts its credibility in global sustainability discussions, encouraging other nations to set ambitious targets. This alignment also fosters collaboration with international organizations, businesses, and civil society groups that share sustainability goals. Such collaboration enhances the effectiveness of sustainability initiatives, accelerating progress towards a more sustainable and equitable global future [8, 9].

5 Impact Assessment

The European Green Deal is a transformative strategy aimed at creating a sustainable and climate-neutral Europe by 2050. This initiative is expected to significantly impact the environment, economy, and society. This section analyzes the anticipated effects of the Green Deal, exploring how it may reshape Europe's environmental landscape, economic structure, and social well-being.

A. *Environmental Impact*

The European Green Deal sets legally binding targets to achieve climate neutrality by 2050, focusing on reducing greenhouse gas emissions through renewable energy adoption, improved energy efficiency, and industrial decarbonization. These actions are essential for curbing global warming and fulfilling international climate commitments, such as the Paris Agreement. The goal of zero pollution under the Green Deal involves reducing industrial emissions, improving agricultural practices, and promoting sustainable transportation to achieve cleaner air, water, and soil. These efforts will lead to stronger, more resilient ecosystems, reduced respiratory diseases, and better protection of aquatic systems and drinking water quality. Biodiversity preservation and restoration are also central to the Green Deal, aiming to protect natural habitats, enhance ecosystem resilience, and promote sustainable use of land and marine resources. Initiatives like reforestation and expanding protected areas will help maintain ecological balance and increase natural carbon sequestration. The Farm to Fork strategy promotes sustainable agricultural practices, including reduced pesticide and fertilizer use, improved soil health, and increased resilience to climate change. It supports organic farming and agroecology, which can lower greenhouse gas emissions from agriculture, protect biodiversity, and ensure sustainable natural resource management [2, 5].

B. *Economic Impact*

The European Green Deal serves two interrelated purposes: it is both an environmental initiative and a comprehensive economic strategy. Its goal is to transform Europe into a sustainable, competitive, and resilient economy. This transition is expected to significantly boost job creation and economic growth, especially in sectors like renewable energy, energy efficiency, sustainable agriculture, and green construction. Key initiatives, such as renovating existing buildings, expanding renewable energy infrastructure, and developing sustainable transport networks, will stimulate economic activity and create substantial employment opportunities. Additionally, investments in green technologies and innovation will further drive economic growth, enhancing Europe's global competitiveness. The Green Deal emphasizes innovation and technological advancement, particularly in research and development areas like advanced energy storage, smart grids, and carbon capture and storage. These innovations will not only mitigate environmental impacts but also establish Europe as a global leader in sustainable solutions, encouraging economic diversification and the development of new industries, which will drive additional economic growth. Additionally, the Green Deal strengthens economic resilience by decreasing dependence on fossil fuels and supporting the shift towards a circular economy. This strategy strengthens Europe's ability to respond to external economic shocks, such as energy price fluctuations and resource scarcity. A diversified and sustainable economy is better equipped to adapt to changing global market conditions and recover more quickly from disruptions. The Green Deal is also expected to generate significant cost savings by improving energy efficiency, reducing waste, and minimizing resource consumption. These measures will lower energy costs and material expenses for both businesses and households. Energy-efficient building renovations will directly contribute to these savings, while the shift to a circular economy will optimize resource use and minimize waste. These cost savings will support economic stability and enhance the standard of living across Europe [3, 4].

C. *Social Impact*

The European Green Deal is committed to ensuring that the transition to a sustainable economic system is both inclusive and equitable, aiming to generate widespread social benefits. A key social outcome of the Green Deal is improved public health. By reducing pollution and enhancing environmental quality, the Green Deal is expected to lead to cleaner air and water, reduced exposure to harmful chemicals, and increased access to green spaces. These improvements will enhance public health outcomes, lower healthcare costs, and improve the overall quality of life for European citizens. The Green Deal also emphasizes social equity and inclusion, ensuring a just transition for all communities, particularly those vulnerable or dependent on carbon-intensive industries. The Just Transition Mechanism offers financial support and training programs to help workers adapt to new roles in the green economy. This approach ensures that the benefits of the green transformation are distributed, leaving no individual or group behind. Additionally, the Green Deal seeks to improve quality of life by promoting sustainable urban development, clean transportation, and energy-efficient housing. These initiatives aim to provide access to affordable, clean energy, sustainable mobility options, and green public spaces, enhancing the livability and resilience of urban areas [7, 10].

6 Challenges and Opportunities

While the European Green Deal offers a comprehensive and ambitious plan to achieve sustainability, climate neutrality, and economic resilience by 2050, its successful implementation faces several challenges. However, these obstacles also present opportunities for innovation, leadership, and global influence. This section explores the key challenges and opportunities associated with the European Green Deal.

A. *Challenges*

Implementing the Green Deal will require substantial financial investment, estimated at over €1 trillion over the next decade. This funding is essential for infrastructure upgrades, renewable energy projects, building renovations, and other green initiatives. Mobilizing such investment, especially in the context of economic disruptions like the COVID-19 pandemic, is a significant challenge. Collaboration among governments, businesses, and financial institutions is crucial to secure the necessary resources while balancing economic growth and fiscal stability. Without adequate funding, achieving the Green Deal's ambitious targets may be difficult, potentially leading to delays and reducing the effectiveness of climate action efforts. Although the Green Deal promotes environmentally friendly technologies, key innovations, such as large-scale carbon capture and storage (CCS) and advanced energy storage solutions, are still in early development stages. Developing these technologies requires substantial research and development, as well as support for startups and SMEs engaged in innovation. The pace of technological advancement will be critical to meeting the Green Deal's climate and energy goals. Without accelerated innovation, Europe may struggle to maintain its leadership in the global shift towards a sustainable economy. The transition to a sustainable economy will likely encounter resistance from various stakeholders. Industries heavily reliant on fossil fuels, such as coal, oil, and natural gas, may oppose regulations that threaten their business models. Communities dependent on these industries for employment and economic stability may resist changes that could lead to job losses or economic decline. Addressing these concerns through effective communication, financial support, and retraining programs is essential to garner public and political support for the Green Deal. Ignoring the socio-economic impacts could hinder the necessary changes, slowing the transition to sustainability. The Green Deal spans multiple policy areas, including energy, transportation, agriculture, and waste management. Coordinating these policies across the diverse EU member states, each with unique economic priorities, regulatory frameworks, and political contexts, poses a significant challenge. Ensuring policy consistency and alignment requires robust governance, effective communication, and cooperation between national and regional authorities. Inconsistent policy enforcement or conflicting priorities could impede the Green Deal's objectives, highlighting the need for unified and collaborative strategies. The shift towards renewable energy and electric vehicles depends on the availability of critical raw materials, such as lithium, cobalt, and rare earth elements. Reliance on these materials raises concerns about supply chain security, geopolitical risks, and the environmental impact of mining and extraction. Developing sustainable supply chains, recycling programs, and alternative materials is crucial to address these challenges and support the scalability of green technologies. Without resolving supply chain issues, the transition to renewable energy and sustainable transport could face significant barriers, undermining the Green Deal's effectiveness. These challenges underscore the complexity of implementing the European Green Deal and the need for a comprehensive approach that includes financial investment, technological innovation, stakeholder engagement, and effective policy coordination. Overcoming these obstacles is essential to achieving the Green Deal's ambitious goals, paving the way for a sustainable and climate-neutral Europe [7–10].

The European Environment Agency's (EEA) 2020 'State of the Environment' report highlights that Europe urgently needs to change direction to tackle the alarming rate of biodiversity loss, the escalating impacts of climate change, and the overconsumption of natural resources. Despite some progress over the past decades, Europe is not on track to meet its 2030 and 2050 environmental goals without significant and immediate action. While public awareness, technological innovations, and initiatives like the European Green Deal provide a basis for hope, the report stresses that incremental changes are insufficient. Instead, Europe must radically scale up efforts to protect nature, reduce greenhouse gas emissions, and shift towards sustainable production and consumption patterns, focusing on key sectors such as food, energy, and mobility. Enhanced policy implementation, improved coordination, and rethinking investments towards sustainable practices are crucial to ensuring a socially fair transition to a climate-neutral and prosperous future [12].

B. *Opportunities*

The European Green Deal positions the EU as a global leader in climate action and sustainable development. By setting ambitious goals and implementing comprehensive policies, Europe can serve as a model for other nations, influencing international climate negotiations and encouraging the adoption of similar measures. This leadership role enhances the EU's diplomatic influence and solidifies its position in global sustainability and environmental protection discussions. The Green Deal demonstrates that economic growth and environmental stewardship can coexist, offering an example for other countries to follow. As a pioneer in climate policy, Europe can lead the international agenda, advocating for stronger commitments and collaboration to address global climate challenges. The Green Deal provides a unique opportunity to stimulate economic growth by promoting innovation and developing emerging industries. Investments in renewable energy, energy efficiency, and sustainable technologies can create new markets and business opportunities, driving economic activity. European companies leading in green technology will gain a competitive edge, establishing themselves as global leaders in the green economy. This focus on innovation not only promotes economic growth but also strengthens Europe's economic resilience by reducing dependence on volatile fossil fuel markets and enhancing the capacity to adapt to economic changes. Creating high-quality jobs in green sectors supports economic stability and long-term prosperity. Implementing the Green Deal will bring significant social benefits, including improved public health, a higher quality of life, and a more equitable society. Cleaner air and water, increased access to green spaces, and sustainable urban development will contribute to healthier, more liveable communities. These improvements address health issues related to pollution and urban density, reducing healthcare costs and improving overall well-being. Moreover, the Green Deal's emphasis on a just transition ensures vulnerable populations receive the support they need, reducing inequality and fostering social cohesion. By prioritizing social equity, the Green Deal promotes inclusive growth, ensuring all societal groups benefit from the transition to a sustainable economy. These social benefits may also strengthen public support for sustainability initiatives, creating a positive feedback loop that encourages further environmental action. Building a sustainable and climate-neutral economy enhances resilience to environmental and economic shocks, including extreme weather events, resource scarcity, and market volatility. By decreasing dependence on fossil fuels and advancing a circular economy, the European Green Deal helps shield Europe from external shocks, thereby securing long-term stability. This resilience is crucial for safeguarding economic prosperity and social well-being amid global uncertainties. A sustainable economy is better equipped to withstand disruptions, adapt to new challenges, and recover from crises. Implementing sustainability measures strengthens resilience, protecting the environment and maintaining robust economic systems capable of supporting future generations. The Green Deal offers a valuable opportunity for citizens to actively engage in the transition to a sustainable future. Public awareness campaigns, educational programs, and community initiatives can foster a culture of sustainability, encouraging individuals to adopt environmentally responsible behaviours and make informed choices. Empowering citizens to participate in sustainability efforts enhances the Green Deal's effectiveness and fosters a shared sense of responsibility for the planet. Public involvement in sustainability initiatives not only raises awareness but also drives behavioural change, leading to more sustainable consumption patterns and stronger support for environmental policies. Widespread public engagement is essential for achieving the Green Deal's goals, as it is broad participation and commitment that will drive the large-scale changes needed for a sustainable future [2–4, 7, 8] (Table 1).

C. Additional Measures to Support the EU 2050 Climate Goals

In order to achieve the EU's objective of climate neutrality by 2050, it is possible that airlines will be required to operate low-carbon emission aircraft within specified timeframes. This approach would markedly diminish the aviation industry's carbon footprint and stimulate the development and adoption of innovative, low-emission technologies. Biodiesel, a renewable fuel derived from plant oils and animal fats, represents a vital transitional solution for reducing CO_2 emissions, particularly within the transportation sector. By increasing the use of biodiesel, conventional fossil fuels

Table 1 Overview of historical trends, future outlooks and likelihood of achieving policy goals [12]

Theme	Past trends and outlook		Prospects of meeting policy objectives/targets		
	Past trends (10-15 years)	Outlook to 2030	2020	2030	2050
Protecting, conserving and enhancing natural capital					
Terrestrial protected areas	Improving	Mixed	Largely on track		
Marine protected areas	Improving	Mixed	Largely on track		
EU protected species and habitats	Mixed	Mixed	Largely not on track		
Common species (birds and butterflies)	Deteriorating	Deteriorating	Largely not on track		
Ecosystem condition and services	Deteriorating	Mixed	Largely not on track		
Water ecosystems and wetlands	Mixed	Mixed	Largely not on track		
Hydromorphological pressures	Deteriorating	Mixed	Largely not on track		
State of marine ecosystems and biodiversity	Mixed	Deteriorating	Largely not on track		
Pressures and impacts on marine ecosystems	Mixed	Deteriorating	Largely not on track		
Urbanisation and land use by agriculture and forestry	Deteriorating	Deteriorating			Largely not on track
Soil condition	Deteriorating	Deteriorating	Largely not on track		
Air pollution and impacts on ecosystems	Mixed	Mixed	Partially on track	Partially on track	
Chemical pollution and impacts on ecosystems	Mixed	Deteriorating	Largely not on track		
Climate change and impacts on ecosystems	Deteriorating	Deteriorating	Largely not on track		
Resource-efficient, circular and low-carbon economy					
Material resource efficiency	Improving	Mixed	Largely on track		
Circular use of materials	Improving	Mixed		Partially on track	
Waste generation	Mixed	Mixed	Partially on track		
Waste management	Improving	Improving	Partially on track		
Greenhouse gas emissions and mitigation efforts	Improving	Mixed	Largely on track	Largely not on track	Largely not on track
Energy efficiency	Improving	Mixed	Partially on track	Largely not on track	Largely not on track
Renewable energy sources	Improving	Mixed	Largely on track	Largely not on track	Largely not on track
Emissions of air pollutants	Mixed	Mixed	Largely on track	Partially on track	
Pollutant emissions from industry	Improving	Mixed	Partially on track		
Clean industrial technologies and processes	Improving	Mixed	Partially on track		
Emissions of chemicals	Mixed	Deteriorating	Largely not on track		
Water abstraction and its pressures on surface and groundwater	Improving	Mixed	Largely not on track		
Sustainable use of the seas	Mixed	Mixed	Partially on track		
Safeguarding from environmental risks to health and well-being					
Concentrations of air pollutants	Improving	Mixed	Largely not on track	Largely on track	
Air pollution impacts on human health and well-being	Improving	Mixed		Largely on track	
Population exposure to environmental noise and impacts on human health	Mixed	Deteriorating	Largely not on track		
Preservation of quiet areas	Mixed	Mixed	Largely not on track		
Pollution pressures on water and links to human health	Mixed	Mixed	Largely not on track		
Chemical pollution and risks to human health and well-being	Mixed	Deteriorating	Largely not on track		
Climate change risks to society	Deteriorating	Deteriorating	Partially on track		
Climate change adaptation strategies and plans	Improving	Improving	Partially on track		

Indicative assessment of past trends (10-15 years) and outlook to 2030	Indicative assessment of prospects of meeting selected policy objectives/targets
Improving trends/developments dominate	Year — Largely on track
Trends/developments show a mixed picture	Year — Partially on track
Deteriorating trends/developments dominate	Year — Largely not on track

Note: The year for the objectives/targets does not indicate the exact target year but the time frame of the objectives/targets.

can be replaced, which will result in a substantial reduction in greenhouse gas emissions and an enhancement of the circular economy. In addition to advancing carbon capture and storage technologies, the EU could prioritize investments in research on hydrogen-powered engines for heavy-duty transport and aviation. Furthermore, urban greening initiatives, such as vertical gardens and green roofs, provide ecological advantages by reducing CO_2 emissions while also improving air quality and enhancing urban microclimates.

7 Conclusions

The European Green Deal seeks to make Europe the first continent to achieve climate neutrality by 2050. This holistic strategy combines environmental sustainability with economic development and social inclusion, tackling the pressing issues of climate change, environmental degradation, and socio-economic inequality through a cohesive approach. By aligning with the United Nations Agenda 2030 and the Sustainable Development Goals (SDGs), the Green Deal sets ambitious goals for Europe while also contributing to global sustainable development initiatives. The Green Deal's strategic priorities—climate action, clean energy transition, circular economy, sustainable agriculture, biodiversity conservation, and zero pollution—provide a clear roadmap for achieving a sustainable future. Its implementation relies on policy frameworks, financial investments, technological innovation, stakeholder engagement, and monitoring mechanisms, ensuring that the transition to sustainability is effective and equitable for all. While the Green Deal offers significant opportunities for economic growth, technological innovation, and improved quality of life, it also faces substantial challenges. Successful implementation will require careful management of economic costs, technological barriers, political resistance, regulatory complexities, and resource constraints. Overcoming these challenges demands collaboration among governments, businesses, civil society, and citizens, along with a sustained commitment to innovation, education, and public engagement. Impact assessments indicate that the Green Deal could deliver substantial environmental, economic, and social benefits, including reductions in greenhouse gas emissions, improved air and water quality, biodiversity preservation, job creation, and enhanced public health. These benefits will improve the well-being of European citizens and reinforce Europe's role as a global leader in climate action and sustainable development.

In conclusion, the European Green Deal represents a bold and necessary response to the urgent challenges posed by climate change and environmental degradation. Its success depends on the collective efforts of policymakers, businesses, communities, and individuals. By advancing sustainability, innovation, and social equity, Europe can lead the way towards a more resilient, prosperous, and sustainable future, setting an example for other regions to follow.

References

1. European Commission (2019) The European Green Deal (COM (2019) 640 final). EUR-Lex. https://eur-lex.europa.eu/legal-content/EN/TXT/HTML/?uri=CELEX:52019DC0640&from=EN
2. European Commission (2019) The European Green Deal. COM 640. https://commission.europa.eu/document/daef3e5c-a456-4fbb-a067-8f1cbe8d9c78_de
3. European Commission, The European Green Deal. https://commission.europa.eu/strategy-and-policy/priorities-2019-2024/european-green-deal_de
4. Almihat MGM, Kahn MTE, Aboalez K, Almaktoof AM (2022) Energy and sustainable development in smart cities: an overview. Smart Cities 5(4):1389–1408. https://doi.org/10.3390/smartcities5040071
5. European Environment Agency (2019) SOER 2020—Europe's environment: State and outlook report. https://www.eea.europa.eu/highlights/soer2020-europes-environment-state-and-outlook-report
6. European Commission (2019) The European Green Deal: New growth strategy for Europe. https://ec.europa.eu/commission/presscorner/detail/de/ip_19_6691
7. Soergel B, Kriegler E, Weindl I et al (2021) A sustainable development pathway for climate action within the UN 2030 agenda. Nat Clim Chang 11:656–664. https://doi.org/10.1038/s41558-021-01098-3
8. United Nations, Sustainable Development Goals. https://sdgs.un.org/goals
9. United Nations General Assembly, Transforming our world: The 2030 agenda for sustainable development, A/RES/70/1, 2015. https://www.un.org/Depts/german/gv-70/band1/ar70001.pdf
10. Österreich B (2023) Umsetzung der Agenda 2030 in und durch Österreich 2020–2022. Bundeskanzleramt Österreich, Vienna
11. Edtstadler K, Schallenberg A, Kocher M, Gewessler L, Rauch J (2023) Österreich und die Agenda 2030: Bericht zur Umsetzung der Agenda 2030 in und durch Österreich im Zeitraum 2020–2022, Bundeskanzleramt Österreich
12. European Environment Agency (2019) The European environment—State and outlook 2020: Knowledge for transition to a sustainable Europe. Luxembourg: Publications Office of the European Union. https://doi.org/10.2800/96749

Artificial Intelligence Applications

A Hesitant Fuzzy SWARA Methodology for Big Data Maturity Assessment

Emine Elif Nebati and Biset Toprak

Abstract Big data analytics empowers organizations to make faster, more accurate, and cost-effective decisions compared to their competitors. While the potential benefits of big data are widely recognized, effective integration into long-term corporate planning remains a challenge. This study evaluated the big data maturity of a company operating in the food and retail sectors using the DELTTA (data, enterprise, leadership, targets, technology, and analysts) model. Given the complexity of big data maturity assessment, the study incorporated hesitant fuzzy (HF) numbers into the Stepwise Weight Assessment Ratio Analysis (SWARA) method to determine the relative importance of each maturity dimension. Subsequently, expert responses to the Big Data Readiness Assessment Survey were analyzed to evaluate the company's maturity level across all dimensions. The Hesitant Fuzzy SWARA application revealed 'data' as the most critical dimension, while 'enterprise' was identified as the least important. Furthermore, survey findings highlighted the company's strength lies in its 'Analysts and Data Scientists,' offering a notable advantage in big data management. Based on these insights, strategic recommendations were formulated to strengthen the company's big data capabilities, streamline operational processes, and enhance decision-making efficiency.

Keywords Big data · Big data maturity model · Readiness assessment · Hesitant fuzzy SWARA · Case study

E. E. Nebati (✉) · B. Toprak
Department of Industrial Engineering, Istanbul Sabahattin Zaim University, İstanbul, Turkey
e-mail: emine.nebati@izu.edu.tr

B. Toprak
e-mail: biset.toprak@izu.edu.tr

N. M. Durakbasa and K. G. Gülen (eds.), *Sustainable Green Conversion*, Lecture Notes in Mechanical Engineering, https://doi.org/10.1007/978-3-031-83583-4_6

1 Introduction

Technological advancements over the past decade have fueled a significant increase in data generation and storage, sourced from various devices capturing data. As data analysis and technology progresses, the notion of big data has emerged, contributing to the efficiency and effectiveness of industries [1]. The term "big data" encompasses information assets characterized by their high volume, high velocity, and high variety, necessitating cost-effective and innovative information processing methods to facilitate enhanced insight, decision-making, and process automation" [2]. Big data analytics and integration offers unprecedented benefits for decision-making across both public and private sectors. Processing vast quantities of data by machine learning, visual analytics, and natural language processing (NLP), empowers organizations to make faster, more accurate, and cost-effective decisions than their competitors. Data-driven decisions can optimize every phase of product or service development, leading to higher customer satisfaction, revenue, and cost reductions [3–6].

While the potential benefits of Big Data for organizations are widely recognized, there remains a significant need for comprehensive guidance on effectively integrating Big Data initiatives into long-term organizational planning [4, 7]. Typically, a Maturity Model (MM) or Maturity Assessment Model (MAM) is used to evaluate an organization's capabilities and outline development steps, focusing on key dimensions relevant to the company's structure [4, 8]. In this context, the Big Data Maturity Model (BDDM) functions as a framework aimed at assisting organizations with their big data endeavors and guides structuring processes, technologies, and analytics to facilitate success in creating value from data [6, 7, 9]. By utilizing such models, organizations can assess their current capabilities, identify areas for improvement, and implement targeted adjustments to enhance the effectiveness of their Big Data initiatives [6].

The purpose of this research is to assess the big data maturity within a company operating in the food and retail sectors. By employing the DELTTA model proposed by Davenport [3], the company's maturity level across six distinct dimensions will be evaluated, and its strengths and weaknesses will be revealed. Given the inherent complexity of big data maturity assessment, which involves multiple dimensions, determining the relative importance of each dimension within the model is critical [1]. To overcome this challenge, this study will integrate hesitant numbers into the Stepwise Weight Assessment Ratio Analysis (SWARA) method to calculate the relative importance of each maturity dimension. Subsequently, expert responses to the Big Data Readiness Assessment Survey [3] will be analyzed to determine the company's maturity level across all dimensions.

The remainder of the paper is organized as follows: The subsequent section provides a literature review, comparing existing Big Data Maturity Models (BDMMs) based on their purpose, level count, and dimensions. Additionally, it presents related studies on big data maturity and highlights limitations in current BDMMs. Sections 3 and 4 introduce the research methodology and the case study application, respectively. Finally, Section 5 presents the findings and offers recommendations for future research.

2 Literature Review

Maturity models for the adoption of Information and Communication Technologies (ICT) and big data in organizations have evolved significantly since the 1970s [5, 10]. Following the introduction of the first Big Data Maturity Model (BDMM) by the Data Warehousing Institute (TDWI), which was designed to evaluate the maturity of Big Data and Big Data analytics programs, numerous BDMMs have been developed over the past decade. Table 1 provides a comparative overview of the currently available Big Data Maturity Models (BDMMs).

Big Data Maturity Models are generally grouped into three main categories: descriptive models, prescriptive models, and comparative models [8]. Descriptive models solely evaluate an organization's current maturity level with respect to a particular technology or capability without providing recommendations to increase the organization's proficiency in big data maturity. Prescriptive models go a step further by not only evaluating the organization's current maturity level but also offering actionable recommendations and steps to enhance its proficiency in Big Data maturity. Comparative models involve analyzing the positions of similar organizations within the big data maturity model and creating a road map by comparing them with similar organizations. To construct a comparative big data maturity model, a reference group comprising multiple organizations is essential for benchmarking purposes [4, 5, 11].

Recent studies based on big data maturity assessment are presented as follows:

Drus and Hassan [12] evaluated various big data maturity models, focusing on their generic, design, and application attributes. The aim was to assist organizations in selecting the most suitable BDMM effectively and competently. Peña et al. [1] proposed a fuzzy ELECTRE framework to evaluate big data maturity, specifically in small- and medium-sized enterprises (SMEs) within the healthcare sector. Their study utilized the TDWI methodology [13] to identify criteria for assessing big data management maturity in SMEs across infrastructure, data management, organization, governance, and analysis dimensions. Carvalho et al. [10] examined the deficiencies in existing maturity models for health analytics and underscored the obstacles to the adoption of Data Analytics (DA) in Hospitals Information Systems (HIS). As a solution, they introduced a one-dimensional maturity model in the domain of Data Analytics, named HISMM-DA. This model comprises six stages and is a component of the broader HISMM framework. Okuyucu and Yavuz [5] reviewed big data maturity models explicitly developed for governmental adoption of big data. They identified two major models: Klievink et al. [14] and Kuraeva [15]. While Klievink et al.'s [14] model assesses big data maturity at the organizational level, Kuraeva's

Table 1 Currently available big data maturity models [5, 11]

Assessment model	Purpose of use	Number of levels	Maturity dimensions
TDWI big data maturity model	Comparative	5	Data management, infrastructure, analytics, organization governance
Big data business maturity model index	–	5	Organization, business process, and organization's situation
IDC maturityscapebig data and analytics	Comparative	5	Intent, data technology, people, process
Maturity model for big development	Prescriptive	4	Data management, strategy, efficiency, effectiveness, new solutions, transformation, data and analytics, security and policy, and partnership
Enterprise architecture maturity assessment tool	Prescriptive	4	Technology, staffing, business focus, Big Data management and governance, data type and quality
DELTA model	Descriptive	5	Data, enterprise, leadership, targets, analysts and data scientists
DELTTA model	Descriptive	5	Data, enterprise, leadership, targets, technology, analysts and data scientists
Big data maturity assessment	Descriptive	4	Business need, technology platform, operating model, analytics, and information management
Big data maturity framework	Prescriptive	4	Technical/infrastructure, data availability and governance, data-driven, decision-making culture, organization and resources, and sponsorship
Big data maturity model	Prescriptive	6	Vision, strategy, value and metrics, governance, trust and privacy, people and organization, data sources, data management, analyticsand visualization
A maturity model for big data and analytics IBM	Descriptive	4	Business strategy, information, analytics, culture and operational execution, architecture and governance
Zakat big data maturity model	–	5	Organization, leadership, data governance and integration, and analytics
The big data temporal maturity model	–	5	Data/knowledge, IT solutions, functionalities
Hortonworks big data maturity model	–	4	Sponsorship, data and analytics, technology and infrastructure, organization and skills; and process management
Big data maturity model	–	6	Strategic alignment, data, organization, governance, information technology

(continued)

Table 1 (continued)

Assessment model	Purpose of use	Number of levels	Maturity dimensions
A Maturity model for big data and analytics IBM	–	5	Organization, governance, data management, strategy, value and metrics, and trust and privacy
A maturity model for big data analytics in airline network planning	Comparative	6	Strategic alignment, organization, data, information technology

[15] model aims to evaluate big data maturity as part of the overall government adoption process at the state level. Limpeeticharoenchot et al. [9] developed a dynamic Big Data Maturity Model (BDDM) using latent class analysis (LCA) tailored for SMEs in Thailand. Their web-based model incorporated dimensions such as organizational and attitude actors, information technology, technology, and people readiness. This model allows respondents to compare their business with others at similar or higher maturity levels. De Mauro et al. [7] implemented the Consensual Big Data Maturity Assessment System (CBDAS) in a multinational leader company in the consumer goods sector. Through interviews with managers, they assessed dimensions including "data strategy," "integrated architecture," "IT infrastructure", "human data interface," "analytical human workforce," "integrated organization," "data-friendly corporate culture," and "data-reliant business process." The study found "data friendliness" to be the most critical dimension, followed by "integrated organization" and "data strategy," with overall maturity levels on data domains appearing relatively high. Mamak [16] measured the big data maturity of a large company operating in various sectors. Initially, a fuzzy DEMATEL method was employed to calculate the weights and interrelationships of the six dimensions outlined in the Davenport's BDMM [3]. Subsequently, Davenport's BDMM model was applied to the company, and the results were integrated with the DEMATEL findings. As a result, the "data" dimension emerged as the company's strongest aspect. The company demonstrated high performance in data access, integration, and quality. Following the "data"; "enterprise", "targets", "data scientists and analysts", "technology", and "leadership" dimensions were ranked in order of importance.

Malaracia et al. [6] created an evaluation system based on the Analytic Hierarchy Process (AHP) and assessed the big data maturity of sample Italian companies spanning various industries and scales. In their two-stage model, the initial step involved assessing eight factors contributing to big data success through a questionnaire completed by participants. Subsequently, these factors were assigned weights using the AHP methodology. The outcomes of the AHP-weighted scores revealed "Data friendliness" as the most influential factor, followed by "Integrated Organization" and "Analytical Skills." Al-Sai et al. [11] systematically reviewed Big Data maturity models from 2007 to 2022. Their analysis of 15 existing models underscores the deficiency in comprehensive details such as documentation and dimensions. Furthermore, they highlighted that there is a lack of information regarding the development process, as well as their validation and evaluation. The study highlights that existing Big Data Maturity Models (BDMMs) require further research and development, as many of these models currently lack comprehensive evaluation tools and visual aids to effectively assess and interpret outcomes.

3 Methodology

In this study, a two-stage research methodology is proposed. In the first stage, the six basic dimensions of the DELTTA model [3]—data, enterprise, leadership, targets, technology, and analysts—will be evaluated by decision-makers using the Hesitant Fuzzy SWARA method. In the second stage, Big Data Readiness Assessment Survey [3] will be applied to the managers in a company, and the current maturity level of the company will be evaluated. As we mentioned earlier, big data maturity models are grouped in three categories: descriptive, comparative, and prescriptive models. The DELTTA Model, proposed by Davenport [3], is one of the descriptive models. This model enables organizations to assess their strengths and/or weaknesses. Organizations develop their analytical capabilities according to "Data", "Enterprise", "Leadership", "Targets", "Technology", "Analysts and Data Scientists" dimensions. The stages in the DELTTA model can be used as a benchmark for capabilities and improvements by organizations that want to enhance their maturity level [17]. Therefore, the DELTTA model will be used in this study. In the first version of the model, called DELTA (data, enterprise, leadership, targets, and analysts) model, Davenport [18] evaluated the technology dimension under the title of data. Then, the DELTTA model (data, enterprise, leadership, targets, technology, and analysts), was developed by Davenport [3] by considering technology under a separate title.

A. Hesitant Fuzzy SWARA

The hesitant fuzzy SWARA method enables decision-makers to prioritize criteria based on their specific context and current circumstances. This approach is particularly useful in situations where uncertainty or ambiguity exists, as it incorporates hesitant fuzzy numbers to account for varying expert opinions and incomplete information. Hesitant fuzzy sets (HFSs), introduced by Torra and Narukawa [19] and Torra [20], offer a flexible and robust framework for addressing uncertainties inherent in decision-making processes. Recently, the hesitant fuzzy SWARA method has been used in various fields, such as education, health, and transportation [21–23]. The hesitant fuzzy SWARA method involves steps similar to those of the classical SWARA method. The main difference between the two methods lies in the integration of fuzzy numbers into the decision-making process. Once the criteria have been established, the process follows the steps outlined in Fig. 1 [24, 25].

Step 1. Define the objectives and criteria of the decision problem; identify the decision-makers.

Step 2. Determine the most important criteria according to expert opinions.

- $\bar{t}_j = \sum_{k=1}^{p} t_{jk}/p$ (1)

Step 3. Determine the relative importance of the criteria according to the most important criteria.

- Decision makers indicate the relative importance of each criterion with respect to the most important criterion by using undecided fuzzy linguistic terms from the set S. The decision maker can express his/her opinion with a single linguistic term or with two or more linguistic terms using the conjunction 'between'.

- S = $\begin{cases} S_0 \text{Nothing}, S_1 \text{ Very strongly less important}, S_2 \text{ Strongly less important}, \\ S_3 \text{ Very slightly less important}, S_4 \text{ Slightly less important}, S_5 \text{ Equally important}, \ S_6 \text{ Very High}, \ S_7 \textit{ Fairly high} \end{cases}$

Step 4. Integrate expert opinions.

- $\text{HFLWA}(H_1, H_2, \ldots, H_p) = \int_{k=1}^{p} (W_k H_k) = U_{s_{a_k}} \in H_K \left\{ S_{\sum_{K=1}^{p} w_K a_K} \right\}$ (2)

Step 5. Calculate criterion scores.

- The criterion score is calculated (S_j) and he linguistic term sets are reduced to a single value.
- $\text{P}(H_s) = \frac{1}{L} \sum_{l=1}^{L} \frac{\varphi_1}{\#H-1}$ (3)

Step 6. Criteria coefficients and importance vector are calculated.

- $K_J = \begin{cases} 1 & j = 1 \\ S_J + 1 & j > 1 \end{cases}$ (4)

Step 7. Determine criterion weights.

- Criteria weights are calculated by normalising the importance vector
- $W_J = \frac{T_J}{\sum_{J=1}^{n} T_J}$ (5)

Fig. 1 Steps of hesitant fuzzy SWARA application (Ren et al., 2019)

4 Application

The objective of this study is to assess the big data maturity of a company operating in the food and retail sectors. Using the DELTTA model, the company's maturity will be evaluated across six key dimensions—data, enterprise, leadership, targets, technology, and analysts—to identify its strengths and weaknesses. Decision-makers involved in the study include experts, assistant experts, and engineers who work as data analysts within the company. The research is conducted in two stages. In the first stage, the Hesitant Fuzzy SWARA method is employed to determine the importance weights of the six main dimensions of the DELTTA model. This stage involves consultation with three experts. In the second stage, the company's big data maturity level is assessed using the Big Data Readiness Assessment Survey [3].

A. Determination of Criteria Weights with Hesitant Fuzzy SWARA

The steps of the Hesitant Fuzzy SWARA method were applied as follows:

Step 1. As proposed in the DELTTA model [3], the criteria were determined as "Data", "Enterprise", "Leadership", "Targets", "Analysts and Data Scientists", and "Technology".

Step 2. Decision makers determined the criteria in order of importance. The evaluations of the three decision-makers were combined by geometric mean. The most important factor was determined as "Data". In the following steps, other criteria were evaluated according to the "Data". The final ranking obtained using Eq. (1) is given in Table 2.

Step 3. Three decision-makers compare each criterion with "data" using Table 3. The criteria values are given in Table 3 (Table 4).

Step 4. For each criterion, the opinions of the decision-makers are integrated with the given aggregation operator in Eq. (2). Table 5 shows the aggregation values.

Step 5. The criteria score is calculated (S_j) using the integrated decision-maker opinions and the linguistic term sets are reduced to a single value.

Table 2 Order of importance of criteria according to decision-makers

Criteria	Data	Enterprise	Leadership	Targets	Technology	Analysts and data scientists
DM-1	1	6	5	4	2	3
DM-2	1	5	6	4	3	2
DM-3	1	6	5	4	3	2
Combined final ranking	1,00	5,67	5,33	4,00	2,67	2,33
	1	**6**	**5**	**4**	**3**	**2**

Table 3 Linguistic terms

Scale			
s0	0	s4	4
s1	1	s5	5
s2	2	s6	6
s3	3	s7	7

Table 4 Relative importance levels of criteria according to "data" criterion

Criteria	DM-1	DM-2	DM-3
Analysts and data scientists	{s6}	{s5}	{s5}
Technology	{s6, s7}	{s6, s7}	{s6, s7}
Targets	{s7}	{s7}	{s7}
Leadership	{s6, s7}	{s7}	{s6, s7}
Enterprise	{s7}	{s7}	{s7}

Table 5 Integrated values obtained with the aggregation operator

Criteria	Integrated value
Analysts and data scientists	{5,333; 5,333}
Technology	{6,000; 7,000}
Targets	{7,000; 7,000}
Leadership	{6,333; 7,000}
Enterprise	{7,000; 7,000}

Step 6. The most important factor was included in the calculations, the criterion scores obtained in the previous step were converted into criterion coefficients determined using Eq. (4) and the criterion coefficients were converted into an importance vector using Eq. (5) (Tables 6 and 7).

Step 7. Importance vector values are normalized by Eq. (6) and importance weights are obtained. The importance weights of all factors are given in Table 8.

The Hesitant Fuzzy SWARA method revealed that 'data' is the most critical dimension with an importance weight of 0.47, while 'enterprise' is the least significant dimension with a weight of 0.018 (see Table 8). Data management ensures that analysts can easily access and integrate data. Current and reliable data is the basis for all analysis and decision-making. Without high-quality data, the reliability of analyses is reduced. Data management plays a critical role in ensuring data quality and integrity. The effectiveness of big data analytics depends on collecting and integrating data from different sources. A well-designed data management infrastructure includes technical solutions for storing, processing, and analyzing data. This infrastructure is essential for the scalability and flexibility of data. This has a direct impact

Table 6 Criteria scores

Criteria	Score
Analysts and data scientists	0,7619
Technology	0,9286
Targets	1,0000
Leadership	0,9524
Enterprise	1,0000

Table 7 Criteria coefficients and significance vector

Criteria	Coefficient	Significance vector
Data	1,0000	1,0000
Analysts and data scientists	1,7619	0,5676
Technology	1,9286	0,2943
Targets	2,0000	0,1471
Leadership	1,9524	0,0754
Enterprise	2,0000	0,0377

Table 8 Criteria weights obtained by the hesitant fuzzy SWARA method

Criteria	Weights	Ranking
Data	0,4712	1
Analysts and data scientists	0,2675	2
Technology	0,1387	3
Targets	0,0693	4
Leadership	0,0355	5
Enterprise	0,0178	6

on the success of big data projects. In contrast, the 'enterprise' dimension, which typically relates to the organizational structure and culture, is important but does not have the same direct technical impact on the success of big data analytics projects.

B. Evaluating the Big Data Maturity of the Company with the DELTTA Model

In the second step, the company's big data maturity was evaluated based on expert responses to the Big Data Readiness Assessment Survey [3] presented in Table 9. Responses were collected using a 5-point Likert scale for each of the six dimensions. Then, the responses for each of the six dimensions were averaged by dividing the total score by the number of questions within each dimension (see Table 10).

According to survey results, as presented Fig. 1, 'Analysts and Data Scientists' is the strongest dimension of the company. This criterion was ranked second according to the Hesitant Fuzzy SWARA method (see Table 8). This finding indicates that the company has a significant advantage in big data management. The "data" dimension is the weakest dimension of the company. Given the critical role of big data technologies in promoting sustainable competition and effective decision-making, the company should prioritize significant investments in data management, as the hesitant fuzzy SWARA analysis highlights the importance of 'data' dimension. Suggestions to the company are as follows:

Cloud-based data storage solutions can provide scalability as data volumes increase. For example, cloud providers such as Google Cloud offer flexible solutions for storing and processing large amounts of data. Agreements can be made with trusted data providers or marketplaces to access external data. Such services can help the company to analyze its environment better and gain a competitive advantage. Data quality can be improved by data cleaning, validation, and fact-checking processes. These processes ensure that data is complete, accurate, and current. Data quality and correct data use should be emphasized in training programs. Data security and confidentiality should be ensured, and data breaches and leaks should be prevented. These recommendations can help the organization to address big data management weaknesses, improve efficiency, and make more accurate and effective decisions (Fig. 2).

Table 9 Big data readiness assessment survey [3]

Data
We have access to very large, unstructured, or fast-moving data for analysis
We integrate data from multiple internal sources into a data warehouse or mart for easy access
We integrate external data with internal to facilitate high-value analysis of our business environment
We maintain consistent definitions and standards across the date we use for analysis
Users, decision makers, and product developers trust the quality of our data
Enterprise
We employee a combination of big data and traditional analytics approaches to achieve our organization's goals
Our organization's management ensures that business units and functions collaborate to determine big data and analytics priorities for the organization
We structure our data scientists and analytical professionals to enable learning and capabilities sharing across the organization
Our big data and analytics initiatives and infrastructure receive adequate funding and other resources to build the capabilities we need
We collaborate with channel partners, customers, and other members of our business ecosystem to share big data content and applications
Leadership
Our senior executives regularly consider the opportunities that big data and analytics might bring to our business
Our senior executives challenge business unit and functional leaders to incorporate big and analytics into their decision making and business processes
Senior executives in our organization utilize big data and analytics to guide both strategic and tactical decisions
Non-executive level managers in our organization utilize big data and analytics to guide their decisions
Oure process for prioritizing and deploying our big data assets (data, people, software and hardware) is directed and reviewed by senior management
Targets
We prioritize our big data efforts to high-value opportunities to differentiate us from our competitors
We consider new products and services based on big data as an aspect of our innovation process
We identify internal opportunities for big data and analytics by evaluating our processes, strategies, and marketplace
We regularly conduct data-driven experiments to gather data on what works and what doesn't in our business
We evaluate our existing decisions supported by analytics and data to assess whether new, unstructured data sources could provide better models
Technology
We have explored or adopted parallel computing approaches (e.g., Hadoop) to big data processing

(continued)

Table 9 (continued)

Data
We are adept at using data visualization to illuminate a business issue or decision
We have explored or adopted cloud-based services for processing data and doing analytics
We have explored or adopted open-source software for big data and analytics
We have explored or adopted tools to process unstructured data such as text, video, or images
Analysts and data scientists
We have a sufficient number of capable data scientists and analytics professionals to achieve our analytical objectives
Our data scientists and analytics professionals act as trusted consultants to our senior executives on key decisions and data-driven innovation
Our data scientists and analytical professionals understand the business disciplines and processes to which big data and analytics are being applied
Our data scientists, quantitative analyst, and data management professionals operate effectively in teams to address big data and analytics projects
We have programs (either internal or in partnership with external organizations) to develop data science and analytical skills in our employees

Table 10 Scores of the six maturity dimensions

Dimensions	Score
Data	3,70
Enterprise	4,38
Leadership	4,30
Targets	3,90
Technology	4,10
Analysts and data scientists	4,8

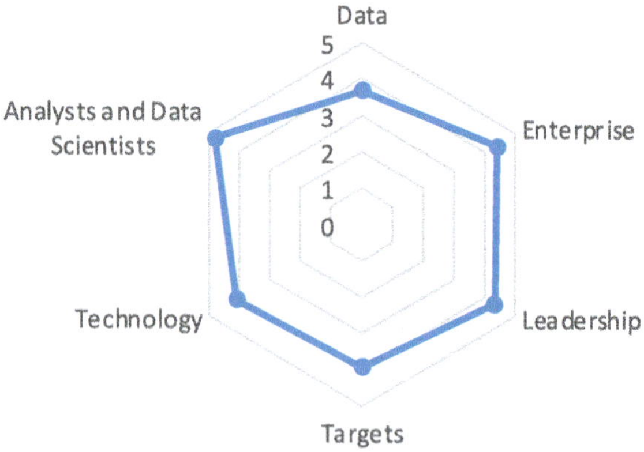

Fig. 2 Big data maturity levels of the company

5 Conclusion

Big data management is a critical component for modern businesses. By effectively harnessing and utilizing vast amounts of data, organizations can gain a competitive edge, improve efficiency, enhance customer satisfaction, and make more informed decisions. This study focused on evaluating the big data maturity level within a company operating in the food and retail sectors. Using Davenport's DELTTA model, the company's maturity was assessed across six key dimensions: "Data", "Enterprise", "Leadership", "Targets", "Technology", and "Analysts and Data Scientists".

A two-stage methodology was employed in the study. In the first stage, the hesitant fuzzy SWARA method was used to determine the importance weights of the six dimensions. The results revealed that "Data" is the most critical dimension, while "Enterprise" is the least significant. In the second stage, the Big Data Readiness Assessment Survey, based on Davenport's model, was conducted to assess the company's big data maturity level. According to the survey results, "Analysts and Data Scientists" was the company's strongest dimension, indicating a significant advantage in big data management. However, "Data" was identified as the weakest dimension of the company, suggesting a critical area for improvement.

The findings underscore the need for the company to prioritize investments in data management to enhance its overall big data maturity. Specific recommendations include adopting cloud data storage, integrating external data sources, implementing practices to improve data quality, focusing on data security, and providing employee training on data quality and utilization. By addressing these recommendations, the organization can strengthen data management capabilities, increase overall big data maturity, and leverage data-driven insights to improve decision-making, derive innovation, and gain competitive advantage.

References

1. Peña A, Bonet I, Lochmuller C, Tabares MS, Piedrahita CC, Sánchez CC Chiclana F (2019) A fuzzy ELECTRE structure methodology to assess big data maturity in healthcare SMEs. Soft Comput 23:10537–10550
2. Gartner IT Glossary (2020). Big data. https://www.gartner.com/en/information-technology/glossary/big-data
3. Davenport TH (2014) Big data at work: dispelling the myths. Harvard Business Review Press, Uncovering the Opportunities
4. Comuzzi M, Patel A (2016) How organisations leverage big data: a maturity model. Ind Manag Data Syst 116(8):1468–1492
5. Okuyucu A, Yavuz N (2020) Big data maturity models for the public sector: a review of state and organizational level models. Transform GovMent: People Process Policy 14(4):681–699
6. Malacaria S, De Mauro A, Greco M, Grimaldi M (2023) An application of the analytic hierarchy process to the evaluation of companies data maturity. SN Comput Sci 4(5):696
7. De Mauro A, Greco M, Grimaldi M, Malacaria S, Mignacca B (2022) Toward the implementation of a consensual maturity model for big data in consumer goods companies. In: Proceedings

of the international forum on knowledge asset dynamics, knowledge drivers for resilience and transformation (IFKAD, 2022), pp 2380–402
8. Al-Sai ZA, Abdullah R (2019) A review on big data maturity models. In: 2019 IEEE Jordan international joint conference on electrical engineering and information technology (JEEIT 2019). IEEE, pp 156–161
9. Limpeeticharoenchot S, Cooharojananone N, Chavarnakul T, Charoenruk N, Atchariyachanvanich K (2022) Adaptive big data maturity model using latent class analysis for small and medium businesses in Thailand. Expert Syst Appl 206:117965
10. Carvalho JV, Rocha Á, Vasconcelos J, Abreu A (2019) A health data analytics maturity model for hospitals information systems. Int J Inf Manag 46:278–285
11. Al-Sai ZA, Husin MH, Syed-Mohamad SM, Abdullah R, Zitar RA, Abualigah L, Gandomi AH (2022) Big data maturity assessment models: a systematic literature review. Big Data Cogn Comput 7(1):2
12. Drus SM, Hassan NH (2017) Big data maturity model–a preliminary evaluation. In: Proceedings of the 6th International Conference on Computing and Informatics (ICOCI 2017), Kuala Lumpur
13. Halper F, Krishnan K (2013) TDWI big data maturity model guide: interpreting your assessment score. TDWI benchmark guide 2014:2013
14. Klievink B, Romijn B-J, Cunningham S, de Bruijn H (2017) Big data in the public sector: uncertainties and readiness. Inf Syst Front 19(2):267–283
15. Kuraeva A (2016) Big data analysis influence on public administration processes. Master's thesis, Higher school of economics university, Moscow. www.hse.ru/en/edu/vkr/182647584
16. Mamak F (2022) Büyük veri olgunluk ölçümü: DEMATEL temelli bir vaka analizi. Master's thesis. Ibn Haldun University, Turkey
17. Davenport T (2018) DELTA plus model & five stages of analytics maturity: a primer. Ebook. International institute for analytics, Portland, Oregon, USA
18. Davenport TH, Harris JG, Morison R (2010) Analytics at work: smarter decisions, better results. Harvard Business Press
19. Torra V, Narukawa Y (2009) On hesitant fuzzy sets and decision. In: 2009 IEEE international conference on fuzzy systems. IEEE, pp 1378–1382
20. Torra V (2010) Hesitant fuzzy sets. Int J Intell Syst 25(6):529–539
21. Mardani A et al (2020) A novel extended approach under hesitant fuzzy sets to design a framework for assessing the key challenges of digital health interventions adoption during the COVID-19 outbreak. Appl Soft Comput 96:106613
22. Kamali Saraji M, Streimikiene D, Ciegis R (2022) A novel Pythagorean fuzzy-SWARA-TOPSIS framework for evaluating the EU progress towards sustainable energy development. Environ Monit Assess 194(1):42
23. Kaya SK, Erginel N (2020) Futuristic airport: a sustainable airport design by integrating hesitant fuzzy SWARA and hesitant fuzzy sustainable quality function deployment. J Clean Prod 275:123880
24. Ren R, Liao H, Al-Barakati A, Cavallaro F (2019) Electric vehicle charging station site selection by an integrated hesitant fuzzy SWARA-WASPAS method. Transform Bus Econ 18(2)
25. Ecer B (2021) Kararsız Bulanık Dilsel Terimler Temelli SWARA Yöntemi ile Ülke Değerlendirme Kriterlerinin Ağırlıklandırılması. Bulanık Çok Kriterli Karar Verme Yöntemleri-Ms Excel® ve Software Çözümlü Uygulamalar, Ankara: Nobel Akademik Yayıncılık, pp 217–229

Development of Agriculture in Turkey and Changing Dynamics: Factor Analysis Approach

Hasan Hüseyin Çelebi and Ayşe Paksoy

Abstract This study investigates the factors affecting agricultural productivity in Turkey from 1990 to 2023 using factor analysis. As the global population increases, understanding the variables that influence crop yields becomes crucial for sustaining the food supply. The research examines data such as average temperature, rainfall, inflation rates, and fuel prices, sourced primarily from the Turkish Statistical Institute. By applying factor analysis, the study groups these variables into meaningful clusters that reveal the underlying correlations between them and agricultural outputs like wheat, barley, rice, and corn. The results indicate that certain economic and environmental factors, such as consumer inflation and annual average temperature, consistently influence multiple crops. This research contributes a new perspective to the literature by identifying the key factors that drive agricultural productivity in Turkey, providing insights that can inform future agricultural policies. The study concludes with suggestions for further analysis involving a broader range of variables to enhance understanding of the complex dynamics in Turkish agriculture.

Keywords Agriculture · Factor analysis · Turkey · Wheat · Data analysis

1 Introduction

With the increasing population around the world, the nutrition of food products is an important factor. Agriculture is an important study that has a wide historical period in human history. Along with agriculture, it provides support to humanity in matters such as feeding the majority of the population, providing food to the elements in the livestock sector, and increasing soil fertility [1].

H. H. Çelebi (✉)
Industrial Engineering, Istanbul Kültür University, Istanbul, Turkey
e-mail: h.celebi@iku.edu.tr

A. Paksoy
Industrial Engineering, Istanbul University-Cerrahpaşa, Istanbul, Turkey
e-mail: ayse.paksoy@iuc.edu.tr

N. M. Durakbasa and K. G. Gülen (eds.), *Sustainable Green Conversion*, Lecture Notes in Mechanical Engineering, https://doi.org/10.1007/978-3-031-83583-4_7

Agriculture in Turkey has changed over the years in line with factors such as the rapidly increasing population, changes in eating habits, changes in soil fertility, and the number of farm animals. Economically, agricultural production has a positive relationship regarding exports and imports, raw material supply, and food prices [2].

The product that emerges as a result of agriculture is called a crop. Crop productivity varies depending on variables such as average temperature throughout the country, richness of underground and surface water resources, amount of precipitation, etc.

In this study, considering the data affecting agriculture in Turkey between 1990 and 2023, it is aimed to group the variables that provide variability in the same structure by clustering the factors affecting agricultural productivity in a smaller number with the factor analysis method. It has been determined that there are few studies on agriculture in the Turkish region as broad as the study, and this study brings a new perspective to the literature in this respect. In the following parts of the study, the studies in the field of agriculture were examined with the factor analysis method in the literature.

2 Literature Review

Agricultural productivity is of paramount importance for ensuring food security and the long-term sustainability of the agricultural sector. A review of studies in industrial engineering and manufacturing fields reveals a number of innovative approaches and technologies that have the potential to enhance agricultural productivity.

The literature review presented in Table 1 provides a comprehensive overview of the existing research on agricultural productivity and related topics. The studies in question examine a number of different factors by conducting a productivity analysis on a variety of geographical regions and agricultural products.

The noteworthy historical interconnection between agricultural (Ag) and non-agricultural (NonAg) economic expansion and its impact on the advancement of the agricultural sector is meticulously delineated by George [3]. A robust correlation has been identified between the GDP per capita of the agricultural sector and grain yield in a number of countries, regions, and globally, as evidenced by an examination of time series data. Furthermore, the data has demonstrated a correlation between the aforementioned variables and the GDP per capita of the non-agricultural sector. Analyses of the data have also revealed comparable relationships when the data is examined in relation to groups of countries based on income, debt, and development level. The revitalization of the agricultural industry is contingent upon two factors: strong GDP growth in agriculture and the possibility of higher rewards for all market players. This will necessitate a high level of contract compliance within the market.

A subsequent study proposed a new paradigm for analyzing variations in agricultural labor productivity, taking into account structural, land density, and land productivity effects. This method permitted the data to be disaggregated redundantly from a number of aggregation layers. A data envelopment analysis model was constructed

Table 1 Literature review on agricultural productivity

Researcher(s)	Year	Subject of the article
Xu et al.	2023	The objective of this study is to examine regional agricultural labor productivity in China in the context of improving urban–rural integration
Baležentis et al.	2021	Analysis of changes in agricultural labor productivity
Pokharel et al.	2021	The article's focus is an examination of the productivity of agricultural cooperatives in the United States
Dai	2024	The article's subject is an analysis of the potential of smart systems and sensors to enhance agricultural productivity. The study concentrates on the ways in which these technologies can be employed to enhance the efficiency and yield of agricultural practices
Zhu et al.	2023	It examines the subject of productivity growth measures in the context of European Union (EU) agriculture, with a specific focus on those that have been adjusted for energy-related greenhouse gas (GHG) emissions
Ngepah et al.	2023	The article's subject is an examination of the influence of government agricultural expenditure on the growth of agricultural total factor productivity (TFP) in South Africa. The study is concerned with elucidating the manner in which government expenditure in the agricultural sector affects the overall growth in productivity in the country
Baležentis et al.	2021	This article presents an extended by-production model for consideration and application
Şenol Çelik	2012	The article's primary focus is an examination of the development of provinces in Turkey, with a particular emphasis on the quantities of crops produced. The study employs factor analysis to examine and elucidate the variations and trends in crop production across different provinces
Serez et al.	2022	It evaluates the value of agricultural lands in a region affected by the Yortanlı dam and the factors affecting these values
Valizadeh et al.	2021	The creation and validation of an index for the assessment of agricultural sustainability was the main goal of the current investigation
Yaghoubi et al.	2022	Its goal is to investigate and examine the difficulties posed by precision agricultural methods in the province of Semnan's pistachio farms
Petrescu	2014	The analysis of labor productivity levels in agriculture in Romania's various development regions is the article's main topic. Evaluating the differences in labor productivity between the nation's various regions is the study's goal
Başer et al.	2023	The article addresses beef sustainability in the context of a range of global agricultural challenges, including shrinking farmland, soil contamination, climate change, water scarcity and rural poverty

(continued)

Table 1 (continued)

Researcher(s)	Year	Subject of the article
Iqbal et al.	2018	The subject of this article is the risk factors confronting cotton farmers in the Punjab province of Pakistan. The objective of the study is to document the aforementioned risks, assess the awareness of these risks among growers, understand the perceived impacts of these risks, and examine the strategies employed by farmers to manage these risks at the farm level
Dash et al.	2022	The article's focus is on the utilization of the Extra-trees classifier machine learning model to ascertain the socioeconomic factors that are significantly correlated with agricultural productivity. The study entails the training of the model on a dataset comprising a range of socioeconomic variables, with agricultural productivity designated as the target variable
Singh et al.	2016	This article addresses two key issues: firstly, the identification of the variables influencing Punjab's agricultural development; and secondly, the analysis of the regional differences resulting from these variables
Gaytancıoglu et al.	2007	The subject of the article is the analysis of wheat production in the Thrace region
Özyazıcı et al.	2013	The objective of this article is to present the results of an assessment of the physical and chemical characteristics of the soil in the tea-growing regions of the Rize province, which has been conducted using factor analysis as the methodological technique
Gedikli et al.	2014	This article presents an analysis of the corn yield in the Samsun province's Alaçam, Bafra, and Terme districts. The objective of this study is to ascertain the impediments pertaining to plant protection and the factors influencing the utilization of pesticides in agricultural practices

with the objective of identifying potential increases in agricultural labor productivity resulting from optimized input usage and output production. The analysis employed the logarithmic mean Divisia index. Consequently, if complete technical efficiency were to be achieved, the frontier analysis indicated that agricultural worker productivity may increase by approximately 45% on average [4].

The objective of the study undertaken by Serez [5] was to determine the value of agricultural land in the region and the factors that influence this value in the event of the operationalization of the Yortanlı dam. The parcels constituted the basis for this initial analysis, and the land values were determined using the market value and income technique. In order to ascertain the factors influencing the respective values of the parcels, a hedonic pricing analysis and factor analysis were employed. The results of the model indicate that the distance of the parcel from the district center and settlement area, in addition to its rocky terrain, is associated with a reduction in land value. It has been established that the value of the land in question increases in direct proportion to both the size of the parcel in question and the distance from the city center and the dam.

Table 2 Correlation matrix

TABLE II
CORRELATION MATRIX

	CORRELATION MATRIX																			
	Wheat	**AAT**	**Population**	**CI**	**WI**	**Rainfall**	**FP**	**AFCA**	**TAGP**	**NRF**	**TNATCPAP**	**AAD**	**Drought**	**Barley**	**Corn**	**Rice**	**Rye**	**Oat**	**Einkorn Wheat**	**Millet**
Wheat	1	-0.269	0.183	-0.153	-0.157	-0.518	-0.073	0.336	0.302	-0.399	0.137	-0.389	-0.268	0.691	0.376	0.678	0.494	0.61	0.207	0.059
AAT	-0.269	1	0.593	0.564	0.511	0.232	0.438	-0.045	0.299	-0.271	0.592	-0.54	0.48	-0.458	0.387	-0.654	-0.246	0.057	-0.229	0.327
Population	0.183	0.593	1	0.743	0.689	-0.173	0.808	0.534	0.741	-0.556	0.951	-0.812	-0.087	-0.401	0.285	-0.166	-0.254	-0.062	-0.471	0.609
CI	-0.153	0.564	0.743	1	0.988	-0.097	0.928	0.207	0.393	-0.207	0.804	-0.387	-0.138	-0.686	-0.184	-0.292	-0.64	-0.424	-0.522	0.426
WI	-0.157	0.511	0.689	0.988	1	-0.194	0.919	0.272	0.335	-0.109	0.774	-0.313	-0.223	-0.707	-0.253	-0.284	-0.679	-0.421	-0.545	0.432
Rainfall	-0.518	0.232	-0.173	-0.097	-0.194	1	-0.249	-0.701	0.091	0.178	-0.3	0.238	0.833	-0.161	0.255	-0.401	-0.012	-0.358	0.234	-0.37
FP	-0.073	0.438	0.808	0.928	0.919	-0.249	1	0.433	0.356	-0.32	0.877	-0.471	-0.346	-0.688	-0.269	-0.218	-0.701	-0.354	-0.727	0.625
AFCA	0.336	-0.045	0.534	0.207	0.272	-0.701	0.433	1	0.4	-0.072	0.521	-0.405	-0.585	-0.185	0.045	0.148	-0.169	0.122	-0.46	0.552
TAGP	0.302	0.299	0.741	0.393	0.335	0.091	0.356	0.4	1	-0.251	0.555	-0.566	0.197	-0.083	0.615	0.053	0.156	-0.109	0.032	0.154
NRF	-0.399	-0.271	-0.556	-0.207	-0.109	0.178	-0.32	-0.072	-0.251	1	-0.554	0.783	0.145	-0.252	-0.195	-0.179	-0.282	-0.31	0.102	-0.381
TNATCPAP	0.137	0.592	0.951	0.804	0.774	-0.3	0.877	0.521	0.555	-0.554	1	-0.708	-0.258	-0.49	0.052	-0.276	-0.389	-0.031	-0.633	0.767
AAD	-0.389	-0.54	-0.812	-0.387	-0.313	0.238	-0.471	-0.405	-0.566	0.783	-0.708	1	-0.023	-0.072	-0.528	-0.132	-0.182	-0.276	0.094	-0.338
Drought	-0.268	0.48	-0.087	-0.138	-0.223	0.833	-0.346	-0.585	0.197	0.145	-0.258	-0.023	1	0.048	0.653	-0.333	0.215	0.017	0.447	-0.452
Barley	*0.691*	*-0.458*	*-0.401*	*-0.686*	*-0.707*	*-0.161*	*-0.688*	*-0.185*	*-0.083*	*-0.252*	*-0.47*	*-0.72*	*0.048*	*1*	*0.402*	*0.718*	*0.884*	*0.623*	*0.731*	*-0.485*
Corn	0.376	0.387	0.285	-0.184	-0.253	0.255	-0.269	0.045	0.615	-0.195	0.052	-0.528	0.653	0.402	1	0.09	0.586	0.435	0.492	-0.23
Rice	0.678	-0.654	-0.166	-0.292	-0.284	-0.401	-0.218	0.148	0.053	-0.179	-0.276	-0.132	-0.33	0.718	0.09	1	0.502	0.179	0.486	-0.457
Rye	0.494	-0.246	-0.24	-0.64	-0.679	-0.012	-2.701	-0.168	0.156	-0.282	-0.389	-0.182	0.215	0.884	0.586	0.502	1	0.478	0.816	-0.473
Oat	0.691	0.057	-0.062	-0.424	-0.421	-0.358	-0.354	0.122	-0.109	-0.31	-0.31	-0.276	0.017	0.623	0.435	0.179	0.478	1	0.142	-0.17
Einkorn Wheat	0.207	-0.229	-0.471	-0.522	-0.545	0.234	-0.727	-0.46	0.032	0.102	-0.633	0.094	0.447	0.731	0.492	0.486	0.816	0.142	1	-0.876
Millet	0.059	0.327	0.609	0.426	0.432	-0.37	0.625	0.552	0.154	-0.381	0.767	-0.338	-0.452	-0.485	-0.23	-0.457	-0.473	0.17	-0.876	1

A separate study employed factor analysis to examine the evolution of agricultural production in Turkey's provinces. The findings indicate that the Central Anatolia Region is primarily engaged in the cultivation of grains, legumes, and sugar beets, whereas the Mediterranean Region is predominantly focused on the production of fruits and vegetables. The Mediterranean and Southeastern Anatolia regions are renowned for their corn and melon production, while the Marmara region is particularly renowned for its stone and pome fruits. The Eastern Anatolia Region is a significant producer of mulberries and apricots. Training is a fundamental aspect of various agricultural endeavors. It is of the utmost importance to consider the distinctive characteristics of the regions in question, including their vegetation, land structure, and climate, throughout the production process. It is recommended that efforts be made to enhance output, effectiveness, and revenue by identifying the optimal areas for agricultural production. Upon achieving these objectives, a notable enhancement to the country's economic standing is foreseen [6].

Yaghoubi [7] observed the challenges associated with the implementation of precision agriculture in pistachio orchards in the Semnan region. A stratified random sampling approach was employed to achieve this objective. The survey constituted the primary research instrument employed during the field study phase. The exploratory factor analysis yielded four principal components, namely: investment and support; structural; supply and development of human resources; and, finally, organizational and program problems.

In a separate investigation, an additional tree classifier machine learning model was employed to ascertain socioeconomic characteristics. It was found that the target variable may be able to identify socioeconomic variables that are associated with agricultural productivity. A test sample comprising the socioeconomic variables that influence farmers' agricultural productivity in Sambalpur District, Odisha State, India, was employed to evaluate the study. Two significant findings emerged from the investigation. Firstly, the method employed for the identification of socioeconomic characteristics and agricultural productivity proved to be an effective means of establishing the interrelationships between the two. Secondly, it was demonstrated that the proposed collective learning-based approach was an effective method for calculating agricultural productivity [8].

The objective of the follow-up study, which employed factor analysis, was to devise and validate an index for assessing agricultural sustainability. The population under investigation consisted of wheat growers in Iran. Principal component factor analysis identified four elements or criteria for the indicators of agricultural sustainability. The four factors identified as being of particular importance were social fairness and well-being, resilience, stability and adaptation, and productivity and efficiency. Following the confirmatory factor analysis of the factor structure of the agricultural sustainability measurement index, the model's fit indices demonstrated that the measurement structure of the index was an appropriate representation of the underlying data. To conclude, despite the challenges that researchers, decision-makers and policymakers face in improving and capturing agricultural sustainability, the agricultural sustainability measurement index and processes developed and validated have the potential to inform and guide agricultural policies, as well as facilitate comparisons between different regions in an agriculturally rational manner [9].

In a separate study, the productivity of American agricultural cooperatives was evaluated utilizing the Malmquist Productivity Index (MPI). The analysis was conducted at two-year intervals and under a number of scenarios with varying returns to scale. The findings indicate that augmented investment in technology and enhanced managerial efficacy can result in augmented productivity [10]. A comparative analysis of worker productivity in different development zones in Romania's agricultural industry revealed a decline in productivity. The results of the factor analysis indicated that the primary factor contributing to the observed decline in productivity was a reduction in the value of agricultural output [11].

The sustainability of agricultural systems is facing significant challenges due to a multitude of global agricultural issues, including the loss of arable land, soil contamination, climate change, water scarcity, and rural poverty. A number of investigations have been conducted with the objective of addressing these concerns. In one study, a method for assessing the environmental, social, and economic sustainability of beef cattle farms was developed. Furthermore, the investigation examined the influence of farm size on the overall sustainability of these farms [12]. A further study examined the numerous risks faced by cotton farmers in Punjab, Pakistan. The findings indicate that the majority of farmers adopt risk mitigation strategies in a manner that is consistent with their comprehension of these strategies. Furthermore, they perceive output declines to be associated with various threats [13]. Significant elements, including land resources, farm inputs, productivity, and diversity were shown to be the main causes of regional differences and to have a major impact on the observed variation in development indices, according to a study on Punjab's agricultural development [14]. Furthermore, a study conducted in the tea-producing regions of Rize, Turkey, examined the chemical and physical characteristics of the soil and identified several factors that contribute to soil heterogeneity, with human activity being a particularly significant factor in coastal areas [15].

In a study on corn production in Turkey's Samsun province, Gedikli [16] observed that a number of variables, including field rent, operating costs, and the age of the family head, exert a significant influence on the usage of pesticides. Furthermore, the utilization of pesticides is contingent upon the availability of subsidies. Balezentis [17] put forth an enhanced co-production model and its application to elucidate the interrelationship between production and downstream technologies that result in pollution.

In examining strategies and effects of government spending to increase agricultural productivity, studies such as Xu [18] and Ngepah [19] are of particular interest. Similarly, a study by Zhu [20] investigates the effects on technical efficiency and energy-based greenhouse gas emissions. Furthermore, a study by Dai [21] investigates the role of smart systems and sensors in increasing agricultural productivity and the implementation challenges associated with precision agriculture. Conversely, a study by Gaytancıoğlu [22] examines crop productivity and production processes in specific regions. These studies evaluate the effectiveness of various strategies and technologies in increasing the efficiency of agricultural production processes, providing valuable insights that inform current practices in the sector.

3 Methodology

A. Factor Analysis

A popular multivariate statistical method for reducing a large set of interrelated variables into a smaller number of more significant, comprehensible, and independent factors is factor analysis. By organizing variables that are related to one another in a multidimensional space, factor analysis seeks to extract fewer new and independent variables. Similarly, factor analysis is a technique for reducing the dimensionality of a dataset and identifying any potential dependency structures. The first step in factor analysis is to create a covariance or correlation matrix. The principal component analysis warnings are duly considered when selecting the optimal matrix. Nevertheless, the correlation matrix is typically the superior option. In the process of factorizing a correlation matrix for the purpose of conducting a factor analysis, a variety of methods are employed. The aforementioned techniques are employed to elucidate latent structures within the dataset and to ascertain the elements that give rise to these patterns. Principal factors analysis (PAF), principal components analysis (PCA), least squares, maximum likelihood, extended least squares, and unweighted least squares (ULS) represent a selection of these techniques [23].

A statistical technique employed for the purpose of examining the connections between observable variables is referred to as factor analysis. The initial stage of the methodology entails the calculation of the correlation between the observed variable values, which transforms the original variables into a correlation matrix. The correlation matrix allows the relationships between variables to be identified and understood. In the second stage, the unrotated factors are derived using factor derivation from the correlation matrix. Principal component analysis (PCA) is the most commonly employed method for factor derivation, although alternative approaches may also be utilized. The objective of this approach is to identify a limited set of factors that collectively account for the shared variance between the variables. The third step comprises the application of a rotation technique for the purpose of converting the unrotated factors into rotated factors. The objective of this technique is to enhance the interpretability of the components. The two principal methods of rotation are oblique and orthogonal. Conversely, the former is typically the preferred option, as it enables a more straightforward structure while maintaining the elements' independence.

Prior to its utilization, factor analysis is contingent upon the satisfaction of several prerequisites. Primarily, the dataset must exhibit linear relationships and a normal distribution. Furthermore, the information must be quantitative. It is essential that the dataset be free of anomalies and heteroscedasticity issues. It is imperative that the independent variables are not multicollinear in the final analysis. The fulfillment of these requirements ensures the validity and dependability of the factor analysis results [24].

Fig. 1 Factor analysis stages

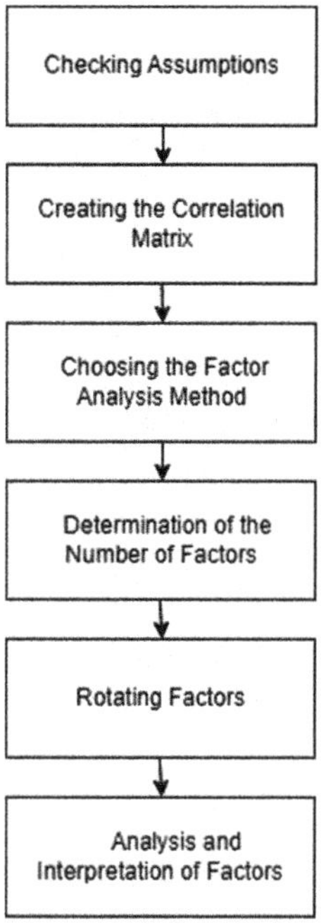

The stages of the factor analysis process are illustrated in Fig. 1. The method of factor analysis commences with the validation of the underlying assumptions, after which the dataset is evaluated for its suitability for analysis. Subsequently, the correlations between the variables are illustrated in a correlation matrix. The number of factors to be employed is determined subsequent to the identification of the analytical method. The objective of factor rotation is to enhance the interpretability of the results. The final phase comprises a comprehensive examination and evaluation of the factors. This procedure is fundamental to the comprehension of complex datasets.

In the field of factor analysis, two principal methodologies are typically employed. An exploratory factor analysis is conducted when the researcher is uncertain about the number of factors that the measurement tool is measuring. The objective is to gain insight into the characteristics of the components that the measurement tool is designed to assess, rather than to test a specific hypothesis. The specific type of analysis is known as confirmatory factor analysis.

4 Implementation

Data that collaborated mostly from the Turkish Statistical Institute (TURKSTAT) have conducted individual factor analysis on SPSS due to agricultural products. When conducting factor analysis, the varimax rotation method is used in software.

At every single result, the correlation matrix is observed and checked to see if there are any statistical data that cannot be used on factor analysis; there was no data that could not be used on factor analysis.

From Table 2, the data that represents the correlation between each of variable that used in factor analysis is shown. Correlation between products will not be meaningful at the part of the analysis; however, correlation among the products and the variables that affect the productivity is going to make some observations, such as if one product and the variable are going to have a strong correlation at the factor analysis stage, it is expected to be in the same group at the factor analysis stage. For instance, Einkorn wheat and fuel prices have a strong correlation with a − 0,727 value, so we expect that they are going to be in the same group when we perform individual factor analysis.

Data has been grouped if they are at the same component number, and we gave every component a symbol. If the symbol in corresponding data is "*" then that data is at the first component, " − " symbol is at the second component, "/" symbol is at the third component, and " + " is at the fourth component. Only one analysis revealed 4 factors, while all other analyses revealed 3 factors.

From Table 3, wheat is at component 2nd with total annual groundwater potential (TAGP), number of farmers (NRF), and agricultural areas. Barley in component 1st with annual average temperature (AAT), consumer inflation (CI), wholesale inflation (WI), fuel prices (FP), and total number of animals that can produce animal products (TNATCPAP). Corn is in the 3rd component with total annual groundwater potential (TAGP), rice is in the 1st component with annual average temperature, consumer inflation, whole inflation, fuel prices, and TNATCPAP. Rye is at 1st component with annual average temperature, consumer inflation, whole inflation, fuel prices, and TNATCPAP. Oat is in the 3rd component with a few registered farmers and agricultural areas (decare). Einkorn wheat is in the 1st component with annual average temperature, consumer inflation, wholesale inflation, fuel prices, and TNATCPAP. Millet is in the 3rd component with drought, rainfall, and annual fertilizer amounts. Also, we can conclude that consumer inflation, whole inflation, annual average temperature fuel prices, and TNATCPAP have the same impact on agriculture in Turkey and each other because they are mostly placed at the same component. Also, drought, rainfall, and annual fertilizer consumption amounts have similar impacts on agriculture and each other.

5 Conclusions

In this paper, it is aimed to find the correlation between agricultural products and the variables that affect the productivity of these products. Finding this correlation and using factor analysis to group these elements has a high impact on understanding the nature of agriculture. While collecting data, mostly Turkish Statistical Institute sites were used. At the analysis stage, mostly average annual temperature, consumer inflation, wholesale inflation, fuel prices, and the total number of animals that can

Table 3 Individual factor analysis

	AAT	Drought	Population	CI	WI	Rainfall	FP	AFCA	TAGP	NRF	TNATCPAP	AAD
Wheat (–)	*	/	*	*	*	/	*	/	–	–	*	–
Barley (*)	*	/	–	*	*	/	*	/	–	–	*	–
Corn (/)	*	–	*	*	*	–	*	–	/	+	*	+
Rice (*)	*	/	–	*	*	/	*	/	–	–	*	–
Rye (*)	*	/	–	*	*	/	*	/	–	–	*	–
Oat (/)	*	-	*	*	*	-	*	–	+	/	*	/
Einkorn Wheat (*)	*	/	–	*	*	/	*	/	–	–	*	–
Millet (/)	*	/	–	*	*	/	*	/	–	–	*	–

produce animal products are in the same group with the agricultural products such as wheat, barley, rice, rye, and Einkorn wheat. The other 3 products are grouped in the 3rd group with other variables that affect productivity.

From the findings from factor analysis, we can conclude that the product that was grouped in the 1st group has a high correlation between the variables that were placed in the 1st group. The variability of these elements is going to have an effect on the products; some of these variables have a negative correlation with the resulting products, while others have a positive correlation.

At the next stage of this study, the number of variables that have an impact on agriculture can be increased. As we apply factor analysis one by one based on agricultural products, one more factor analysis that takes every variable and products can be made and also correlation between variables can be examined. After this application, the differences between the previously performed factor analysis and the newly performed factor analysis can be compared.

References

1. Ronaghi M, Ronaghi MH (2021) Investigating the impact of economic, political, and social factors on augmented reality technology acceptance in agriculture (livestock farming) sector in a developing country. Technol Soc 67:101739. https://doi.org/10.1016/j.techsoc.2021.101739. ISSN 0160-791X
2. Yılmaz H, Özyakar HD, Dağ MM (2024) Exploratory factor analysis of manure utilization for sustainable dairy farming: evidence from crop-dairy farming systems in Turkey. Waste Manag Bull 1(4):164–171. https://doi.org/10.1016/j.wmb.2023.10.010. ISSN 2949-7507
3. George T (2020) A new look at agricultural development and the non-agriculture economy in low-income countries. Glob Food Sec 26:100449
4. Baležentis T, Li T, Chen X (2021) Has agricultural labor restructuring improved agricultural labor productivity in China? A decomposition approach. Socio-Econ Plan Sci 76:100967
5. Serez BS, Engindeniz S, Örük G (2022) Tarım arazisi değerlerini etkileyen faktörlerin analizi: Yortanlı baraj havzası örneği. Türk Tarım ve Doğa Bilimleri Dergisi 9(2):320–329
6. Çelik, Ş (2012) Türkiye'de İllerin Bitkisel Üretiminin Faktör Analizi İle İncelenmesi. Yuzuncu Yıl University. J Agric Sci 22(2):69–76. [3] Azibi R, Vanderpooten D (2002) Construction of rule-based assignment models. Eur J Oper Res 138:274–293
7. Yaghoubi M, Niknami M (2022) Challenges of precision agriculture application in pistachio orchards: factor analysis from Iranian agricultural experts perspective. Tekirdağ Ziraat Fakültesi Dergisi 19(3):473–482
8. Dash PB, Naik B, Nayak J, Vimal S (2022) Socio-economic factor analysis for sustainable and smart precision agriculture: an ensemble learning approach. Comput Commun 182:72–87. [5] (2018) The ISPR2018 website. http://www.ispr2018.org/
9. Valizadeh N, Hayati D (2021) Development and validation of an index to measure agricultural sustainability. J Clean Prod 280:123797
10. Pokharel KP, Featherstone AM (2021) Examining the productivity growth of agricultural cooperatives: the biennial malmquist index approach. J Co-Op Organ Manag 9(2):100148
11. Petrescu IE (2014) Factor analysis of labor productivity in agriculture in terms of sustainable development
12. Başer U, Bozoğlu M (2023) The impact of farm size on sustainability of beef cattle farms: a case study of the Samsun province Turkey. Int J Agric Sustain 21(1):2253647
13. Iqbal MA, Ping Q, Zafar MU, Abbas A, Bashir MK, Ali A, Kousar R (2018) Farm risk sources and their mitigation: a case of cotton growers in Punjab. Pak J Agric Sci 55(3)

14. Singh P, Grewal IS, Javed M, Kaur A (2016) Statistical evaluation of agricultural development in Punjab-a factor analytic approach. Indian J Econ Dev 12(1):33–40
15. Özyazıcı MA, Sağlam M, Dengiz O, Erkoçak A (2014) Çay tarımı yapılan topraklara yönelik faktör analizi ve jeoistatistik uygulamaları: Rize ili örneği
16. Gedikli O, Uzundumlu AS, Tozlu G (2015) Mısır üretiminde kimyasal ilaç kullanımını etkileyen faktörlerin analizi: Samsun ili örneği. ÇOMÜ Ziraat Fakültesi Dergisi 3(1):1–8
17. Baležentis T, Blancard S, Shen Z, Štreimikienė D (2021) Analysis of environmental total factor productivity evolution in European agricultural sector. Decis Sci 52(2):483–511
18. Xu B, Tu J, Kai W, Chen X (2023) Regional diversity and evolution of agricultural labor productivity in China. J Chin Hum Resour Manag 14:90–105
19. Ngepah N, Sunge R (2023) Agricultural expenditure and agricultural total factor productivity growth in South Africa. AIMS Agric Food 8(2):637–661
20. Zhu N, Streimikis J, Yu Z, Balezentis T (2023) Energy-sustainable agriculture in the European Union member states: overall productivity growth and structural efficiency. Socioecon Plann Sci 87:101520
21. Dai S (2022) Improve the design intelligence through wireless networks to increase agricultural production in provincial countries. Wirel Commun Mob Comput 2022(1):5991783
22. Gaytancıoğlu SKO (2007) Türkiye'de buğdayda uygulanan tarım politikaları ve Trakya bölgesi buğday üreticilerinin sorunları. Tekirdağ Ziraat Fakültesi Dergisi 4(3):249–259
23. Saraçlı S (2011) Faktör analizinde yer alan döndürme metotlarının karşılaştırmalı incelenmesi üzerine bir uygulama. Düzce Üniversitesi Sağlık Bilimleri Enstitüsü Dergisi 1(3):22–26
24. Yılmaz E (2018, Temmuz 19) Faktör analizi nedir? Nasıl uygulanır? *Veri Bilimi Okulu.* https://www.veribilimiokulu.com/faktor-analizi-nedir-nasil-uygulanir/

Enhancing Stock Price Prediction with Extreme Learning Machine and Multi-Indicator Fusion: A Comparative Study

Elham Pashaei and Binnur Gürül

Abstract This study presents a novel approach to stock price prediction by enhancing the Extreme Learning Machine (ELM) with multi-indicator fusion. Our method extracts nine technical indicators from historical stock price data, combining them to significantly enhance forecasting accuracy. The dataset includes stock data from NYSE and NASDAQ spanning several years, covering eight firms: GS, MSFT, AAPL, IBM, CTSH, BAC, ORCL, and HAL. Our approach is rigorously evaluated through a comprehensive comparative study using six key performance metrics: Average Relative Value (ARV), Mean Absolute Error (MAE), Symmetric Mean Absolute Percentage Error (SMAPE), Root Mean Squared Error (RMSE), Mean Squared Error (MSE), and R^2, applied across multiple datasets. We compare the effectiveness of our ELM-based approach with several well-known machine learning models, including Artificial Neural Networks (ANN), Extreme Gradient Boosting (XGBoost), Long Short-Term Memory (LSTM), and Multi-Layer Perceptron with Backpropagation (MLP-BP). The results show that the ELM model, enhanced with a new combination of technical indicators, outperforms competing methods in prediction accuracy and robustness. This study highlights the efficacy of ELM in stock price forecasting and contributes to the growing research on integrating diverse data sources with machine learning techniques for financial predictions.

Keywords Stock price prediction · Extreme learning machine (ELM) · Multi-indicator fusion · Comparative study · Financial forecasting models

E. Pashaei (✉)
Software Engineering Department, Istanbul Gelisim University, Istanbul, Türkiye
e-mail: elhpasha@iu.edu

B. Gürül
Industrial Engineering Department, Istanbul Gelisim University, Istanbul, Türkiye
e-mail: bgurul@gelisim.edu.tr

N. M. Durakbasa and K. G. Gülen (eds.), *Sustainable Green Conversion*, Lecture Notes in Mechanical Engineering, https://doi.org/10.1007/978-3-031-83583-4_8

1 Introduction

Making well-informed investing decisions and creating successful market strategies depend on accurate stock price prediction. Reliable forecasts empower investors to maximize returns and minimize risks, offering a competitive edge in the financial markets [1]. However, because financial markets are so intricate and unpredictable, forecasting stock values is a difficult undertaking by nature. Numerous factors, such as market sentiment, corporate performance, and economic conditions, affect stock prices, making accurate forecasting more challenging [2].

To address these challenges, a variety of methods have been developed, including machine learning (ML) algorithms [3, 4], deep learning models [1, 5] conventional statistical techniques [6, 7], and hybrid methods [8, 9]. Recent advancements in stock price prediction have focused on leveraging deep learning architectures and hybrid models. For instance, Wenjie Lu et al. [10] proposed a CNN-BiLSTM-AM method combining convolutional neural networks (CNN), bi-directional long short-term memory (BiLSTM), and an attention mechanism (AM) to predict stock closing prices. This hybrid approach effectively identified key features influencing stock prices, achieving superior mean absolute error (MAE) and root mean square error (RMSE) metrics on 1000 trading days of the Shanghai Composite Index. Similarly, Pengfei Yu and Xuesong Yan [11] developed a deep neural network (DNN) model for stock price prediction by utilizing long short-term memory networks (LSTM) and time series phase-space reconstruction (PSR). Their results demonstrated higher prediction accuracy compared to conventional ML algorithms, underscoring the DNN's ability to handle the complex, nonlinear nature of financial data. Xuan Ji et al. [12] extended this approach by integrating traditional financial indices with social media data using deep learning. They employed Doc2Vec to process text features from social media, and wavelet transforms to decompose time series data, which enhanced the model's predictive capability by incorporating investor sentiment along with financial variables. Further advancements include the work by Hadi Rezaei et al. [13], who introduced hybrid models like CEEMD-CNN-LSTM and EMD-CNN-LSTM, combining empirical mode decomposition techniques with CNN and LSTM architectures. These models were shown to outperform their standalone counterparts by effectively analyzing different frequency spectra of time series data, thereby enhancing the prediction accuracy of financial markets. Another notable approach by Wu et al. [14] employed a graph-based CNN-LSTM model that incorporates leading indicators such as options and futures, which improved the prediction accuracy of stock prices by utilizing a sequence array as an input to CNN for feature extraction, followed by LSTM for temporal prediction. Billah et al. [5] empirically analyzed LSTM networks for time-series predictions, reporting a mean accuracy of 89.7% across multiple financial datasets, and a reduction of 23.4% in MAE when compared to conventional forecasting techniques. Additionally, Ma et al. [2] introduced VGC-GAN, a multi-graph convolutional adversarial network that generates multiple correlation graphs from historical stock data, using a Generative Adversarial Network (GAN) framework combined with Multi-Graph Convolutional Networks and Gated Recurrent Units (GRU). This approach enhanced predictive performance by deeply exploring inter-stock correlations and temporal dependencies, validated on several real datasets. Lastly, Minrong Lu and Xuerong Xu [1] proposed an effective Time-series Recurrent Neural Network (TRNN) that addressed the inefficiencies of traditional RNN and LSTM models by incorporating sliding windows and data compression techniques. Their TRNN model demonstrated improvements in both accuracy and computational efficiency.

These methods vary in their ability to model complex patterns and adapt to market changes. Besides, feature engineering has a significant impact on raising the prediction accuracy of ML and DL-based techniques [3]. Technical indicators, such as the Simple Moving Average (SMA), Relative Strength Index (RSI), and Exponential Moving Average (EMA), are commonly used to provide insights into stock price trends and market conditions. These indicators help capture various aspects of price movements, enhancing the model's ability to forecast future trends [15, 16].

Despite the progress made with existing ML and DL models, predicting stock prices remains a formidable challenge because of the many influencing elements and the constantly changing structure of financial markets. Existing methods, while effective in certain contexts, often face limitations in adapting to sudden market shifts, capturing intricate dependencies among diverse data sources, and optimizing the balance between model complexity and interpretability. Moreover, the reliance on feature engineering and the need for preprocessing steps like frequency decomposition or sentiment analysis can add complexity and reduce the generalizability of models.

Given these challenges, there is a clear need for a novel approach that can seamlessly integrate multiple indicators, handle the nonlinearity and high volatility of stock data, and provide robust predictions with minimal preprocessing.

In this study, we introduce a novel methodology termed ELM-MIF, which synergistically combines Extreme Learning Machine (ELM) with multi-indicator fusion (MIF) to improve predictive accuracy. This approach utilizes nine technical indicators derived from historical stock data of eight companies. In our experimental study, we benchmark the ELM-MIF model against several prominent ML techniques, including LSTM, Extreme Gradient Boosting (XGBoost), Artificial Neural Networks (ANN), and Multi-Layer Perceptron with Backpropagation (MLP-BP). Performance is assessed using six critical evaluation metrics: Coefficient of Determination (R^2), Average Relative Value (ARV), Symmetric Mean Absolute Percentage Error (SMAPE), Mean Squared Error (MSE), MAE, and RMSE. The findings demonstrate that the ELM-MIF model performs better than other models at predicting stock prices, consistently surpassing the comparison models in prediction accuracy, highlighting its robustness and potential in financial forecasting.

2 Preliminaries

A. Extreme Learning Machine

ELM is a prominent ML method utilized for both regression and classification tasks, particularly within Single Hidden-Layer Feedforward Neural Networks (SLFNs). In SLFNs, ELM significantly simplifies the process of the training model by fixing the weights of the hidden layer neurons, which are randomly assigned and remain unchanged during training. This contrasts with traditional training mechanisms, such as gradient descent or backpropagation, which involve iterative adjustments of both hidden layer and output layer weights, often requiring multiple epochs and extensive computational resources. ELM's approach eliminates the need for these iterative adjustments, providing a more efficient and rapid training process. Its advantages include reduced training time and computational complexity, making it particularly effective for handling complex data. Recent research [17–19], has demonstrated ELM's effectiveness in stock price prediction, highlighting its capability to achieve high forecasting accuracy while maintaining computational efficiency.

B. Technical Indicators

Technical indicators are quantitative tools for analyzing changes in stock prices and forecasting future trends. They help investors and analysts assess market conditions, identify trading signals, and make informed decisions [15, 20, 21]. Some of the most important indicators include the SMA, EMA, William's R%, Stochastic %K, Moving Average Convergence Divergence (MACD), Signal Line, RSI, Stochastic %D, and Rate of Change (ROC).

The SMA smoothes out short-term swings by calculating the average closing price over a predetermined period of days. The EMA provides a more responsive trend indicator by giving current prices more weight. The *RSI* gauges the speed and change of price movements, identifying overbought or oversold conditions. The *MACD* measures the difference between two EMAs, indicating potential buy or sell signals. A 9-day EMA of the MACD is called the Signal Line, and it is used to further refine trading signals. The *Stochastic%K* and *Stochastic%D* indicators measure price momentum relative to historical price ranges, aiding in identifying potential reversal points. The *ROC* quantifies the percentage change in price over a given period, reflecting the strength of price movement. Finally, *William's R%* assesses overbought and oversold levels by contrasting the present price with previous highs and lows. Their formulas are detailed in Table 1, providing the mathematical foundations for these indicators.

Table 1 Selected technical indicators

Technical indicators	Description	Formula
SMA	Simple moving average (5 days)	$SMA_t = \frac{1}{t}\sum_{i=1}^{t} C_i$ Where C_i is the price at day i, and t is the number of periods
EMA	Exponential moving average (5 days)	$EMA_t = C_t \times M + EMA_{t-1} \times (1 - M)$; $M = \frac{1}{n+1}$
RSI	Relative strength index	$RSI = 100 - 100/(1 + RS)$; $RS = AverageLoss/AverageGain$
MACD	Moving Average Convergence Divergence	$MACD = EMA_{12} - EMA_{26}$ where EMA_{12} and EMA_{26} are the 12-day and 26-day EMAs
Signal	A 9-day EMA of the MACD line	$SignalLine = EMA_9(MACD)$
Stochastic %K	Measures the current price relative to the range of prices over a specific period (usually 14 days)	$\%K = \frac{currentprice - LL}{HH - LL} \times 100$ where LL and HH are the lowest and highest prices over a specific period (usually 14 days)
Stochastic %D	A 3-day moving average of %K, is used to smooth out %K and confirm buy or sell signals	$\%D = SMA_3(\%K)$ where $SMA_3(\%K)$ is the 3-day simple moving average of $\%K$.
ROC	Rate of change	$\frac{Price(t)}{Price(t-n)} \times 100$
William's R%	A momentum indicator that measures overbought and oversold levels	$William'sR\% = \frac{\text{HH -current price}}{\text{HH -LL}} \times (-100)$

3 Proposed Stock Prediction Model

The proposed ELM-MIF method enhances stock price prediction by leveraging historical stock data to derive nine technical indicators, which are then analyzed using an ELM neural network. We utilize four historical stock price data points—the highest price, lowest price, opening price, and closing price—to compute these nine technical indicators, as detailed in Table 1. These indicators were selected based on their relevance to stock trend prediction, supported by prior research [14–16, 22].

The dataset consists of 13 features in total: four historical data points and nine technical indicators, where closing price is the label. These features are normalized to a range of [0,1], and their importance for prediction is illustrated in Fig. 1, highlighting how each contributes to forecasting accuracy. The data is split into training and testing subsets. ELM is used to train the SLFN model, to optimize the weights between the output and hidden layers. The framework of the suggested ELM-MIF approach is depicted in Fig. 2, showcasing the integration of historical data, technical indicators, and the ELM model in the stock price prediction process.

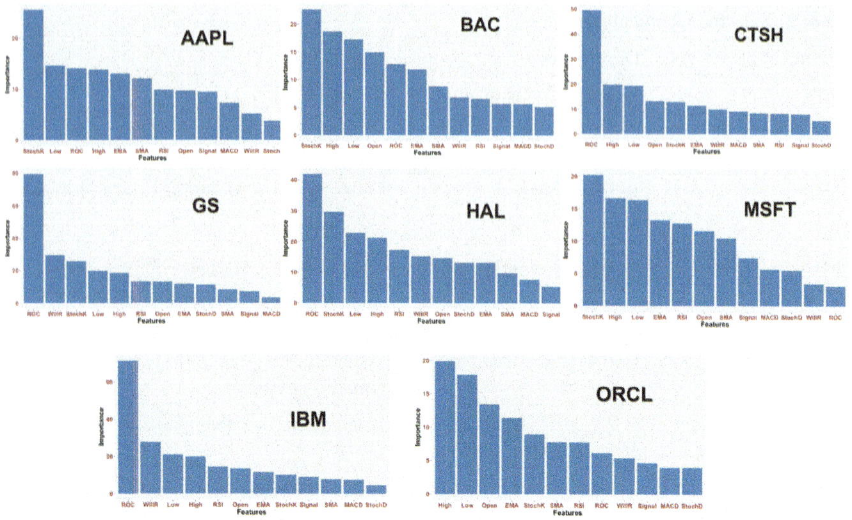

Fig. 1 Feature importance based on indicator values for companies: the y-axis represents features of importance, with the top variable being the most significant

4 Experimental Results

A. Dataset

Daily stock data from frequently traded companies were used to evaluate the hybrid ELM-MIF model. The analysis included eight firms—GS, MSFT, AAPL, IBM, CTSH, BAC, ORCL, and HAL—drawn from the NYSE and NASDAQ. We retrieved historical stock data from Yahoo Finance (https://finance.yahoo.com/), covering a period of more than ten years, from January 2009 to May 30, 2024. Each data sample comprises the date, trading volume, highest price, opening price, lowest price, and closing price. For the study, the dataset was refined to include the opening, closing, lowest, and highest prices. Additionally, the ELM-MIF model received inputs in the form of nine technical indicators that were obtained from historical data. The data was segmented, allocating 80% for training and 20% for testing. Details of the dataset are provided in Table 2.

B. Evaluation Metrics and Experimental Setup

To assess the suggested method's effectiveness, we use six performance metrics: MAE, MSE, RMSE, SMAPE, ARV, and R^2.

MSE quantifies the mean squared difference between actual values (y) and predicted values ($\hat{y}$), with smaller values signifying superior model performance:

$$MSE = \frac{1}{n}\sum_{i=1}^{n}(\hat{y}_i - y_i)^2 \tag{1}$$

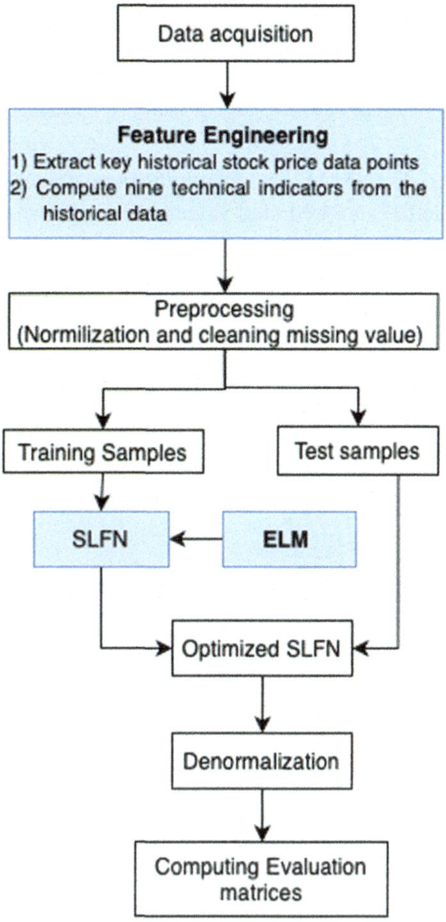

Fig. 2 The suggested model for stock prediction

Table 2 Description of dataset

Total number of samples	Period for training	Number of samples used for training (80%)	Period for testing	Number of samples used for testing (20%)
3877	2009–01-02/ 2021–04-29	3102	2021–04-30/ 2024–05-29	775

The MAE quantifies the average absolute deviation between predicted and actual values, serving as a direct indicator of accuracy:

$$MAE = \frac{1}{n}\sum_{i=1}^{n}|\hat{y}_i - y_i| \quad (2)$$

RMSE, derived as the square root of MSE, accentuates larger errors and provides performance measurement in the original data's units:

$$RMSE = \sqrt{MSE} \tag{3}$$

The SMAPE calculates the percentage error normalized by the sum of the absolute actual and predicted values, offering a relative measure of accuracy:

$$SMAPE = \frac{1}{n}\sum_{i=1}^{n}\frac{|\hat{y}_i - y_i|}{|\hat{y}_i| + |y_i|} \times 100\% \tag{4}$$

ARV assesses the variance of residuals (errors), reflecting the dispersion of prediction errors. It is given by:

$$AVR = \frac{1}{n-1}\sum_{i=1}^{n}(\hat{y}_i - y_i - \overline{r})^2 \tag{5}$$

where $\overline{r}$ is the mean of the residuals.

The R^2 calculates the proportion of the dependent variable's variance and a better model fit is indicated by higher R^2 value:

$$R^2 = 1 - \frac{\sum_{i=1}^{n}(\hat{y}_i - y_i)^2}{\sum_{i=1}^{n}(y_i - \overline{y})^2} \tag{6}$$

where $\overline{y}$ represents the actual values' mean.

For the experimental setup, various R packages were utilized, including "xgboost," "elmNNRcpp," "Keras," and "caret." The suggested model has an output layer and one hidden layer with twelve neurons. For comparison, we used an MLP-BP, which includes a single hidden layer with 15 neurons, and an ANN, which has two hidden layers with 15 and 5 neurons, respectively. The MLP-BP and ANN models both employ the sigmoid activation function.

Additionally, LSTM and XGBoost models were evaluated. For the XGBoost model, instead of using all 13 features, we created lagged features (lag1 and lag2) from the closing price. Lag features capture past values of the time series, providing temporal context for the model. The LSTM model was configured with a single time step, focusing on learning sequential dependencies from the data.

C. Performance Assessment and Comparative Analysis

Several criteria were used to assess and compare the suggested ELM-MIF model's performance with some benchmark models. Table 3 demonstrates that the proposed model consistently attains low MAE, MSE, and RMSE across stock data from GS, MSFT, AAPL, IBM, CTSH, BAC, ORCL, and HAL, reflecting its high accuracy and minimal prediction errors. The model also shows low SMAPE and ARV, with high R^2 values, indicating its effectiveness in capturing stock price variance. Figure 3

Table 3 Performance metrics for proposed ELM-MIF predictive model applied to stock data from GS, MSFT, AAPL, IBM, CTSH, BAC, ORCL, and HAL

Dataset	MSE	MAE	RMSE	SMAPE	ARV	R^2
AAPL	0.002988	0.042020	0.054660	0.036190	0.000016	1
BAC	0.000085	0.006949	0.009251	0.022580	0.000003	1
CTSH	0.000244	0.011660	0.015640	0.018060	0.000003	1
GS	0.005888	0.058570	0.076730	0.020740	0.000007	1
HAL	0.000165	0.009834	0.012880	0.031090	0.000003	1
MSFT	0.002525	0.038580	0.050250	0.021100	0.000002	1
IBM	0.000814	0.021040	0.028540	0.014900	0.000002	1
ORCL	0.000290	0.012160	0.017040	0.020400	0.000003	1

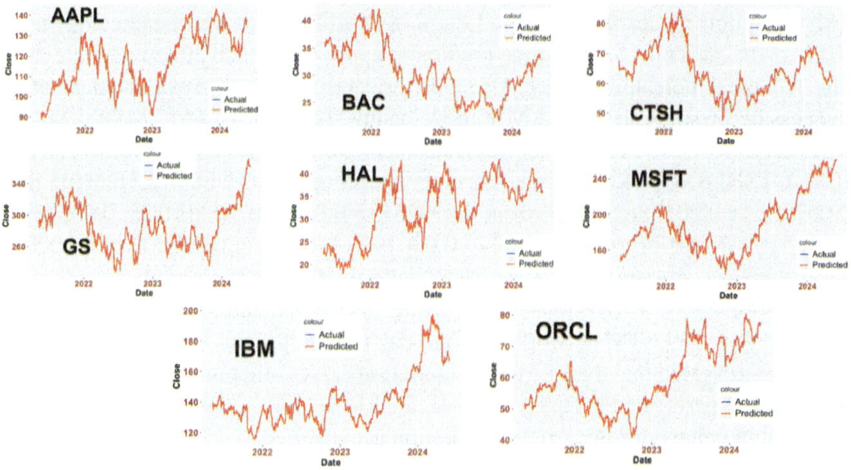

Fig. 3 Comparison of actual and predicted closing prices using ELM-MIF on Test data

illustrates the actual versus predicted closing prices using the model, providing a visual representation of the model's ability to forecast stock prices.

Table 4 presents a comparison of RMSE values across different models, including MLP-BP, ANN, XGBoost, LSTM, and the proposed ELM-MIF model. The ELM-MIF model consistently achieves the lowest RMSE values across all datasets, outperforming all models. Specifically, the ELM-MIF model demonstrates significantly lower RMSE values compared to MLP-BP, which exhibits higher errors, and ANN, which shows the highest RMSE values, indicating poorer performance. Although XGBoost is effective, it still has higher RMSE values than ELM-MIF. The LSTM model, while competitive, is outperformed by the ELM-MIF model across all datasets. This comprehensive evaluation underscores the ELM-MIF model's effectiveness in minimizing prediction errors and accurately capturing stock price movements, establishing it as the most accurate and reliable model in this study.

Table 4 Performance comparison of the suggested approach (ELM-MIF) with other machine learning methods using RMSE metric

Methods	Datasets							
	AAPL	BAC	CTSH	GS	HAL	MSFT	IBM	ORCL
MLP-BP	4.1998	0.6122	0.4092	3.8292	0.3034	4.3418	0.5821	1.0758
ANN	5.3850	2.1837	1.4558	5.7210	0.7275	6.7130	1.5485	2.2741
XGBoost	5.3241	1.1590	2.3392	9.2004	2.4301	6.0336	3.4078	1.8806
LSTM	1.4952	0.3403	0.5916	3.1600	0.3184	3.8153	1.3047	1.4608
ELM-MIF	**0.0546**	**0.0092**	**0.0156**	**0.0767**	**0.0128**	**0.0502**	**0.0285**	**0.0170**

5 Conclusion

This study introduces the Extreme Learning Machine with Multi-Indicator Fusion (ELM-MIF) as a novel stock price prediction approach, which effectively integrates nine technical indicators extracted from historical stock data. We evaluated the effectiveness of the suggested ELM-MIF model using six metrics: MAE, MSE, SMAPE, RMSE, ARV, and R^2. The model was tested across eight datasets: GS, MSFT, AAPL, IBM, CTSH, BAC, ORCL, and HAL, demonstrating results with MSE ranging from 0.002988 to 0.005888, RMSE from 0.009251 to 0.076730, MAE from 0.006949 to 0.058570, SMAPE from 0.018060 to 0.036190, ARV from 0.000003 to 0.000016, and R2 values consistently at 1 across all datasets. Comparative analysis with other methods—namely a traditional MLP using backpropagation, an ANN with two hidden layers, an LSTM network with one-day time steps, and XGBoost with lagged features—revealed that the ELM-MIF model surpasses other methods in terms of all metrics, and performed robustly. Overall, the ELM-MIF method showed promising potential in enhancing stock price prediction accuracy.

References

1. Lu M, Xu X (2024) TRNN: an efficient time-series recurrent neural network for stock price prediction. Inf Sci (N Y) 657:119951. https://doi.org/10.1016/J.INS.2023.119951
2. Ma D, Yuan D, Huang M, Dong L (2024) VGC-GAN: a multi-graph convolution adversarial network for stock price prediction. Expert Syst Appl 236:121204. https://doi.org/10.1016/J.ESWA.2023.121204
3. Akhtar MM, Zamani AS, Khan S, Shatat ASA, Dilshad S, Samdani F (2022) Stock market prediction based on statistical data using machine learning algorithms. J King Saud Univ Sci 34(4):101940. https://doi.org/10.1016/J.JKSUS.2022.101940
4. Kumar D, Sarangi PK, Verma R (2022) A systematic review of stock market prediction using machine learning and statistical techniques. Mater Today Proc 49:3187–3191. https://doi.org/10.1016/J.MATPR.2020.11.399
5. Billah MM, Sultana A, Bhuiyan F, Kaosar MG (2024) Stock price prediction: comparison of different moving average techniques using deep learning model. Neural Comput Appl 36(11):5861–5871. https://doi.org/10.1007/S00521-023-09369-0/FIGURES/13

6. Khashei M, Hajirahimi Z (2019) A comparative study of series arima/mlp hybrid models for stock price forecasting. Commun Stat Simul Comput 48(9):2625–2640. https://doi.org/10.1080/03610918.2018.1458138
7. Kobiela D, Krefta D, Król W, Weichbroth P (2022) ARIMA vs LSTM on NASDAQ stock exchange data. Procedia Comput Sci 207:3836–3845. https://doi.org/10.1016/J.PROCS.2022.09.445
8. Rao KV, Ramana Reddy BV (2023) HM-SMF: an efficient strategy optimization using a hybrid machine learning model for stock market prediction 24(2). https://doi.org/10.1142/S021946782450013X
9. Rao KV, Reddy BVR (2024) OFML-SMF: optimal feature selection with hybrid machine learning classifier for stock market price forecasting using social media and secondary data sources. Multimed Tools Appl 83(2):4703–4729. https://doi.org/10.1007/S11042-023-15461-3/FIGURES/7
10. Lu W, Li J, Wang J, Qin L (2021) A CNN-BiLSTM-AM method for stock price prediction. Neural Comput Appl 33(10):4741–4753. https://doi.org/10.1007/S00521-020-05532-Z/TABLES/3
11. Yu P, Yan X (2020) Stock price prediction based on deep neural networks. Neural Comput Appl 32(6):1609–1628. https://doi.org/10.1007/S00521-019-04212-X/TABLES/12
12. Ji X, Wang J, Yan Z (2021) A stock price prediction method based on deep learning technology. Int J Crowd Sci 5(1):55–72. https://doi.org/10.1108/IJCS-05-2020-0012
13. Rezaei H, Faaljou H, Mansourfar G (2021) Stock price prediction using deep learning and frequency decomposition. Expert Syst Appl 169:114332. https://doi.org/10.1016/J.ESWA.2020.114332
14. Wu JMT, Li Z, Herencsar N, Vo B, Lin JCW (2023) A graph-based CNN-LSTM stock price prediction algorithm with leading indicators. Multimed Syst 29(3):1751–1770. https://doi.org/10.1007/S00530-021-00758-W/TABLES/15
15. Kumar Chandar S (2021) Grey wolf optimization-Elman neural network model for stock price prediction. Soft comput 25(1):649–658. https://doi.org/10.1007/s00500-020-05174-2
16. Chung H, Shin KS (2018) Genetic algorithm-optimized long short-term memory network for stock market prediction. Sustainability (Switzerland) 10(10). https://doi.org/10.3390/su10103765
17. Zhang F (2023) RETRACTED: extreme learning machine for stock price prediction. Int J Electr Eng Educ 60:3972–3985. https://doi.org/10.1177/0020720920984675/ASSET/0020720920984675.FP.PNG_V03
18. Li X et al (2016) Empirical analysis: stock market prediction via extreme learning machine. Neural Comput Appl 27(1):67–78. https://doi.org/10.1007/S00521-014-1550-Z/TABLES/10
19. Novykov V, Bilson C, Gepp A, Harris G, Vanstone BJ (2023) Empirical validation of ELM trained neural networks for financial modelling. Neural Comput Appl 35(2):1581–1605. https://doi.org/10.1007/S00521-022-07792-3/TABLES/6
20. Long W, Lu Z, Cui L (2019) Deep learning-based feature engineering for stock price movement prediction. Knowl Based Syst 164:163–173. https://doi.org/10.1016/J.KNOSYS.2018.10.034
21. Nti IK, Adekoya AF, Weyori BA (2019) A systematic review of fundamental and technical analysis of stock market predictions. Artif Intell Rev 53(4):3007–3057. https://doi.org/10.1007/S10462-019-09754-Z
22. Ntakaris A, Kanniainen J, Gabbouj M, Iosifidis A (2020) Mid-price prediction based on machine learning methods with technical and quantitative indicators. PLoS ONE 15(6):e0234107. https://doi.org/10.1371/JOURNAL.PONE.0234107

How Artificial Intelligence Technology Affects Productivity

Elizaveta Semenova and Mikhail Komarov

Abstract The article investigates the impact of large language models (LLMs), such as ChatGPT, on productivity within the digital product development industry. The research highlights the transformative role of LLMs in enhancing task completion speed, job satisfaction, and reducing fatigue, particularly for junior employees and less experienced professionals. Through a survey-based approach, the study identifies that the practical application of LLMs is more beneficial in stages like coding, code review, and bug fixing, while their effectiveness diminishes in more creative or planning-intensive phases. Despite the positive correlation between LLM usage and productivity improvements, the study underscores the lack of significant empirical data across diverse organizational settence.

Keywords Large language models (LLM) · Artificial intelligence (AI) · Productivity · Digital product development · ChatGPT

1 Introduction

Artificial Intelligence (AI) is often perceived as a universal technology that is being rapidly and massively adopted and impacting various sectors of the economy. At the centre of this technological breakthrough are Large Language Models (LLMs), such as GPT-4 and its successors, which demonstrate exceptional capabilities in natural language processing and generation. The main feature of this technology is the creation of new and complementary modes of production that can increase productivity over time. Hence, it can be assumed that the adoption of artificial intelligence technologies, especially its machine learning component, will enable firms to create new business opportunities and increase productivity.

E. Semenova (✉) · M. Komarov
Graduate School of Business, HSE University, Moscow, Russia
e-mail: Elansemenova@edu.hse.ru

M. Komarov
e-mail: mkomarov@hse.ru

N. M. Durakbasa and K. G. Gülen (eds.), *Sustainable Green Conversion*, Lecture Notes in Mechanical Engineering, https://doi.org/10.1007/978-3-031-83583-4_9

However, at the moment, the questioning of whether artificial intelligence (AI) is the new engine of productivity growth is controversial among scholars. This is due to the lack of empirical data from companies.

The relevance of the study stems from the rapid development of AI technologies and the growing need for organisations to improve productivity and competitiveness. Big language models have already shown their effectiveness in various domains, but their impact on the processes of organisations is still poorly understood.

Since digital product development requires a high degree of creativity, accuracy and speed, this makes it an ideal area for AI applications. Large language models can automate routine tasks, assist in code generation, creating technical documentation and even in the interface design process. The implementation of such technologies promises not only to speed up development processes, but also to improve the quality of final products. Therefore, this study will be based on the data collected from this industry.

2 Overview of AI Technology

A. *Definition and classification of artificial intelligence*

Advances in science and new technologies mimicking human intelligence have made the concept of Artificial Intelligence (AI) ambiguous and problematic. However, understanding AI is essential to interpreting its impact on business processes. AI is interpreted differently across various fields, and debates over its definition continue. Many existing definitions are not universal and often simplify or focus on specific aspects, like machine learning or automated decision-making.

Despite the absence of a universally accepted definition, AI definitions share commonalities. S. Legg and M. Hutter [1] analyzed key AI definitions, highlighting similarities and differences. The European Commission defines AI as intelligent systems that analyze the environment and act autonomously, whether as software (e.g., voice assistants, image recognition) or hardware (e.g., robots, drones). The Association for the Advancement of AI [2] defines it as the scientific understanding of thinking mechanisms and their embodiment in machines. AI levels vary and don't necessarily match human intelligence, encompassing deep learning, reasoning, and human–machine interaction. Most experts divide AI into weak and strong AI. (Shown in Fig. 1).

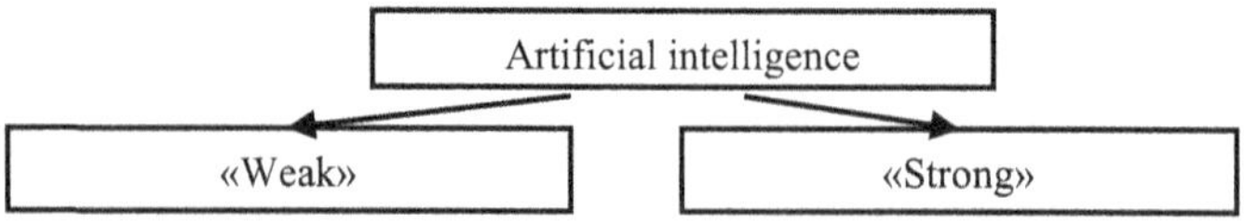

Fig. 1 Classification of AI

Weak AI (or narrow AI) refers to systems designed for specific intellectual tasks, often outperforming humans in these areas. A. Kaplan and M. Hanline [3] describe it as analytical intelligence tailored to particular functions. Modern AI systems and robots are examples of weak AI.

Strong AI (or general AI) is software with cognitive abilities akin to human intelligence, capable of independently solving a wide range of tasks. P. Wang, B. Goertzel, and S. Franklin [4] describe it as thinking machines resembling humans. Search Enterprise AI defines it as generalized cognitive abilities that can handle unfamiliar tasks without human intervention [5].

Confusion often arises from equating AI with methods like machine learning and neural networks. Machine learning involves data analysis techniques for decision-making, including supervised, unsupervised, and reinforcement learning. Artificial neural networks are mathematical models inspired by biological neurons.

In modern society, AI primarily exists as weak AI, designed for specific tasks. Its systemic nature ensures software and hardware interaction for data collection, interpretation, and decision-making.

B. *Classification of AI in terms of applied technologies*

Artificial intelligence technologies can be classified in terms of their functions. The classification is presented in Table 1.

C. *Large language models*

In artificial intelligence, creating technology that understands human language is a significant challenge, known as Natural Language Processing (NLP). NLP advancements are widely used in virtual assistants and spam filters. Recently, large language models (LLMs) like GPT-3 have made significant progress, handling various NLP tasks with impressive capability.

NLP enables computers to interpret, analyze, and understand human language, playing a crucial role in processing text and speech, including handling dialects, slang, and errors common in everyday language. Companies use NLP to automate tasks such as processing documents, analyzing customer feedback, running chatbots, and classifying text.

NLP enhances customer interactions by enabling chatbots to sort queries, answer common questions, and escalate complex issues, reducing costs and improving customer satisfaction. Industries like insurance, law, and healthcare use NLP to automate data redaction and protect privacy. NLP-powered chatbots and voice bots create more human-like interactions, optimizing customer experience while cutting operational costs.

Marketers utilize NLP to analyze customer sentiment from written feedback, gaining insights into customer emotions. NLP combines computational linguistics, machine learning, and deep learning to process human language. It employs various approaches:

- Supervised learning: Training software on labeled data to process new, unknown data, such as sorting documents by specific labels.

- Unsupervised learning: Using statistical models to detect patterns in unlabeled data, like text message autocomplete.
- Natural Language Understanding (NLU): Analyzing sentence meaning and interpreting polysemous words.
- Natural Language Generation (NLG): Generating human-like text from keywords, allowing chatbots to interact like customer service representatives.

In the pre-training phase, a language model (LLM) is trained on vast text datasets to learn language structure and rules. Recent LLMs, like GPT-3, have been pre-trained on large datasets from the internet, including web pages, messages, and books from over 50 million domains.

Table 1 Artificial intelligence technologies

Technology	Description [6, 7]	Example [8]
Rule-based systems	The systems operate on the basis of predefined logical rules developed by experts. Systems operate on the basis of if–then logic (IF–THEN). They require careful manual programming and cannot adapt to new data without human intervention	DENDRAL: An expert system used for chemical analysis to predict molecular structure
Machine learning	Machine learning involves creating algorithms that can learn and make predictions from data. Machine learning algorithms use linear regression methods, decision trees, SVM (support vector method), clustering and neural networks	Many of the advanced recommendation systems, such as Google's YouTube
Deep learning	Deep learning is a subset of machine learning. Deep learning uses multi-layer neural networks such as super-accurate neural networks (CNNs) for image processing and recurrent neural networks (RNNs) for sequential data processing	Autonomous vehicles: control and navigation of autonomous vehicles
Robotics	Robotics combines mechanical and electronic systems with AI to perform tasks that require interaction with the physical world. Robotics uses sensors, and AI algorithms to perceive, plan and execute actions	Robot couriers, cleaners
Computer vision	Computer vision includes methods and technologies for processing and analysing visual information. This technology is based on object recognition, image segmentation and classification, and CNN model for analysing visual data	Facial recognition system
Natural language processing, NLP	NLP includes algorithms and methods for processing and analysing human language. Machine learning and deep learning techniques are used to understand and generate text	Chatbots and virtual assistants
Neural networks	Neural networks consist of many interconnected nodes (neurons) organised into layers. Each neuron accepts input data, handles them and transmits them to the next layer. Neuron network trained weight on the basis of feedback (error)	Image and video processing, face and object recognition

These datasets are processed by a Transformer, a deep neural network effective with sequential data. Transformers use an encoder-decoder architecture: the encoder extracts meaning from input text, converting it into a vector, and the decoder generates text from that vector. A key feature is the self-feedback mechanism, which helps the model focus on important words in a sentence and allows for parallel processing, making training faster.

After pre-training, the model can be fine-tuned on specific tasks, improving performance. Methods like Quick Tuning and One-shot tuning involve providing the model with examples to guide its output. However, LLMs have limitations:

- Lack of contextual understanding: LLMs rely on statistical patterns rather than reasoning and can produce nonsensical answers.
- Hallucinations: LLMs sometimes generate inaccurate information, like fabricated quotes.
- Biases and limited knowledge: LLMs may internalize biases from training data and have limited knowledge of post-2023 events.

D. *Overview of large language models*

Large language models require significant resources, limiting their development to major tech companies like Google and OpenAI. For example, training GPT-3 cost nearly $5 million. Once developed, these models may be released under closed or open-source licenses, allowing others to customize them for specific applications.

This does not require significant cost and computational resources, so startups often work with an optensor LLM. Such modified models are called forks.

Examples of popular basic models that have become the basis for open source LLMs include:

- LLaMA and Meta's LLaMA 2, the latter of which is co-developed with Microsoft.
- BLOOM (BigScience Large Open-Science Open-Access Multilingual Language Model), created by the BigScience project with Hugging Face.
- GPT-2, published by OpenAI a few years ago when it planned to develop exclusively open-source solutions.
- Falcon is the latest development from the Technology Innovation Institute (TII) of Abu Dhabi, UAE.
- Google's T5 family of models.

The most used LLM models are shown in Fig. 2.

3 Approaches to Assessing the Impact of Artificial Intelligence on Performance

The literature presents different perspectives on AI's role in organizational performance. In 'The Economics of Artificial Intelligence' [14], AI is viewed as a production element that enhances productivity by generating new ideas, technologies, and

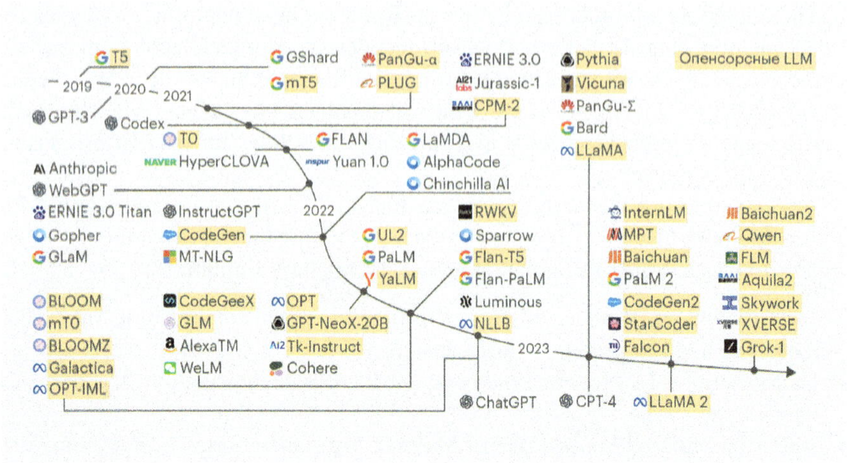

Fig. 2 LLM development timeline with a size of more than 10 billion parameters [13]

solving complex problems. E. Brynjolfsson [15] sees AI as intangible capital, arguing that investing in AI can increase productivity, though the effects may not be immediate and may require additional process adaptation and investment.

Some economists [17, 18] are more cautious, citing challenges in developing new ideas, and barriers to accessing the knowledge and data needed for AI to be effective. Damioli [19] studied the impact of AI on productivity through AI-related patents filed by 5,257 companies from 2000 to 2016, finding that AI patents positively influence productivity, especially in smaller firms.

The study 'Artificial Intelligence and Firm-Level Productivity' [20] used data from European Commission AI adoption surveys and linked AI adoption to improved productivity. Another study, 'Quantifying the Impact of AI on Productivity and Labour Demand' [21], analyzed U.S. Census microdata and AI patents, finding that AI innovations increase output per worker and wage inequality, though AI adoption is limited to a few leading companies.

Overall, the literature highlights diverse methodologies and results but also notes a lack of empirical evidence supporting any definitive conclusions.

A. *A review of research on the impact of large language models on organisational performance*

As of May 2024 [22], a literature review identified only one empirical study on the impact of large language models (LLMs) on performance. This online experiment involved 444 college-educated professionals from various fields, including marketing, consulting, data analysis, and management. Participants completed two profession-specific tasks such as press releases, reports, and emails, taking 20–30 minutes each.

Participants were split into two groups:

- The experimental group (50%) could use ChatGPT between tasks.
- The control group used the LaTeX Overleaf editor.

Researchers measured performance, time spent, job satisfaction, self-efficacy, and attitudes towards automation. They monitored task completion times and used independent experts to evaluate task quality.

Key findings:

- 92% of the experimental group registered on the OpenAI platform, and 81% used ChatGPT for the second task.
- The experimental group's task completion time decreased by 37%, and expert evaluations improved by 0.45 standard deviations.
- In the control group, first and second task scores correlated at 0.49, while the experimental group's correlation dropped to 0.25, indicating improved performance with ChatGPT.
- ChatGPT slightly increased self-esteem by 2% points and job satisfaction by 4% points.
- After the experiment, 33% of the experimental group continued using ChatGPT at work.

The study concluded that ChatGPT improved performance, job satisfaction, and self-efficacy in professional tasks. The digital product development process and the impact of AI on it.

Before analyzing the impact of AI technologies, particularly large language models (LLMs), on digital product development, it's essential to understand the digital product development process and its stages. The Institute of Electrical and Electronics Engineers defines this process as a systematic, disciplined, and quantifiable approach to software development, operation, and maintenance.

Klopper R. et al. [23] outline the key stages of the software development process:

- Planning: Defines project scope, objectives, and resources. LLMs can assist in task planning, assigning employees, and forecasting project timelines and budgets more accurately.
- Requirements Analysis: Involves understanding customer needs. LLMs, such as ChatGPT, can automate requirement elicitation, improving stakeholder communication [24].
- Product Design and Engineering: Creates a software plan based on requirements. LLMs can generate and optimize designs by analyzing system models, ensuring quality.
- Coding and Code Review: Involves writing and reviewing code. LLMs enhance this stage through code completion, bug detection, and fragment prediction [25].
- Testing and Integration: Ensures software functionality. LLMs can automate testing and write testing strategies [26].
- System Support and Bug Fixing: LLMs provide technical support, troubleshooting, and best practice recommendations.

- Documentation Generation: Produces final documents describing the development process.

In the further part of the work highlights the limited empirical evidence on AI's impact on performance. Performance metrics like lines of code or team productivity vary across industries. The literature review shows that AI affects productivity at different stages of digital product development, but due to the novelty of AI, academic opinions on its impact vary, shaping this study's research design.Analysis of the impact of artificial intelligence technologies on the productivity of organisations

A. *Research Hypotheses*

The rapid growth of Artificial Intelligence technologies, especially AI-driven language models such as ChatGPT, has generated considerable interest in the possibilities of using them in the digital product development process. However, the impact of Big Language Models on performance during development remains poorly understood. Determining the extent to which this technology affects performance can provide valuable information for organisations looking to optimise their processes and drive innovation in a competitive environment.

As mentioned earlier, to identify the factors of influence of Artificial Intelligence technology on the performance of organisations, specifically the performance of organisations from the digital product development industry Hypotheses to be tested as well as the expected results are demonstrated in Table 2.

B. *Research design*

The research design includes a survey of employees involved in the digital product development process. The survey consists of different types of questions designed to test each hypothesis Table 3.

To obtain representative data, the survey consisted of different types of questions. A description of the questions, a link to the questionnaire and the survey results can be provided on request.

C. *Sample formation and data collection*

The questionnaire was filled in by respondents who used LLM in their work at least once. The questionnaire was posted on the service 'Yandex. Forms' service and was available by clicking on the link [27]. The link to the questionnaire was sent out via professional chats of employees in the digital product development industry. The respondents who took the questionnaire were representatives of different industries, with different levels of experience and involved in different stages of development. This ensures a diversity of views among the participants and allows for representative, comprehensive data.

The data collection period started on 01.05.2024 and ended on 10.06.2024. Over 3000 messages were sent to potential participants. As a result, 156 responses were received from professionals willing to share their experience and opinions on the topic.

Table 2 Research hypothesis and expected results

No	Hypothesis	Expected result	Source
1	Less experienced employees benefit more from the use of LLMs (Large Language Models) than those with higher levels of expertise	Higher estimates of LLM impact among respondents with less experience	Junior developers may use LLMs to get guidance, best practices, or to understand complex algorithms, allowing them to adapt faster. (Kumari et al. 2018)
2	Employees with lower qualifications benefit more from LLMs	Higher estimates of LLM impact among respondents with lower qualifications (Junior) compared to Middle and Senior	Junior developers may use LLM to get guidance, best practices or to understand complex algorithms, allowing them to adapt faster. (Kumari et al. 2018)
3	There will be no differences in the impact on productivity for employees of large organisations and small firms / freelancers	Identical assessment of influence LLM among respondents from different categories of categories of companies	Interviews with developers
4	The impact of LLM will vary for different stages of software development	Different estimates of the impact of LLM for different stages	Section 'Understanding the Software Development Process and the Transformational Impact of ChatGPT' and (Pressman R. S. (2010))
5	Professionals using the paid version of the LLM will record a more positive impact on their productivity	Higher overall productivity impact score reported by participants with the paid version of the LLM	Interviews with developers
6	Younger participants will record a greater impact of the LLM as they tend to be more active in adopting new technologies in their lives	Higher overall productivity impact score indicated by younger participants	
7	The impact of LLM on development productivity is higher in teams where the use of this technology is actively supported by the manager (however, it is assumed that active support is rare)	Higher estimates of the impact of LLM among respondents whose managers are aware of and support the use of LLM	

Table 3 Research design

No	Type of question	Hypothesis tested
1	Demographic questions (age, gender, length of service, job title, position level, type of employment, company size and main industry of the company)	H1, H3, H6
2	Closed-ended questions (use of LLM, frequency of use)	H6
3	Likert scale questions (impact on productivity, speed of task completion, quality of work, error reduction, problem solving abilities)	H1, H2, H5, H6, H7
4	Stage-specific Likert scale questions (impact of LLM on each stage of software development)	H4, H5
5	Yes/No question (manager's awareness of using LLM)	H7

D. *Data analysis*

The collected survey data were carefully analysed using statistical techniques such as: correlation analysis, analysis of variance (ANOVA) and description of mean values.

Correlation analysis was used to examine the relationships between various continuous variables such as duration of LLM use, its effect on productivity and other relevant factors. By calculating correlation coefficients, the strength and direction of these relationships were assessed, providing valuable insights into how different variables are related to each other.

Analysis of variance (ANOVA) was used to investigate the impact of LLM for different groups: different experience, occupation and industry of the company. By analysing the differences between groups, we were able to identify whether there were significant differences in productivity outcomes based on these categorical factors.

In addition, mean, mode, and median values were used to understand the differences in influences on the different stages of software development.

E. *Justification of the method of analysis used*

A quantitative method was chosen for this study for a number of reasons. Firstly, quantitative methods of analysis allow the collection of fairly large samples compared to qualitative methods.

Secondly, the results are considered less skewed as it is relatively easy to compare them with other similar studies if they are published in the future.

The survey is the main method used in this study to obtain relevant data and information. In creating this format, attention was paid to simplicity and clarity. Likert scale was chosen as the main method.

The study considers different industries within which digital products are developed, as well as different companies and tasks, therefore the introduction of metrics such as 'Number of lines per hour' etc. was rejected due to the fact that the amount of work is not homogeneous in different environments.

The Likert scale, on the other hand, allows for relative assessments. The value of this scale lies in a range of ranks from 1 (strongly disagree) to 5 (strongly agree), i.e. 5 degrees, and this approach is widely used in management research papers.

F. *Validity and reliability of the research conducted*

Several measures were taken to ensure the validity and reliability of the study:

- Designing a survey with clear, concise and relevant questions to accurately elicit the views of the respondents;
- Selecting respondents who have experience of using LLMs;
- Ensuring respondents are anonymised to encourage honest and unbiased responses;
- Using rigorous statistical methods such as ANOVA and correlation analysis to identify significant patterns and relationships in the data.

G. *Limitations*

Despite the careful development of the research methodology and design, there are certain limitations that may affect the generalisability and applicability of the results.

Sample size and representativeness

The sample size of this study may not be large enough to provide a fully representative picture, which could potentially limit the generalisability of the results to the wider population of professionals in the digital product development industry.

Self-completion of the survey

The survey is based on respondents completing the survey on their own without further explanation, which may cause some bias and inaccuracies. Participants may be influenced by socially desirable behaviour, which may lead them to provide answers that they believe are expected or acceptable, rather than expressing their true opinions. In addition, respondents may interpret survey questions differently or may not accurately remember specific details related to their experiences.

Time constraints

This survey captures respondents' experiences with MML at a specific point in time. As a result, the findings may not reflect changes in the impact of MML on performance over time or the dynamic nature of digital product development processes and technologies. Further research will be required to track the dynamics of MML's impact on productivity and to understand the long-term implications of its use.

Subjectivity in data analysis

The subjective opinions and assumptions of researchers can affect their interpretation of survey data and the identification of patterns and relationships. Although statistical tools such as t-tests and correlation analysis can reduce subjectivity in data analysis, there is a possibility that the results may be skewed by the subjective opinions and assumptions of the author of the paper.

H. *Summary of key findings*

Descriptive statistics

Age of Respondents:

The most common age group is 22–25 years, representing 20% of the total sample (32 respondents). The 45 and above age group is the smallest, with 4 respondents (3.6%).

Time in the Digital Product Development Industry:

The largest group of respondents has 1–3 years of experience (27.4%). Those with over 20 years of experience account for 11.3%.

Job Roles:

The top three roles are Product Manager (26.4%), Developer (17%), and Business Analyst (17%), making up over 60% of the sample.

Occupational Level:

Most respondents are at the Senior level (45.3%), with Trainees being the smallest group (4.7%).

Organization Size:

Half of the respondents work in large organizations (50%), and 25.5% work in medium-sized companies.

Industry:

IT companies dominate at 34%, followed by finance and economics (13.2%) and metallurgy (12.3%). Freelancers are minimally represented (1.9%).

4 Summary

The most frequent respondent group is aged 22–25 (20%) and at the Senior level (45.3%). Most respondents work in large (50%) or medium-sized (25.5%) organizations, with IT being the leading industry (34%).

Descriptive statistics of the dependent variables

In order to assess the impact of language patterns, several aspects of its impact on productivity and specific stages of software development were assessed, which were later used as dependent variables in the analysis. This section will study the scaled variables to which it is appropriate to apply mean scores and assess the significance of the results obtained.

Thus, this study has a number of limitations that should be considered when interpreting the results Table 4.

Table 4 Parameters of dependent variables affecting performance: description and statistics

Description	Average	No	Std. deviation
Assessing the overall impact of using language models on your performance at work	3,61	106,00	0,76
LLM help me complete my work tasks faster	3,72	106,00	0,81
Using LLM helps me to be less tired when performing my work tasks	3,41	106,00	0,79
Using LLM allows me to complete more tasks during the work day	3,52	106,00	0,89
Using LLM gets me out of mundane tasks and helps me concentrate on more complex work	3,49	106,00	0,97
I am less tired during the working day when using LLM	3,40	106,00	0,86
Using LLM models reduces the number of errors in my work	3,24	106,00	0,96
Using LLM improves the quality of my work	3,27	106,00	0,89

Evaluation of the overall impact of using language models on your productivity at work: 3.61 is a very good score that shows the overall high productivity of implementing language models in the workflow.

Help you to complete work tasks quickly: a score of 3.72 shows that the models help you to complete work tasks quite well.

Allow less fatigue in the realisation of work tasks: a mean score of 3.41 shows a positive effect in the distribution of effort during working hours.

Increase the number of tasks during the working day: 3.52 out of 5 is a positive score, therefore, it can be concluded that the use of GPT allows to realise more tasks.

Getting rid of routine tasks and helping to concentrate on more complex work: 3.49 in line with the previous answers shows the usefulness of using models to differentiate tasks by level of complexity.

Reduction of fatigue during the working day: 3.40 is above average, indicating a positive effect of using models on reducing fatigue during the working day.

Reduction of work errors: 3.24 is not the highest score presented, indicating a moderate opinion on the number of errors.

Improving the quality of work: 3.27 is identical to the previous indicator, which indicates a moderate impact on improving the quality of work.

At this stage we do not consider the question of how involved the respondents are in the process of realisation of each stage of development, and what they give their assessment of time saving at each stage on a five-point scale. This is analysed in detail in Sect. 3.5. Here we will limit ourselves to a summary table showing the general picture (Table 5).

Table 5 Software development phases and the influence of language models on them: the big picture

	Planning	Requirements analyzing	Designing and developing architecture	Coding	Code-review	Deployment	Support and bug-fixing	Documentation	Meetings
Valid	59	64	47	55	49	46	51	65	70
Missing	47	42	59	51	57	60	55	41	36
Mean	2,07	2,03	2,23	2,91	2,90	2,52	2,59	2,18	2,57
Std. deviation	0,94	1,07	0,91	1,08	1,14	1,31	1,15	1,27	0,99

Thus, the use of language models in work showed high effectiveness, with a score of 3.61 for the overall effect on productivity. Models significantly help in completing tasks quickly (3.72) and have a positive effect on effort allocation and fatigue reduction (3.41). They also facilitate the completion of more tasks (3.52) and help to focus on complex work (3.49). The effects on reducing errors (3.24) and improving the quality of work (3.27) were moderate.

Correlations to test the relationship between the effect of language patterns on job performance and respondent groups.

After preliminary analyses, correlations related to the relationship about the effect of language on performance were identified. All possible variations of respondent groups were examined. Kendall's coefficient was used for the correlation analysis, as the rating scale represents ordinal values. Correlation analysis did not reveal any clear strong correlations between the evaluation of the impact of language patterns on performance at work according to developer level, work experience and company size. A moderate correlation was only observed between 'evaluation of the overall impact of using language models on job performance' and 'help in quickly completing work tasks' and the frequency of using language models.

It can be concluded that the more often an employee works with MML models, the more highly he/she evaluates their overall impact on improving productivity and, in particular, on helping to complete work tasks quickly.

ANOVA test to find the difference between groups.

In order to determine the difference in the evaluation of aspects of perceived usefulness of language models by different groups of respondents, we will use a one-factor analysis of variance. For this purpose, we will apply the ANOVA test because it allows us to compare the mean values of 3 or more groups to see if there is a statistically significant difference between them.

The Anova tests performed did not reveal any statistical significance between the factors analysed.

Evaluation of the use of MML in individual stages of development.

After understanding the differences in groups' general perceptions of the impact of language models on different aspects of work, it was time to analyze the responses relating to the different stages of digital product development. For this purpose, means, medians and modes were analyzed as this best suit the aims of the study (Table 6).

Thus, at the stage of team meetings language models are used most often and help to save time.

Based on the analysis, we can conclude that language models are of practical use at the stages of coding, code-review, support and bug-fixing and team meetings. In these stages at mean score, mode and median are close to each other and equal to 3.

In the planning, requirements analysis and documentation stages, the use of language models is not appropriate. This is most likely due to the need for a personalised approach and the lack of templates for applying artificial intelligence.

A verification test to determine the level of creativity of each stage of development.

According to one of the hypotheses, the use of artificial intelligence in the format of language models is applicable to less creative stages of software development. The

Table 6 Software development phases and the influence of language models on them

	Planning	Requirements analysis	Dezigning and developing architecture	Coding	Code-review	Deployment	Support and bug-fixing	Documentation	Meetings
Valid	59	64	47	55	49	46	51	65	70
Mean	2,07	2,03	2,23	2,91	2,90	2,52	2,59	2,18	2,57
Median	2,00	2,00	2,00	3,00	3,00	2,00	3,00	2,00	3,00
Moda	1,00	1,00	2,00	3,00	3,00	1,00	2,00	1,00	3,00
Std. deviation	0,94	1,07	0,91	1,08	1,14	1,31	1,15	1,27	0,99
Minimum	1,00	1,00	1,00	1,00	1,00	1,00	1,00	1,00	1,00
Maximum	4,00	5,00	5,00	5,00	5,00	5,00	5,00	5,00	5,00

respondents were asked to choose the most creative stages. Based on the frequency analysis of multiple answers, the following table was obtained.

More creative stages: Software architecture design and development, Software deployment, Coding, Requirements gathering.

Less creative stages: Planning, Code-review, Documentation, Support and Bug-fixing.

When cross-analysing pts. 3.5. and 3.6. we can state that the hypothesis is partially confirmed. To three of the four less creative stages it is reasonable to apply language models to save time. The exception was the documentation stage.

Among the creative stages, the exception is the Coding stage, where the majority of respondents noted that the use of large language models saves time. Perhaps, this situation is due to the fact that all respondents evaluated the creativity of the stages, and only the respondents involved in each of the stages evaluated the time saving from language models.

I. *The results*

The sample analysed is predominantly made up of participants aged between 19 and 44. The top 3 job roles are Product Manager (26.4%), Developer (17%) and Business Analyst (17%). This adds up to over 60% of the total sample. The vast majority of respondents are at the Senior level (45.3%). The majority of respondents work in large (50%) and medium-sized (25.5%) organisations, with IT being the main industry (34%).

In terms of using large language models at work, over 80% of respondents used this technology, with over 62% doing so at least 3–4 days per working week, indicating that the digital product development industry is highly interested in this technology.

That said, the more often an employee uses MML models, the higher they rate its impact on improving productivity and helping them complete tasks. Correlation analysis showed no strong correlations between the evaluation of the impact of language models on productivity and the level of the developer, work experience or company size. A moderate correlation was only found between 'evaluation of the overall impact of language models on productivity' and 'help in performing tasks quickly' and the frequency of using these models. This may be due to the fact that employees need time to learn how to ask the right questions of LLMsitems to get satisfying answers.

Employees in small enterprises are much higher in characterising the overall productivity of using language models.

Trainees report maximum usefulness (4.6) of language models. The Seniors give a relatively low rating of 3.54 for help in performing work tasks.

Based on the analysis, it can be concluded that language patterns have practical usefulness at the stages of coding, code-review, support and bug fixing and team meetings.

At the planning, requirements analysis and documentation stages, the use of language models is not practical. This is most likely due to the need for a customised approach and the lack of templates for applying artificial intelligence.

Thus, analysing the results supports the following findings. Less experienced employees and lower-skilled professionals benefit more from the use of large language models (LLM), which is supported by hypotheses H1 and H2. The influence of LLMs varies across different stages of software development, which is consistent with hypothesis H4. Younger employees show greater impact from the use of LLM as they are more active in adopting new technologies, which is supported by hypothesis H6. However, hypothesis H3, which suggested that there were no differences in productivity impact between employees in large and small organisations, was not confirmed. Differences in productivity between users of the paid and free version of the LLM, as well as the impact of supervisor support on LLM use were found but were not statistically significant, as indicated by hypotheses H5 and H7 respectively.

5 Conclusion

Based on an understanding of the phenomenon of Artificial Intelligence and in particular Big Language Model technology, a literature review of approaches to assessing the impact of AI technologies on organisational performance was conducted. However, it was found that there is a lack of empirical evidence for research on the impact of AI on productivity. And commonly accepted metrics for assessing performance are not universal as they vary greatly from project to project and industry to industry. Literature analysis showed that AI technologies affect performance at different stages of digital product development, but the factors that differentiate the impact are different, as the opinions of the sheets on this issue vary due to the novelty of the technology. These aspects formed the basis for the research design. As a result of the analysis of the theoretical framework of the impact of artificial intelligence technologies on the productivity of organizations in the digital product development industry, a research design was developed. The study contained 7 hypotheses, 4 of which were confirmed, 1 was disproved, and 2 had a correlation, but it was not statistically significant.

The findings of this study indicate that large language models (LLMs), such as ChatGPT, do positively impact productivity in the digital product development process. Specifically, LLMs enhance task completion speed, reduce fatigue, and improve job satisfaction, particularly among junior employees and those with less experience. The research shows that LLMs are most beneficial in the coding, code review, and bug-fixing stages of development, where routine tasks can be automated, allowing employees to focus on more complex work. However, their effectiveness is less pronounced in more creative stages such as planning, requirements analysis, and documentation, where a personalized approach is essential.

Despite these promising results, several limitations must be acknowledged. The impact of LLMs on productivity is influenced by a wide range of contributing factors, including employee experience, company size, and the stage of the development process. Additionally, the time frame for evaluating the effectiveness of LLMs is still too short to draw definitive conclusions. As AI technologies continue to evolve, further longitudinal studies are needed to fully understand their long-term implications for productivity across different industries and organizational contexts.

In conclusion, while LLMs hold significant potential to enhance productivity in specific areas of digital product development, their overall effectiveness remains contingent upon various contextual factors. The rapid pace of AI advancements necessitates ongoing research to capture the dynamic nature of these technologies and their impact on organizational performance.

This study will be extended to identify the initial factors in initiating the use of LLMs in work and to identify the effects on the economic performance of organizations.

References

1. Legg S, Hutter M (2007) A collection of definitions of intelligence. In: Goertzel B, Wang P (eds) Advances in artificial general intelligence: concept, architectures and algorithms. Amsterdam, IOS Press, vol 157, pp 17–24
2. AI overview: broad discussions of artificial intelligence. AI topics. http://aitopics.org/topic/aio verview
3. Kaplan A, Haenlein M (2019) Siri, Siri in my hand, who's the fairest in the land? On the interpretations, illustrations and implications of artificial intelligence. Kaplan A, Heinlein M Bus HorizS 62(1)
4. Wang P, Goertzel B, Franklin S (2008) Artificial general intelligence. In: 2008: Proceedings of the first AGI conference. Wang P, Goertzel B, Franklin S, Washington DC. IOS Press 507
5. Léon E, Djou C (2018) Métamorphose des managers: à l'ère du numérique et de l'intelligence artificielle. Pearson
6. Brynjolfsson E, McAfee A (2014) The second machine age: work, progress, and prosperity in a time of brilliant technologies. WW Norton & Company
7. Jarrahi MH (2018) Artificial intelligence and the future of work: human-AI symbiosis in organizational decision making. Bus HorizS 61(4):577–586
8. Léon E, Dejoux C (2018) Métamorphose des managers: à l'ère du numérique et de l'intelligence artificielle. Pearson
9. Principles of NLP natural language learning. https://arxiv.org/abs/2306.04980. Accessed 12 Apr 2024
10. Liu J, Liu A, Lu X, Welleck S, West P, Bras RL, Choi Y, Hajishirzi H (2021) Generated knowledge prompting for commonsense reasoning. https://arxiv.org/pdf/2110.08387.pdf
11. Alammar J (2020) How GPT-3 works: visualizations and animationsf
12. OpenAI's GPT-3 language model: a technical overview. OpenAI's GPT-3 language model: a technical overview (lambdalabs.com). Accessed 10 May 2024
13. Zhao WX, Zhou K et al. (2023) A survey of large language model. Carnell University. Accessed 15 Mar 2024
14. Aghion P, Jones BF, Jones CI (2019) Artificial intelligence and economic growth. Kap. 9 i A. Agrawal, J. Gans og A. Goldfarb (red) The economics of artificial intelligence: an agenda
15. Brynjolfsson E, Rock D, Syverson C (2019) Artificial intelligence and the modern productivity paradox. The economics of artificial intelligence: an agenda
16. Corrado C, Hulten C, Sichel D (2013) Measuring capital and technology: an expanded framework. In Corrado C, Haltiwanger J, Sichel D (eds) Measuring capital in the new economy. Chicago, IL, 2005; online edn, Chicago Scholarship. Accessed 21 Feb 2013
17. Gordon RJ (2014) The demise of US economic growth: restatement, rebuttal, and reflections. NBER working paper no. 19895, Cambridge, MA

18. Gordon RJ (2018) Why has economic growth slowed when innovation appears to be accelerating? NBER working paper no. 24554, Cambridge: mA
19. Damioli G, Van Roy V, Vertesy D (2021) The impact of artificial intelligence on labor productivity. Eurasian Bus Rev 11:1–25
20. Czarnitzki D, Fernández GP, Rammer C (2022) Artificial intelligence and firmlevel productivity. ZEW–Leibniz centre for European economic research discussion paper no. 22–005
21. Alderucci D, Branstetter L, Hovy E, Runge A, Zolas N (2019) Quantifying the impact of AI on productivity and labor demand: evidence from U.S. Census Microdata. Center for AI Analysis of Patents Carnegie Mellon University and NBER
22. Noy S, Zhang W (2023) experimental evidence on the productivity effects of generative artificial intelligence. Working paper, MIT
23. Klopper R, Gruner S, Kourie D (2007) Assessment of a framework to compare software development methodologies proceedings of the 2007
24. Nair S, Lyytinen K, Kannabiran G (2018) AI in requirements engineering. IEEE Softw 35(5):69–75
25. Kumari V, Kulkarni S (2018) Use of artificial intelligence in software development life cycle requirements and its model. Int Res J Eng Technol
26. Ch R The role of AI in software development: trends, statistics, and growth. https://aloa.co/blog/the-role-of-ai-in-software-developmenttrends-statistics-and-growth
27. Online questioner. https://forms.yandex.ru/cloud/665386a8505690b7ba82ef66/

Unveiling the Transformative Influence of ChatGPT on Service Sector

Melike Zehir and Mustafa Kemal Yılmaz

Abstract The study examines the implementation of a new technological setting in the business environment, focusing on the service sectors that are most likely to be affected by ChatGPT. Initially potential use cases in the selected sectors are identified, then the use case-specific efficacy of ChatGPT-4 in generating coherent responses by devising a series of questions are assessed. According to the results, some service sectors will be affected from ChatGPT more than the others as it influences managerial, and operational work differently at the strategic, and functional levels. In e-commerce and digital marketing, financial and healthcare sectors, ChatGPT provides directly or potentially applicable outputs, beyond sole suggestions. In education and management consulting sectors, the outputs are directly applicable. In the IT and cybersecurity sectors, the majority of outputs are definitive solutions, while in the tourism, travel, and hospitality sectors, many outputs serve as suggestions that guide users toward potential solutions. The study also provides valuable insights about use of ChatGPT in overcoming the challenges faced by firms in the service sector.

Keywords ChatGPT · Generative AI · Management · Use cases · Service sector

1 Introduction

Generative Artificial intelligence (AI) has taken assistive technology to a new level by providing powerful capabilities to nontechnical users, and offering a wide range of applications for firms [1]. ChatGPT (Chat Generative Pre-trained Transformer) is one of the most revolutionary AI-powered chatbot launched in December 2022. It enables users to discuss with the AI by inputting prompts [2]. By enabling the automation of many tasks, it produces human-like replies and interactions to queries.

M. Zehir (✉) · M. K. Yılmaz
Ibn Haldun University, Istanbul, Türkiye
e-mail: melike.zehir@ihu.edu.tr

M. K. Yılmaz
e-mail: mustafa.yilmaz@ihu.edu.tr

N. M. Durakbasa and K. G. Gülen (eds.), *Sustainable Green Conversion*, Lecture Notes in Mechanical Engineering, https://doi.org/10.1007/978-3-031-83583-4_10

Goldman Sachs (2023) [3] predicts that generative AI could boost global GDP by 7% (nearly USD 7 trillion) and increase productivity by 1.5% over the next decade, while OpenAI projects of ChatGPT are expected to generate USD 1 billion in 2024.

ChatGPT has the potential to create new opportunities and challenges for companies, transforming working environment and the interaction of people with technology. The system employs transformer-based neural networks to make predictions from given inputs. By leveraging both natural language processing and machine learning algorithms, it can accurately and quickly interpret and respond to user queries. This allows for text-based interactions between the user and the artificial intelligence system and it can be applied to produce various content i.e., text, audio, images and videos [4]. It enhances creativity and innovation in services by effective integration of human and machine intelligence and skills [5]. Many firms across different sectors, including e-commerce, healthcare, finance, education, tourism, use it for several purposes. It offers numerous advancements over traditional chatbots. Unlike other search engines, it interprets conversational questions in natural language, and learns from its interactions with users over time.

Emerging digital technologies lead to transformative changes in the interactions of users with service systems [6]. Many firms integrate GPT to exemplify its transformative potential. Microsoft Bing uses ChatGPT-4 to enhance the user experience of its search engine, while Stripe, a financial platform, deploys this tool to protect chatrooms from fraudulent activities [7]. Morgan Stanley adopted ChatGPT-4 to enhance the efficiency of its in-house technical support operations while Salesforce use it to furnish tailored suggestions [8]. Microsoft Office 365 integrate ChatGPT-4, enabling the creation of entire PowerPoint presentations through a single text prompt, along with the ability to summarize lengthy documents in Word [9].

ChatGPT may have far-reaching implications for companies in different sectors. For instance, educational institutions may employ it to provide customized teaching assistance to the students, and students can get quick answers for their queries. Entertainment companies become more interactive through virtual assistants that understand user preferences. ChatGPT offers better suggestions based on the preferences of users, enabling them to explore new content that people may enjoy doing. Financial institutions may provide customers automated advice on investment decisions based on real-time data analysis and the risk profiles of clients. Healthcare companies may offer smarter patient care solutions by using predictive analytics that can detect patterns in medical data. Mass media may produce news by using the features of ChatGPT. GPT-enabled chatbot also allows employees to complete complex tasks more efficiently without spending much time navigating complicated menus. Thus, anyone that knows coding or not, may use this AI tool, paving the way for new innovations across sectors.

Although there are many studies held on ChatGPT, studies that investigate the applications of ChatGPT in businesses are notably limited. Most of these studies have a rather narrow focus, primarily concentrating on the enumeration of ChatGPT's advantages and drawbacks. This limited scope leaves a significant gap, particularly on how ChatGPT functions in service-sectors. The present study fills this gap in three dimensions. First, it goes beyond the advantages and disadvantages of ChatGPT and explores its potential applications in service-sector firms. This approach provides a comprehensive view, enabling stakeholders to make more informed decisions. Second, it deepens the discourse by examining the utility of ChatGPT across different sub-sectors and using sector-specific questions that consider unique sectorial dynamics. Finally, this study provides valuable insights for service-sector firms on the potential use of ChatGPT to improve corporate performance and manage complicated issues.

We structure the remaining parts of the study as follows. Section 2 reviews the literature. Section 3 provides the methodology. Section 4 presents evidences on the potential uses of the ChatGPT across selected service sectors. Section 5 argues the limitations and challenges of ChatGPT. Section 6 discusses the ability of ChatGPT-4 to address the use cases in the selected service sectors. Finally, the last section concludes and discusses the implications of the study.

2 Conceptual Framework and Literature Review

ChatGPT allows computers to understand human speech and respond in a natural-sounding way. It relies on a type of neural network that learns from past conversations and additional information provided by users. This learning approach makes it more accurate than other chatbots that use fixed rules for responding. ChatGPT is also designed to quickly adapt to new topics and different types of conversations based on user input. With the use of reinforcement learning, the system improves its performance over time as it processes new data.

ChatGPT fits for real-world applications. It reduces mistakes caused by misunderstandings or wrong interpretations on either side during interactions. There is an increasing number of studies that explore the impact of ChatGPT on diverse areas, ranging from education [10] to health [11], media [12] to tourism [13], environmental science [14] to academia and libraries [15], supply chain [16] to intelligent driving [17], human resource management [18] to real estate valuation [19], finance [20] to organizational behavior [18], advertisement [21] to intellectual property law [22].

In the management field, ChatGPT serve as a new context for the operations of firms. The employment of algorithms significantly advances the optimization of decision-making [23, 24]. ChatGPT also play a vital role in knowledge management. Recent studies show that ChatGPT can be used for information retrieval [25], text mining [26], and document summarization [27]. ChatGPT also provides innovative opportunities for shaping customer-firm relationship, by offering quick and customized answers to the inquiries. Firms can provide 24/7 customer services by using ChatGPT. In this sense, ChatGPT improves customer interaction by providing uninterrupted services, and responding to inquiries on real-time. These experiences increase the loyalty of customers [28, 29]. However, the results may vary from one sector to another. Thus, a close look into how the experiences of customers with services provided by ChatGPT influence their relationship should be explored at sectorial level.

Most of the studies held in management field so far have provided preliminary insights into the application of ChatGPT in supply chain [16], digital leadership and technology integration [30], human resource management [18], organizational behavior [31], corporate performance [1] and technology acceptance [32]. Frederico (2023) [16] focused on how ChatGPT plays a role in enhancing customer satisfaction, operational efficiency, and cost reduction. ChatGPT optimizes communication with customers and suppliers, automates tasks like tracking shipments and managing orders, and supports data analysis throughout the supply chain process [33, 34]. Karakose et al. (2023) [29] test the quality of the outputs provided by this tool for detailed prompts about digital leadership and technology integration in terms of accuracy, and concisenes. Budhwar et al. (2023) [18] employed diverse perspectives to observe the applications of regenerative AI in human resources management. Dasborough (2023) [31] examined the impact of ChatGPT on organizational behavior. Kopplin (2023) [32] explored the acceptance of ChatGPT in co-working spaces. Table 1 presents a summary of the selected studies.

Although ChatGPT have significantly contributed to firms there are relatively limited studies held on the service sectors. Most of the prior studies focus on health and education sectors. Thus, there is a need to conduct an in-depth analysis to reveal the quality of the outputs provided by ChatGPT in the service sectors. In this context, several questions should be answered. Can the user follow the output of ChatGPT as supportive advice or guidance aiding to decision-making? Can the user directly use the output, empowering the user's skills, speed, and productivity? Is the output of ChatGPT directly integrated with the use cases, or is there a need for further efforts for data integration, system access, and authorization to implement the output? The answers to these questions represent an important gap to be filled in the literature.

3 Methodology

ChatGPT offers various services in different sectors. However, a comprehensive investigation is necessary to identify the most suitable use cases and evaluate their adaptability. This study reviews the primary service sectors that could integrate ChatGPT into their business based on prior studies. It presents the possible cases in which ChatGPT can be used in the selected service sectors. For the use cases, prompts demanding clear, solid and detailed examples are prepared. We assign various tasks to ChatGPT-4, the most recent version of the selected regenerative AI tool, including planning based on scenario, making predictions, generating reports, creating algorithms, and producing content. It is explicitly requested that ChatGPT-4 provide clear and specific answers. After analyzing the responses of ChatGPT-4 to these tasks, we pose additional follow-up questions, and assign more tasks to get deeper into the subject matter. This process also involves evaluating the proficiency of ChatGPT-4 in handling these complex situations. Finally, the outputs of GPT-4 are categorized according to their applicability based on the criteria that if the provided output,

Table 1 The prominent relevant works in the literature on generative AI applications in business fields

Category	Authors	Sector	Method	Focus area
Marketing	Paul et al. (2023) [35]	Various	Conceptual	The influence of ChatGPT on consumers, its advantages and potential pitfalls
	Dwivedi et al. (2023) [36]	Hospitality and tourism	Conceptual	Practices, challenges, and establishment of a research plan for implementing generative AI
	Gursoy et al. (2023) [37]	Hospitality and tourism	Conceptual	Transformative impact of ChatGPT, its benefits, challenges, and potential to reshape customer interactions, decision-making, and service delivery processes
Finance	Hofert (2023) [38]	Finance	Correlation analysis	ChatGPT capabilities in addressing potential issues related to correlations in financial and insurance risk modeling
Human resource management	Budhwar et al. (2023) [18]	Various	Conceptual	Employment relations, employee well-being, engagement, human resource management outcomes
Organizational behaviour	Dasborough (2023) [31]	–	Conceptual	The effects of ChatGPT on organizational behavior
Technology integration	Karakose et al. (2023) [30]	Education	Comparative analysis	Digital leadership and teachers' technology integration
Supply chain	Frederico (2023) [16]	Logistic	Conceptual	Research on ChatGPT and the gap in supply chain area
	Burian (2023) [33]	Logistic	Conceptual	Applications that can enhance supply chain management
	Nunez (2023) [34]	Logistic	Conceptual	Main advantages of ChatGPT

- Is a suggestion or guidance that the user should follow to adapt to its own processes,
- Is a direct solution that can be used without any further adaptation efforts,
- Is a solution that can be easily integrated into the existing processses if the required authorization and/or availabilities (such as access to past social media post data for sentiment analysis of a company or access to financial information of a customer to access their credibility and other) are provided to the tool.

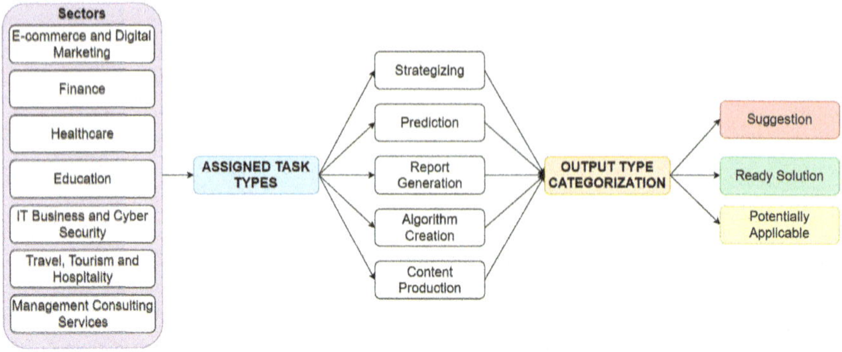

Fig. 1 The details of the methodology employed in the present study

The responses of ChatGPT-4 are categorized by consulting with industry experts to ensure a comprehensive analysis. This allowed us to clarify the extent to which the answers provided by ChatGPT-4 are applicable to the specific questions that we asked. The results provide valuable insights for companies and market professionals in the selected service sectors to prioritize sector-specific use cases for effective regenerative AI deployment, integration and utilization. Figure 1 depicts the research methodology.

4 The Potential Uses of Chatgpt in the Selected Service Sectors

ChatGPT offers an automated way to manage customer relationships and improve internal operations for companies. It provides predictive analytics to come up with new products to meet the needs of customers, enabling firms to develop new strategies on real-time, giving them more control over marketing campaigns, and allowing to optimize their operations. Firms integrate ChatGPT into their workflow to provide quick solutions to the clients to enhance customer satisfaction. They develop distinctive experiences for new clients, boost customer engagement, and build trust with the aid of ChatGPT. They also access new markets, run efficient marketing campaigns, and set up closer relationship with existing and prospective clients [2]. ChatGPT aids to identify what makes a firm successful by examining its marketing tactics, clientele, and product attributes. It makes recommendations for the best channels a firm may use for a campaign and modify campaign tactics to improve outcomes in real-time. The following part discuss the potential uses of ChatGPT on the selected service sectors.

A. E-Commerce and Digital Marketing

Firms that engage in e-commerce heavily depend on customer service. Customer support could be costly and time-consuming with traditional methods i.e., phone calls, emails, while by the capabilities of ChatGPT, firms can quickly and accurately respond to the queries of customers, and offer a more interactive encounter, without incurring additional cost, improving customer experience [39]. Companies may craft personalized offers based on data collected during conversations to increase online sales. ChatGPT also help e-commerce firms better understand the needs of users that speak different languages. This ability allows companies to access to new markets and increase sales abroad. Other benefits of integrating ChatGPT into an e-commerce business can be summarised as follows:

- Sales and upselling: ChatGPT can act as virtual sales representatives, suggesting related products or upselling opportunities, thus increasing average order value.
- Lead generation and conversion: ChatGPT can engage website visitors and convert them into leads by providing information and guding them through the sales process.
- Reduced cart abandonment: By adressing customer concerns promptly and providing real-time support, GPT chatbots can help reduce cart abandonment rates.
- Integration and customization: GPT chatbots can easily be integrated with the existing e-commerce platforms and be customized to match the brand's style. They enhance word-of-mouth referrals, and improve consumer satisfaction in shopping [40]. They also generate customized recommendations, i.e., product suggestions, promotional content, shopping insights, based on the consumer's past behavior and preferences [41, 42].
- Reduced costs: ChatGPT-driven chatbots and conversational agents has the potential to enhance the quality of services and boost productivity [43]. They offer a cost-effective mean of customer service and assistance, managing multiple interactions simultaneously without human intervention [41].
- Enhanced advertising strategies: ChatGPT enables the creation of compelling promotional content that deeply resonates with consumers, ultimately resulting in elevated levels of engagement and improved conversion rates [44].

ChatGPT has the potential to make revolutionary changes in digital marketing. Firms may develop effective digital marketing campaigns, and engage with the target audience by using the capabilities of this AI-powered tool. ChatGPT may help firms create materials for social media updates, blog entries, and may recommend text in marketing materials for facilitating the work of digital marketers. It can identify the preferences of customers by analysing big data. By this way, firms may better understand what customers want, provide quick solutions to their problems, offer proactive suggestions, and may execute successful campaigns.

ChatGPT speed up response times, facilitate customer support, reducing the workload of customer representatives. In this sense, ChatGPT acts as an efficient alternative for call centers. Firms save time and offer a customized experience by automating customer services on real-time. ChatGPT can also be used with analytics to make digital marketing campaigns more effective. It increases efficiency, and improve customer engagement, leading to better decision-making, cost savings and greater scalability when entering into new markets.

B. Finance

ChatGPT allow financial institutions to cut the costs of customer services and to respond customers more quickly and accurately. It leads to automate tasks like checking account balances and answering account queries, allowing customer representatives to focus on more complex issues. It also enables financial institutions to engage with each customer based on their past interactions over time. By this way, customers could get tailored recommendations and personelized offers for their financial needs in line with their risk tolerance, and investment background [45]. Thus, ChatGPT assist financial institutions in making prudent decisions, and fostering customer loyalty and trust.

ChatGPT can help detect fraudulent activities to prevent financial losses in financial institutions by analyzing the patterns of customer interactions. Financial companies can employ this tool to inspect client information, produce credit ratings, and make better lending decisions. In this context, ChatGPT can analyze the credit records of a borrower, and market conditions for mitigating the risk associated with an investment or loan. It also ensures that financial institutions comply with regulations, and generate regular reports. However, marketing communication in these institutions may not solely rely on ChatGPT, but human involvement is still needed to verify the trustworthiness of personalised offers. Finally, since ChatGPT is powered by machine learning algorithms that are constantly improved through continuous training with new datasets, the accuracy level in these institutions stays high in responding to new questions, making sure that the clients of the institutions always get reliable answers quickly.

Financial intermediaries may use ChatGPT to select the best stocks that will outperform traditional investment funds and to give more accurate investment advice to their customers. In a recent study held by Lopez-Lira and Tang (2023) [46] the authors found that ChatGPT outperforms traditional sentiment analysis methods and could predict the stock price movements of specific companies more accurately. For instance, JP Morgan uses IndexGPT, its trademark product, that tap cloud computing software using AI for analyzing and selecting securities tailored to the customers' needs. Morgan Stanley use ChatGPT to get answers to the queries of financial advisors. The engineers of Goldman Sachs use ChatGPT to create codes.

C. Healthcare

ChatGPT has a significant impact on healthcare sector. It can speed up and improve many tasks in healthcare business, such as keeping medical records and taking care of patients. For instance, ChatGPT could be employed for data entry into electronic health records, saving time, increasing information accuracy, and helping doctors and nurses to more quickly and simutaneously access patient records. Additionally, the system can make it easier for hospital employees to talk to each other by simplifying routine tasks like scheduling appointments and filing documents. This allows staff to spend more time on important jobs. ChatGPT can also be tailored for early detection of disease outbreaks by regularly updating its training data from diverse sources [11]. This will improve the AI's ability to spot infectious disease patterns and trends. To reduce errors and biases in ChatGPT's responses, it's crucial to use accurate, diverse, and high-quality data for training, and to update the model regularly. It's also beneficial to have human experts check the information generated by the AI. To make ChatGPT useful for healthcare professionals and researchers who speak different languages, it's important to train it with multilingual datasets and add translation features. Working with language experts to adapt the system will enhance its accessibility.

Healthcare companies, particularly those focusing on telemedicine services by virtual visits to doctors over video conferencing platforms, benefit more from ChatGPT. Patients could get direct and real-time physical and emotional care from virtual assistants powered by ChatGPT, making easier for them to get answers for their treatment plans. This opportunity may reduce the stress of people, particularly those that face with difficult and urgent health issues. Furthermore, ChatGPT could reduce expenditures associated with physical visits due to the lack of additional personnel or physical space compared to traditional medical facilities, making ideal to use ChatGPT services for rural people whose access to quality medical care is limited.

However, when dealing with patient care, it's crucial to manage conversations carefully to protect medical privacy. Using ChatGPT's features, healthcare providers can more easily handle tasks like filling out patient forms, all while making sure they meet legal standards during each interaction. Finally, healthcare experts can use ChatGPT to help diagnose and treat diseases more accurately. They can collect data from conversations with patients to spot patterns that will help them make better decisions.

D. Education

ChatGPT has the most significant impact on education sector, leading to several opportunities and challenges for different parties. On the one side, it provides students personalized learning experiences and allows them to write assignments. On the other side, teachers and academicians can use ChatGPT to make course plans and shape their lectures for each student based on their skills, and give them immediate feedback. ChatGPT also enables teachers to grade the assignments more quickly. Thus, it offer benefits that improve teacher-student interactions and class productivity in all grades. ChatGPT may produce unique academic articles by employing an incredible capacity to fast search online and powerful writing abilities [2]. On the institutional level, ChatGPT could be employed as an automated virtual assistant that can answer questions about course content and campus services in an educational institution.

However, ChatGPT leads some challenges in education sector. Since it processes information from text input it reduces the originality of work, and tends to be uncreative. Therefore, students should master critical, and creative thinking through learning and making practices. Another challenge of ChatGPT is that if the training data contains biases the results of the search for an assignment may contain inaccurate outcome. Therefore, educational institutions should address these challenges to ensure that the use of ChatGPT in education is ethical, and effective. ChatGPT also affect educational technology firms that provide homework help as a service, as well as digital and physical textbook rentals. For instance, the stock of EdTech Chegg has dropped by – 50% as ChatGPT is making their business obsolete. The company has been hit hard by the recent advancements in AI as more students turned to ChatGPT for studying and learning. Its net profit margin dropped by 92% in 2022.

E. IT and Cyber Security

A closely related field to digital transformation is the use of ChatGPT in the IT business and cyber security. ChatGPT has significant effects on software development in IT companies. Developers use it to create text generation and code completion systems, making software applications more interactive and user-friendly. ChatGPT is also a valuable tool for proofreading and bug fixing. Depending on the programming language, it can show sample code structures, and give developers more time to work on improving core features. NLP capabilities of ChatGPT allow it to analyze existing source codes more accurately than humans. Finally, it can generate new source codes based on samples that already exist. By this way, developers can create more sophisticated chatbots that can respond to user queries.

ChatGPT has a significant impact in the field of cyber security to detect and prevent cyber-attacks. It helps identify phishing emails and detect malware, where it can pinpoint malicious code by analyzing the language used in the code [47]. Moreover, ChatGPT can be used to generate secure passwords that are difficult to guess. It helps debugging programs and highlightes cyber vulnerabilities, improving the development of software with enhanced cyber security [48]. ChatGPT can also be employed in developing strategies to combat cybercrime [49]. It can improve the cybersecurity skills of workforce and minimize negative impacts of future similar attacks [50].

F. Tourism, Travel and Hospitality

The implementation of ChatGPT in the tourism, travel and hospitality sectors have dramatically changed how companies in these sectors operate to enhance customers experience [51]. Firms use ChatGPT to meet the requirements of travellers that search for information, construct itineraries, select suitable timetables, and evaluate alternative services in travelling. Travellers can get quick information to shape travel experiences. Meanwhile, AI-generated texts are used to improve back-end operations and marketing campaign. When ChatGPT is combined with voice assistants it will offer smarter services [52].

On the supply side, ChatGPT can assist travel and tourism firms with customer phasing. By allowing travellers to use AI to create a trip itinerary and find hotels, the time-consuming process of sifting through numerous options can be reduced. ChatGPT can also provide travel recommendations for activities, hotels, restaurants, and other value-added travel services, as well as give advice on local customs, and visa requirements. It can assist with booking and reservation inquiries and provide information about availability and pricing. It offers personalized recommendations based on the interests of travellers. ChatGPT can evaluate alternative routes and transportation options for travellers and travel agencies. By offering multilingual support, it can communicate with travellers in multiple languages, helping to improve their experience. Finally, it can generate marketing content and detailed descriptions for attractions, hotels, and facilities, and develop destination content on websites and social media to attract more visitors.

Some examples may provide evidence on the adoption of ChatGPT by travel companies. In February 2023, Singapore-based online travel agency OTA) Trip.com launched TripGen, allowing the company's mobile app users to get recommendations on flights, hotels, and other travel services. Kayak, owned by Booking Holdings and Expedia announced the introduction of plugins powered by ChatGPT, while eDreams ODIGEO started a partnership with Google's Bard. The rapid implementation of these tools by major OTAs shows the significant interest in using Generative AI for trip planning.

G. Management Consulting Services

ChatGPT has significant impact on consulting services. It enables consultants to provide quicker, and more personalized services and insightful advice to the clients by collecting, analyzing and interpreting large data volumes. Chatbots can conduct surveys, collect feedback, and provide support, while virtual assistants can automate repetitive tasks and provide recommendations [47]. ChatGPT allows consultants to collaborate and share knowledge more efficiently, creating new opportunities for innovation in the consulting sector.

Table 2 presents human insight and manual work needed in different activities performed by the management consultants, and summarizes the potential impact of ChatGPT on these activities. In Table 2, the tasks that need human insight are not repetitive and still require the expertise of consultants. However, certain activities would become more efficient by using ChatGPT. For example, ChatGPT can help streamline research work. However, posing the right questions and understanding the interplays between the answers would need human insight (Fig. 2).

5 Limitations and Challenges of ChatGPT

Although ChatGPT offers remarkable advantages, it has also some limitations. One of them is the potential for bias in its responses [35]. AI and chatbot models frequently face allegations of exhibiting bias and discrimination [41]. Thus, ChatGPT has the potential to reinforce biases inherent in its training data, leading to prejudiced results and it is imperative to consider both the training data and prompts employed to ensure that the outputs of the model remain impartial and equitable [48]. To minimize these problems, curating the training data carefully and continually monitoring the

Table 2 The potential impact on ChatGPT on human insight and manual work required activities performed by the management consultants

Activity	Tools used	Human insight and ingenuity	Repetitive and manual work	Impact of ChatGPT
Problem identification	Interviews, surveys, meetings	High	Low	Low
Data collection and analysis	Surveys, financial statements, data visualization software	Medium	Medium	Medium
Research and benchmarking	Industry reports, research databases, benchmarking software	Medium	Low	Medium
Solution development	Mind-mapping software, prototyping tools, business modeling software	High	Low	Medium
Recommendation and implementation	Project management software, communication tools	Medium	Medium	Medium
Monitoring and evaluation	Performance metrics, surveys, data visualization software	Low	Medium	High
Communication and presentation	Presentation software, Communication tools	High	Low	Low
Project management	Project management software, scheduling tools	Medium	Medium	Medium

Source www.datadesign.fi

responses of ChatGPT to identify and correct potential biases is essential. Furthermore, the responses of ChatGPT are restricted by the domain knowledge it has acquired through its training data. Thus, it may need help for highly specialized topics. Another drawback of ChatGPT is its tendency to generate information that might not be accurate. This becomes particularly important when consumers seek reliable insights about products, as doubts can arise regarding the reliability of the AI [49].

ChatGPT need more emotional intelligence. In human conversation, ChatGPT may experience difficuties in recognizing and responding to emotional clues. This can result in insensitive responses that may frustrat users. Therefore, it may be essential to

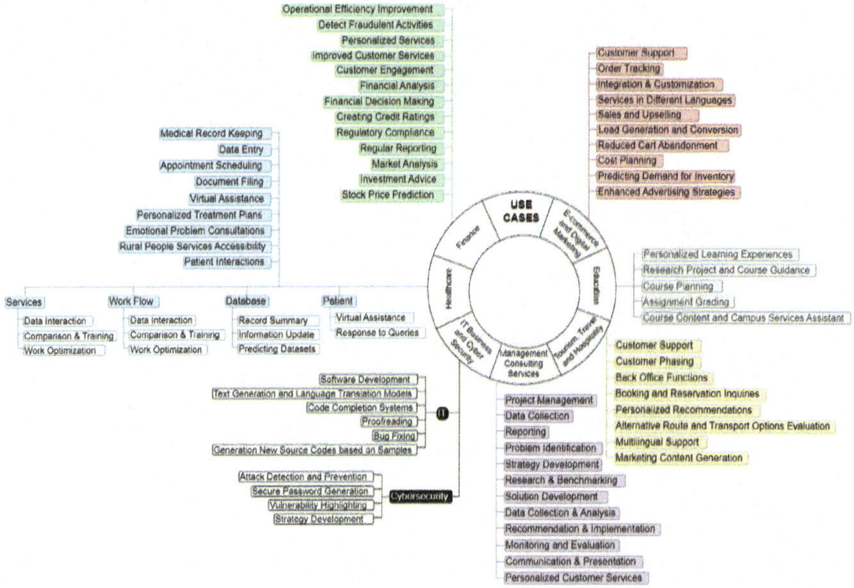

Fig. 2 The detailed identified subfields ChatGPT can serve under each considered sector

incorporate additional programming to help ChatGPT better understand and respond to emotional clues for creating an empathetic user experience. ChatGPT has also the limitation of offering limited dialogue options to users that can restrict their ability to engage in meaningful conversations. It generates responses based on statistical patterns of its training data, but it can sometimes misenderstand the context of the prompts it receives, leading to inappropriate answers [50]. A closely related issue to this limitation is that ChatGPT may be unable to understand the context of a conversation and thus, may give inaccurate answers. This could result in negative interactions with consumers that may harm the brand reputation.

Other challenges of ChatGPT include privacy and security. ChatGPT has the capability to gather data from interactions with consumers, raising concerns related to privacy [51]. ChatGPT also possesses the capability to create convincing and high-quality malicious content such as phishing emails and social engineering attacks [52]. Although safeguarding measures are implemented, they can be bypassed by altering the wording of the queries. This issue highlights the complexity of maintaining security in the AI systems. Though many companies assess ways in which ChatGPT could be used in the business environment some firms ban the use of ChatGPT due to security reasons. For instance, Samsung restricted the use of ChatGPT for its employees after the misuse of this chatbot. The company reported that some employees uploaded sensitive codes to ChatGPT that lead to concern in security risk considering the leaks of crucial information [53, 54]. For similar reason, US investment bank JP Morgan restricted the use of ChatGPT among its staff earlier in 2023 [54], while Amazon warned its employees not to upload confidential information to ChatGPT. However,

Samsung is also looking for ways in which generative AI could be used safely to enhance employee productivity and efficiency. Despite these restrictions, ChatGPT can help engineers and software developers in companies to generate computer codes to speed up their tasks, as this is the case in Goldman Sachs.

6 The Ability of Chatgpt-4 to Address the Use Cases in the Selected Service Sectors

This section explores the ability of ChatGPT-4, the latest version of the regenerative AI tool, to address some use cases in the selected service sectors. We asked ChatGPT-4 to provide solid, detailed, and deterministic answers for the use cases. The responses are then evaluated by taking the opinions of industry experts from considering the type and applicability of the output. Table 3 presents the outcomes of inquiries related to the use cases in each sector. The results include a classification of the content types of the responses and an overview of the types of outputs generated by ChatGPT-4. In the categorization of the output type, "suggestion" refers to the provision of recommendations that guide the user to solution, but not providing a direct solution; "ready solution" represents direct use of the outputs for the use case; "potentially applicable" means the output can be adapted to real world cases, if certain authorizations or data-system integration is provided such as access bank user account details in finance sector.

In this study, we provided only a sample of the questions and answers directed to and generated by ChatGPT-4 for the selected sectors due to large size of the responses and answers. The remaining questions and their answers can be provided by the authors upon requests.

The findings of the exploratory study in Table 3 show that, in e-commerce and digital marketing, financial and healthcare sectors, ChatGPT-4 can provide directly or potentially applicable outputs beyond suggestions. In education and management consulting sectors, in the all use cases, ChatGPT-4 provided directly applicable, and satisfactory outputs. In the IT and cybersecurity sector, most of the outputs, and in the tourism, travel and hospitality sectors some outputs are suggestions, rather than direct solutions, guiding the users towards solutions.

7 Conclusion

ChatGPT has produced remarkable changes in the business environment and the interactions of people since its release in December 2022. It offers companies significant opportunities like enhancing customer satisfaction, saving time and money, and gaining a deeper understanding of customer behavior by a natural conversational platform. The machine learning capabilities of ChatGPT enable it to be updated by

Table 3 The types of the outputs ChatGPT-4 provided for the explored detailed use cases in different sectors

Sectors	Use case no	Use case detail	Type of the output		
			Suggestion	Ready solution	Potentially applicable
E-commerce and digital marketing	1	Build **smart product recommendation engines** to personalize user experiences in e-commerce sector			+
	2	Assist in **monitoring brand mentions and sentiment** across social media platforms		+	
	3	Help **segmenting customers** based on various metrics for targeted marketing			+
	4	Assist in implementing **visual search features** for products			+
	5	Use data to **optimize pricing strategies** for different products		+	
	6	**Predict** inventory **demand**			+
Finance	1	Generate **reports and summarize market analysis** and **potential market movement**		+	
	2	Assist customers in performing basic **banking transactions** using **natural language commands**			+
	3	**Assess loan applications** and help **determine creditworthiness** by analyzing financial and historical data			+
Healthcare	1	Provide **initial guidance to individuals** based on the **symptoms**, directing them to **appropriate care**		+	
	2	Periodically ask patients for **updates on their condition** and **relay** that information **to** provide **healthcare**			+

(continued)

Table 3 (continued)

Sectors	Use case no	Use case detail	Type of the output		
			Suggestion	Ready solution	Potentially applicable
	3	Analyze **public data** to **identify potential outbreak** of diseases			+
	4	Generate **possible treatments** based on guidelines to assist doctors, though which they make final decision		+	
Education	1	**Generate lesson plans** or offer suggestions for activities		+	
	2	Guide students on how to **approach a research project** and suggest sources to **structure it**		+	
	3	Offer pros and cons **for educational decisions**, such as implementing a new curriculum		+	
	4	Assist in learning new topics by giving **examples**		+	
IT and cybersecurity	1	Assist in **generating or updating technical documentation**, manuals, and best practices guideline	+		
	2	Offer insights into **code quality, readability**, and potential **security flaws**		+	
	3	Suggest optimization strategies for **allocating hardware and software resources** in IT scenarios	+		
	4	Generate **incident response reports during a security event**	+		

(continued)

Table 3 (continued)

Sectors	Use case no	Use case detail	Type of the output		
			Suggestion	Ready solution	Potentially applicable
	5	**Prioritize vulnerabilities** based on textual descriptions and available data		+	
	6	Create **cyberattack detection algorithms**		+	
Management consulting services	1	Provide insights into **different types of business models** and **adaptations**		+	
	2	Outline **strategic objectives** and **action plans**		+	
	3	Run **SWOT analysis** for a firm		+	
Tourism, travel and hospitality	1	**Offer suggestions** for sightseeing, dining, and other activities based on user preferences		+	
	2	Offer **immediate scripted advice** in emergency cases until professional help is available	+		
	3	**Translate** content in **multiple languages**		+	

new data generated from customer engagements, social media platforms and other sources to provide precise responses to the needs of customers.

This study examines how ChatGPT affects firms across the selected service sectors. At the strategic level, ChatGPT assists in data collection and analysis, offering valuable insights for managers to make better decisions by using educated reasoning. At this level, ChatGPT may influence knowledge management by facilitating the spread of information within a firm. This opportunity offers new ways for employees to generate innovative ideas. At the functional level, ChatGPT improves customer support service, and shape customers experiences and relationship with the firm, increasing operational efficiency. Finally, at the administrative level, ChatGPT may facilitate the organization of work time, the simplification of procedures, the automation and scheduling of tasks like document reviews, report generation.

On the sectorial basis, ChatGPT provide different advantages to firms. It has a positive effect on e-commerce and digital marketing operations. Firms can quickly process customer queries and provide customized responses to them, building strong relationships with customers, leading to increase in sales. In healtcare sector, ChatGPT enables companies to provide automated patient support services that can reduce costs and improve outcomes for patients through better diagnosis and faster response to medical concerns. In education business, ChatGPT assists students with course material, writing assignments, answering their queries, allowing them to focus more on research projects. It also helps teachers and academicians in evaluation and grading, moving their emphasis from fixing mistakes to developing more creative writing and innovative abilities. Financial institutions may use ChatGPT to enable customers to interact with their accounts without having to go to a branch, among many other things. In travel, tourism and hospitality sectors, ChatGPT assist travellers by generating answers about destinations, and transportation and providing personalized recommendations for activities, hotels, restaurants, and other value-added travel services based on the preferences of travellers. Finally, ChatGPT makes the job of management consultants much interesting. It allows them to collaborate more efficiently, creating new opportunities in the consulting sector.

The findings indicate a diverse landscape for the applicability of ChatGPT-4 across service sectors. In e-commerce, digital marketing, financial services, and healthcare, ChatGPT-4 shows a strong propensity for delivering directly applicable and impactful outputs, suggesting a ripe opportunity for firms to integrate AI-driven solutions for getting significant operational benefits. In the education and management consulting sectors, the tool offers directly applicable and satisfactory solutions for the use cases, indicating a potential to augment human expertise. However, for the IT and cybersecurity sector, as well as tourism, travel, and hospitality sectors, ChatGPT-4 plays the role of a suggestive guide rather than a solution provider.

A. Implications of the study

ChatGPT offers multiple benefits, i.e., reduced costs, improved user experience, increased productivity, for companies in service sectors. Companies that leverage this technology are likely to gain significant competitive advantages and they could seize their market share by offering better customer services. ChatGPT also can facilitate better allocation of employees to tasks, increasing operational efficiency. Therefore, companies should invest in this technology to pinpoint the right models of using these solutions to enhance corporate performance. Moreover, given the varied level of applicability, firms should carefully consider how to strategically deploy ChatGPT-4, taking into account strengths and weaknesses of their industry requirements. This approach is crucial for maximizing the potential benefits of AI adoption.

However, companies should be aware of the fact that the performance of ChatGPT is limited to making predictions based on the data it is provided. It needs creativity, empathy, and other human-specific abilities. Users should verify the information from reliable sources before relying on the responses of ChatGPT. They should also adapt this technology by incorporating it into their corporate values. They should identify the parts of the business where ChatGPT could have the most immediate impact and implement a mechanism to monitor it, given the fact that AI-powered chatbots evolve quickly. Finally, companies should exercise caution to ensure that private, sensitive, and secret business information remains where it belongs. Therefore, policymakers should assess the possible consequences of ChatGPT and work out to introduce new regulations and standards to ensure the privacy and secracy in the corporate world.

B. Limitations of the study and future research

We acknowledge that the analyses in this study are limited to the available knowledge on this technology. Further research is necessary to ensure the reliability of the solutions offered by ChatGPT. The ethical implications of incorporating ChatGPT into large endeavors should also be investigated to reveal its potential effects. Future studies may also explore whether ChatGPT leads to better employee performance. Finally, although ChatGPT improves operational efficiency and optimizes workflows in the service sectors, the close task collaboration between humans and machines should be explored to manage any anxiety that may arise within firms.

References

1. Chu MN (2023) Assessing the benefits of ChatGPT for business: an empirical study on organizational performance. IEEE Access 11:76427–76436. https://doi.org/10.1109/ACCESS.2023.3297447
2. Sharma P, Dash B (2023) Impact of big data analytics and ChatGPT on cybersecurity. In: 2023 4th International conference on computing and communication systems (I3CS). IEEE, pp 1–6. https://doi.org/10.1109/I3CS58314.2023.10127411. Accessed on 20 Oct 2023
3. Sachs G (2023) Generative AI could raise global GDP by 7%. https://www.goldmansachs.com/intelligence/pages/generative-ai-could-raise-global-gdp-by-7-percent.html
4. Korzynski P (2023) Generative artificial intelligence as a new context for management theories: analysis of ChatGPT. CentL Eur Manag J 31(1):3–13. https://doi.org/10.1108/CEMJ-02-2023-0091
5. Ameen N, Viglia G, Altinay L (2023) Revolutionizing services with cutting-edge technologies post major exogenous shocks. Serv Ind J 43(3–4):125–133. https://doi.org/10.1080/02642069.2023.2185934
6. Wirtz J, Hofmeister J, Chew PY, Ding X (2023) Digital service technologies, service robots, AI, and the strategic pathways to cost-effective service excellence. Serv Ind J 1–24. https://doi.org/10.1080/02642069.2023.2226596
7. Stripe (2023) Stripe and OpenAI collaborate to monetize OpenAI's flagship products and enhance stripe with GPT-4. https://stripe.com/en-gb-fr/newsroom/news/stripe-and-openai. Accessed on 15 Nov 2023
8. Yakar TJ (2023) GPT-4 and salesforce potential features. https://www.apexhours.com/gpt-4-salesforce-potential-features/. Accessed on 20 Nov 2023
9. Microsoft (2023) How we can use GPT-4 to continually improve AI models in dynamics 365 and power platform. https://cloudblogs.microsoft.com/dynamics365/it/2023/03/20/how-we-can-use-gpt-4-to-continually-improve-ai-models-in-dynamics-365-and-power-platform/. Accessed on 20 Nov 2023
10. Lo CK (2023) What is the impact of ChatGPT on education? A rapid review of the literature. Educ Sci 13(4):410. https://doi.org/10.3390/educsci13040410
11. Ray PP, Majumder P (2023) AI tackles pandemics: ChatGPT's game-changing impact on infectious disease control. Ann Biomed Eng 1–3. https://doi.org/10.1007/s10439-023-03239-5
12. Feng Y, Poralla P, Dash S, Li K, Desai V, Qiu M (2023) The impact of ChatGPT on streaming media: a crowdsourced and data-driven analysis using Twitter and Reddit. In: 2023 IEEE 9th Intl conference on big data security on cloud (Big data security). IEEE, International conference on high performance and smart computing, (HPSC) and IEEE international conference on

intelligent data and security (IDS). IEEE, pp 222–227. https://doi.org/10.1109/BigDataSecurity-HPSC-IDS58521.2023.00046
13. Carvalho I, Ivanov S (2023) ChatGPT for tourism: applications, benefits and risks. Tour Rev. https://doi.org/10.1108/TR-02-2023-0088
14. Agathokleous E, Saitanis CJ, Fang C, Yu Z (2023) Use of ChatGPT: what does it mean for biology and environmental science? Sci Total Environ 888(2023):164154. https://doi.org/10.1016/j.scitotenv.2023.164154
15. Houston AB, Corrado EM (2023) Embracing ChatGPT: implications of emergent language models for academia and libraries. Tech Serv Q 40(2):76–91. https://doi.org/10.1080/07317131.2023.2187110
16. Frederico GF (2023) ChatGPT in supply chains: initial evidence of applications and potential research agenda. Logistics 7(2):26. https://doi.org/10.3390/logistics7020026
17. Gao Y, Tong W, Wu EQ, Chen W, Zhu G, Wang FY (2023) Chat with ChatGPT on interactive engines for intelligent driving. IEEE Trans Intell Veh 8(3):2034–2036. https://doi.org/10.1109/TIV.2023.3252571
18. Budhwar P, Chowdhury S, Wood G, Aguinis H, Bamber GJ, Beltran JR, Varma A (2023) Human resource management in the age of generative artificial intelligence: perspectives and research directions on ChatGPT. Hum Resour Manag J 33(3):606–659. https://doi.org/10.1111/1748-8583.12524
19. Cheung KS (2023) Real estate insights unleashing the potential of ChatGPT in property valuation reports: the red book compliance chain-of-thought (CoT) prompt engineering. J Prop Invest Financ. https://doi.org/10.1108/JPIF-06-2023-0053
20. Neilson B (2023) Artificial intelligence authoring financial recommendations: comparative Australian evidence. J Financ Regul. https://doi.org/10.1093/jfr/fjad004
21. Lee C, Kim J, Lim JS (2023) How does fact-check labeling impact the evaluations of inadvertently placed brand ads? Soc Sci J 1–17. https://doi.org/10.1080/03623319.2023.2216965
22. Buckingham L, Williams M (2023) I, robot: Is IP law ready for the age of AI?:(Thaler and other provocations to our existing systems). Aust Intellect Prop J 33(3):140–162
23. Cristofaro M (2017) Herbert Simon's bounded rationality: its historical evolution in management and cross-fertilizing contribution. J Manag Hist 23(2):170–190. https://doi.org/10.1108/JMH-11-2016-0060
24. Sieja M, Wach K (2019) The use of evolutionary algorithms for optimization in the modern entrepreneurial economy: interdisciplinary perspective. Entrep Bus Econ Rev 7(4):117–130
25. Guo J, Fan Y, Pang L, Yang L, Ai Q, Zamani H, Cheng X (2020) A deep look into neural ranking models for information retrieval. Inf Process Manag 57(6):102067. https://doi.org/10.1016/j.ipm.2019.102067
26. Kumar S, Kar AK, Ilavarasan PV (2021) Applications of text mining in services management: a systematic literature review. Int J Inf Manag Data Insights 1(1):100008. https://doi.org/10.1016/j.jjimei.2021.100008
27. El-Kassas WS, Salama CR, Rafea AA, Mohamed HK (2021) Automatic text summarization: a comprehensive survey. Expert Syst Appl 165:113679. https://doi.org/10.1016/j.eswa.2020.113679
28. Prentice C, Nguyen M (2020) Engaging and retaining customers with AI and employee service. J Retail Consum Serv 56:102186. https://doi.org/10.1016/j.jretconser.2020.102186
29. Karakose T, Demirkol M, Yirci R, Polat H, Ozdemir TY, Tülübaş T (2023) A conversation with ChatGPT about digital leadership and technology integration: comparative analysis based on human–AI collaboration. Adm Sci 13(7):157. https://doi.org/10.3390/admsci13070157
30. Sung EC, Bae S, Han D-ID, Kwon O (2021) Consumer engagement via interactive artificial intelligence and mixed reality. Int J Inf Manag 60:102382. https://doi.org/10.1016/j.ijinfomgt.2021.102382
31. Dasborough MT (2023) Awe-inspiring advancements in AI: the impact of ChatGPT on the field of organizational behavior. J Organ Behav 44(2):177–179. https://doi.org/10.1002/job.2695

32. Kopplin CS (2022) Chatbots in the workplace: a technology acceptance study applying uses and gratifications in coworking spaces. J Organ Comput Electron Commer 32(3–4):232–257. https://doi.org/10.1080/10919392.2023.2215666
33. Burian J (2023) Does ChatGPT offer real value to the supply chain? https://www.iiot-world.com/industrial-iot/connected-industry/does-chatgpt-offer-real-value-to-the-supply-chain/. Accessed on 20 Nov 2023
34. Nunez V (2023) How Chat GPT thinks it can revolutionize the logistics industry. https://www.shiplilly.com/blog/how-chat-gpt-thinks-it-can-revolutionize-the-logisticsindustry/#:~:text=Chatbots%20and%20language%20models%20like,customer%20satisfaction%2C%20and%20reduce%20costs. Accessed on 20 Nov 2023
35. Paul J, Ueno A, Dennis C (2023) ChatGPT and consumers: benefits, pitfalls and future research agenda. Int J Consum Stud 47(4):1213–1225. https://doi.org/10.1111/ijcs.12928
36. Dwivedi YK, Kshetri N, Hughes L, Slade EL, Jeyaraj A (2023) So what if ChatGPT wrote it? Multidisciplinary perspectives on opportunities, challenges and implications of generative conversational AI for research, practice and policy. Int J Inf Manag 71:102642. https://doi.org/10.1016/j.ijinfomgt.2023.102642
37. Gursoy D, Li Y, Song H (2023) ChatGPT and the hospitality and tourism industry: an overview of current trends and future research directions. J Hosp Market Manag 32(5):579–592. https://doi.org/10.1108/IJBM-09-2021-0444
38. Hofert M (2023) Correlation pitfalls with ChatGPT: would you fall for them? Risks 11(7):115. https://doi.org/10.3390/risks11070115
39. Mattas P (2023) ChatGPT: a study of AI language processing and its implications. Int J Res Publ Rev 4(2):435–444. https://doi.org/10.55248/gengpi.2023.4218
40. Konya-Baumbach E, Biller M, von Janda S (2023) Someone out there? A study on the social presence of anthropomorphized chatbots. Comput Hum Behav 139:107513. https://doi.org/10.1016/j.chb.2022.107513
41. Haluza D, Jungwirth D (2023) Artificial intelligence and ten societal megatrends: an exploratory study using GPT-3. Systems 11(3):120. https://doi.org/10.3390/systems11030120
42. Ma L, Sun B (2020) Machine learning and AI in marketing–connecting computing power to human insights. Int J Res Mark 37(3):481–504. https://doi.org/10.1016/j.ijresmar.2020.04.005
43. Balakrishnan J, Dwivedi YK (2021) Role of cognitive absorption in building user trust and experience. Psychol Mark 38(4):643–668. https://doi.org/10.1002/mar.21462
44. Taecharungroj V (2023) What can ChatGPT do? Analyzing early reactions to the innovative AI Chatbot on Twitter. Big Data Cogn Comput 7(1):35. https://doi.org/10.3390/bdcc7010035
45. Mogaji E, Soetan T, Kieu T (2020) The implications of artificial intelligence on the digital marketing of financial services to vulnerable customers. Australas Mark J 29(3):235–242. https://doi.org/10.1016/j.ausmj.2020.05.003
46. Lopez-Lira A, Tang Y (2023) Can ChatGPT forecast stock price movements? Return predictability and large language models. https://doi.org/10.48550/arXiv.2304.07619
47. Haleem A, Javaid M, Singh RP (2023) An era of ChatGPT as a significant futuristic support tool: a study on features, abilities, and challenges. BenchCouncil Trans Benchmarks Stand Eval 2(4):1–8. https://doi.org/10.1016/j.tbench.2023.100089
48. Kalla D, Smith N (2023) Study and analysis of ChatGPT and its impact on different fields of study. Int J Innov Sci Res Technol 8(3):827–833
49. Mijwil M, Aljanabi M (2023) Towards artificial intelligence-based cybersecurity: the practices and ChatGPT generated ways to combat cybercrime. Iraqi J Comput Sci Math 4(1):65–70. https://doi.org/10.52866/ijcsm.2023.01.01.0019
50. Biswas S (2023) Role of ChatGPT in cybersecurity. Available at SSRN 4403584
51. Buhalis D, Harwood T, Bogicevic V, Viglia G, Beldona S, Hofacker C (2019) Technological disruptions in services: lessons from tourism and hospitality. J Serv Manag 30(4):484–506. https://doi.org/10.1108/JOSM-12-2018-0398
52. Buhalis D, Moldavska I (2022) Voice assistants in hospitality: using artificial intelligence for customer service. J Hosp Tour Technol 13(3):386–403. https://doi.org/10.1108/JHTT-03-2021-0104

53. Bloomberg (2023) Samsung bans staff's AI use after spotting ChatGPT data leak. https://www.bloomberg.com/news/articles/2023-05-02/samsung-bans-chatgpt-and-other-generative-ai-use-by-staff-after-leak#xj4y7vzkg. Accessed on 20 Oct 2023
54. Forbes (2023) Samsung bans ChatGPT. https://www.forbes.com/sites/siladityaray/2023/05/02/samsung-bans-chatgpt-and-other-chatbots-for-employees-after-sensitive-code-leak/?sh=5c3c9c6e6078

Industrial Applications

An Approach to On-Site Radiated Emissions Measurement Using a Compact Reference Field Source

Tamerlan Hajiyev

Abstract On-site measurements remain the only option for measuring radiated disturbance from large-sized equipment that cannot be delivered to a standard testing site and/or cannot run anywhere except the operation place. CISPR-16-2-3 standard describes the direct and substitution approaches to the on-site measurements in general, however, obtaining the actual far-field strength values comparable with standard limits still is a challenge due to the list of factors including random wave propagation, unpredictable reflections and interference, available measurement distance comparing to the size of the equipment, and high level ambient noise. This paper describes the method that was developed and applied for on-site measuring of radiated emissions in 30–1000 MHz frequency range using a compact reference *E*-field source to define the wave propagation conditions. The findings show that a random site can be described with a statistically obtained relatively stable path loss value similar to the site attenuation of standard testing sites, and narrow-band ambient noises can be detected as outliers within the statistical analysis.

Keywords Electromagnetic compatibility · On-site · In-situ · Pre-compliance testing · Radiated emissions · Actual values · Path loss · Site attenuation · Testing site · Total correction factor

1 Introduction

The electromagnetic compatibility (EMC) compliance standards for Radiated Emissions require an Open Area Test Site (OATS) or a Semi-Anechoic Chamber (SAC) for performing full-compliance measurements in 30–1000 MHz frequency range. However, there is a wide range of specialized and/or large-sized equipment under test (EUT), like factory engines and machinery, marine and supply equipment, network and data processing enclosures etc., that can't be placed and run in a standard testing

T. Hajiyev (✉)
Radio Engineering and Telecommunications Department, Azerbaijan Technical University, The Republic of Azerbaijan, Baku, Azerbaijan
e-mail: tamerlan.hajiyev.1995@gmail.com

N. M. Durakbasa and K. G. Gülen (eds.), *Sustainable Green Conversion*, Lecture Notes in Mechanical Engineering, https://doi.org/10.1007/978-3-031-83583-4_11

site, thus the only option for evaluating their emissions is the *on-site* measurements (also called *in-situ* measurements).

2 Problem

Arranging on-site measurements remains the subject of ongoing solution research [1–4]. There exist two main issues of on-site measurements shown in Fig. 1:

1. Cancelling the ambient noise (disturbances from other equipment).
2. Measuring the actual values of the emissions.

This paper is related to the 2nd issue. It is caused by random and non-uniform electromagnetic wave propagation environment of the equipment location, frequency resonance and wave reflection-interference issues as well, hence unlike standard-specified testing sites, random on-site measurement area can't be described by constant values of site attenuation.

Therefore, it usually makes the operator perform consistently three measurements of near field, cabling and far-field emissions to obtain a proper vision of the actual far-field emissions [3].

From CISPR 16–2-3 [5] we know that for the basic measurement set "receiver—cable—antenna" running on OATS & SAC the *E*-field strength is calculated the following way:

$$E = V_r + CL + AF,$$

here V_r is the receiver input voltage; *CL* is cable loss; *AF* is antenna factor.

Besides the strong ambient signals and wave reflection-interference issues, the main obstacle using this equation in a random area is *AF* value which has been measured above the ground plane and takes into account the site attenuation of the ground plane. However, $(CL + AF)$ sum can be replaced by some *Total Correction Factor (TCF)* that could include both of antenna factor and site attenuation by measuring some reference *E*-field.

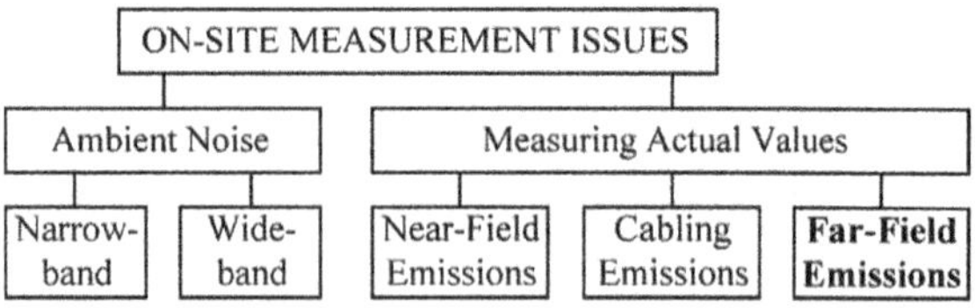

Fig. 1 Classification of on-site radiated emissions measurement issues

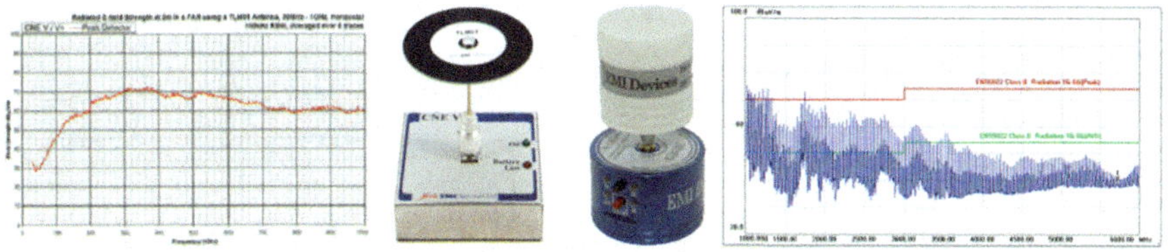

COMPARISON NOISE EMITTER COMB GENERATOR

Fig. 2 Compact reference signal sources

3 Method Description

The method proposed in this paper describes an approach to measure actual values of the far-field emissions in a way similar to the standard method. It requires basic spectrum analyser, allows to check conditions to obtain stable *TCF* value and provides the spectrum graph, where the dots with strong narrow-band ambient signals are rejected and interpolated with next dots.

The mentioned *reference E-field source (REF)* could be a *Comparison Noise Emitter (CNE)* or a *Comb Generator (CG)* shown in Fig. 2—these are compact devices used by EMC testing laboratories for verification, proficiency testing and measurement set check.

4 Measurement Procedure

The following steps describe the procedure for a single antenna placement. In case of multiple placement measurements, angle sector or all around the EUT, the procedure would be repeated for each point.

A. *Measurement set arranging*

Measurement distance should meet Rayleigh's far-field criterion [6]:

$$d \geq \frac{2D^2}{\lambda} = \frac{fD^2}{150},$$

here D is largest dimension of the physical aperture of the antenna, m; λ is wavelength, m; f is frequency, MHz.

If the chosen d isn't the same with REF's calibration distance $d \neq d_{cal}$, then the E-values of the source could be recalculated for the specific d value [7]:

$$E = E_{cal} + 20\,\lg(d_{cal}/d).$$

This measurement is performed using the set shown in Fig. 3, before assembling the equipment under test (EUT) or in front of it. Measurement height for the REF is

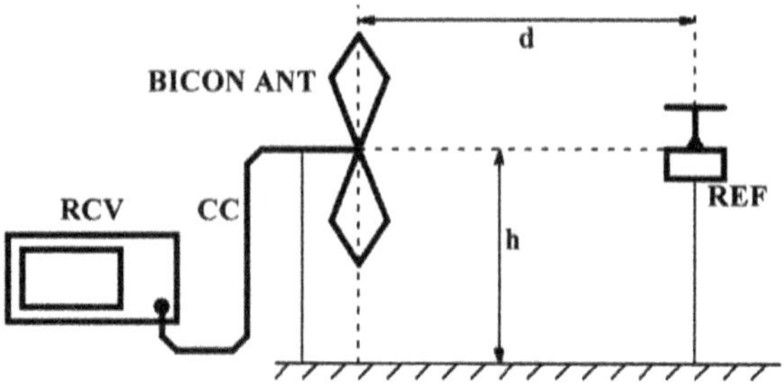

Fig. 3 TCF measuring set: RCV—measuring receiver; CC—coaxial cable; BICON ANT—biconical antenna; REF—reference *E*-field source

chosen as half of the height of the EUT $h = 0{,}5h_{EUT}$, however the receiving antenna should scan height for max value.

Following equation is the total correction factor:

$$TCF = AF + CL + SA = E_{REF} - V_{REF},$$

here E_{REF} and V_{REF} are calibrated and measured values of the reference *E*-field.

This equation is given for a biconical antenna, thus it doesn't take into account antenna phase centre correction.

It's recommended to control environment temperature and humidity during the measurements to ensure *TCF* would remain relatively stable for a while, and the receiver should be set to the same RBW and VBW as it is noted in the calibration certificate of the REF.

The actual measured emissions strength is then:

$$E_{SIG} = V_{SIG} + TCF,$$

Here V_{SIG} is the received signal value.

B. *Input data arranging*

The *TCF* is measured several times (samples taken), the input and processed data is arranged in arrays the following way:

$$TCF_{ij} = E_{REF_i} - V_{RCF_{ij}};$$

$$\begin{bmatrix} TCF_{11} & TCF_{12} & \dots & TCF_{1j} \\ TCF_{21} & TCF_{22} & \dots & TCF_{2j} \\ \dots & \dots & \dots & \dots \\ TCF_{i1} & TCF_{i2} & \dots & TCF_{ij} \end{bmatrix} = \begin{bmatrix} E_{REF_1} \\ E_{REF_2} \\ \dots \\ E_{REF_i} \end{bmatrix} - \begin{bmatrix} V_{REF_{11}} & V_{REF_{12}} & \dots & V_{REF_{1j}} \\ V_{REF_{21}} & V_{REF_{22}} & \dots & V_{REF_{2j}} \\ \dots & \dots & \dots & \dots \\ V_{REF_{i1}} & V_{REF_{i2}} & \dots & V_{REF_{ij}} \end{bmatrix}.$$

i–frequency number (vertical);

j–sample number (horizontal);

n–number of frequencies;

$m \geq 10$–number of samples;

E_{REF_i}–actual value of reference signal;

$V_{REF_{ij}}$–received value of reference signal.

C. *Calculating average TCF value for each frequency*

The average *TCF* value is calculated after removing the outlier values that are caused by sudden ambient noises and reflection-interference issues. Outliers are detected with *one-sided Grubbs' outlier test* which finds one outlier from each *sample* set related to a single frequency [8].

Input average *TCF* and its standard deviation values:

$$TCF_{AVG_i} = \frac{1}{m}\sum_{j=1}^{m} TCF_{ij};$$

$$sd_i = \sqrt{\sum_{j=1}^{m}\left(TCF_{ij} - TCF_{AVG_i}\right)^2/(n-1)};$$

Grubbs' test value for each TCF_{ij} and critic value:

$$G_{ij} = \frac{\left|TCF_{ij} - TCF_{AVG_i}\right|}{sd_i};$$

$$G_{cr} = \frac{m-1}{\sqrt{m}}\sqrt{\frac{t^2\left(\frac{\alpha}{2m}; m-2\right)}{m-2+t^2\left(\frac{\alpha}{2m}; m-2\right)}}.$$

here $\alpha = 0{,}05$ is accepted significance level; $t_{\left(\frac{\alpha}{2m}; m-2\right)}$ is upper critical value of the Student's t-distribution.

Outlier detection rule, logical variable and number of outliers for each *sample* set:

$$TCF'_{ij} = \begin{cases} TCF_{ij} & if \quad G_{ij} \leq G_{cr} \\ N/A & if \quad G_{ij} > G_{cr} \end{cases};$$

$$x_i = \begin{cases} 0 & if \quad G_{ij} \leq G_{cr} \\ 1 & if \quad G_{ij} > G_{cr} \end{cases};$$

$$x = \sum_{i=1}^{m} x_i.$$

The processed *TCF* average value, standard deviation and coefficient of variance are calculated only for TCF'_{ij} values, without detected *single* outlier:

$$TCF'_{AVG_i} = \frac{1}{m-1}\sum_{j=1}^{m} TCF'_{ij};$$

$$sd'_i = \sqrt{\sum_{j=1}^{m}\left(TCF'_{ij} - TCF_{AVG'_i}\right)^2 / (n-2)};$$

$$CV'_i = sd'_i / TCF_{AVG'_i}.$$

D. *Cancelling strong ambient narrow-band signals*

It's obvious that a narrow-band ambient signal which has been almost strong or stronger than the reference signal and has been received within most of samples, can't be detected by the outlier test. Since its attribute is $TCF_{ij} \leq 0$ it's reasonable to set some minimal *gap* value that would distinguish such a signal.

Logical variable describing the gap test result, number of rejected points and output TCF average value:

$$y_i = \begin{cases} 0 & if \ \ \min\{TCF_{i1}; TCF_{i2}; \ldots; TCF_{ij}\} > GAP \\ 1 & if \ \ \min\{TCF_{i1}; TCF_{i2}; \ldots; TCF_{ij}\} \leq GAP \end{cases};$$

$$y = \sum_{i=1}^{m} y_i;$$

$$TCF''_{AVG_i} = \begin{cases} TCF'_{ij} & if \ \ y_i = 0 \\ N/A & if \ \ y_i = 1 \end{cases}$$

The gap value could be taken approximately as the expanded uncertainty from the REF's calibration certificate $GAP \approx 5 - 10dB$.

E. *Processing the signal*

Since TCF''_{AVG_i} values have been defined, the measured signal value V_{SIG_i} can be processed to obtain the actual emissions values E_{SIG_i} and the spectrum graph:

$$E_{SIG_i} = V_{SIG_i} + TCF''_{AVG_i}$$

The points of $TCF''_{AVG_i} = N/A$ are not shown on the graph but get linearly interpolated with the next points with real values.

5 Data Presentation

This section describes the practical result applying the proposed method. *TCF* was measured in 30–1000 MHz band using a spectrum analyser, a biconical antenna, and a CNE as the reference signal source (RBW = VBW = 100 kHz). The CSV-files from the spectrum analyser were processed with MS Excel.

The spectrum graph consists of $n = 1000$ dots. Total $m = 11$ sample sets have been taken in the same single position in two days as shown in Table 1. Each measurement has taken about 30 s long in MAX HOLD mode.

Results are shown in Fig. 4. The number of outlier detected frequencies is $x = 46$, number of rejected strong ambient noise frequencies is $y = 55$ with $GAP = 8dB$. Rejected ambient signals are FM broadcast in 87–108 MHz band, handheld radios in 156–174 MHz band, and GSM cellular network in 800–900 MHz band.

Significant fact is that for 95% of the frequencies the coefficient of variance of *TCF* didn't exceed $CV_i' \leq 0{,}1$ and for 88% of the frequencies standard deviation of *TCF* didn't exceed $sd_i' \leq 1dB$, this demonstrates satisfactory stability of *TCF* in the site.

One important note is that for effective *gap* test it's worth to listen to the ambient noises as long as possible to catch most of the suddenly appearing nearby noise sources (Fig. 4d shows such a missed strong signal at 195 MHz). Since limited battery supply of CNE it should not run longer than 0,5–1 min for each measurement, hence after short V_{CNE} measuring the ambient noise V_{AMB} should be "listened" for couple of minutes, so the maximum number of ambient noises stronger than CNE signal could be distinguished with equation.

$$V_{REF} = \max\{W_{CNE}; V_{AMB}\}.$$

Table 1 Measurement conditions

#	Day–time	t, °C	rh, %
1	I–16:29:51	21,8	50
2	I–16:47:26	23,0	50
3	I–16:50:39	23,1	50
4	II–08:45:07	22,3	52
5	II–08:58:53	25,0	51
6	II–09:28:59	25,1	53
7	II–09:45:11	25,1	53
8	II–09:46:54	25,2	54
9	II–09:52:49	25,3	53
10	II–09:57:30	25,3	53
11	II–10:09:02	25,3	53
[Avg ± dev]		24,2 ± 2,4	52 ± 2

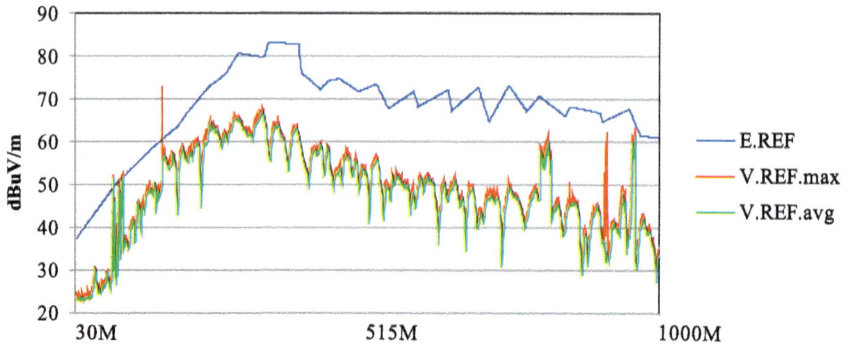

a) Input data: calibrated reference field strength E_{REF} and measured values V_{REF}.

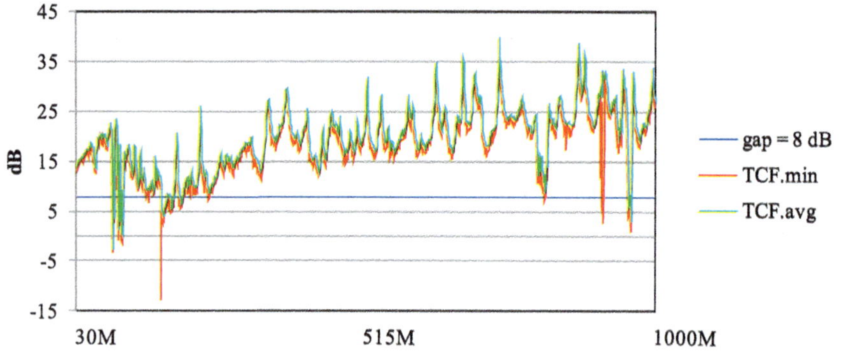

b) *TCF* values. The minimum values close to zero and lower are ambient signals stronger than E_{REF}.

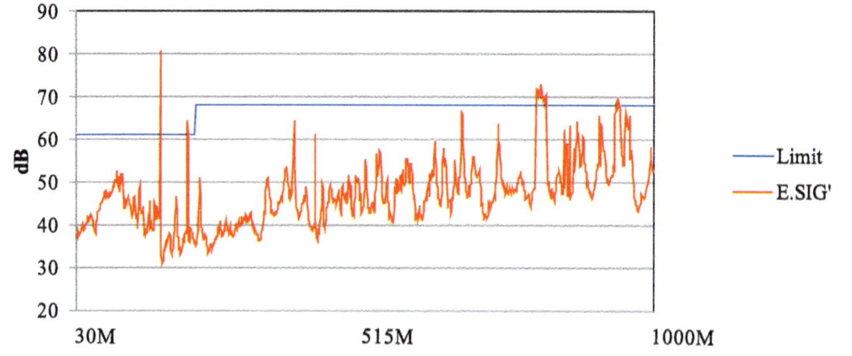

c) Example on measuring actual values of radiated emissions with outlier test but without *gap* rejection $E'_{SIG} = V_{SIG} + TCF'_{AVG}$.

Fig. 4 Graphs obtained performing the proposed method **a** Input data: calibrated reference field strength E_{REF} and measured values V_{REF} **b** *TCF* values. The minimum values close to zero and lower are ambient signals stronger than E_{REF} **c** Example on measuring actual values of radiated emissions with outlier test but without *gap* rejection $E'_{SIG} = V_{SIG} + TCF'_{AVG}$. **d** Example on measuring actual values of radiated emissions with outlier test and *gap* rejection $E''_{SIG} = V_{SIG} + TCF''_{AVG}$

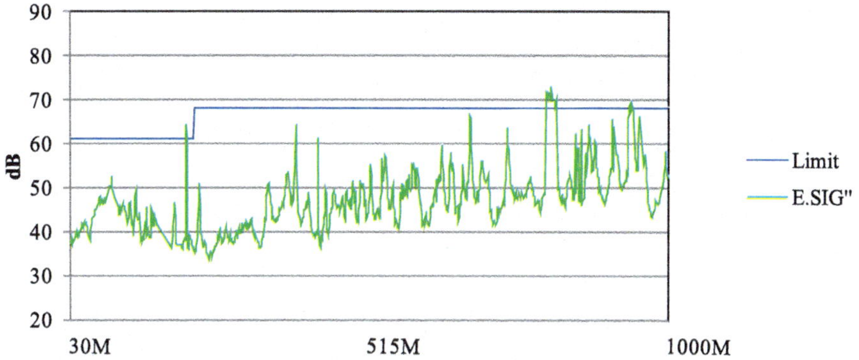

d) Example on measuring actual values of radiated emissions with outlier test and *gap* rejection $E''_{SIG} = V_{SIG} + TCF''_{AVG}$.

Fig. 4 (continued)

6 Conclusions

As result of the research reported in this paper it was found out, that in some conditions attenuation of a random testing site can remain stable for the period enough for performing on-site radiated emissions measurements and obtaining the actual values of far-field radiated emissions, therefore the proposed method can be used for on-site measurements and pre-compliance testing in general. Due to the potential of this method, there will be further research on expanding frequency range and cancelling ambient noise.

References

1. Catrysse J, Vanhee F, Knockaert J, Hendrickx I, Beauvois V (2008) In situ testing of large machines: alternative method for radiated emission measurement. In: IEEE international symposium on electromagnetic compatibility 2008. Detroit, MI, USA, pp 1–6
2. Solé-Lloveras J, Azpúrua MA, Homar AM, Yoshioka Y, Silva F (2023) Efficient in situ assessment of radiated emissions using time-domain measurements. In: international symposium on electromagnetic compatibility – EMC Europe 2023. Krakow, Poland, pp 1–6
3. Zhang M (2023) In-situ radiated emission testing of large systems installations. In: compliance magazine. https://incompliancemag.com/in-situ-radiated-emission-testing-of-large-systems-installations
4. Wyatt K (2021) Pre-compliance testing large systems in-situ. EDN. https://www.edn.com/pre-compliance-testing-large-systems-in-situ

5. CISPR 16–2–3:2014. Specification for radio disturbance and immunity measuring apparatus and methods—Part 2–3: methods of measurement of disturbances and immunity—radiated disturbance measurements. Clause 7.3.1
6. Kraus J (2002) Antennas for all applications. McGraw Hill, p 832
7. CISPR 32:2012. Electromagnetic compatibility of multimedia equipment—emission requirements. Annex C.2.2.4
8. Grubbs' test for outliers (2012) Engineering statistics handbook. National institute of standards and technology (NIST). https://www.itl.nist.gov/div898/handbook/eda/section3/eda35h1.htm

Artificial Intelligence Management in Mechatronic Systems in the Automotive Industry: Privacy and Sustainability Challenges

Țîțu Aurel Mihail, Bâlc Emanuel, and Bâlc Daniel

Abstract This paper explores the integration of artificial intelligence (AI) in mechatronic systems in the automotive industry, focusing on major challenges related to data privacy and sustainability. As AI becomes an essential element in autonomous vehicles and automated manufacturing processes, the proper management of personal data and the reduction of the environmental impact of AI technologies are critical. The paper analyses the risks related to the security of data collected by smart vehicles, emphasizing the need for protection measures according to international regulations. Additionally, it highlights the energy impact of AI and its potential to contribute to sustainability by optimizing processes and reducing waste. Finally, the paper emphasizes the importance of intelligent AI management to balance technological innovation with data protection and environmental responsibility.

Keywords Mechatronic systems · Artificial Intelligence (AI) · Automotive industry · Data privacy · Sustainability

1 Introduction

With the rapid evolution of the manufacturing industry towards the Industry 5.0 model, human-centered, sustainable and resilient technologies have become a key priority. Artificial intelligence (AI) plays a major role in this process, opening up new opportunities for additional functionality and innovative features in the industrial

Ț. A. Mihail (✉)
Faculty of Engineering, Industrial Engineering and Management Department, "Lucian Blaga" University of Sibiu, Sibiu, Romania
e-mail: mihail.titu@ulbsibiu.ro

B. Emanuel
Lucian Blaga University of Sibiu, Sibiu, Romania
e-mail: emybalc@yahoo.ro

B. Daniel
National University of Science and Technology POLITEHNICA Bucharest, Buchares, Romania

N. M. Durakbasa and K. G. Gülen (eds.), *Sustainable Green Conversion*, Lecture Notes in Mechanical Engineering, https://doi.org/10.1007/978-3-031-83583-4_12

landscape. At the same time, the Industry 4.0 paradigm, focused on technology and automation, is still prevalent, which creates a context in which industrial artificial intelligence must evolve to meet both technological and human requirements.

In the automotive industry, mechatronic systems, which combine mechanics, electronics and software, are the backbone of modern vehicles. By using AI, these systems can be managed more efficiently, becoming smarter, capable of self-diagnosis and self-adaptation. However, the use of AI in mechatronic systems is not without its challenges, especially in terms of data privacy and process sustainability. Proper AI management is crucial to maximize technological benefits while maintaining a balance between efficiency, security, and environmental impact. Mechatronic systems in the automotive industry have become increasingly complex, incorporating advanced technologies that allow precise control and automation of various vehicle functions. From automated braking systems and stability control, to energy management in electric vehicles, mechatronics plays a key role in ensuring safety and performance. However, as these systems become more complex, they also require more sophisticated management, and that's where AI comes in.

The integration of AI into mechatronic systems has enabled a significant evolution in terms of the ability to self-diagnose and optimize in real time. For example, AI can analyse data collected from sensors to predict failures or optimize energy consumption based on road conditions and driving style. This approach not only improves the operational efficiency of the vehicle, but also contributes to a reduced environmental impact through a more rational use of resources.

Artificial intelligence has become an essential component in the management and optimization of mechatronic systems. In the context of Industry 4.0, the focus is on automation and massive data collection, and AI plays a crucial role in analysing and capitalizing on this data [1]. As the industry evolves towards Industry 5.0, AI will facilitate the transition from technology-focused manufacturing to one that also integrates social and sustainability aspects [2]. In this sense, AI offers three major opportunities for the automotive industry: collaborative intelligence, self-learning, and collective intelligence. They allow vehicles and mechatronic systems to learn and adapt in real time, collaborating with both other vehicles and human operators. For example, an AI-equipped vehicle can learn from data collected from other vehicles in traffic to improve performance and safety.

With all the benefits that AI brings to the management of mechatronic systems, there are also significant challenges, especially in terms of data privacy and sustainability. In an era where smart vehicles collect huge amounts of data, including sensitive data on user behaviour, protecting this information becomes essential. Legislations such as the GDPR impose strict standards on the collection, storage, and use of personal data, forcing manufacturers to implement AI solutions that comply with these regulations.

At the same time, sustainability is becoming an imperative requirement for the automotive industry. AI can significantly contribute to reducing negative environmental impacts by optimizing resource consumption, but this requires careful management of how AI is implemented. Some systems can become too complex and energy-intensive, thus increasing costs and the ecological footprint. In this regard, the design of AI systems must be well adapted to the specific industrial context, avoiding the unnecessary use of advanced algorithms in simple scenarios and vice versa.

With the transition to Industry 5.0, the automotive industry is moving towards a human-centric and society-centric production model [3]. This means that AI will no longer only be used to optimize technological efficiency, but also to create a more collaborative and safer work environment for employees. Collaborative robots, equipped with AI, will work side by side with humans, performing repetitive and laborious tasks, while human operators will focus on strategic and personalized decisions.

Another central aspect of Industry 5.0 is sustainable development. This involves implementing circular processes that reduce waste and improve resource efficiency. In this context, AI can play a crucial role, facilitating the recycling and reuse of vehicle components and optimizing manufacturing processes to reduce energy consumption and materials used. The integration of artificial intelligence into mechatronic systems in the automotive industry represents a huge opportunity to improve the performance, safety and sustainability of modern vehicles. However, the success of these implementations depends largely on how AI is managed and adapted to the real needs of industry and society. The challenges related to data privacy and the sustainability of AI implementations must be carefully addressed to ensure a balanced transition to Industry 5.0, where technology and human requirements coexist harmoniously. Therefore, effective AI management in mechatronic systems is essential not only to achieve technological performance goals, but also to ensure a sustainable and responsible future for the automotive industry. AI is not only an optimization tool, but also an essential component of a more resilient, human-oriented and environmentally friendly industrial system.

2 Integrating Artificial Intelligence into Mechatronic Systems: Opportunities and Challenges in the Automotive Industry

Technology is advancing at an accelerated pace, and AI is playing a crucial role in revolutionizing the automotive industry. By integrating AI into vehicle mechatronic systems, new possibilities for innovation have opened up, especially in autonomous vehicles and other advanced mobility solutions. However, despite promising opportunities, implementing AI comes with important challenges, including data privacy, sustainability, and effective management of technology. This chapter explores the impact of AI on mechatronic systems, providing an overview of the opportunities and challenges facing the automotive industry.

One of the most obvious and revolutionary applications of AI in the automotive industry is the development of autonomous vehicles [4]. These vehicles use sophisticated AI algorithms to process real-time data collected from sensors, cameras, radars, and GPS. This information is used to make autonomous decisions, such as avoiding obstacles, adjusting speed, and optimizing routes. AI significantly improves road safety by reducing the risk of accidents, anticipating hazards, and reacting quickly to changes in the environment. A study by Waymo revealed that autonomous vehicles powered by AI reduce accidents by 94%, compared to human-driven vehicles [5]. Companies in the automotive sector are investing heavily in the development of AI technologies to improve the performance of autonomous vehicles. For example, AI can optimize energy consumption based on road conditions or adapt driving style to user preferences. Tesla's Full Self-Driving system further highlights AI's potential to optimize energy consumption by tailoring vehicle performance to environmental conditions [6]. Moreover, these vehicles are becoming more efficient in navigating complex environments such as busy urban traffic, which improves both the user experience and the efficiency of transport [7].

AI is also widely used in manufacturing and assembly processes in the automotive industry. Mechatronic systems in factories are optimized with the help of AI to monitor and improve production lines, minimizing downtime and manufacturing defects. For example, General Motors implemented AI-driven predictive maintenance, reducing production downtime by 10% and improving equipment efficiency [8].

AI-equipped robots can learn from errors and continuously improve production processes, contributing to better resource management and reducing industrial waste. For example, AI can be used to analyse data from machines and equipment in real-time, detecting anomalies or malfunctions before they become major problems. This type of predictive maintenance reduces operating costs and extends the life of equipment, providing a competitive advantage to companies that adopt these technologies.

With the increasing use of AI in vehicles and factories, data privacy and security are becoming a major challenge. Autonomous vehicles and connected systems collect massive amounts of data, including sensitive information about users, locations, and driving behaviours. This data is essential for the optimal functioning of AI, but protecting it is essential to prevent security breaches and abuse. Additionally, businesses must comply with strict privacy regulations such as the GDPR in the European Union, making it necessary to implement robust cybersecurity measures. Non-compliance has led to significant fines, as demonstrated by a major EU-based automaker fined €50 million for improper handling of personal data in 2022 [9]. The lack of clear protocols for data protection can lead to a decrease in consumer confidence in autonomous vehicles and other advanced automotive technologies.

While AI offers important opportunities to optimize energy efficiency and reduce environmental impact, there is also a downside. Studies have shown that training large neural networks generates as much carbon emissions as five average cars over their lifetime [10]. Training and running complex AI models requires significant computing power, which translates into high energy consumption and carbon emissions [11]. Deep learning algorithms are particularly resource-intensive, requiring large and energy-intensive data centers for processing. This dual relationship between AI and sustainability has raised concerns about AI's contribution to climate change.

While AI can help combat the climate crisis by optimizing industrial processes and monitoring the environment, its cost in terms of carbon footprint cannot be ignored. Thus, a major focus is on developing more energy-efficient AI models that can deliver performance without an excessive environmental footprint. For instance, NVIDIA's development of energy-efficient AI chips has reduced energy consumption in training by 40% [12].

Another major challenge in integrating AI into mechatronic systems is the complexity of the implementation process. While AI offers clear benefits, there are risks to overusing or inappropriately using these technologies. In some cases, AI systems are oversized for simple tasks, leading to unnecessary resource consumption and increased costs. In other cases, AI can be underutilized in complex scenarios, where existing solutions are not advanced enough to deliver reliable results. This discrepancy between AI capabilities and application requirements can compromise both product quality and operational efficiency. For example, a complex AI algorithm implemented in a simple production scenario can increase complexity without bringing real benefits. Conversely, using inappropriate AI solutions in complex scenarios can lead to costly errors and even production losses.

As the automotive industry evolves towards the vision of Industry 5.0, the integration of AI into mechatronic systems will continue to become increasingly sophisticated. Human-centric technologies such as collaborative robots and autonomous vehicles will become more common, and AI will play a crucial role in ensuring the safety, efficiency, and sustainability of these systems. At the same time, continuous investments will be needed in the development of AI solutions that are both high-performance and environmentally friendly. In the future, AI will need to be integrated in a way that balances technological efficiency with sustainability and privacy requirements, helping to create a greener and more responsible automotive industry. The integration of AI into mechatronic systems in the automotive industry offers tremendous opportunities for innovation, safety, and efficiency. However, there are major challenges related to data privacy, sustainability, and the complexity of implementation. Managing these challenges will be key to ensuring a sustainable and secure future for the automotive industry, where AI not only improves technological performance, but also meets social and environmental requirements.

3 Data Privacy and Sustainability in Automated Processes: The Role of Intelligent AI Management

In a world where automated processes and AI technology are becoming an integral part of automotive mechatronic systems, the challenges of data privacy and sustainability are more acute than ever. AI brings multiple benefits in the management and optimization of vehicles and production processes, but at the same time, it raises essential issues related to the security of personal information and the ecological impact of the technologies used. Intelligent AI management can solve these challenges by providing innovative solutions that balance technological requirements with the needs of data protection and reduction of the ecological footprint.

The collection and use of data in AI mechatronic systems is essential for modern vehicles, whether we are talking about driver assistance systems, autonomous vehicles or automated manufacturing processes [13]. These systems depend on the collection and analysis of huge volumes of data, which are used to make real-time decisions, optimize vehicle performance, and improve the efficiency of production lines. For example, AI can analyze data from sensors to detect mechanical malfunctions or adjust vehicle settings based on road conditions. However, this reliance on data opens the door to significant privacy risks. The data collected from vehicles and users may include sensitive information such as location, driving behavior, personal preferences, or even biometric information. Managing this data in a responsible and ethical way is essential to protect users' rights and prevent abuse.

As AI becomes increasingly integrated into the automotive industry, the risk of cybersecurity breaches also increases. Connected and autonomous vehicles are vulnerable to cyberattacks, and hackers could exploit these vulnerabilities to gain access to users' personal data or even take control of vehicles. Toyota faced a major cybersecurity breach in 2020, where data from 3.1 million connected vehicles were compromised, exposing how vulnerable these systems can be [14]. To mitigate such risks, intelligent AI management must prioritize strong encryption and anonymization techniques, alongside transparent data handling practices. Also, the collection of data without the explicit consent of users can lead to serious violations of data protection legislation. In this context, intelligent AI management must include rigorous cybersecurity and data protection measures. It is essential to implement advanced encryption and data anonymization technologies, as well as audit and control procedures that ensure compliance with international regulations. Users must also be clearly informed about how their data is collected and used, giving them the opportunity to manage their privacy preferences.

In addition to privacy challenges, the use of AI in automated processes also raises sustainability concerns. While AI can help optimize resource consumption and reduce waste in manufacturing processes, it requires significant amounts of energy to operate. Advanced AI models, especially those based on deep learning, require massive computing power, which leads to high energy consumption and, implicitly, significant carbon emissions [15]. Training AI algorithms often involves the use of large, energy-intensive data centers, which contribute to increasing the carbon footprint. AI's energy consumption can be significant, with data centers supporting AI systems consuming as much as 1% of the world's energy [16]. Moreover, AI-driven manufacturing processes, though more efficient, often rely on energy-intensive computing resources. A recently study revealed that training a state-of-the-art AI model could produce 284 tons of CO_2, equivalent to the lifetime emissions of five cars [11].

However, AI also has the potential to become a powerful tool for promoting sustainability. In the automotive industry, AI can help develop more energy-efficient vehicles that reduce fuel consumption and CO_2 emissions. For example, AI algorithms can optimize energy management in electric vehicles, extending the life of batteries and reducing the need for frequent recharging. AI-powered systems in electric vehicles, like Tesla's energy optimization algorithms, have improved battery efficiency by 25%, extending driving range while reducing environmental impact [17]. Moreover, AI can be used to monitor and manage manufacturing processes in a sustainable way, reducing waste and the use of natural resources. In vehicle manufacturing, AI can analyze data from assembly lines to identify opportunities for optimization, minimizing material and resource losses. Additionally, AI-driven predictive maintenance in automotive plants can extend the lifespan of machinery by 20%, reducing the need for resource-intensive manufacturing of replacement parts [18].

To minimize the negative impact of AI on the environment, companies must adopt intelligent AI management that focuses on developing energy-efficient solutions. One solution is to use machine learning algorithms that consume less power. In many cases, simpler algorithms can provide results similar to those of deep learning models, but with significantly less resource consumption. Also, deploying renewable energy sources to power data centers and the necessary AI infrastructure can help reduce the carbon footprint. At the same time, companies can invest in research and development to create more efficient AI models that require fewer resources to train and run.

Intelligent AI management plays a crucial role in balancing privacy and sustainability challenges [19]. In this regard, companies in the automotive industry must develop strategies that ensure both the protection of user data and the reduction of environmental impact. A first step is to implement transparent data collection and use policies that comply with international regulations and give users control over their personal information.

In terms of sustainability, intelligent AI management must include measures to optimize energy consumption and reduce carbon emissions associated with the operation of AI technologies. This can be achieved by choosing more efficient AI models and investing in green computing technologies. Additionally, companies can develop initiatives to educate and raise awareness among users about the impact of AI on the environment and the importance of data privacy [20]. Through proactive and transparent management, the automotive industry can reap the benefits of AI without compromising data security or process sustainability. Data privacy and sustainability are key challenges in integrating AI into automated processes in the automotive industry. Intelligent AI management can solve these challenges by adopting rigorous data protection measures and developing energy-efficient solutions. In this way, AI can contribute to the development of safer, more efficient and environmentally friendly vehicles, balancing technological needs with social and ecological requirements.

4 Conclusions

The integration of artificial intelligence into mechatronic systems in the automotive industry represents a technological revolution that brings multiple benefits, but also significant challenges. As the industry evolves towards automation and digital optimization, AI is becoming indispensable in managing processes, improving vehicle performance, and creating safer and more efficient systems. However, the success of implementing AI in this context depends on how two critical issues are addressed: data privacy and sustainability.

The challenges of data privacy have become a central concern with the increase in the number of smart and connected vehicles. AI collects and processes huge volumes of sensitive data, from users' location to their driving behaviour. This massive data collection opens the way to security risks, making it necessary to implement rigorous measures to protect and secure personal information. At the same time, international regulations such as GDPR impose strict standards that companies must adhere to in order to ensure transparency and give users control over their data.

Sustainability is another major challenge in the process of automation with the help of AI. On the one hand, AI can help reduce environmental impact by optimizing resource consumption and energy efficiency, both in factories and in vehicle use. For example, AI can improve energy management in electric vehicles, reducing battery consumption and maximizing efficiency. On the other hand, training and running complex AI algorithms requires intensive computing resources, which leads to high energy consumption and, implicitly, high carbon emissions.

Intelligent AI management in mechatronic systems in the automotive industry must address these two challenges simultaneously. In terms of privacy, a robust security infrastructure is required, including data encryption, anonymization, and clear procedures for collecting and using information. Transparency towards users and compliance with current regulations are essential to maintain trust in new technologies. To reduce their environmental impact, companies need to adopt energy-efficient solutions, explore less intensive AI models, and use renewable energy sources to power the necessary infrastructure. The development of "green" AI technologies will not only help improve sustainability, but will also bring long-term economic benefits by reducing the costs associated with resource consumption.

In conclusion, artificial intelligence has a huge potential to transform the automotive industry, bringing innovative solutions and optimizing mechatronic processes. However, success depends on smart and balanced management that ensures the protection of user data and minimizes the impact on the environment. As the industry continues to evolve, companies must adopt proactive strategies to meet these challenges, thereby ensuring the sustainable and ethical development of AI technologies.

References

1. Lee J, Davari H, Singh J, Pandhare V (2018) Industrial artificial intelligence for industry 4.0-based manufacturing systems. Manuf Lett 18:20–23
2. Leng J, Zhong Y, Lin Z, Xu K, Mourtzis D, Zhou X, Zheng P, Liu Q, Zhao JL, Shen W (2023) Towards resilience in industry 5.0: a decentralized autonomous manufacturing paradigm. J Manuf Syst 71:95–114
3. Leng J, Sha W, Wang B, Zheng P, Zhuang C, Liu Q, Wuest T, Mourtzis D, Wang L (2022) Industry 5.0: prospect and retrospect. J Manuf Syst 65:279–295
4. Nahavandi D, Alizadehsani R, Khosravi A, Acharya UR (2022) Application of artificial intelligence in wearable devices: opportunities and challenges. Comput Methods Programs Biomed 213:106541
5. Di Lillo L, Gode T, Zhou X, Atzei M, Chen R, Victor T (2024) Comparative safety performance of autonomous- and human drivers: a real-world case study of the waymo driver. Heliyon 10(14):e34379
6. Constantinides P, Monteiro E, Mathiassen L (2024) Human-AI joint task performance: learning from uncertainty in autonomous driving systems. Inf Organ 34(2):100502
7. Shaterabadi M, Jirdehi MA, Mehrjerdi H, Karimi H (2024) Chapter 5—artificial intelligence for autonomous vehicles: comprehensive outlook. In: handbook of power electronics in autonomous and electric vehicles. Academic Press, pp 59–79
8. Johnson R, Smith P (2022) General motors predictive maintenance data. Automot Eng J 18:78–90
9. Zhao Y, Li Z, Lv S (2024) Enhancing AI system privacy: an automatic tool for achieving GDPR compliance in NoSQL databases. Comput, Mater Contin 80(1):217–234
10. Strubell E, Ganesh A, McCallum A (2019) Energy and policy considerations for deep learning in NLP. J Artif Intell Res 55:123–145
11. Gaur L, Afaq A, Arora G, Khan N (2023) Artificial intelligence for carbon emissions using system of systems theory. Eco Inform 76:102165
12. Zhang Y, Li J, Sun W, Zhang J, Peng Y, Chen Z, Zheng X, Peng L (2023) Intelligent patrol terminal of transmission line based on AI chip accelerated calculation. Energy Rep 9:190–198
13. Nahavandi D, Alizadehsani R, Khosravi A, Acharya UR (2024) Chapter 5—the role of automotive artificial intelligence. The real case for driverless mobility, Elsevier, pp 111–130
14. Rahman MH, Wuest T, Shafae M (2023) Manufacturing cybersecurity threat attributes and countermeasures: review, meta-taxonomy, and use cases of cyberattack taxonomies. J Manuf Syst 68:196–208
15. Schoch A, Refflinghaus R, Zivkovic P (2023) Deep Learning based predictive testing strategy in the automotive industry. Procedia CIRP 118:1108–1113
16. Alijoyo FA (2024) AI-powered deep learning for sustainable industry 4.0 and internet of things: enhancing energy management in smart buildings. Alex Eng J 104:409–422
17. Gu J, Wu Z, Song Y, Nicolescu A-C (2024) A win-win relationship? New evidence on artificial intelligence and new energy vehicles. Energy Econ 134:107613
18. Mallioris P, Aivazidou E, Bechtsis D (2024) Predictive maintenance in Industry 4.0: a systematic multi-sector mapping. CIRP J Manuf Sci Technol 50:80–103
19. Bâlc S (2018) Particulars of educational management in the context of global changes, In: Proceedings of the 7th international rais conference on social sciences. Tampa, Florida, USA, pp 203–207
20. Bâlc S (2018) Evolution of educational management and its impact on society, In: Proceedings of the 8th international rais conference on social sciences. Washington, DC, USA, pp 110–115

Blockchain and the Environment: Impacts Across Key Industries

Gokcen Bas

Abstract The blockchain technology has sparked innovations across various sectors, including finance, supply chain, mining, fashion industry, transportation, and industrial manufacturing systems. The literature along with the real-life blockchain applications showcase the potential for improving transparency, efficiency, traceability and security of operations. This study provides an overview of the use cases from various sectors that provide valuable tools for improving sustainability and positive impact on environmental management due to reductions in waste, reduction in carbon emission, improved supply chain traceability, and transparency enabling ethical sourcing. The study further focuses on the manufacturing industry as specific with integrated use of Internet of Things (IoT) together with the Blockchain technology to assess the future of smart factories. The environmental impact of blockchain applications, though minimised by the use of renewable energy resources and various mining algorithms, cannot be ignored and are assessed to be as the main indicator of the future environmental integrity of industrial applications.

Keywords Environmental impact · Sustainability · Blockchain · Manufacturing industry · Internet of things (IoT)

1 Introduction

The Blockchain technology offers several opportunities for the strategies and practices of the modern manufacturing industry. While offering innovative solutions for transparency, efficiency, security, and traceability, the application of blockchain technologies in the multiple sectors, including finance, supply chain, manufacturing and energy require a careful assessment of environmental impacts and ecological footprint.

This study starts with the definition of the Blockchain technology. Further it provides an overview of the applications and studies covering the environmental,

G. Bas (✉)
CypherX, Larnaca, Cyprus
e-mail: gb@cypherx.tech

N. M. Durakbasa and K. G. Gülen (eds.), *Sustainable Green Conversion*, Lecture Notes in Mechanical Engineering, https://doi.org/10.1007/978-3-031-83583-4_13

sustainability and innovation aspects. The global use cases are summarized. The study focuses on the manufacturing industry as specific with integrated use of Internet of Things (IoT) together with the Blockchain technology as the next step. The European Union incentives are introduced as the final part of the study to provide a policy and regulatory roadmap for further developments.

2 The Blockchain Technology

A blockchain functions as a distributed and immutable ledger, where data is recorded in blocks linked together cryptographically. In the context of interconnected devices, each device can act as a node in the network, contributing to the validation and addition of new blocks to the chain. This decentralized approach ensures data integrity and transparency, as any attempt to alter information would require consensus from the majority of the network. The blockchain's inherent security and traceability make it ideal for managing and securing data generated by a multitude of devices in a trustless environment.

The Fig. 1 depicts a simplified blockchain structure. Each block contains a unique hash and the hash of the previous block, forming a chain where any alteration to a single block would invalidate all subsequent blocks, thus ensuring data integrity. The "Genesis Block" is the first block in the chain and has no previous hash.

The Fig. 2 illustrates a decentralized network of nodes (X, Y, Z, W) where each node maintains a copy of the ledger. Consensus, represented in the center, is the process these nodes use to agree on the validity of transactions and updates to the ledger, ensuring everyone has the same tamper-proof record.

3 The Environmental Impact of Blockchain Applications in Modern Industries

The integration of blockchain technologies in the modern industries offer several opportunities such as security and fraud protection, improving product, process and quality information transparency and auditability, administrative efficiency and automation. The strategy and practice of innovative solutions to increase transparency, efficiency, and security, may provide operational benefits as well as mitigating the environmental impacts.

The literature review identified various projects to adopt the blockchain technology as an alternative to improve their operations. However, adopting these practices, it's necessary to assess the environmental impact with further consideration of computational power, leading to higher energy consumption along with energy source carbon footprint. As the blockchain technology evolves, it's essential to establish an approach of balancing between innovation and environmental sustainability.

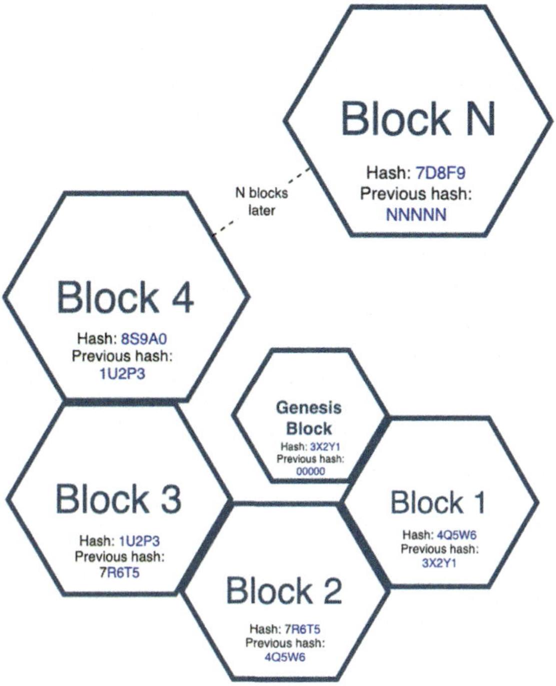

Fig. 1 A simplified blockchain structure

Table 1 provides an review of some studies on the blockchain technology applications applied in various industries.

4 The Internet of Things (IoT) in the Manufacturing Industry

IoT in the manufacturing industry provides benefits through the use of sensors, cameras, and smart equipment that optimize the operation and the maintenance, by communicating with each other, sharing data, automating repetitive tasks, and making decisions based on artificial intelligence and machine learning programming. The digital transformation with IoT integration in manufacturing facilitates predictive maintenance, smart operations, energy optimization and efficiency increase by minimizing downtime [15, 16].

Blockchain applications in the IoT enabled manufacturing systems offer unique features in traceability, efficiency, security. The convergence of blockchain and IoT is considered promising for creating transparent, secure, and decentralized production systems [17, 18]. Some use cases of the Blockchain technologies' potential applications in manufacturing industry are as below:

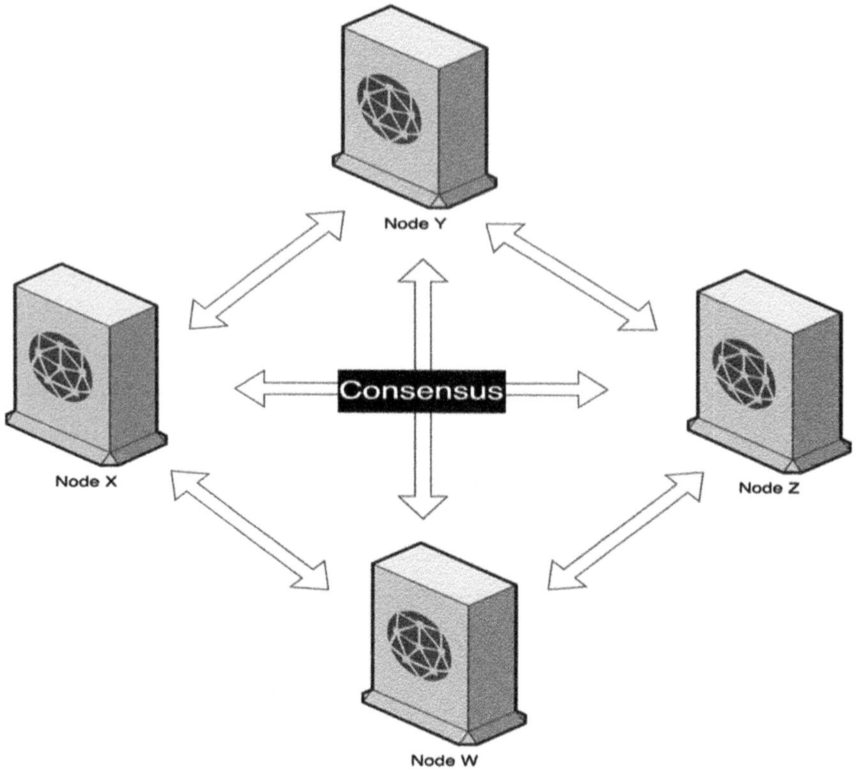

Fig. 2 A decentralized network of nodes (X, Y, Z, W) where each node maintains a copy of the ledger

- The supply chain traceability of raw materials, parts, and end products allows for accountability and sustainable operation practices [19]. The real time tracking e.g., temperature monitoring, or other crucial information ensuring the goods are handled appropriately reduces the waste in the supply chain [20].
- The IoT devices can securely exchange data without a central unit, which reduces the risk of cyber-attacks and unauthorized access to critical manufacturing data increasing security [21].
- IoT integrated devices can operate autonomously based on the blockchain based smart contracts, which enables the system to work smart [22]. The smart manufacturing systems, which is called smart factory operations, enhance operational sustainability and efficiency by enabling waste reduction and resource use optimization through automatic reordering of materials made by the production line machine when the inventory reaches a threshold [23].

Blockchain applications in the IoT integrated manufacturing industry provide valuable tools for mitigating the environmental impacts. Further advances coupled

Table 1 Overview of the blockchain application studies in various industries

Source	Overview of the industrial studies		
	Business	Description	Environmental impact
[1]	Supply chain	Coffee supply chain traceability and transparency study	Environmental sustainability: no deforestation, safe handling and storage of chemicals, water treatment system, renewable energy e.g., solar panels
[2]	Supply chain	Assessment of blockchain technology and smart contracts with potential application to sustainable supply chain management	Environmental dimension of sustainability can be used in blockchain integrated supply chain applications, further enabling the UN's sustainable development goals by ensuring traceability
[3]	Supply chain	Developing a model to monitor supply chain performance and optimize both emission levels and operational costs in a synchronised way	Carbon emissions can be reported, and compliance to carbon taxation policy can be ensured, representing a significant potential to reduce emissions, as well as the costs due to the emissions
[4]	Supply chain	Explorative qualitative research to lead to sustainable social and environmental impacts	Carbon footprints can become procurement conditions, incentivising greener practices, sustainable consumption
[5]	Supply chain	Performance assessment of blockchain in supply chain management on the domains of environment, economy, customer and information	The environmental performance of the corporate is as follow: 1-green image/marketing designing more sustainable products, 2-corporate responsibility including renewable energy grids, sustainable food production and e-waste monitoring, 3-waste/emission reduction by applying low-carbon product design and production, 4-product life cycle to track the product environmental effects, 5-nonrenewable resource reduction, 6-green packaging, 7-green supply chain
[6]	Fashion industry	Investigation of operational transparency and environmental efforts in the fashion industry	The integration of information disclosure and traceability plays a significant role in consumers' demand, affecting the environmental impacts in both manufacturing and supply chain of the fashion products

(continued)

Table 1 (continued)

Source	Overview of the industrial studies		
	Business	Description	Environmental impact
[7]	Mining	Evaluation of the recent advances in the mining industry contributing to transparency and sustainability	It's possible to assure ethical mineral sourcing, accountability for every participant in the supply chain leading to all environmental obligations of auditing during the operations and after the mine closure to enable reforestation, soil remediation, and water management
[8]	Sustainable development	Digital economy structuring for sustainable development	Efficiency, transparency, and traceability throughout the supply chain enable better resource management and waste management. Minimizing environmental footprints
[9]	Sustainable operations management	A proposal of a firm-regulator-consumer interaction framework to achieve sustainability	Incentives by the regulators motivate green technology selection, production of energy-saving products, and decision on emission reduction. Incentives by the customers motivate green product development/design, low-carbon effort, green product pricing, emission reduction strategy, environmental R&D decisions, green marketing and sustainability investment
[10]	Airport industry	Sustainability performance study for operations management with a focus on decision-making processes in supply-chain management	The flights' punctuality reduces the fuel consumption and negative environmental effects The flexible pre-departure planning reduces apron and taxiway congestion, reduces the ground movement and handling resources
[11]	Digital economy	Assessment the structure of the digital economy by integrating artificial neural network and blockchain technology	Structure improvement on adverse environmental effects, waste reduction, and environmental sustainability

(continued)

Table 1 (continued)

Source	Overview of the industrial studies		
	Business	Description	Environmental impact
[12]	Manufacturing	Identifying the potentials to enhance environmental development goals performance in manufacturing firms	Transformation to green practices in production and its supply chain with compliance of sustainable development goals will further environmentally friendly practices such as green purchasing, eco-design, sustainable material and energy use, green warehousing, sustainable transportation
[13]	Electric vehicles	Electric vehicle battery state-of-health estimation and anomaly detection	The proposed charging data collection strategy can provide an efficient and reliable data basis for tracking the battery health, battery aging, secondary use, and recycling of power batteries
[14]	Waste management	Assessment of innovation for management of recyclable plastic waste	The waste management can be optimized by monitoring waste volumes and contamination, improving the efficiency of recycling markets and reducing environmental impact

with initiatives, responsible resource use and policy making will have a positive effect as observed already by the use cases.

5 Sustainable Blockchain Technology Applications in the European Union

The European Commission has been promoting blockchain for sustainability focusing on fostering sustainable economic development, addressing climate change, and supporting the European Green New Deal, which aims towards becoming climate-neutral by 2050 [24, 25].

There are use cases set as example by the European Commission e.g., certification and validation processes, data sharing management, registration and data notarization processes, financial management and support, internal management process, authentication of self-sovereign digital ID services, tracking of goods and assets along the supply chain, payments and international transactions, governance and voting, procurement management, service integration, managing copyright and intellectual property rights, improving cybersecurity, predicting and planning, monitoring policy implementation [26]. The use cases implemented in Georgia, Malta, Sweden, Switzerland, Luxemburg, and the Netherlands resulted with quantitative and qualitative benefits such as cost savings, efficiency, increased accountability and incorruptibility, and environmental benefits such as reduced energy consumption [26].

The long-term sustainability of blockchain technology will be determined by their implementations, which align with the global sustainability goals, such as the United Nations' Sustainable Development Goals (SDG) [27].

6 Environment Impact of Blockchain Technology

The blockchain technology and cryptocurrencies have garnered significant attention due to its energy consumption and carbon footprint [28]. It's crucial to lay out the data, identify the causes and effects, and discuss potential regulatory and legal changes rather than forming a biased and non-scientific perspective. This study aims to address the environmental aspects and the broader implications for sustainability highlighting the innovative nature of the existing operations still transforming.

The energy consumption associated with cryptocurrency mining requires careful consideration. By harnessing the benefits of blockchain applications while minimizing their ecological footprint depends on the factors as listed below:

- The mechanisms of mining: Mining in cryptocurrencies involves validating and adding them to the blockchain.
  - Hash rate: The total computational power used in mining correlates with the energy consumption.
  - Network difficulty: As more miners join the network, the difficulty increases, correlating with higher energy consumption.
  - Hardware efficiency: Advances with application-specific integrated circuits (ASICS) provides efficiency, though the energy consumption remains high.
- The consensus mechanism: Proof of Stake (PoS) or Proof of Work (PoW) require different computational power, determining the expected energy consumption.
- The geographical distribution of mining operations
  - Location: Places abundant with clean energy sources such as water, wind, or sun, enable reduction of the carbon footprint.
  - Energy costs: High energy costs can lead to the alternative of self-sufficient power plant use.
  - Renewable energy sources: Integrating renewable energy sources into mining can mitigate the carbon footprint.
- The local regulations and incentives: Regions with favorable regulations and incentives for adopting cleaner sources, or allowing carbon credit purchase have an impact on the carbon footprint.

7 Conclusions

The future research is required to continue as the development of technologies such as artificial intelligence and blockchain is transforming. Implementation of these innovative technologies has challenges of scalability, high energy consumption. These challenges are being addressed and worked on as the technologies mature and the adoption increases. The implications of environment and social impacts need to be addressed contributing to more ethical, responsible and sustainable operations.

There are various industry implementations already providing data for decision making about the environmental impact of the blockchain applications. The adoption of innovative technologies presents both opportunities and challenges for environmental sustainability. It is essential to implement solutions improving the carbon footprints. Therefore, the focus for the future work shall be improving energy efficiency by various mining algorithms, application of specific integrated circuits, using renewable energy resources, and reducing the carbon emission to ensure the benefits of the innovative technologies that will be realized without compromising environmental sustainability.

References

1. Bager SL, Singh C, Persson UM (2022) Blockchain is not a silver bullet for agro-food supply chain sustainability: insights from a coffee case study. Curr Res Environ Sustain 4:100163. https://doi.org/10.1016/j.crsust.2022.100163
2. Saberi S, Kouhizadeh M, Sarkis J, Shen L (2019) Blockchain technology and its relationships to sustainable supply chain management. Int J Prod Res 57(7):2117–2135
3. Manupati VK, Schoenherr T, Ramkumar M, Wagner SM, Pabba SK, Singh IR (2019) A blockchain-based approach for a multi-echelon sustainable supply chain. Int J Prod Res 58(7):2222–2241. https://doi.org/10.1080/00207543.2019.1683248
4. Köhler S, Pizzol M, Sarkis J (2021) Unfinished paths—from blockchain to sustainability in supply chains. Front Blockchain 4:720347. https://doi.org/10.3389/fbloc.2021.720347
5. Hong L, Douglas NH (2021) Blockchain performance in supply chain management: application in blockchain integration companies. Ind Manag Data Syst 121(9):1969–1996. https://doi.org/10.1108/IMDS-10-2020-0598
6. Guo S, Sun X, Lam HKS (2023) Applications of blockchain technology in sustainable fashion supply chains: operational transparency and environmental efforts. IEEE Trans Eng Manage 70(4):1312–1328. https://doi.org/10.1109/TEM.2020.3034216
7. Onifade M, Adebisi JA, Zvarivadza T (2024) Recent advances in blockchain technology: prospects, applications and constraints in the minerals industry. Int J Min Reclam Environ 38(7):497–533. https://doi.org/10.1080/17480930.2024.2319453
8. Hong Z, Xiao K (2024) Digital economy structuring for sustainable development: the role of blockchain and artificial intelligence in improving supply chain and reducing negative environmental impacts. Sci Rep 14:3912. https://doi.org/10.1038/s41598-024-53760-3
9. Hong Z, Zhang H, Gong Y, Yu Y (2021) Towards a multi-party interaction framework: state-of-the-art review in sustainable operations management. Int J Prod Res 1–37. https://doi.org/10.1080/00207543.2021.18943
10. Di Vaio A, Varriale L (2020) Blockchain technology in supply chain management for sustainable performance: evidence from the airport industry. Int J Inf Manag 52:102014

11. Cai T, Hong Z (2024) Exploring the structure of the digital economy through blockchain technology and mitigating adverse environmental effects with the aid of artificial neural networks. Front Environ Sci 12:1315812. https://doi.org/10.3389/fenvs.2024.1315812
12. Jan A, Salameh AA, Rahman HU, Alasiri MM (2023) Can blockchain technologies enhance environmental sustainable development goals performance in manufacturing firms? Potential mediation of green supply chain management practices. Bus Strat Environ 33(3):2004–2019. https://doi.org/10.1002/bse.3579
13. Jin R, Wei B, Luo Y, Ren T, Wu R (2021) Blockchain-based data collection with efficient anomaly detection for estimating battery state-of-health. IEEE Sens J 21(12):13455–13465. https://doi.org/10.1109/jsen.2021.3066785
14. Bułkowska K, Zielińska M, Bułkowski M (2024) Blockchain-based management of recyclable plastic waste. Energies 17(12):2937. https://doi.org/10.3390/en17122937
15. Qin J, Liu Y, Grosvenor R (2016) A categorical framework of manufacturing for industry 4.0 and beyond. Procedia CIRP 52:173–178
16. Wan J, Tang S, Li D, Wang S, Liu C, Abbas H, Vasilakos AV (2016) A manufacturing big data solution for active preventive maintenance. IEEE Trans Industr Inf 13(4):2039–2047
17. Yaga D, Mell P, Roby N, Scarfone K (2019) Blockchain technology overview. National institute of standards and technology
18. Reyna A, Martín C, Chen J, Soler E, Díaz M (2018) On blockchain and its integration with IoT. Challenges and opportunities. Futur Gener Comput Syst 88:173–190
19. Kamble SS, Gunasekaran A, Dhone NC (2020) Industry 4.0 and lean manufacturing practices for sustainable organizational performance in Indian manufacturing companies. Int J Prod Res 58(5):1319–1337
20. Feng T (2017) A supply chain traceability system for food safety based on HACCP, blockchain and Internet of things. In: 2017 international conference on service systems and service management. Dalian, pp 1–6. https://doi.org/10.1109/ICSSSM.2017.7996119
21. Christidis K, Devetsikiotis M (2016) Blockchains and smart contracts for the internet of things. IEEE Access 4:2292–2303
22. Makhdoom I, Abolhasan M, Ni W, Anwar A, Jamalipour A (2019) Blockchain's adoption in IoT: the challenges, and a way forward. J Netw Comput Appl 125:251–279
23. Kang J, Yu R, Huang X, Maharjan S, Zhang Y, Hossain E (2019) Enabling localized peer-to-peer electricity trading among plug-in hybrid electric vehicles using consortium blockchains. IEEE Trans Ind Inform 13(6):3154–3164
24. The European commission website (2022). https://digital-strategy.ec.europa.eu/en/policies/blockchain-climate-action
25. The European commission website (2022). https://commission.europa.eu/strategy-and-policy/priorities-2019-2024/european-green-deal_en
26. Bosch JM, Tangi L, Burian P (2022) European landscape on the use of blockchain technology by the public sector. Publications office of the European union, Luxembourg, JRC131202. https://doi.org/10.2760/18007
27. The European commission website (2015). https://www.undp.org/sustainable-development-goals
28. Treiblmaier H (2023) A comprehensive research framework for bitcoin's energy use: fundamentals, economic rationale, and a pinch of thermodynamics. Blockchain Res Appl 4(3):100149. https://doi.org/10.1016/j.bcra.2023.100149

Challenges in STEM Teaching at Engineering Education

András Horváth and Gabriella Farkas

Abstract Science, technology, engineering, and mathematics (STEM) is an umbrella term used to group together the distinct but related technical disciplines of science, technology, engineering, and mathematics. STEM education aims to prepare students with the skills needed in the twenty-first century, particularly in areas that require significant technological and scientific advances. This form of education emphasizes problem-solving skills, critical thinking and practical applications. A questionnaire survey was conducted on STEM education at Óbuda University, Hungary, to explore the place of STEM education at the university. The focus was on exploring university instructors' knowledge and experience regarding STEM education, how open they are to further learning about different educational technologies, what results they have achieved using STEM-based teaching methods, and what tools they use for teaching STEM subjects. The student survey questions centred on student satisfaction with STEM education. The purpose of the survey was to determine how, if the respondents are satisfied with STEM teaching and learning, this type of educational practice can be further deepened in the future. The results indicated that both instructors and students are open to the multidisciplinary approach to STEM subjects. To achieve this, greater emphasis needs to be placed on the development of teaching methods for instructors, as well as the application of modern technological tools in education.

Keywords STEM · Engineering education · Complex approach · New trends in STEM teaching · Teacher professional development

A. Horváth (✉) · G. Farkas
Bánki Donát Faculty of Mechanical and Safety Engineering, Óbuda University, Budapest, Hungary
e-mail: horvath.andras@bgk.uni-obuda.hu

G. Farkas
e-mail: farkas.gabriella@bgk.uni-obuda.hu

N. M. Durakbasa and K. G. Gülen (eds.), *Sustainable Green Conversion*, Lecture Notes in Mechanical Engineering, https://doi.org/10.1007/978-3-031-83583-4_14

1 Introduction

Educating engineering students with a practical mindset is key to future technological development and innovation. One way of achieving this is through integrated teaching of Science, Technology, Engineering and Mathematics (STEM) subjects. Universities need to work out how to integrate it effectively so that students are best prepared to succeed in the labour market. In addition, improving the effectiveness of STEM education can contribute to increasing global competitiveness and economic growth.

Individuals must address the demands of modern society, which requires individuals with the ability to solve problems, produce and evaluate scientific evidence, work in teams and, above all, understand and act upon the world and the phenomena that make it up [1]. In this context, the STEM (science, technology, engineering, and mathematics) approach has emerged as a proposal to strengthen education in these skills [2]. Although this approach is increasingly popular, it is challenging to establish a unified concept [3, 4].

The following outlines the current characteristics of technical education, which has fundamentally changed based on industrial demands. It is crucial to assess the present situation within our higher education and technical institution. Understanding the opinions and preparedness levels of instructors, as well as the needs of students, is necessary to address the challenges that industry-based practical applications pose for university educators [2].

The traditional methods of education, which current instructors experienced, have been radically transformed with the advent of the information society. The 21st -century technological revolution has significantly altered the industry and, concurrently, the channels through which modern society acquires information. Previously, knowledge acquisition relied almost exclusively on textbooks and written materials, whereas today, gaining new knowledge increasingly depends on electronic sources. As the societal approach to information dissemination has evolved, the traditional book-based education must also adapt to continue effectively transmitting knowledge [5].

Instructors now face the challenge of conveying knowledge using a teaching method that has emerged in the past two decades, contrasting with the methodologies they originally learned. Another challenge is the rapid advancement of current industrial technology, such as the expanding use of artificial intelligence in the industry, which evolves almost daily [2, 6].

Historically, instructors were primarily responsible for knowledge acquisition, continuous monitoring, and transmission, forming the basis of the teaching profession. This responsibility has now expanded to include openness to interdisciplinarity, active collaboration with other scientific fields, and the ability to use and transmit knowledge through evolving IT tools and channels. Additionally, instructors must keep their IT skills up to date.

Current expectations require instructors to pass on interdisciplinary knowledge using new types of tools [7]. They must also meet students' needs and learning readiness. The effectiveness of knowledge transfer is measured by students' understanding and satisfaction, with assessment results and institutional surveys reflecting these aspects.

Institutions and students expect instructors to employ new educational tools that align with current standards. Instructors must integrate their scientific expertise with knowledge of new technological tools, often involving daily familiarity and proficient use of IT resources. As IT tools and systems continue to evolve and become more complex, instructors are required to engage in ongoing learning to stay proficient.

The quality of university education is dependent on organizational elements, students' competences, and the effort and commitment of academic teachers that are influenced by motivation, competences, and personal views about what good teaching practice is. Therefore, the professional development of teaching staff at university is important [8].

Our research aims to understand how instructors at Bánki Donát Faculty of Mechanical and Safety Engineering, Óbuda University, Hungary evaluate their own effectiveness in teaching through new educational tools and methods. This is important because the instructor embodies the expectations of the institution where they teach. The other goal is to get student outcomes and feedback regarding the STEM integrated teaching activities existed at our university.

2 Material and Methods

During the research, questionnaire surveys was conducted for both instructors and students. For the instructors, basic questions regarding the field of education, gender, position, and university teaching experience were asked at the beginning of the questionnaire to allow for more detailed, grouped analysis. The following questions (and answers) were posed to the instructors:

- QT01 Would you consider additional training or professional development to better incorporate current STEM trends into your teaching? (1. Not at all 2. Not too much 3. Maybe 4. Yes, probably 5. Yes 6. Yes, definitely)
- QT02 How would you rate your ability to follow and implement current trends in the STEM (e.g. ICT, team works, project based learning, e.t.c.) ? (1. Very poor ability 2. Less poor ability 3. Poor ability 4. Good ability 5. Very good ability 6. Excellent ability)
- QT03 Do you think the current availability of literature limits the quality of your STEM teaching? (1. Very limiting 2. Limiting 3. A little bit limiting 4. Does not limit much 5. Does not limit two much 6. Does not limit at all)
- QT04 How well do students achieve good results in combined STEM subjects? (1. Not at all, 2. Very little, 3. Little 4. Solid progress, 5. Sufficiently progress, 6. Very much 7. I do not combine STEM)

- QT05 Is the teaching more difficult for students when using STEM? (Based on student feedback) (1. Much simpler 2. Simpler 3. Demanding 4. More demanding 5. I can´t judge)
- QT06 Which pedagogical approaches do you use in your STEM teaching?
  A. Traditional teaching where the teacher gives information, and students learn from it. Traditional direct instruction (lessons are focused on the delivery of content by the teacher and the acquisition of content knowledge by the students).
  B. Teaching with experiments (experiments are used in the classroom to explain the subject matter)
  C. Project-/Problem-based approach (students are engaged in learning through the investigation of real-world challenges and problems)
  D. Inquiry-Based Science Education (students design and conduct their own scientific investigations)
  E. Collaborative learning (students are involved in joint intellectual efforts with their peers or with their teachers and peers)
  F. Formative assessment, including self-assessment (student learning is constantly monitored and ongoing feedback is provided; students are provided with opportunities to reflect on their own learning).

For the students, basic information was collected regarding their field of study, type of studies (undergraduate, graduate, PhD), the number of completed semesters, as well as their gender and age. The following questions (and possible answers) were asked of the students:

- QS01 How many subjects focused on a combination of math, problem solving, technology and science did you have in your studies so far? (none -some – most of them – all of them)
- QS02 In your education, do you often deal complex problems that are similar to problems in practice? (1. Not at all, 2. Mostly no, 3. Rather not, 4. Rather yes, 5. Mostly yes, 6. Yes, definitely)
- QS03 What portion of courses/subjects do you use digital technologies (PC, tablet, mobile phone) (1 to 10)
- QS04 I see myself in STEM a career (in the future) (1. I don't see myself in a STEM career 2. Maybe I see myself in a STEM career 3. I slightly plan to occasionally pursue a STEM career 4. I strongly plan to occasionally pursue a STEM career 5. I have already taken steps towards my STEM career 6. I am in a STEM career)
- QS05 I would appreciate more chances to learn STEM. (1. I am not interested to learn about STEM. 2. There are not opportunities 3. There are not enough opportunities 4. There are limited opportunities 5. There are opportunities 6. There are many opportunities 7. There are extremely many opportunities).

The questions were sent to the participants through an online form, with a two-week period allowed for completion. The questionnaires were distributed to both students and instructors at the Bánki Donát Faculty of Mechanical and Safety Engineering, Óbuda University. The faculty consists of 102 instructors and approximately 2,000 students. The goal was for about 50% of the instructors and about 10% of the students to complete the questionnaire to draw appropriate conclusions.

3 Results and Discussion

A. *Evaluation of the teachers' questionnaire*

A large proportion of the teachers (49 teachers, approx. 50%) who responded to the survey teach in engineering, technology and mathematics. Some of the respondents do not teach STEM subjects, but economics, quality assurance or safety science. There is a significant gender difference in the number of lecturers, with many respondents (71.40%) being male, which is in line with the proportion at Óbuda University. Most respondents teach as associate professors, but many assistant professors with PhD and without PhD also appear (Fig. 1). Within the teaching field, there is a significant number of experienced teachers, with no responses with less than one year of experience.

The QT01 and QT02 questions are dealing with the professional development and experience in STEM education. Some teachers train regularly, and some do not. Just over half of respondents can only keep up with current trends in STEM. Nearly 16% think they would not like to receive further training on the use of the STEM approach, 1/3 are undecided on this issue, the remaining respondents are open for development (Fig. 2). The responses to the question on following and implementing current STEM trends reveal a range of confidence levels. Most respondents rate their ability as "Good" (40.8%) or "Poor" (28.6%), suggesting a general comfort with current trends. However, only a small proportion rated their ability as "Very Good" (8.2%) or "Excellent" (6.1%), indicating that while many feel capable, few see themselves at the highest level of proficiency. Additionally, 16.3% of respondents rated their ability as very poor and less poor, highlighting a need for further support and development in this area (Fig. 3).

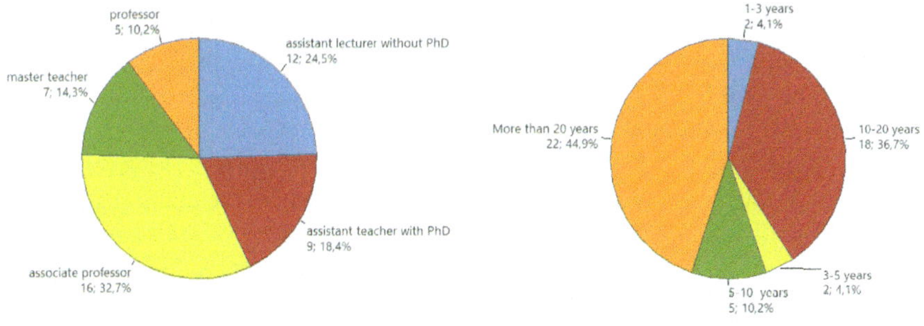

Fig. 1 Teacher respondents' answers to the question position and teaching experience at university

Fig. 2 QT01 results would you consider additional training or professional development to better incorporate current STEM trends into your teaching?

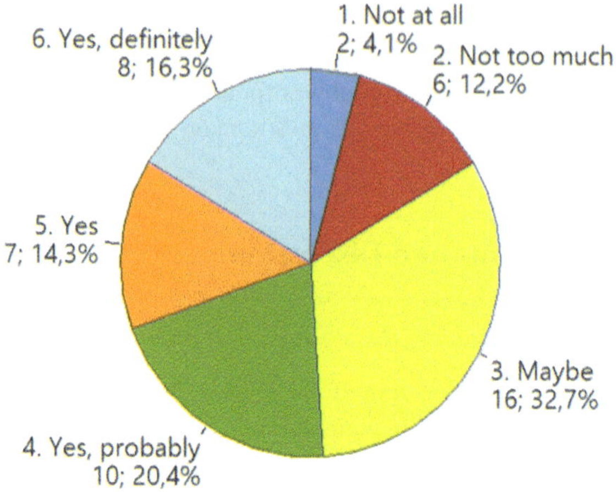

Fig. 3 QT02 results how would you rate your ability to follow and implement current trends in the STEM?

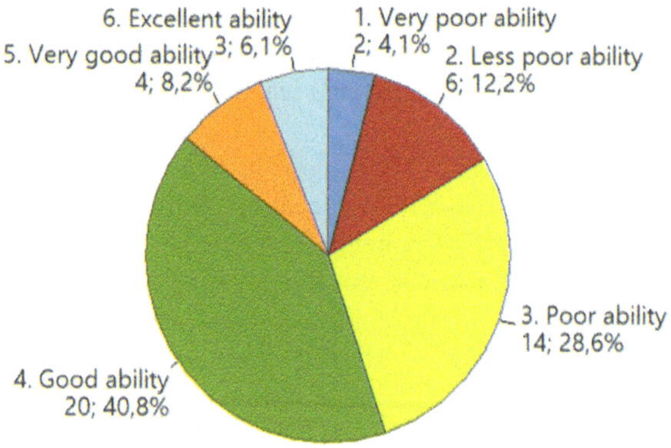

The third question is referred to the material support. Literature and other teaching materials are available according to half of the respondents. There are some teachers who believe that the possible lack of literature does not affect STEM education (Fig. 4).

The student achievement is the topic for QT04 and QT05. In combined STEM subjects, students perform better according to many respondents. Of those who were able to judge, around two-thirds think it is easier to teach with STEM education (Fig. 5). The responses regarding the difficulty of teaching students using STEM tools show diverse opinions. A majority find it "Simpler" (46.9%) or "Much simpler" (6.1%), suggesting that STEM tools often make teaching easier. However, 30.6% view it as "Demanding," indicating that for some, the process remains challenging. No respondents felt that it was "More demanding," and 16.3% could not judge the impact (Fig. 6). Overall, while most see STEM tools as a simplification of teaching, there is a notable portion who still find it challenging.

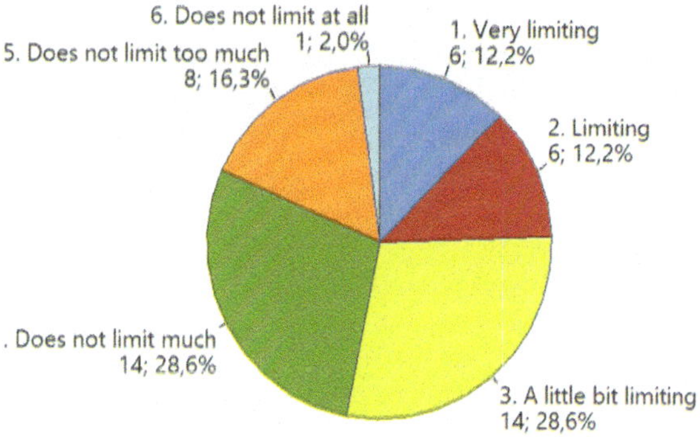

Fig. 4 QT03 results do you think the current availability of literature limits the quality of your STEM teaching?

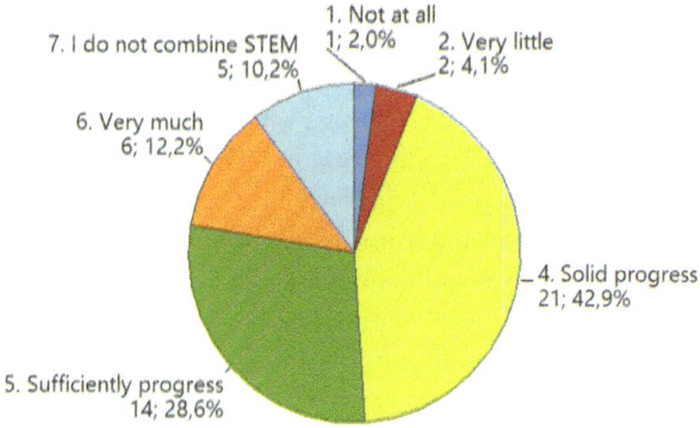

Fig. 5 QT04 results how well do students achieve good results in combined STEM subjects?

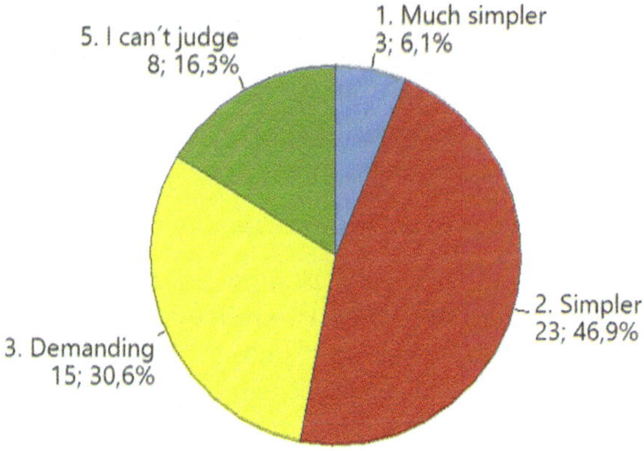

Fig. 6 QT05 results is the teaching more difficult for students when using STEM?

The last question (QT06) is related professional development and experience in STEM. The pedagogical approaches used in STEM teaching show a variety of methods. The most used approach is *Traditional teaching* (29 respondents), where the teacher delivers content and students absorb it. *Inquiry-Based Science Education* is also widely utilized (26 respondents), allowing students to design and conduct their own investigations. *Teaching with experiments* and *Collaborative learning* each have

Fig. 7 QT06 results which pedagogical approaches do you use in your STEM teaching?

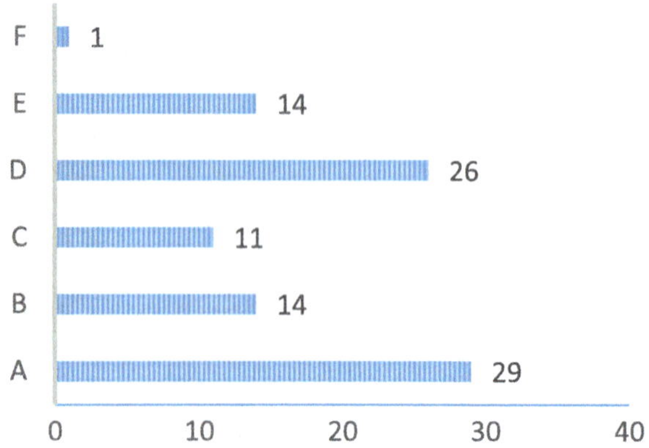

14 respondents, indicating that these methods are used to engage students through practical applications and peer interaction. The *Project-/Problem-based approach* is employed by 11 respondents, focusing on real-world challenges. *Formative assessment* is the least used, with only 1 respondent incorporating it, highlighting a potential area for growth in monitoring and reflecting on student learning (Fig. 7).

The responses to the questions did not show any differences by category (gender, position, educational experience).

The comments highlighted that some teachers face challenges in using STEM tools due to time constraints or software limitations. Nevertheless, positive student feedback suggests that these tools facilitate teaching and boost motivation. Targeted training for instructors would benefit the faculty. Additionally, more information on the system's elements and methodology is needed. Concerns were raised about the emphasis on paper-based, practice-oriented education in a curriculum increasingly focused on reducing fundamental subjects, such as mechanics and mathematics. There is a risk that students may become proficient in software usage but struggle to critically evaluate results unless the evaluation is left to the software itself.

B. *Evaluation of the students' questionnaire*

A total of 207 students from the Bánki Donát Faculty of Mechanical and Safety Engineering, Óbuda University completed the questionnaire. The majority are mechanical engineering students, but there are also students of mechatronics, security engineering and cyber engineering (Fig. 8). 93% of the respondents are BSc students and 90% are male. 1/3 of the respondents have completed 1–2 semesters and 1/3 have completed 3–4 semesters, which means that most respondents are in the middle of their bachelor's studies (Fig. 8). The age distribution is more likely to be under 23, but there are also students up to 36 years old.

The majority of students' subjects are in science, engineering and maths. The University offers engineering and problem-solving related subjects, although there are some students who say they do not (4.3%) (Fig. 9). The responses (QS02, Fig. 10) deal with complex problems similar to those in practice reveal a range of experiences. A combined total of 48.6% (7.2% "Not at all," 20.2% "Mostly no," and 21.2% "Rather

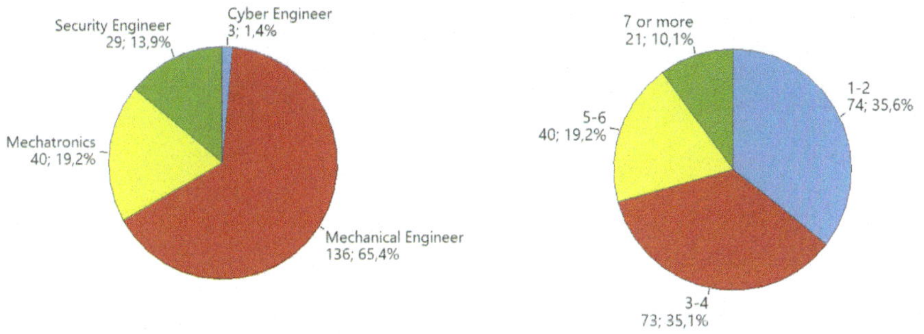

Fig. 8 Student respondents' answers to the question field of studies and finished number of semesters at university

not") report limited engagement with such problems. In contrast, 50.4% of respondents (30.3% "Rather yes," 16.8% "Mostly yes," and 4.3% "Yes, definitely") feel that their education does involve practical problems, though only a small percentage find this to be a definitive aspect of their learning. This suggests that while many encounters practical problems, there remains a significant portion who do not.

The third question is related to the technology used throughout STEM education. The use of digital technologies (PC, tablet, mobile phone) across courses/subjects varies widely (Fig. 11). The highest portion of respondents (37) reported using digital technologies in all their courses (10). The use decreases gradually, with 25 respondents using them in 7 courses, and 24 in 8 courses. Fewer respondents reported lower levels of usage, with the least number using them in only 1 course (6). Overall, there is a noticeable trend towards extensive use of digital technologies, especially in the majority of courses.

More than half of students prioritise STEM education in their further studies (Fig. 12). Only a small proportion of students (2.4%) do not choose a STEM field as their future career, with the majority envisaging some form of STEM career.

Fig. 9 QS01 results how many subjects focused on a combination of math, problem solving, technology and science did you have in your studies so far?

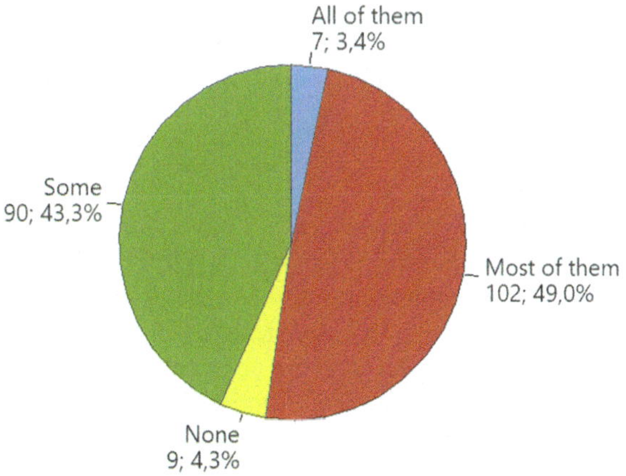

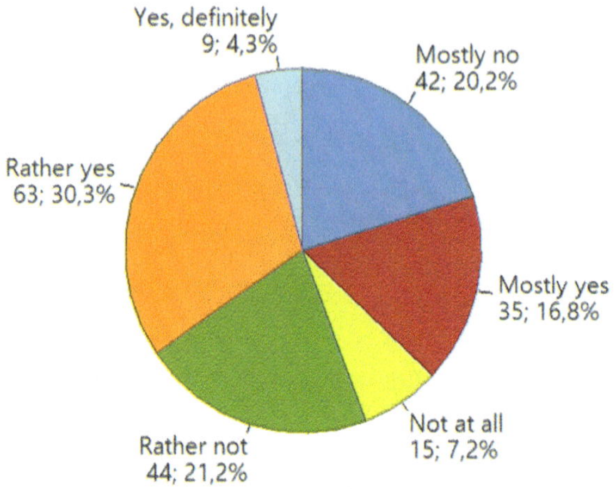

Fig. 10 QS02 results in your education, do you often deal complex problems that are similar to problems in practice?

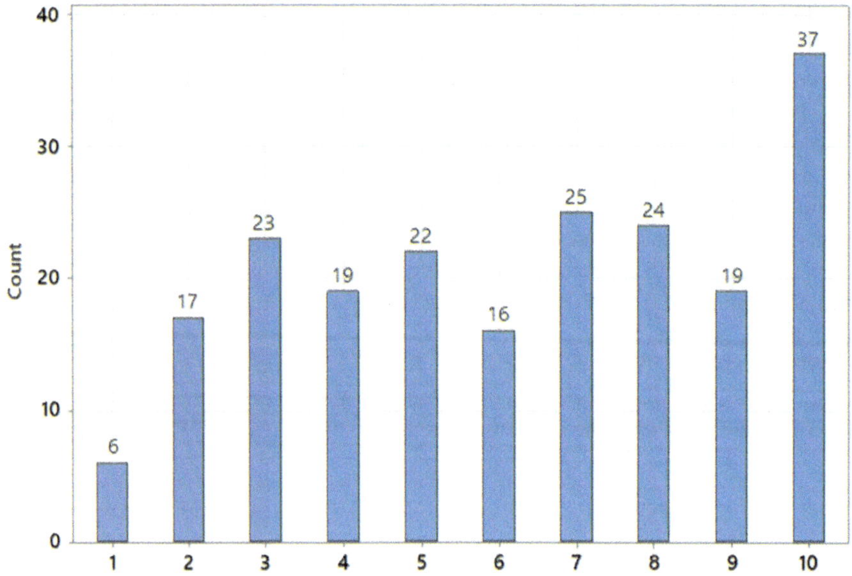

Fig. 11 QS03 results what portion of courses/subjects do you use digital technologies? (1–10)

Nearly 10% of students are not interested in STEM education at all, 1/3 say there are few opportunities to learn about STEM, and more than half of students perceive that they have opportunities to participate in STEM education. Students are open to learning innovative technologies and are able to think logically (Fig. 13).

Comments reveal that the university curriculum did not cover essential skills related to STEM education. Although students are enrolled in practical courses, most of their learning has occurred through presentations rather than firsthand experiences. This situation contradicts the university's promotion of its programs as being practical

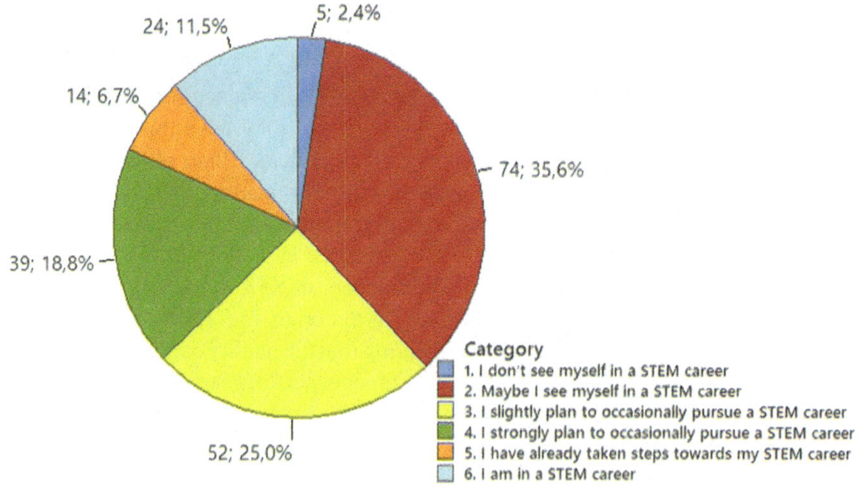

Fig. 12 QS04 results I see myself in STEM a career (in the future)

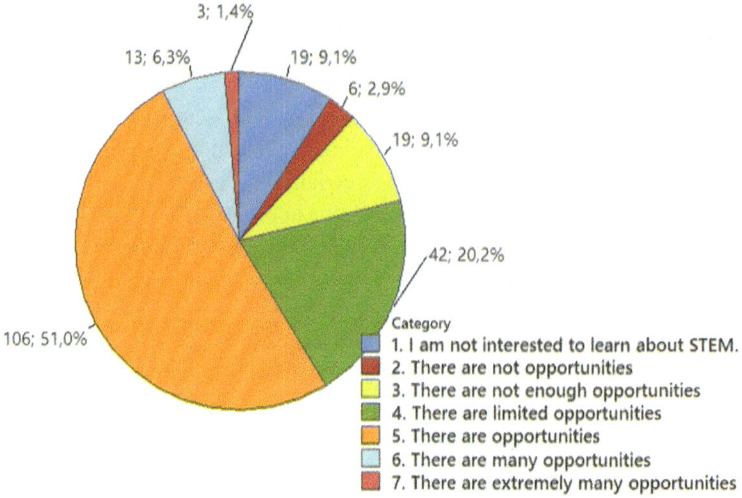

Fig. 13 QS05 results I would appreciate more chances to learn STEM

and application focused. Additionally, there is a strong demand for incorporating more visual tools and elements into teaching. This suggests that the current methods may not fully engage students or support their learning needs, highlighting a gap between the advertised and actual educational experience.

4 Conclusion

Most teacher who responded feel able to teach STEM (science, technology, engineering, mathematics) subjects, but few teachers feel they have the right level of knowledge about this new way of teaching. This indicates that, although there is general satisfaction with the effectiveness of current teaching, there is a need for further improvement to keep up to date with current teaching practice.

The results of the research also show that there are large differences between the willingness and perception of the skills of teachers in STEM education. The literature and general availability of other teaching materials in STEM education elicit different opinions among respondents, indicating that the quality of access to teaching materials varies, but that students in STEM subjects perform better according to the responses, suggesting the effectiveness of multidisciplinary approaches in developing problem solving and critical thinking. The use of STEM tools is seen by most respondents as a simplifying factor in teaching, highlighting the need for additional support and training.

Innovative methods such as project-based learning and research-based teaching are beneficial for students, but time and technological constraints make their effective use difficult. Positive feedback from students suggests that modern educational tools are motivating, but there is a risk that overuse of software can weaken students' critical thinking.

Students encounter practical problems during their learning, but these do not always become a dominant part of their studies. This suggests that practical application is not yet fully integrated into the curriculum, despite the importance that a large proportion of students would attach to this approach. However, the widespread use of digital technologies is becoming more widespread in most courses, which is a positive sign for the integration of technological tools, but to a varying extent across subjects. The implementation of modern technological tools (e.g. VR glasses) already available in engineering education is essential for the effective use of STEM.

According to the students' responses, the current teaching methods of the university do not fully meet the expectations of the practical and application-oriented nature of the knowledge taught. This suggests a need for continuous STEM training of teachers in engineering higher education. This could include the development of in-service training materials, practical training for trainers and the development of a system for sharing good practice.

Our research shows that a paradigm shift is needed in technical higher education to acquire the engineering knowledge required by industry. A STEM interdisciplinary approach is an effective way to enable engineering students to acquire knowledge, skills, attitudes and autonomous decision-making.

Acknowledgements We would like to express our thank to the International Visegrad Fund (IVF) for their generous support through the "Visegrad Grant+" program for our project, "Improving STEM Teaching Process using Digital Transformation - DigSTEM." Our heartfelt thanks go to the dedicated colleagues and participants who have contributed their time, expertise, and enthusiasm to this initiative.

References

1. García Y, González RD, Burgos Oviedo F (2017) Actividades STEM in la formación inicial de profesores: nuevos enfoques didácticos para los desafíos del diglo XXI [STEM activities in initial teacher training: new didactic approaches for the challenges of the 21st century]. Rev Electrónica Diálogos Educ Electronic Mag Educ Dialogues 18(2017):37–112
2. Rodríguez CMA, González-Reyes RA, Ballen AB, Merchán MAM, Barrera EAL (2024) Characterization of STEM teacher education programs for disciplinary integration: a systematic review. Eurasia J Math Sci Technol Educ 20(3):em2408
3. Breiner JM, Harkness SS, Johnson CC, Koehler CM (2012) What is STEM? A discussion about conceptions of STEM in education and partnerships. Sch Sci Math 112(1):3–11
4. Martín-Páez T, Aguilera D, Perales-Palacios FJ, Vílchez-González JM (2019) What are we talking about when we talk about STEM education? A review of literature. Sci Educ 103(4):799–822
5. Saat RM, Fadzil HM, Adli DSH, Awang K (2021) Stem teachers' professional development through scientist-teacher-students partnership (Stsp). J Pendidik IPA Indones 10(3):357–367
6. Guzmán-Duque A, Chalarca-Guzmán L (2024) Educational innovation in higher education with the use of management simulators for the development of competencies. IntechOpen. https://doi.org/10.5772/intechopen.112771
7. Brouwer N, Niemelä M, Maciejowska I, Grecea Ş, Tarallo O, Russo V (2024) Ambassadors of professional development in teaching and learning in STEM higher education. Chem Teach Int 6(3):201–216
8. Brouwer N, Maciejowska I, Lis A, Machado C, Grecea S, Kärkkäinen J, Niemelä M, Kranjc K, Podlipnik Č, Prashar S, Russo V (2020) The need for STEM continuous professional development at European universities. VIRT&L-COMM 21

Developing a Tool for HSE and Social Performance Assessment in Healthcare Facilities

Fatemeh Masoudian, Shiva Faeghi, and Kunibert Lennerts

Abstract Healthcare facilities play a critical role in providing care in certain level of quality and ensuring the safety and well-being of patients, staff, and the environment. However, the lack of standardized and comprehensive tools to assess and evaluate Health, Safety and Environment (HSE) practices and social performance hsecnders effective monitoring and improvement in these crucial areas. This paper aims to address this problem by proposing the development of a simple and straightforward tool specifically tailored to assess HSE and social performance in healthcare facilities. The proposed tool integrates various metrics and indicators to evaluate the HSE practices and social performance of healthcare facilities in the form of a checklist. It encompasses different components, including employee health, environmental management, facility maintenance, and infrastructure, legal and managerial compliance, patient safety, safety and social responsibility. The proposed tool enables healthcare facilities to conduct regular self-assessments, identify areas for improvement, and track progress over time. This tool provides possibility to benchmark HSE and social performance against industry standards and identify best practices. Furthermore, the tool facilitates data-driven decision-making, enabling healthcare facility administrators and stakeholders to effectively allocate resources and prioritize initiatives that have the greatest impact on HSE and social performance. The development and implementation of a standardized tool for assessing HSE and social performance not only contributes to improving the quality of care and patient outcomes, but also promotes a culture of accountability and continuous improvement within healthcare organisations.

F. Masoudian (✉) · S. Faeghi · K. Lennerts
Institute of Technology and Management in Construction, Karlsruhe Institute of Technology, Karlsruhe, Germany
e-mail: fatemeh.masoudian@kit.edu

S. Faeghi
e-mail: shiva.faeghinezhad@kit.edu

K. Lennerts
e-mail: kunibert.lennerts@kit.edu

N. M. Durakbasa and K. G. Gülen (eds.), *Sustainable Green Conversion*, Lecture Notes in Mechanical Engineering, https://doi.org/10.1007/978-3-031-83583-4_15

Keywords Assessment tool · Healthcare facilities · Occupational health and safety performance · Social performance

1 Introduction

Healthcare facilities play a critical role in society, by providing essential medical services and ensuring the well-being of patients, staff, and the surrounding community. As the demand for sustainability in healthcare increases, so does the need for comprehensive assessment tools to evaluate the Health, Safety, and Environment (HSE) practices and the social performance of these facilities. These assessments are crucial for identifying areas for improvement, ensuring compliance with regulations, and promoting a culture of continuous enhancement.

Traditionally, healthcare facilities have focused on patient outcomes and clinical performance measures, often overlooking the importance of HSE practices and broader social responsibilities. However, it is essential to recognize that the success of a healthcare facility is not solely determined by medical outcomes but also by the integration of effective HSE practices and social engagement.

Currently, the lack of standardized and comprehensive tools specifically tailored to assess HSE and social performance of healthcare facilities poses significant challenges. The lack of such tools hampers the ability of healthcare facilities, both hospitals and laboratories, to monitor and improve critical aspects related to the well-being and safety of patients, staff, and the environment.

This paper aims to address this gap by proposing the development of a specialized tool for HSE and social performance assessment in healthcare facilities. The tool will integrate various metrics and indicators to comprehensively assess the different dimensions of HSE practices and social engagement in healthcare settings. It will serve as a practical and user-friendly resource in healthcare facility administrators, enabling them to identify areas for improvement, track progress, and implement targeted interventions.

The proposed tool covers a range of key parameters, including employee health, environmental management, facility maintenance, and infrastructure, legal and managerial compliance, patient safety, safety and social responsibility. By taking a holistic approach to HSE and social performance assessment, as well as enabling an easy and rapid self-assessment, this tool will enable healthcare facilities to consider all aspects of their operations and their impact on various stakeholders.

Implementing a standardized tool for HSE and social performance assessment brings numerous benefits. Firstly, it will provide healthcare facilities with a clear understanding of their current performance levels, enabling them to compare themselves against industry benchmarks and identify areas that require immediate attention. Secondly, it will facilitate the development of targeted action plans and the allocation of resources to address identified gaps. Finally, it will foster a culture of accountability, and continuous improvement within healthcare facilities, ultimately leading to improved patient outcomes and a safer, more socially responsible healthcare environment.

The paper is structured as follows. Section 2 reviews globally recognized references and relevant literature on assessment tools, HSE, and social aspects. Section 3 describes the methodology used to develop the assessment tool. The components of the assessment tool and related items are described in Sect. 4 the and the verification and validation process are explained in Sect. 5.

2 Literature Review

As of the objective of the work to prepare an internationally accepted assessment tool, all globally recognized references, standards and guidelines published by International Standard Organization (ISO), World Bank and World Health Organization (WHO), United Nation guides and some other references reviewed.

Type of reviewed references is categorized as below: World Bank.

- o Environmental and Social Framework [6]
- o Environmental, health, and safety general guidelines [7]

  - World Health Organization

- o Safe management of waste from health-care activities (2nd edition) [8]
- o Patient safety assessment manual (2nd edition) [9]
- o Essential environmental health standards in health-cares [10]
- o WHO Best practices for injections and related procedures toolkit [15]
- o Medical equipment maintenance programme overview [11]

  - UN

- o The 2030 Agenda for Sustainable Development [18]

  - International Atomic Energy Agency

- o Radiation Protection and Safety of Radiation Sources: International Basic Safety Standards [13]

  - European Commission

- o Occupational health and safety risks in the healthcare sector [19]

  - ISO Standards

- o ISO 22320:2018, Security and resilience—Emergency management—Guidelines for incident management [25]
- o ISO 14001:2015, Environmental management systems—Requirements with guidance for use [21]

- o ISO 45001:2018, Occupational health and safety management systems—Requirements with guidance for use [20]
- o ISO 22301:2019, Security and resilience—Business continuity management systems—Requirements [24].

A. *Assessment tool*

Assessment tools play a vital role in measuring compliance across different domains. Developing a reliable and valid assessment tool requires a systematic approach that follows established guidelines and best practices [1]. The importance of assessment tools and their creation, relates to their validity, reliability, and practicality. In diverse range of studies, including education, psychology, healthcare, and business, robust psychometric properties, appropriate administration procedures, and alignment with the intended purpose of assessment is essential.

In the healthcare ecosystem, there is a lack of an efficient and comprehensive assessment tool to evaluate the HSE and social performance of a healthcare facility [2]. To design such an assessment tool, the target audience, purpose, and context of the assessment should be considered. Studies suggest that incorporating clear instructions and providing examples or rubrics can enhance user understanding and accuracy [3].

To decide the components of an assessment tool, a pool of different items should be gathered that adequately represent the construct being assessed [4]. While ISO (International Organization for Standardization) itself does not have a specific standard solely dedicated to creation of checklists, it provides guidelines for the creation of checklists in certain standards. Checklists are commonly used by various certification bodies auditing ISO management system standards to facilitate compliance and improve processes. ISO 19011:2018 [23] provides guidance on the auditing of management systems and suggests that auditors may use checklists as an aid during audits to ensure the systematic and thorough examination of processes, procedures, and controls.

The validity and reliability of assessment tools is critical to obtaining meaningful and trustworthy results. Content validity can be established through expert review and alignment with the intended outcomes [1].

Pilot testing involves administering the assessment tool to a small target group to gather data on item performance and identify potential problems. This step helps assess the clarity, difficulty, and discriminative power of the components.

Validity refers to the extent to which an assessment tool measures the construct it is intended to measure [5]. Different types of validity, such as face, content, and construct validity, should be considered [1]. Content validity ensures that the assessment components adequately cover the construct.

B. *Hse*

The integration of HSE management practices with ISO 45001 and ISO 14001 enables organizations to foster a holistic approach to managing risks, ensuring compliance, and promoting sustainable practices. Numerous studies highlight the inherent synergies between HSE management and ISO 45001 and ISO 14001. Integration of HSE within these standards allows organizations to align their occupational health and safety and environmental efforts, leading to improved risk management, reduced incidents, and enhanced environmental performance. Indeed, based on the United Nations (UN) Sustainable Development Goals (SDGs), these aspects should be considered simultaneously to obtain a sustainable society. In order to manage the HSE aspects of the checklist, globally recognized references [6–25] were selected and reviewed and a content analysis of these principles, rules, and guidelines to prepare a checklist was performed.

C. *Social aspects*

Healthcare facility management encompasses a range of activities that have a direct impact on patient care and staff well-being. Promoting equitable access to healthcare services is crucial in healthcare facility management. Addressing socio-economic disparities, reducing barriers to healthcare, and ensuring equitable distribution of resources contribute to social sustainability and align with the UN commitment to universal health coverage and SDG 3 [25]. It should be noted that safety and satisfaction are of paramount importance in healthcare facilities. Implementing effective patient safety protocols, prioritizing patient-centred care, creating a healthy and safe work environment, promoting work-life balance and ensuring adequate training are all among the UN's emphasis on social responsibility.

In order to manage the social aspects of the checklist, specific references [6–25] were selected and a content analysis of these principles, rules, and guidelines from a social perspective was carried out.

3 Methodology

Assessment tools in the form of checklists are valuable instruments for evaluating compliance, identifying gaps, and driving improvement within healthcare facilities. Once the objectives of the assessment have been determined, the criteria or standards that form the basis of the assessment tool should be identified. These may include industry-specific regulations, domestic policies in each country, best practices, or quality management standards.

Based on a thorough review of the selected criteria and standards a comprehensive understanding of the requirements and expectations has been ensured. As mentioned in the literature review chapter, all fundamental references on HSE and social aspects related to healthcare facilities are studied.

The identified criteria have been broken down into different components. These components represent the key areas and aspects that need to be assessed within healthcare facilities. For each component, clear and measurable objectives have been defined that reflect the desired outcomes or compliance requirements.

Then checklist items that correspond to each component were then formulated. The authors made every effort to use clear language and avoid ambiguity to ensure accurate assessment. The checklist components and items were organized into a logical and coherent structure. Related items grouped together under appropriate components.

4 Components of the Assessment Tool

The purpose of this chapter is to introduce the basic building blocks that make up the proposed assessment tool. Although an item may belong to more than one component, only one component, which in the opinion of the authors is more appropriate, has been selected for better analysis in further research. These components align with the goals and objectives of assessment, whether it is for self-assessment, rapid evaluations, or comprehensive analysis.

The evidence and observation for each item could be a document showing the compliance, or an interview with a recognized accountable person in the healthcare facility or a site visit and observation of the assessors themselves (Table 1).

5 Tool Verification and Validation

The verification process in creating an assessment tool involves ensuring that the tool is accurate, reliable, and aligned with the intended purpose. The process of defining the requirements and specifications for the assessment tool and identifying the objectives, target population, constructs to be measured, and desired outcomes was done based on the needs of the healthcare facilities.

The assessment tool was designed in form of a checklist. The prepared components and items were reviewed carefully by the authors, as the experts in healthcare facility to make sure regarding the accuracy, relevance, and alignment of the tool with the defined requirements. During verification process, the checklist was improved from both the simplicity of the items and coverage of the components. Following the verification steps, it was ensured that the developed checklist could act as a rapid assessment tool to evaluate the HSE and social performance of healthcare facilities and could provide reliable and valid results.

After expert confirmation of the accuracy and sufficiency of the tool, the validation process began. The validation process is an essential step in development an assessment tool. Validity refers to whether the assessment tool measures what it sets out to measure, aligns with the intended purpose and accurately captures the desired constructs or variables. To establish validity, an external team consist of a

Table 1 Checklist and components

Component	Item	References
Employee health	Occupational health training course, trainings for preventions and the required action after occupational hazards for different job groups	[9, 10, 15, 17, 19, 20]
	Individual protection file for staff, including periodic examinations	[6, 17, 19, 20]
	Availability of personal protective equipment (PPE) for personnel including dosimeter and oversight regarding appropriate use	[10, 19, 20]
	Controls and reduction of exposure time for staff in radiation department	[7, 13]
Environmental management	Availability of suitable class ventilation hood	[10, 14]
	Separating wastes, coding and labelling different wastes considering radiation wastes	[7, 8, 10, 21]
	Connection to sewer and its condition, monitoring wastewater disposal indicators	[7, 8, 10, 21]
	Functionality of waste autoclave	[7, 8, 10, 21]
	Optimal energy consumption	[7]
Facility maintenance and infrastructure	Condition of the workplace, door, wall, floor and ceiling	[14]
	Proper air ventilation and circulation, heating, and cooling systems	[7, 14]
	Sufficiency of illuminance and sound in accordance with the occupational exposure limits (OEL)	[7, 19–21],
	Have a preventive maintenance plan	[7, 17]
	Calibration, calibration plan and calibration before use records	[11]
	Equipment technical inspection	[11]
Legal and managerial compliance	Legal permissions for using ionizing radiation generators and magnetic devices	[13]
	Permissions of activity of the healthcare facility and responsible person	[7]
	Existence of environmental and occupational health experts in the healthcare facility	[7]
	Existence of adequate staff to avoid unusual work pressure	[6, 7, 17],
	Considering patient's privacy in providing services to the patients	[9, 18]
	Investigation of medical errors	[9]

(continued)

Table 1 (continued)

Component	Item	References
	Disinfection and sterilization programs	[9, 16]
Patient safety	Existence of policy, program, and patient safety action plan	[9, 17]
	Existence of a patient safety and risk management coordinator with clear responsibilities and required qualification	[9]
	Possibility of conducting initial actions to resuscitate the patient	[7, 9, 16]
	Existence and visibly installation of "code of patients' rights"	[9]
Safety	Availability of essential environment requirements for safety and security of the equipment, including appropriate earthing system, electricity voltage, stabilizers, isolation transformer and backup generators	[15, 19]
	Implementation of a crisis management system	[12, 15, 19]
	Firefighting approval from relevant organizations, existence and inspection of fire alarm and extinguishing system	[12, 15, 19]
	Existence of a mechanism for collecting lessons-learnt in case of accidents and exchanging them	[12, 15, 19]
	Existence and usability of emergency exit doors, displaying appropriate emergency and warning signs for public awareness	[12, 15, 19]
Social responsibility	Accessibility for special groups (elderly, disabled, children and pregnant women and new mothers)	[6, 18]
	Access priority for special groups (senile, disabled, child, pregnant, newborn mothers)	[6, 18]
	Mechanisms for patients, facility personnel, and clients to share/ raise their views, comments, or complaints to the facility's services	[6, 18]
	Instruction for control and prevention of infection disease	[16, 19]

medical doctor[1], an HSE auditor[2] a hospital and a laboratory technical supervisor[3],

[1] Physician with 20 years of experience and acting as Quality Auditor in hospital.

[2] Appointed ISO 9001, ISO 14001, ISO 45001 senior auditor by a TÜV organization, 10 years of experience.

[3] University hospital and laboratory manager, also professor.

were invited and five pilot testing[4] in hospital and laboratories were organized. After these pilot assessments, the checklist was refined and finalized. Based on the findings from the validation process, some items modified or removed. During this period, the opinions of experts and other stakeholders of the testing places were also collected and analysed.

6 Conclusion

The developed comprehensive checklist provides healthcare facilities with an invaluable guide to continuously measure and enhance their performance in the areas of Health, Safety, and Environment (HSE) as well as social responsibility. As a systematic tool, this checklist facilitates rapid self-assessments, enabling healthcare providers to effectively identify strengths and areas for improvement.

The development of the checklist involved a careful review of the key globally recognized references in the field of HSE and social responsibility, ensuring that its components cover the essential aspects for a thorough assessment. Moreover, the checklist's verification and validation steps have been rigorously carried out to provide confidence in its reliability and accuracy. This diligent process guarantees that healthcare facilities can rely on the tool as a credible and robust means of assessing their performance and taking appropriate corrective measures.

As a result of its thorough development and the assurance provided by verification and validation, this tool has a great potential for widespread implementation. Its simple, straightforward, easy-to-understand and uncomplicated language makes it accessible to various healthcare settings, enabling facilities of all sizes to foster a culture of continuous improvement and increased commitment to HSE and social responsibility.

In further work, this checklist could accompany a rating system, so that it could also be used as a maturity model. Indeed, if widely used, the checklist can be used for easy benchmarking against established targets, offering clear insights into the facility's current standing and progress over time.

As this tool, is prepared based on internationally accepted guidelines and standards, to observe national standards in each country, further studies could add national references for local usage. Ultimately, the use of this checklist paves the way for safer, more responsible healthcare practices, benefiting not only the healthcare providers themselves but also the patients and the wider community they serve.

[4] Three university hospitals and two university laboratories in Iran were selected.

References

1. Haladyna TM, Downing SM (2011) Twelve steps for effective test development. Handbook of test development. Routledge, pp 17–40
2. Holmboe ES, Durning SJ, Hawkins RE (2017) Practical guide to the evaluation of clinical competence e-book. Elsevier Health Sciences
3. Boud D, Falchikov N (2006) Aligning assessment with long-term learning. Assess Eval High Educ 31(4):399–413
4. Salvia J, Ysseldyke JE, Bolt S (2007) Portfolio Assessment. Assessment in special and inclusive education, pp 249–267
5. Kane MT (2006) Validation. Educational measurement 4(2):17–64
6. World Bank (2016) Environmental and social framework. https://thedocs.worldbank.org/en/doc/837721522762050108-0290022018/original/ESFFramework.pdf
7. World Bank (2017) Environmental, health, and safety general guidelines. https://documents1.worldbank.org/curated/en/157871484635724258/pdf/112110-WP-Final-General-EHS-Guidelines.pdf
8. World Health Organization (2014) Safe management of waste from health-care activities. https://www.who.int/publications/i/item/9789241548564
9. World Health Organization (2016) Patient safety assessment manual. https://www.who.int/publications/i/item/9789290221203
10. World Health Organization (2008) Essential environmental health standards in health-cares. https://apps.who.int/iris/bitstream/handle/10665/43767/9789241547239_eng.pdf
11. World Health Organization (2011) Medical equipment maintenance programme overview. https://apps.who.int/iris/handle/10665/44587
12. World Health Organization (2018) Communicating risk in public health emergencies: a WHO guideline for emergency risk communication (ERC) policy and practice. https://www.who.int/publications/i/item/9789241550208
13. International Atomic Energy Agency (2014) Radiation protection and safety of radiation sources: international basic safety standards. https://www-pub.iaea.org/MTCD/publications/PDF/Pub1578_web-57265295.pdf
14. World Health Organization (2003) Manual for the laboratory identification and antimicrobial susceptibility testing of bacterial pathogens of public health importance in the developing world: haemophilus influenzae, neisseria meningitidis, streptococcus pneumoniae, neisseria gonorrhoea, salmonella serotype typhi, shigella, and vibrio cholerae. No. WHO/CDS/CSR/RMD/2003.6. World health organization
15. World Health Organization (2012) Strengthening health-system emergency preparedness: toolkit for assessing health-system capacity for crisis management: part 1: user manual. No. E96187. World Health Organization, Regional Office for Europe
16. COVID, WHO. 19: occupational health and safety for health workers: interim guidance. Geneva: world health organization (2021)
17. Geneva WHO (2021) Keep health workers safe to keep patients safe
18. United Nations (2015) Transforming our world: the 2030 Agenda for Sustainable Development. United Nations
19. European commission, Occupational health and safety risks in the healthcare sector
20. Occupational health and safety management systems - requirements with guidance for use. ISO 45001:2018, International organization for standardization (2018)
21. ISO 14001:2015 - Environmental management systems - requirements with guidance for use. ISO 14001:2015, International organization for standardization (2015)
22. ISO 9001:2015 - Quality management systems - requirements. ISO 9001:2015, International organization for standardization (2015)

23. ISO 19011:2018 - Guidelines for auditing management systems. ISO 19011:2018, International organization for standardization (2018)
24. ISO 22301:2019 - Security and resilience - business continuity management systems - requirements. ISO 22301:2019, International organization for standardization (2019)
25. ISO 22320:2018 - Security and resilience — emergency management — guidelines for incident management (2018)

Effects of the New Generation High-Speed Spindle on the Production Performance of Vertical Machining Centers

Çağla Şen, Merve Kurt, Uğur Alev, Cihan Mertöz, Ali Serdar Vanlı, and Anıl Akdoğan

Abstract In order to increase productivity in manufacturing enterprises, a wide variety of applications are made and various support equipment and systems are put into use. In recent years, vertical machining centers have found widespread application in the manufacturing of molds, dies and components in enterprises. The efficient operation of these machines is very important to ensure that mass production is not interrupted and is done efficiently. For this reason, while the developed motor spindles increase the production speeds of vertical machining machines, the quality of the surface roughness they process is also quite satisfactory. In this study, the quality features we achieved in the molds and dies we processed with the new generation motor spindle we use in our CNC vertical machining centers and the satisfactory improvements we experienced in our production performance are presented.

Keywords Vertical machining center · High-speed spindle · Mold production performance

Ç. Şen · M. Kurt · U. Alev · C. Mertöz
Design and Production Department, Mesan Plastik Ve Metal San. A.Ş, İstanbul, Türkiye
e-mail: cagla.sen@mesan.com.tr

M. Kurt
e-mail: merve.kurt@mesan.com.tr

U. Alev
e-mail: ugur.alev@mesan.com.tr

C. Mertöz
e-mail: cihan.mertoz@mesan.com.tr

A. S. Vanlı · A. Akdoğan (✉)
Mechanical Engineering Department, Yildiz Technical University, İstanbul, Türkiye
e-mail: nomak@yildiz.edu.tr

A. S. Vanlı
e-mail: svanli@yildiz.edu.tr

N. M. Durakbasa and K. G. Gülen (eds.), *Sustainable Green Conversion*, Lecture Notes in Mechanical Engineering, https://doi.org/10.1007/978-3-031-83583-4_16

1 Introduction

This study aims to focus on the significant advantages of using high-speed spindle attachments in CNC milling machines, such as reductions in processing times, improvements in surface quality and better adherence to dimensional tolerances. Firstly, the concept of the high-speed spindle attachment is examined from a perspective where a detailed explanation of why this technology is preferred in CNC milling is provided. This information offers a strategic and economic understanding for industry professionals and engineers to make their production processes more effective and optimize part quality in various ways. Simultaneously, a comparison was made by processing the same part with and without the high-speed spindle integrated into a CNC vertical machining center to investigate the effects of the high-speed spindle on parameters such as surface roughness quality, processing time, and tool life improvements. These comparisons aim to increase the efficiency of milling processes and support the more effective use of economic and physical resources. Therefore, the use of high-speed spindles in CNC vertical machining centers represents a strategic step for more competitive, effective and efficient operations in modern production environments.

A CNC vertical machining center is a machine tool that plays a highly critical role in the industry, capable of processing parts with complex geometries with high machining accuracy and efficiency, and continuously evolving with the use of advanced technologies [1]. With the increase in demands for product quality and fast processing, high-speed spindle attachments attract attention in enhancing the machining accuracy and performance of CNC vertical milling machines. High-speed spindles provide significant economic benefits in industries such as automotive, aerospace, and mold-die manufacturing and are increasingly applied [2].

Integrated high-speed spindles in the machine enable access to high speeds that are difficult to reach. Some researchers define high-speed machining (HSM) as machining that exceeds conventional cutting speeds by 5 to 10 times. Parts machined with HSM can be almost stress-free and can be used to process thin-walled workpieces since cutting forces acting on the work piece are lower under HSM conditions [3]. At the same time, it allows for achieving the desired surface quality in the product. CNC high-speed spindles work without the machine's spindle rotating, thus preventing runout and other adverse effects that may arise from the machine over time with use. In the market, there are three types of high-speed spindles: mechanical, pneumatic, and electronic.

Electrical high-speed spindle attachments enhance the quality of the workpiece by reducing tool wear and maintaining machine integrity through advanced control systems that minimize vibrations [4]. They also improve machining efficiency with fast response times. Therefore, they are more reliable in terms of vibration control and deformation prevention. Mechanical high-speed spindles are limited in vibration control and therefore not widely used. Pneumatic high-speed spindles can cause fluctuations in speed due to changes in air pressure, which can trigger undesirable situations [5].

2 Literature Review

Numerous studies have been conducted on machining various materials with high-speed spindles. Some related studies are presented below.

Surface roughness is defined as the geometric error occurring during machining from the movement between the tool and the workpiece, and it is important for evaluating the quality of machined surfaces. Dewei Liu et al. presented a model developed to predict and improve surface roughness in milling. The model examines the effects of the feed per tooth, step-over ratio, corner radius, and minor cutting edge angle on surface roughness. Experiments and analyses determined that surface roughness varies non-linearly with these parameters. The developed model was able to predict surface roughness with higher accuracy compared to the Z-Map prediction model. This study provides an important reference for controlling and optimizing surface quality in CNC vertical machining centers [6].

Rodolfo E. Haber conducted various analyses for monitoring tool wear. The research obtained significant findings by examining cutting force and vibration signals in time and frequency domains for monitoring tool wear. These results contribute to the development of new tool monitoring systems in high-speed machining processes [7].

Rahman et al. evaluated the effectiveness of coated inserts for high-speed machining of titanium. Tests revealed that machining titanium with coated inserts is possible. Especially, coated inserts were found to be useful to machine titanium even at high cutting speeds. The data obtained through the use of mathematical models show that coated inserts are functional in machining titanium. These findings allow for more efficient and effective machining of titanium alloys [3].

Ahmad Hamdan and colleagues focused on optimizing machining parameters using coated carbide tools for high-speed machining of stainless steel. This study aimed to achieve better machining quality by minimizing cutting forces and surface roughness. Results showed a 255% reduction in cutting forces and a 413% improvement in surface roughness performance. These findings indicate that optimization was successful and that significant improvement was achieved in high-speed machining processes using coated carbide tools [8].

2.1 Processing Time

Processing time is critical for the efficient operation of CNC machines. High-speed spindles offer faster machining times, providing flexibility in production and reduced overhead. Additionally, they extend the machine's life by reducing mechanical stress and increase energy efficiency. Buying a high-speed spindle instead of a new machine is an economical solution in terms of cost and time. Expensive and time-consuming processes like grinding and polishing can also be minimized by using a high-speed spindle [9]. The small volume of high-speed spindles can increase efficiency even in limited working environments. Therefore, the use of high-speed spindles in CNC vertical machining centers supports more effective use of economic and physical resources, and represents a strategic step for more competitive and effective operations in modern production environments [10].

2.2 *Surface Quality*

High-speed spindles achieve homogeneous results with smaller chip sizes by processing at high speeds to enhance surface quality. They create a smooth surface by ensuring precise contact with the surface thanks to thin cutting edges. They can even eliminate the formation of BUE (built-up edge) [11]. Advanced models minimize vibrations and increase the thermal stability of the workpiece, preventing deformations. These mechanisms enable high-speed spindle attachments to produce smoother and more homogeneous machined parts [12].

2.3 *Tool Wear*

High-speed spindles play a crucial role in reducing tool wear in CNC vertical machining centers. High-speed machining allows working with lower cutting forces, extending tool life. The stability and precision they provide reduce wear by making more smooth and consistent cuts. Electronic high-speed spindles minimize vibrations and control thermal effects, preventing tool stress and the formation of cracks and wear marks on the surface [12]. Another significant advantage of high-speed machining is minimizing the heat effects on machined parts. Due to the cutting speed and cutting depth, the contact time of the cutting edge is very short. Therefore, there is very little time for heat to spread, which improves tool life [11]. In most cases, the need for coolant may be eliminated. As a result, using high-speed spindles reduces tool wear, lowers costs, and increases the efficiency of the production process [13].

3 Materials and Methods

3.1 *Criteria for Selecting the Processed Part and Material*

The processed part is the core of the zamak injection molds, referred to as the minifix insert, with general dimensions given in Fig. 1. As known from the Al-Zn phase diagrams, the eutectic temperature is approximately 382 °C, and the casting temperature reaches 420 °C. The properties of QR90 mold steel, preferred in zinc die cast

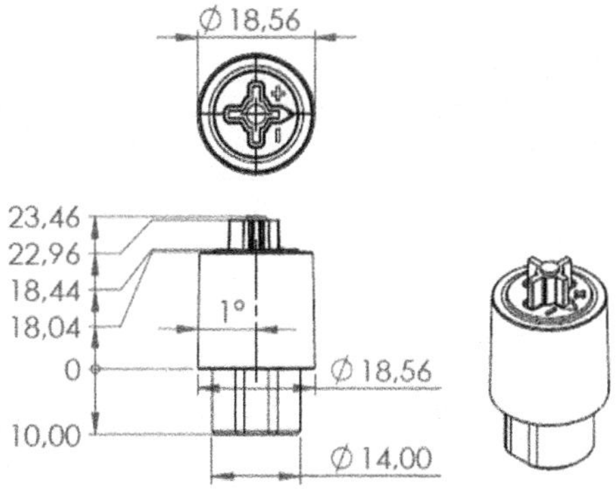

Fig. 1 General dimensions of minifix socket

Table 1 Chemical composition and hardness of QR90 (Uddeholm)

C %	Cr %	Si %	Mo %	Mn %	V %	HRC
0.38	2.60	0.30	2.30	0.80	0.90	46–48

parts exposed to high temperatures and pressures like the minifix insert, are very important. QR90's excellent high-temperature strength and hot hardness properties provide stability during casting, enhancing the part's quality and durability. Its high temper resistance increases its durability against gradual and sudden temperature changes, while its resistance to thermal fatigue maintains the mold's stability. Excellent thermal conductivity ensures homogeneous heat distribution, increasing casting efficiency. Additionally, good hardness and creep properties in longitudinal and transverse directions enhance the mold's durability, while uniform machinability and good heat treatment properties allow for processing the mold as desired. These features clearly explain why QR90 mold steel is preferred in the production of parts exposed to high temperatures like the minifix insert. The table below shows the chemical composition of QR90 mold steel (Table 1).

3.2 Equipment Used in the Experiment

1. High-Speed Spindle (Nakanishi Hes 510): In this study, the NAKANISHI HES 510 high-speed spindle was used on the OKUMA GENOS M560-V-e CNC vertical machining center. The HES510 stands out with its high performance and precision. The high-speed spindle can reach a maximum power of 340 W and a motor speed of 50,000 rpm. The spindle accuracy is within 1 μm, and it can be easily mounted on machining centers and milling machines. Weighing 2.3 kg, this high-speed spindle performs excellently in small-diameter drilling and end milling with φ20 mm or less. The technical drawing of the high-speed spindle

is shown in Fig. 2. The motor speed range is between 5,000 and 50,000 rpm, with a two-digit digital display allowing precise speed adjustments in 1,000 rpm increments. Additionally, it automatically detects the connected motor, so manual adjustment of the maximum motor speed is not necessary. These features provide significant advantages in applications requiring high precision and efficiency with the NAKANISHI HES 510 high-speed spindle (Fig. 3).

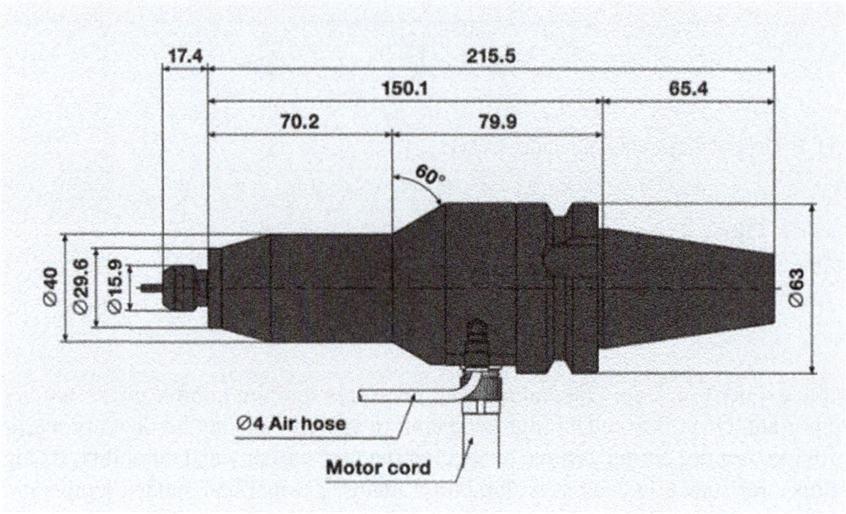

Fig. 2 NAKANISHI HES 510 BT40 technical dimensions

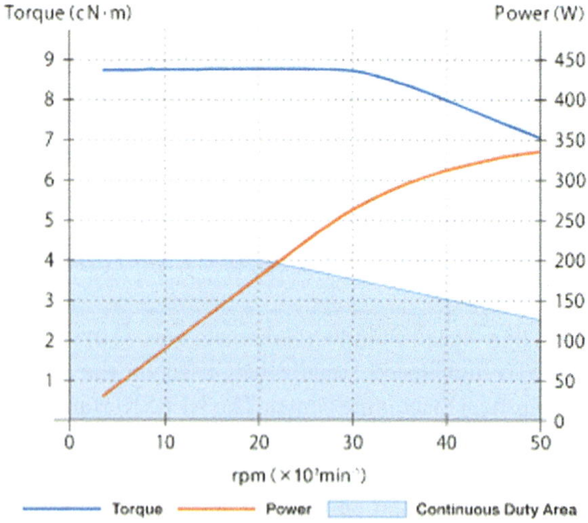

Fig. 3 NAKANISHI HES 510 BT40 torque-power-speed graph

The graph shows the torque and power curves concerning motor speed. While torque is high initially, it decreases as the speed increases, and power increases with speed, reaching a maximum at a certain point before slightly decreasing. This information can be an important guide for optimizing the motor's performance in various speed ranges.

2. CNC Vertical Machining (Okuma Genos M560-V-e): GENOS M560-V-e offers high rigidity and efficiency with its thermally stable, heat-resistant, and precise bridge-type structure. It has a table size of 1,300 × 560 mm, with axis movements of X: 1,050 mm, Y: 560 mm, and Z: 460 mm, and a spindle speed of 15,000 rpm. It has a 32-tool automatic tool changer (ATC) and a motor power of 22/18.5 kW. These features allow the machine to deliver excellent results in high precision and performance applications.
3. Tool Presetting Machine (E460N Nıkken): The E460N tool presetter provides high precision and durability with advanced features. The axis resolution is 1 μm using an optical glass scale and linear transducer. The control speed of the X and Z axes is a maximum of 2 mm/s. The visual inspection system includes a C-MOS sensor, bi-telecentric lens with 30 × magnification, and LED illuminators for tool measurement and cutting inspection. In this study, the diameters of the 2R05 and 1R flat tools used in the finishing process were measured to examine the effect of high-speed spindle machining on tool wear.
4. Surface Roughness Measurement Device (Handysurf+): HANDYSURF+ stands out with its high resolution of 0.0007 μm, a stroke length of 16 mm, and a measurement speed of up to 1 mm/s. It also has a measurement height range of −210 to +160 μm and offers a practical experience with simple use. The environmental temperature was accepted as 20 ± 1 °C in roughness measurements, and the arithmetic average of the obtained roughness values was taken to calculate the average surface roughness (Fig. 4).

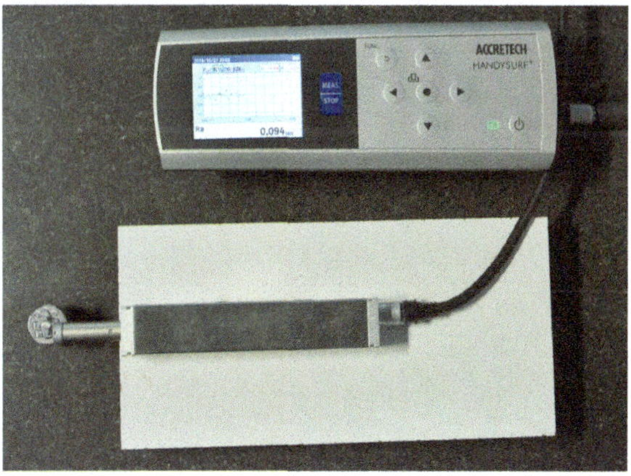

Fig. 4 Handysurf+ Surface roughness measuring device

4 Results and Discussion

The results obtained from this study are discussed under three main headings. The results show that there were improvements in different ratios in processing times, tool wear values, and surface roughness values of the processed surfaces with high-speed spindle systems.

4.1 Processing Time

In this study, inserts were processed both with and without the high-speed spindle on the same machine, with five parts each, and the tools used, speeds, and feeds are shown in the tables below. A new carbide tool was used for both tests, and optimal speeds and feeds were determined according to the standards set by the tool manufacturer. Additionally, the high-speed spindle was not used for roughing in both processes, and the high-speed spindle was activated during the finishing process. The 2R05 and 1R flat tools used in the finishing process are 70 HRC hardness carbide tools (Fig. 5).

In the finishing process without the high-speed spindle, the 1R-Flat tool (9500 rpm, 600 feed, and 0.05 depth of cut) broke twice during the process (Table 2).

As a result, the processing time for five sockets was observed to be 30 h without the high-speed spindle and 12.5 h with the high-speed spindle. The use of the high-speed spindle attachment resulted in a 41.67% reduction in processing time. This increases the production efficiency of the enterprise, making processes more efficient while also reducing production costs.

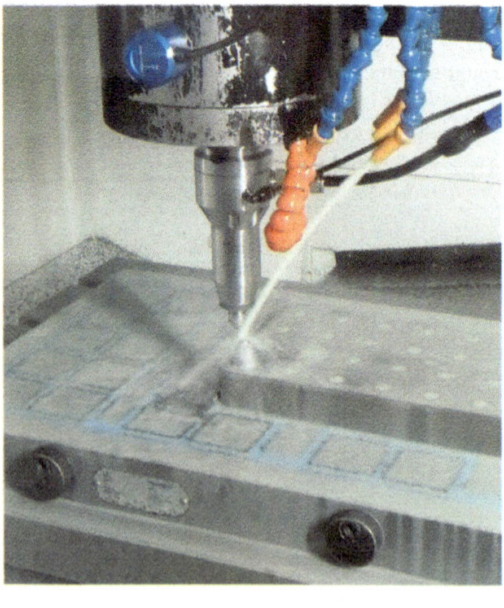

Fig. 5 High speed spindle, cutting tool and part during machining

Table 2 Tools and machine parameters used in standard machining and high-speed spindle machining

	Standard machining				High-speed spindle machining			
	Tool	Speed (rpm)	Feed (mm)	Depth of cut (mm)	Tool	Speed (rpm)	Feed (mm)	Depth of cut (mm)
Roughing	12R1	2500	4000	0.15	12R1	2500	4000	0,15
	10R05	4500	3000	0.1	10R05	4500	3000	0,1
	6R05	5500	3000	0.05	6R05	5500	3000	0,05
	3R05	7500	1500	0.05	3R05	7500	1500	0,05
Finish	2R05 (Moncarb)	9500	600	0.05	2R05	25000	1000	0,05
	1R-DÜZ (Moncarb)	12500	200	0.03	1R-FLAT	30000	400	0,03

4.2 *Tool Wear*

In the 2R05 tool, the tool diameter decreased to 17.42 mm with standard machining, resulting in a 12.9% wear rate. With high-speed spindle machining, the tool diameter decreased to 19.92 mm, resulting in a 0.4% wear rate (Table 3).

Table 3 Tool diameters measured with E460N

Tool	Standard machining	High-speed spindle machining
1R FLAT	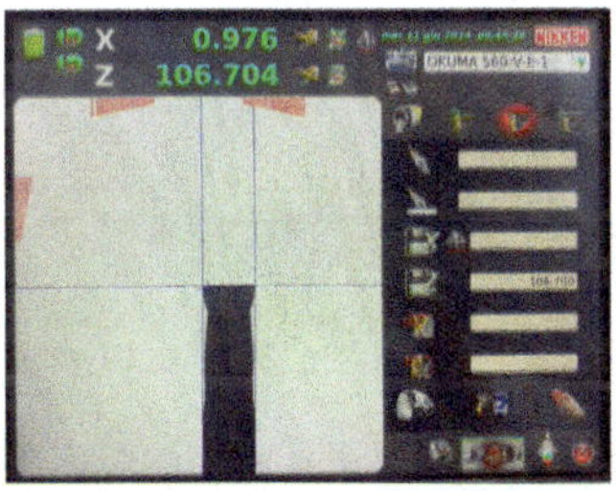	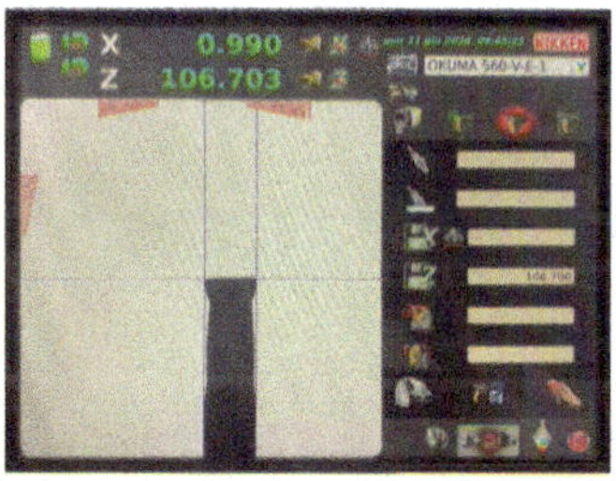
2R05	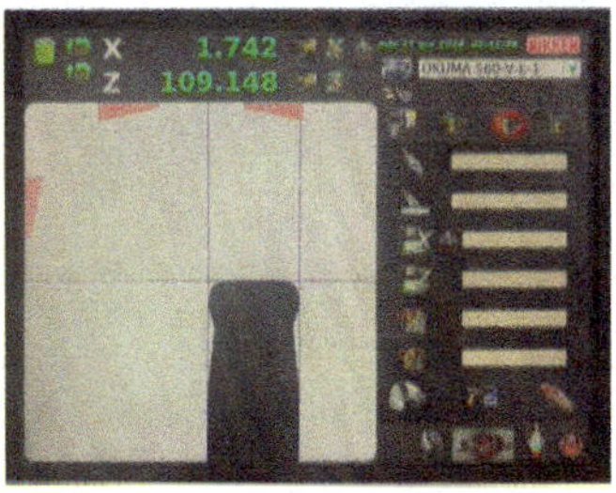	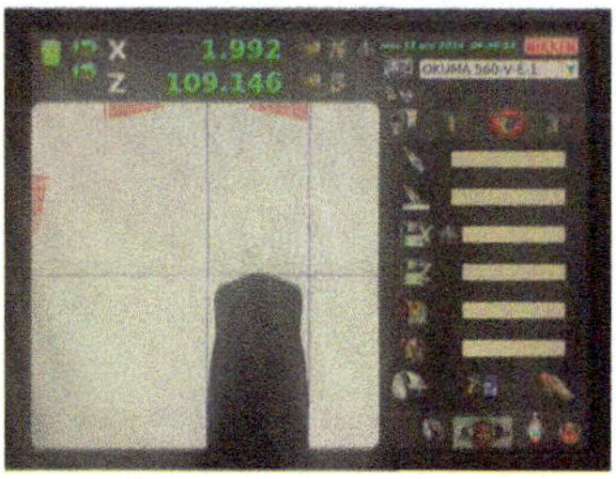

1 In the 1R-Flat Tool, the tool diameter decreased to 9.76 mm with standard machining, resulting in a 24% wear rate. with high-speed spindle machining, the tool diameter decreased to 9.90 mm, resulting in a 1% wear rate.

These data show that the high-speed spindle significantly reduces tool wear. Compared to standard machining, the use of the high-speed spindle extends tool life by reducing tool wear, improving machining quality. This study demonstrates that the use of a high-speed spindle is an effective method for reducing tool wear.

4.3 Surface Roughness

As seen in Table 4, the average surface roughness value of the samples machined with the high-speed spindle is 0.043 μm, while the average surface roughness value of the samples machined without the high-speed spindle is 0.094 μm. In this case, the surface quality of the samples machined with the high-speed spindle is 54.26% better than the samples machined without the high-speed spindle (Fig. 6).

Table 4 Surface roughness values of machined sockets (Ra)

Sample no	High-speed spindle machining (R_a μm)	Standard machining (R_a μm)
1	0.042	0.094
2	0.043	0.096
3	0.043	0.095
4	0.042	0.094
5	0.044	0.093
Average	0.043	0.094

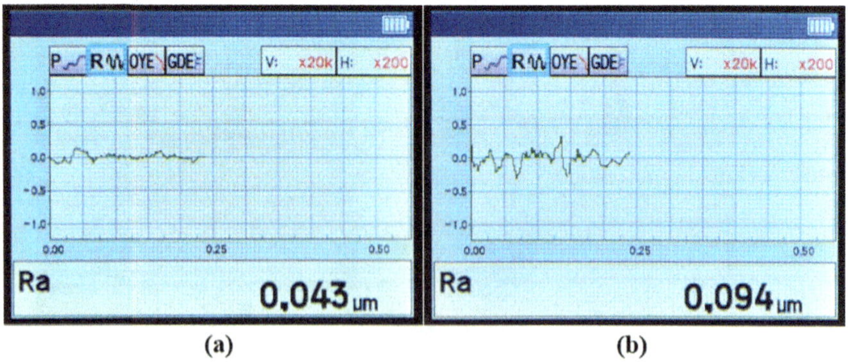

Fig. 6 Surface roughness samples measured with Handysurf+ Surface roughness device (**a**) High-speed spindle machining (**b**) standard machining

The results obtained in this study show that the use of high-speed spindles in CNC vertical machining centers significantly reduces processing times, minimizes tool wear, and improves surface quality. Processes that took 30 h with standard machining were reduced to 12.5 h with the high-speed spindle, achieving a 41.67% time savings. Additionally, tool wear was significantly reduced in machining with the high-speed spindle. In the 2R05 tool, tool diameter wear was 12.9% with standard machining, while it was 0.4% with high-speed spindle machining. In the 1R-Flat tool, tool diameter wear was 24% with standard machining, while it was 1% with high-speed spindle machining. Surface roughness values were also 0.043 μm on average with the high-speed spindle, compared to 0.094 μm with standard machining, representing a 54.26% improvement in surface quality. In conclusion, the use of high-speed spindles in CNC vertical machining centers represents a strategic step for more competitive and effective operations in modern production environments by reducing processing times, tool wear, and improving surface quality.

References

1. Miao H et al (2021) The vibration analysis of the CNC vertical milling machine spindle system considering nonlinear and nonsmooth bearing restoring force. Mech Syst Signal Process 161:107970
2. Lee J, Kim D-H, Lee C-M (2015) A study on the thermal characteristics and experiments of high-speed spindle for machine tools. Int J Precis Eng Manuf 16:293–299
3. Rahman M, Wang Z-G, Wong Y-S (2006) A review on high-speed machining of titanium alloys. JSME Int J, Ser C 49(1):11–20
4. Urbikain G et al (2016) Spindle speed variation technique in turning operations: modeling and real implementation. J Sound Vib 383:384–396
5. Dündar M, Fıçıcı F, Özen F (2016) Mikro Frezeleme Tezgahının Tasarımı ve Kısımları. El-Cezeri 3(3)
6. Liu D et al (2024) Kinematics and improved surface roughness model in milling. Int J Adv Manuf Technol 131(5):2087–2108
7. Haber RE et al (2004) An investigation of tool-wear monitoring in a high-speed machining process. Sens Actuators Phys 116(3):539–545
8. Hamdan, A, Sarhan AAD, Hamdi M (2012) An optimization method of the machining parameters in high-speed machining of stainless steel using coated carbide tool for best surface finish. Int J Adv Manuf Technol 58:81–91
9. Alam S, AU AAP (2012) Surface texture investigation in high speed flat end milling of Ti-6Al-4V. In: Proceedings of international conference on industrial engineering and operations management
10. Salgado DR, Alonso FJ (2009) Optimal mechanical spindle speeder gearbox design for high-speed machining. Int J Adv Manuf Technol 40:637–647
11. Jain A, Bajpai V (2020) Introduction to high-speed machining (HSM). High Speed Machining. Academic Press, pp1–25
12. Abbas AT et al (2021) Fundamental investigation into tool wear and surface quality in high-speed machining of Ti6Al4V alloy. Materials 14(23):7128
13. Wang Y et al (2020) Surface quality improvement in machining an aluminum honeycomb by ice fixation. Chin J Mech Eng 33:1–8

Industry 4.0 and Industry 5.0 Applications

A Qualitative Research on the Market Entry Strategies of Techno-Enterprises' Using Digital Transformation Technologies

Ceren Genç and Dilber Ulaş

Abstract This study aims to understand the market-entry strategies that techno-enterprises employ, utilizing digital transformation technologies to thrive in the new technologies sector. The research question of this study is: "What are the success factors for techno-enterprises?" New digital technologies, such as artificial intelligence, defence, health, space, security, software, SaaS, gaming, fintech, advanced electronic technologies, retail and e-commerce are expected to be the prominent areas of the future in Türkiye and these new technologies can aid start-ups to be effortlessly reached out and distribute their services virtually the whole international markets via online tools. Additionally, personalized medicine, digital health solutions, and initiatives in the field of health and biotechnology stand out and have had a large impact on the rise of start-ups. Digital payment and blockchain technologies are especially developing in the financial sector. We conducted nine techno-enterprises located at Ankara, Balıkesir, Adana and Kastamonu. Data collection involved semi structured interviews, review and analysis of company documents, websites and direct observation of the management operations and processes. The thematic analysis of the data revealed two themes: evolution of a techno-entrepreneur's entry modes and marketing strategies. E-commerce and marketing strategies has provided a way to reach international markets and reduce costs. The results of this study would contribute to techno-entrepreneur's survivability as well as exchanging successful experience among new entrepreneurs. For entrepreneurs who plan to start a business, this study would contribute understanding the skills for initiating a startup.

Keywords Techno-entrepreneur · Digital entrepreneur · Start-up · SME · Incubation center · Entrepreneur · International market entry strategies

---

C. Genç (✉)
Department of Marketing and Advertising, Ankara Medipol University, Ankara, Turkey
e-mail: ceren.genc@ankaramedipol.edu.tr

D. Ulaş
Faculty of Political Science, Department of Business Administration, Ankara University, Ankara, Turkey
e-mail: ulas@politics.ankara.edu.tr

N. M. Durakbasa and K. G. Gülen (eds.), *Sustainable Green Conversion*, Lecture Notes in Mechanical Engineering, https://doi.org/10.1007/978-3-031-83583-4_17

1 Introduction

International market entry strategies refer to the approaches a company adopts to penetrate a new foreign market. To achieve sustainable growth and enhance their competitiveness, companies are seeking opportunities to expand from local to international markets. With the rising use of digital transformation technologies like mobile devices, big data analysis, artificial intelligence, cloud computing, and e-commerce, they are reorganizing their business operations and models. The aim of a market entry strategy is to effectively penetrate a new market and attain long-term growth. Various methods exist for entering international markets, including exports, contractual arrangements, and foreign direct investments. The most suitable method depends on the company's resources, objectives, and the specific characteristics of the market. Each country's market is characterized by distinct cultural, economic, technological, political, legal, geographical, and demographic differences. Therefore, the company's resources, goals, legal and regulatory requirements, market size, market knowledge, estimated growth rate, customer oriented strategy, level of competition, expected costs and risks in the target country market, a unique product, innovation, characteristics of entrepreneur's profile (e.g., their technical and management background, their experiences accumulated over their professional and academic career, and language knowledge) [1], global mindset, ability to identify opportunities, previous experience, acquiring the market knowledge, networks, partnership and experience, should be taken into account. The market entry strategy should be designed to minimize risks and maximize opportunities. The question of which entry strategies to use in different country markets has been a central topic in international business research for many years. In this study, the internationalization literature is examined, the basic features of techno-enterprises entry strategies into international markets are explained, and which factors affect the choice of strategy is emphasized. To help techno-enterprises effectively establish themselves in international markets, a comparative analysis of market entry strategies has been conducted, aiming to provide valuable insights for technology companies seeking expansion into global markets.

2 Literature Review

This section provides a literature review on our study. The word "Technopreneurship" is a combination of the words "Technology" and "Entrepreneurship". The inquiry into how techno-enterprises enter and operate within international markets has emerged as a significant topic in recent years within international business research. The strategic relevance of the entry strategy choice is evident when considering its implications for firms' resources, degree of control, risks and performance.

2.1 The Concepts of Techno-Enterprises

Today, technology is utilized for various purposes, including the production of goods and services, as well as accessing and utilizing information. The concept of entrepreneurship is discussed by different researchers. Joseph A. Schumpeter defines entrepreneur; as a person who innovates, develops untested technologies and production techniques, develops a new product or service, or produces a product using new methods, creating a revolution in this context. In his studies, he sought to elucidate entrepreneurship through the lenses of innovation and technological advancements, concentrating on subjects such as novel products, new services, innovative processes, fresh supply channels, and the generation of new resources to pioneer emerging markets. According to Schumpeter, entrepreneurs are characterized as individuals and institutions that drive change within society. An entrepreneur is viewed as someone who identifies opportunities and, with the requisite time, skill, and effort, can seize these opportunities and convert them into marketable ideas. They create value, shoulder the risks inherent in the competitive market, and reap rewards from these endeavors [2]. Based on the meaning of the concepts of technology and entrepreneur, a techno-entrepreneur is someone who has a technological idea, identifies the problem that the technological idea provides a solution to as an opportunity, manages to bring together the information, material and financial resources that will enable this idea to be realized and transformed into a product, and can manage the R&D project period efficiently. Hence, an entrepreneur is the individual who delivers the technological product they have acquired to ideal customer groups through effective marketing efforts, with the goal of swiftly expanding their business volume through the high added value they generate. The process of techno-entrepreneurship promotes start-ups, which in turn improves the socio-economic outlook of a particular region. A start-up is a young company that has just begun to develop [3]. Startups valued at more than 1 billion dollars are called "unicorns", and Turkish "unicorns" are called "Turcorns". There are currently six unicorn in Türkiye named Peak Games, Getir, Dream Games, Hepsiburada, Trendyol and Insider (Table 1).

In general, the term "technology" denotes the practical application of knowledge to the industrial world, aimed at addressing present challenges or signifies a body of knowledge itself. In addition to major techno-enterprises such as Facebook, Uber, Amazon, Netflix, Alibaba, Google; The quantity of techno-startups emerging with innovative business models, harnessing digital transformation technologies like the Internet of Things (IoT), artificial intelligence, augmented reality (AR), virtual reality (VR), blockchain, and cybersecurity, spans various sectors from health to agriculture to automotive, and is steadily growing day by day. Startups are key players in introducing and scaling radical innovations that replace existing technology. A Born Global company is one that internationalizes a significant portion of its business right from the beginning or within a few years of its establishment. (Spotify, skype, airbnb). Start-up is a technology-oriented business model that develops services or products to solve consumers' problems, carries out this business by using all technological possibilities, and grows in an innovative, fast, and sustainable way. Technology

Table 1 Growth and marketing strategies of unicorns in Türkiye

Peak games	Founded by Sidar Şahin and Hakan Baş in 2010, the game application Peak Games became successful in a short time with the games it made on Facebook. In 2012, it broke records for okey games. Peak Games acquired Saudi Arabia's social gaming giant Kammelna Games and the popular card game Baloot started to be offered on platforms such as Facebook. It was acquired by foreign game studio Zynga in 2020 with a valuation of $ 1.8 billion. Having a young team of 100 people, with an average age of 29, was effective in its success. Marketing team create dozens of messages and pieces of creative content, use advertising partners including Facebook, Google and many others. Through constantly providing data-based feedback, it is the most downloaded game of 2020 on both the App Store and Google Play Store, and local social networking app in Japan. Traditional methods such as TV and billboards are more common even for tech companies. They discovered new games through reading game review websites (https://peak.com/how-we-work/entering-a-unique-market-japan)
Getir	Getir application, which has grown by 5 times in 2017, was founded by Nazım Salur, Serkan Borançılı (founder of Gittigidiyor) and Tuncay Tütek (Entrepreneur). Getir is an organization that can bring thousands of products at a 7/24 fashion, based on an on-demand business model to your desired point (in ten minutes on average) via mobile application. Apart from Türkiye, it entered the UK market in January 2021, and later Amsterdam, Berlin and Paris, and acquired its rival Gorillas. In the UK market, supply and distribution processes were established, personnel were recruited, and initially, five warehouses were maintained. An agreement has been reached with the agency to work with for proper marketing. A payment system that does not exist in Türkiye has been established in the Netherlands. For example, in Istanbul, a city of 16 million people, people go to many different addresses without mixing products. Re-supply of goods, planning their own characteristics, quantities, knowing which warehouses, how many couriers and how many warehousemen will work, and adapting to climatic conditions are all done with data science and technology. Getir decided to leave the French, Italian, Portuguese and Spanish markets and focus on Türkiye (https://fastcompany.com.tr/calisma-hayati/hedef-5-kita/)
Dream games	Founded by former employees of Peak Games Soner Aydemir, İkbal Namlı, Hakan Sağlam, Eren Şengül and Serdar Yılmaz. The startup, which was founded on mobile game development in 2019, became a unicorn with a valuation of $ 2.75 billion in 2022. It is focused on growth in Istanbul and London
Hepsiburada	Founded by Hanzade Doğan in 2000, Hepsiburada was established as an online shopping platform. In e-export, which includes processes such as customs, logistics, ordering and marketing, all processes are gathered in one hand. Sellers on the platform have started to enter the world market. It has more than 100 thousand business partners. Collaborations are being established with successful companies in sectors such as finance, logistics and marketing in Europe, the Middle East and North Africa. Its value increased to 4 billion dollars in 2021

(continued)

Table 1 (continued)

Trendyol	Founded by Demet Mutlu, Trendyol, which started out as an online clothing sales platform, received an investment of 1.5 billion dollars in 2021. Since the beginning of 2022, it has mediated the export of more than 51 thousand of our sellers. With micro exports, it aims to turn 300 thousand of our sellers, 97% of whom are SMEs and tradesmen, into exporters
Insider	Founded by Hande Çilingir in 2012 and developing special software for e-commerce brands, Insider exports to 28 different countries. Artificial intelligence-supported Multichannel Growth Management Platform allows marketers to integrate their digital channels with a single platform, combine user data, predict users' future behavior with a machine learning-supported recommendation engine, and offer personalized experiences. Insider digital marketing experts touch on marketing activities at every stage of the sales cycle, from user acquisition to user conversion (activation), from creating a loyal user base (retention) to profit return (revenue); It enables them to manage important growth metrics such as conversion rate (CR), customer acquisition cost (CAC), customer lifetime value (LTV), average basket amount (AOV). Insider's valuation has increased to over $1 billion in 2022. Serving more than 800 global brands, including industry leaders such as Singapore Airlines, Marks & Spencer, Estée Lauder, Virgin, Samsung, Carrefour, Dominos, Toyota, Newsweek, Avon, MediaMarkt, AVIS, Allianz, BBVA, IKEA and CNN (https://webrazzi.com/2020/07/20/insider-a-32-milyon-dolarlik-yatirim-dogrulandi/)

Source https://upcorn.co/turkiye-nin-unicorn-girisimleri/; https://fastcompany.com.tr/calisma-hayati/hedef-5-kita; https://peak.com/how-we-work/entering-a-unique-market-japan

entrepreneurship must succeed in two basic tasks: ensuring that the technology works according to the needs of the target customer and that the technology can be sold at a profit. For the entrepreneur, making sales while making a profit is usually sufficient. A digital business is an online business venture created, operated, and grown through digital technology. The concept of digital entrepreneurship is a new phenomenon, and both techno-entrepreneur and digital entrepreneur discuss entrepreneurship on a technology or digital basis and can be used in the same sense. A digital entrepreneur is an entrepreneur who uses digital technologies to create value for his customers [4] and offers digital services or products. Examples include developing new digital products or services, creating new platforms or websites, using digital marketing methods to reach new customers, and collaborating with platforms and business partners. When it comes to techno-entrepreneurs, startups should not only come to mind. Existing and large companies can also engage in techno-entrepreneurship, as startups do. Today, in addition to being credit providing institutions, banks are turning to financial technologies such as open banking, service banking and digital banking by using artificial intelligence, advanced analytics, blockchain and 5G technologies.

2.2 *Techno-Enterprises' Market Entry and Entrepreneurial Marketing Strategies to International Markets*

Entry strategies encompass a spectrum, from wholly owned subsidiaries to contractual agreements with independent suppliers, typically categorized into two main types: equity (insourcing) and non-equity (outsourcing) strategies. When a newly established, technology-based start-up with limited resources wants to expand into international markets, deciding which entry strategy to use is an important decision [5–8]. They need to evaluate their options to choose the entry mode that best suits their strategy and goals (Table 2). As seen in Table 2, exporting is typically the easiest way to enter an international market, and therefore most firms begin their international expansion using this model of entry. Exporting is the sale of products and services in different countries that are sourced from the home country. An international licensing agreement allows a foreign company (the *licensee*) to sell the products of a producer (the *licensor*) or to use its intellectual property (such as patents, trademarks, copyrights) in exchange for royalty fees. A strategic alliance involves a contractual agreement between two or more enterprises stipulating that the involved parties will cooperate in a certain way for a certain time to achieve a common purpose. There are several motivations for companies to consider a partnership as they expand globally, including (a) overcoming market entry barriers, government regulations (b) reduce risk and sharing, (c) get the resources (technology, finance, raw materials) (d) joint product development, and (e) low tax in some countries. An acquisition is a transaction in which a firm gains control of another firm by purchasing its stock, exchanging the stock for its own, or, in the case of a private firm, paying the owners a purchase price. Foreign direct investment refers to the formal establishment of business operations on foreign soil the building of factories, sales offices, and distribution networks to serve local markets in a nation other than the company's home country.

New, technological start-ups may export or consider collaborating with a reliable distributor in a market with cultural differences. 362 legally independent companies in the United Kingdom, less than ten years old, established between 1987 and 1997, with at least three employees, participated in the survey, and the responses of 246 high-technology-based start-ups with international activities were considered. It has been determined that start-ups made their first foreign sales to nearby countries (Western Europe, then the USA), while a smaller number of companies expanded into spatially distant markets. The economic size of the market and country-specific opportunities were effective in choosing the target market. When examining entry methods, it was found that 44% of startups favored direct export, 36% opted for distributors, 12% chose agency arrangements, 2% pursued joint ventures, 2% engaged in licensing, and 1% made direct sole investments. Each market entry strategy comes with its own set of advantages and disadvantages (Table 3).

Entrepreneurial marketing refers to marketing created and carried out by entrepreneurs. Hills et al. (2010) emphasize entrepreneurial marketing, particularly through the provision of innovation, creativity, sales expertise, market penetration

Table 2 International market entry strategies

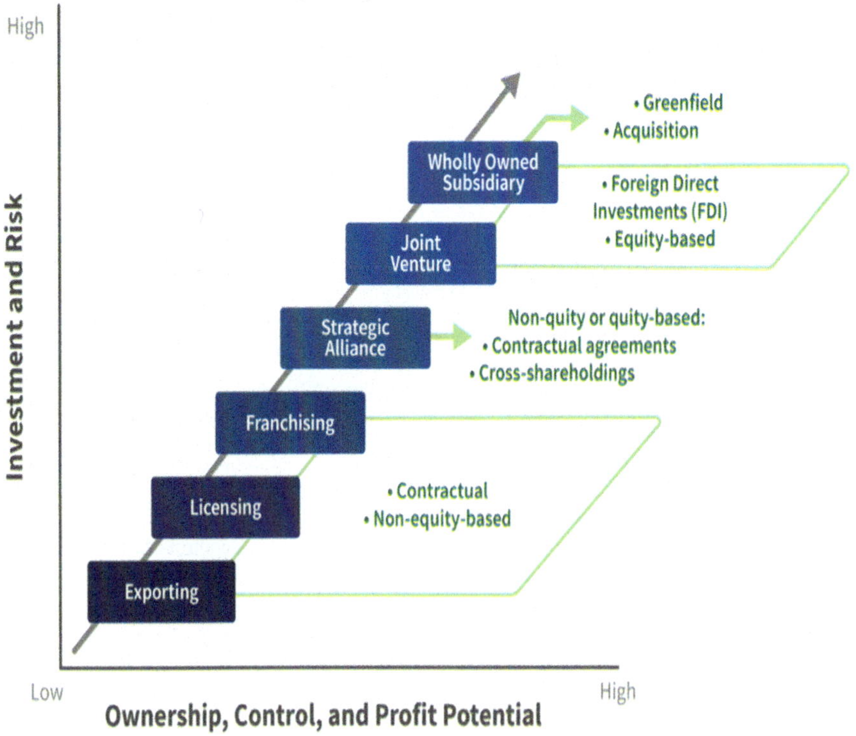

Source Babu John Mariadoss (2017), Core Principles of International Marketing, Open (Access) Textbook, https://ecampusontario.pressbooks.pub/globalmarketing/chapter/6-1-international-entry-modes/-

strategies, networking capabilities, and adaptability. He defined it as a power and orientation that creates perceived customer value through relationships as well as the pursuit of opportunities. According to Covin, entrepreneurial marketing includes providing more comprehensive customer support, placing greater emphasis on product quality, and being more interested in industry and market trends. There are three basic interface areas of marketing and entrepreneurship that constitute entrepreneurial marketing: being change-oriented, opportunistic, and innovative in management approaches [10]. Entrepreneurs first produce an idea and then try to find a market for this idea. In small-scale businesses, the marketing function is carried out intensively by the business owner, and direct communication is established with the target market. As marketing methods; participating in fairs, competitions, investor and entrepreneur meetings, buzz marketing, viral marketing, word of mouth marketing, guerrilla marketing, digital marketing tools (conducting marketing activities via the internet, website, blog, facebook, twitter, google, you tube, instagram, pinterest etc. social media marketing, e- mail marketing, mobile marketing, search engine marketing, content marketing, affiliate marketing), influencer, brand

Table 3 Advantages and disadvantages of international market entry strategies

Type of entry	Advantages	Disadvantages
E-Exporting (B2B-B2C)	Fast-entry, low financial risk, diversified revenue source, with current activities scale up	Low control, low local knowledge, potential negative environmental impact of transportation, cultural and language barriers, transportation costs and complexities, treating the company as a foreigner evaluation
Licensing and franchising	Fast-entry, low-cost, low-risk, low trade barriers, high royalty, shared marketing costs, potential for rapid global expansion, doesn't require factory and staff, preventing imitation	Less control, licensee may become a competitor, the quality of the licensee, legal and regulatory environment (IP and contract law) must be sound, difficulty of protecting information spread, quality/brand control challenges, royalty and franchise fees
Joint venture, partnering and strategic alliance	Overcoming cultural barriers, potential for learning, shared costs reduce investment needed, reduced risk, seen as local entity, access to new markets and customers, rapid market entry. 5 people founded LinkedIn. YouTube, Uber, Airbnb and Facebook started with 3 partners. Instagram, Dropbox and Evernote had 2 partners when they were founded. A significant majority of startups in Türkiye have 2–3 partners	Management difficulty, higher cost than exporting, licensing, or franchising; integration problems between two corporate cultures, risk of sharing sensitive information, lack of full control, dissemination of information, partners rival possibility
Acquisition	Fast entry into a new market, established operations, access to local distribution channels	High upfront cost, integration issues with home office, cultural differences
Greenfield venture (launch of a new, wholly owned subsidiary)	Gain local market knowledge; can be seen as an insider who employs locals; maximum control	High initial investment, high risk due to unknowns, longer time to establish operations, lack of local market knowledge

Source Babu John Mariadoss (2017), Core Principles of International Marketing, Open (Access) Textbook, https://ecampusontario.pressbooks.pub/globalmarketing/chapter/6-1-international-entry-modes/ [9]

ambassador, customer relationship management, brand management, neuromarketing, product placement, neuromarketing, green marketing are used. Entrepreneurs attend various events and fairs which bring customers and investors.

- Participating in Fairs, Competitions, Investor, and entrepreneur meetings: Techno-enterprises that develop innovation should follow fairs and competitions. Fairs are exhibition, communication, and meeting environments where buyers and sellers come together. Techno-enterprises, university and technopark start-ups can demonstrate their projects, can develop their projects, knowledge and ideas, and network. By participating in fairs and exhibitions, techno-enterprises promote their products and services and can get to know their potential competitors as well as their customers.
- Word of Mouth Marketing: Consumers tell their positive or negative product, brand or company experiences to other consumers and people around them (spouse, friend, relative, etc.). It is the verbal distribution and recommendation of information produced by the customer within their personal networks.
- Social media marketing: social media can have a major role in a techno-enterprises market entry strategy. Techno-enterprises can utilize various social media platforms to publish information and content, interact with targeted customers as well as connect with them. Facebook, Linkedin, blogs and microblogs, Twitter, Flicker, videos, compilation boards, online forums, and other social channels. A Web 2.0 social media marketing strategy can be instrumental for start-ups in promoting their products and services, including marketing their vehicles. In this marketing method, a one-on-one relationship is established with customers and negative comments are shared as well as positive ones. Therefore, customer evaluations should be given great importance.
- Viral Marketing: Viral marketing is the spread of content created on the internet like a virus. Viral marketing can be done on the web using social networks.
- Content marketing: It is about creating and distributing meaningful, current, and valid content. The goal is to create content that meets the customer's needs. For this, knowing the target audience and quality factors are very important. Content marketing emphasizes providing value to consumers. Content also serves purposes such as informing, entertaining, emotionally connecting, influencing, educating, building trust, and transforming. In the field of content marketing, SEO application strategies, activities to increase sharing, and content supporting strategies can be used. Prepared contents include blogs, infographics, slides, case studies, whitepapers, e-books, videos, tests, images, etc. Available in a wide variety of formats including:
- Search Engine Marketing (SEM): Search engine marketing is marketing activities carried out to be listed at the top of search engine results pages (SERP). As a historical development, SEM consists of two subheadings: Search Engine Advertising (paid search engine advertising) and SEO (search engine optimization). If a company's website gets an accessible spot on the result page, it can drive more customers to their website. The businesses that master SEO techniques are the ones with the most accessible placements on the search engine page.

- Affiliate marketing: Affiliate marketing is the use of the technopreneurship's network outside its own website, communicating with content creators such as YouTubers, bloggers, influencers who can make sales, and selling goods and services in other digital marketing channels by agreeing on a commission.
- Guerrilla Marketing: Innovative and effective communication techniques such as communication, networking and imagination are used along with creative use of available resources and a highly targeted effect. Techno-entrepreneurs need to actively seek out opportunities. Making sales using online systems, improving sales by giving importance to feedback, experiential marketing where the public will interact, digital guerrilla approaches (storytelling, visuals, social media, pop-up ads, webinars, games, podcasts, applications, and competitions) can be used.
- Email Marketing: Email marketing is a tool that enables businesses to generate and share email messages, commercial information and advertise with their website's guests and customers. It is important to surprise the reader and present innovations. It is convenient, easy to use, and highly effective as a method of engaging large audiences, making sales, and converting customers. Messages are sent to find the target market.
- Mobile Marketing: Mobile marketing is marketing using phones and similar mobile devices and the software and applications on these devices. As examples of mobile marketing applications, SMS/MMS, mobile coupons, mobile games with social sharing that increase user interaction, mobile applications, mobile payment. systems, mobile banking, mobile search optimization, barcodes that direct to a certain site/video when scanned, mobile ticket applications (QR codes), and voice-based mobile advertisements can be given [11].
- Data-Based Marketing: It is a marketing method that includes collecting and evaluating customer data and sending messages according to their lifestyles and behaviors to increase interaction with targeted consumers and improve customer loyalty.
- Brand Ambassador: The person who represents the brand, product, service, or activity of the business, tries to create a reliable image by introducing the brand to the public or a certain target audience, and defends the brand. The brand ambassador tries to increase the awareness and credibility of the techno-startup by creating content such as photos, videos, blog posts about the techno-startup on social media, and by sharing his personal experiences and recommendations.
- Influencer Marketing: Techno-enterprises can also use social media to reach out to key industry influencers and plan promotional social media campaigns in collaboration with them. Influencer marketing is a marketing practice that utilizes well-followed online users who can influence consumers' attitudes and decision-making processes in favor of brands or ideas. Influencers can be named with different concepts such as blogger, youtuber, traveler, influencer and vlogger.
- Customer Relationship Management: is an approach to understanding and influencing customer behavior through meaningful communications to acquire new customers, retain customers, improve customer loyalty and customer profitability. Entrepreneurial marketing focuses on innovative approaches to using existing

customer relationships to create new customer relationships or market shares [12].

- Product placement: basically, it is the placement of brands in various media channels for a fee, with the aim of creating and increasing awareness. Techno-enterprise can apply for product placement support in feature films, documentaries, TV series, animation films regarding the developed product (software, mobile application, digital game, etc.).
- Brand management: is the accurate and effective management of all kinds of message sources such as design, quality, packaging, guarantees offered to the consumer, price, distribution network, stores, sales personnel, advertisements, corporate communication strategies.
- Neuromarketing: It is a field of study that uses medical technologies such as face reading, facial expression analysis, eye tracking, neuroimaging to stimuli to analyze the brain's reactions to the product or service and examines the reactions of consumers to marketing communication messages.
- Green or environmental marketing includes all kinds of environmental and nature-friendly marketing activities aimed at meeting consumer demands and needs. These are studies on the positive or negative effects of marketing activities on environmental pollution, energy consumption and consumption of other resources. In the process of sustainable development, businesses should focus on renewable energy sources that do not harm the environment by making clean production, use environmentally friendly, recyclable raw materials in supply, and adopt green strategies such as green product, green price, green distribution, and green promotion in marketing components. In promotional activities, techno-enterprises that inform their consumers about environmental problems, create environmentally friendly consumer awareness, and develop projects accordingly will reach environmentally friendly customers and provide more customer satisfaction [13].

3 Research Methodology and Findings of Techno-Enterprises Market Entry Strategies

3.1 Sampling, Data Collection and Analysis Methods

Entrepreneurial marketing seeks out new opportunities, with businesses that adopt this strategy showing a creative and innovative marketing approach. Rather than being driven by the market, entrepreneurial marketing continually leads the market through innovation and drives the market by exploring new markets with new products. To delve into our research, we conducted interviews among nine techno-enterprises located at Ankara, Balıkesir, Adana and Kastamonu which are involved in developing, commercializing, or manufacturing (high) technology. These nine techno-enterprises who accepted interview are asked detailed questions through in-depth interview regarding their entry modes and marketing strategies in international markets. The reason for using typical sampling which is a type of purpose sampling is to inform the ones having limited information about the process by interviewing and making use of the sights of enterprises who have an idea of entry modes and marketing strategies. We aimed not to make generalizations. These nine techno-enterprises were called by phone; they were informed about the purpose of the study and asked for an interview. To ensure the research's validity, multiple sources of evidence were used to verify information from existing publications. The findings were gathered from a variety of techno-enterprises. Afterwards we shared our findings with the interviewees. Also, the sights of other researchers working on the same field were consulted. To ensure reliability, an interview form where the research questions are clearly stated was used. The data gathered were quoted directly without any comments. Interviews were conducted with two researchers. The interview questions were previously prepared. We asked the following questions to the techno-enterprises: Are techno-enterprises that have been operating for one year considering entering different country markets? What are the entry strategies to international markets? What integrated marketing communication methods do they use to promote their goods and services once they enter the market?

The research data were summarized in Table 4 based on the themes from the interview form. We assigned a number for each techno-enterprises interviewed and enlisted them. Thus, the statements of nine techno-enterprises were compared, and the collected findings were organized into a table. The techno-entrepreneurs conducted were established between 2017 and 2021.The number of employees in surveyed start-ups is between four and 40. The first question in the interview aimed to discover the entry strategies of techno-enterprises. Next questions are related to the significance of integrated marketing communication methods.

The first techno-entrepreneur in the list stated that 'content marketing, social media marketing and having a company's web page are important for start-up companies'. He said that they received installation support from the state with the support of KOSGEB young entrepreneurs and that Nace codes made their work easier during the establishment phase. Additionally, he said: 'without using digital transformation tools and new generation technologies, start-up companies will fall behind compared to their competitors and help should be sought from mentors'.

Second techno-entrepreneur stated that 'it is impossible to do business and survive in the sector without advertising, customer relationship management, social media and influencers, and it is necessary to get an opinion from at least one mentor'. Moreover, he said: 'In order for the techno-venture to be successful, access to finance is important, taxation should be easier, and the initiative should be started in the early stages. He talked about the importance of angel investors collaborating with start-ups by increasing public policies that support them.

The third techno-entrepreneur stated that they only provide B2B services in the healthcare sector. They also create networks with other companies using the WOM method. They carefully manage the websites and social media of the hospitals they work for rather than their own websites, and that they willingly established themselves outside the technoparks instead of opening an office in the technoparks. He stated that, "Since they are currently working with 47 companies, in some periods

Table 4 International market entry and entrepreneurial marketing strategies of techno-enterprises

		Categories							
		Founding year	First entry to Int. market	Number of team members	Sector	International entry strategy	Use of digital transformation tools	Marketing strategies	What is the business model? Brings solution?
Interviews	Interviewee 1	2021	2022	7	Service (Marketing-advertising)	Trade fairs, B2B (UK)	+	youtube-instagram content marketing, email, WOM, website	To carry out the digital marketing of the companies and to establish graphic design, advertising, and website, and concurrently, to make separate promises for each customer and to meet customer requests and needs. In other words, instead of going to a graphic designer, software developer to set up a website, and an advertiser separately, to provide 3-in-one technology innovations for customers

(continued)

Table 4 (continued)

		Categories							
		Founding year	First entry to Int. market	Number of team members	Sector	International entry strategy	Use of digital transformation tools	Marketing strategies	What is the business model? Brings solution?
	Interviewee 2	2021	2022	5	Software E-commerce	B2B	+	youtube-instagram content marketing, email, WOM, website	Developing and supporting software. With their chatbot service, they offer 24/7 live support and automatic delivery
	Interviewee 3	2017	2019	15	Service (Marketing-Advertising)	B2B	+	WOM, website	To serve only in the health sector and to do every job with new generation technologies in the field of digital marketing and advertising, except for printing, unlike 360-degree advertising

(continued)

Table 4 (continued)

	Categories							
	Founding year	First entry to Int. market	Number of team members	Sector	International entry strategy	Use of digital transformation tools	Marketing strategies	What is the business model? Brings solution?
Interviewee 4	2017	2019	10	Service (Marketing-Advertising)	Trade Fairs, B2B	+	youtube-instagram content marketing, Influencer (collaboration with Instagram and Tiktok influenca	To do video, photography, and advertising business at the same time and to provide digital branding and advertising management services
Interviewee 5	2021	2022	40	Production	Trade Fairs, B2B UK-Germany-Dubai	+	youtube-instagram content marketing, product placement, green marketing, email, WOM, affiliate marketing	Freeze Dry technology is a freeze-drying method. With this method, the water contained in the fruits is removed by sublimation and dried fruit is obtained. The company uses this innovative technology for drying products such as lemons and olives as well as fruits. At the same time, it also obtains freezing with mill dried technology. The company has an e-commerce site and factory related to this innovative technology

(continued)

Table 4 (continued)

		Categories							
		Founding year	First entry to Int. market	Number of team members	Sector	International entry strategy	Use of digital transformation tools	Marketing strategies	What is the business model? Brings solution?
	Interviewee 6	2020	-	20	Service- e-commerce, software (application developer for restaurants	B2B	+	youtube-instagram content marketing, email, WOM, website	Based on his restaurant, an entrepreneur operating in the field of restaurant and animal husbandry developed application software with new generation technologies and added QR codes to his restaurants. It also markets the new generation application software it has developed for its own restaurant for other sectors
	Interviewee 7	2021	–	5	Service (Marketing-advertising)	–	+	instagram-tiktok, green marketing, WOM, email, Influencer (Tiktok)	It is the implementation of all of them under one roof instead of dealing with them separately by using social media and design, video shooting and photography together by carrying out the digital marketing of the companies

(continued)

Table 4 (continued)

		Categories							
		Founding year	First entry to Int. market	Number of team members	Sector	International entry strategy	Use of digital transformation tools	Marketing strategies	What is the business model? Brings solution?
	Interviewee 8	2022	2022	4	Software	B2B	+	Website, WOM	Software development, selling, implementing, developing two projects (learning management system and no code software) abroad
	Interviewee 9	2022	2022	4	Software	B2B		youtube-instagram content marketing, email, WOM, website	Software development, selling

they have freelance colleagues because the number of employees is not sufficient". He said that 'new generation technologies make their lives and jobs easier'. He mentioned about the importance of "seeking help from a mentor from the establishment stage and the process when the idea emerges" and that "start-up companies should open up to international markets as soon as possible" and emphasized that in terms of sustainability, "their priority is economic sustainability."

The fourth techno-entrepreneur said that 'they could not be established in technoparks due to cost, therefore they could not benefit from the incubation center in the technopark' and added that 'digital marketing will be incomplete without digital transformation tools and start-up companies should definitely use social media and new generation technologies'. He stressed the importance of 'to be successful, access to finance and taxation should be easier for the technopreneurship. Additionally, public policy support in the early stages of the initiative, well-established companies being open for cooperation with start-ups as angel customers and reaching big players and potential customers'. Also, he added that 'Environmental sustainability in e-commerce and digital marketing can be achieved through the use of renewable energy'.

The fifth techno-entrepreneur emphasized 'the importance of products dried and frozen with healthier technology without added sugar and additives in terms of environmental, economic and social sustainability' and said that 'dried beet leaves are fed to lambs and there is a solar roof system on the roofs of the factories'. He noted that 'Every company should be aware of green design and the use of solar energy and total water and energy usage. 'In addition, he mentioned "The importance of R&D, production, sales and marketing in the circular economy" to his answers.

The sixth techno-entrepreneur operates in the service sector and emphasizes the 'importance of working with a mentor' while developing application software with new generation technologies, and states that 'both the service sector and e-commerce will become much more important thanks to digital transformation, including the website, and the development of artificial intelligence technologies. Further that, he thought it would make it easier to establish a start-up company. He noticed that 'Innovation expenditures on developing new sustainable products, along with the importance given to research and development, recognizing and valuing the talents of employees, and occupational health and safety, innovation and fair trade are also important in social and economic sustainability'.

The seventh techno-entrepreneur stated that 'they decided to establish a start-up company while receiving marketing and advertising training and received aid from three mentors in this process' and remarked that even though they operate in the field of service, they also present particular importance to green marketing and compliance with carbon footprint and environmental legislation. Additionally, he said, 'Everyone should be made aware of the circular economy by saying that they are aware of all areas of environmental sustainability and that they are and will take part in all studies related to this'.

The eight technoentrepreneur in the table said that 'being established in a technopark has great advantages and they are happy about it'. 'The difference between start-up companies and SMEs and entrepreneurs should also be stated, that there are too many start-ups in the software industry, that it is necessary to open up to international markets during the incubation period (3-year period) by using new generation technologies, and that start-up companies must become limited companies after completing the 3-year process. "It should be transformed," he said. He also noted his foresight: 'there are many developing technoparks in Ankara and there will be more technoparks to be established'.

The last technoentrepreneur said that "We develop software for companies, we have different projects, we managed to enter the Dutch market by advertising on Google and Facebook, we show importance to webinars, e-books and blogs".

4 Conclusion

This study investigated the entry modes and marketing strategies employed by techno-enterprises when entering international markets. Literature studies indicate that there isn't a single universally correct method of internationalization, and this process can't be fully explained by a single theory. The process of each company that achieves internationalization may be different. As emphasized in many studies, the success of techno-enterprises in international markets is related to how accurately entrepreneurial marketing is carried out. Techno-enterprises that want to enter international markets are change-oriented and closely follow rapid changes in technology, digital transformation technologies, and how consumer behavior changes. They adopt a more relaxed approach to assessing consumer behavior. Entrepreneurs engage with individuals they encounter, gathering market, industry, and business insights through conversation and dialogue. They pursue opportunities regardless of available resources. According to Uppsala theory, obtaining market information is still one of the driving forces of internationalization, as companies obtain information through business networks. Creating a network, partnership, customer-oriented strategy, acquiring the market knowledge, mutual trust, and choosing a minority share joint venture strategy may be the right strategy for techno-enterprises. Nowadays, technology and innovation capabilities provide companies with a competitive advantage and help them overcome many challenges. Companies that develop new products and establish a sustainable business model experience a faster growth process. Techno-startups can promote their products and services at technology fairs, as well as create a joint venture strategy by attracting the right investments. They can benefit from incubation centers, competitions, investor, and entrepreneur meetings. Techno-enterprises identifying market opportunities based on technological superiority in the target country market rely on the entrepreneurial behavior of company founders, their networks, experience, and their utilization of entrepreneurial marketing methods to facilitate the internationalization process. Techno-enterprises aim to expand quickly by targeting country markets with significant market opportunities and market size. Despite having fewer financial resources initially compared to large companies, the technology, knowledge, business models, extensive business networks, relationships, carefully chosen partners, and entrepreneurial behavior of the founders play a significant role in the internationalization process. Another issue that caught our attention during the interviews was that their parents had a family business, or their father were entrepreneurs. The limitation of the research is that interviews were conducted with nine newly opened companies that were at a very early stage of entering to international markets and agreed to be interviewed. When entering international markets, techno-enterprises prefer risk-reducing strategies such as B2B exporting, participating in fairs, competitions, investor and entrepreneur meetings strategic cooperation and creating joint ventures, partnerships.

References

1. Felipe B, Cahen F, de Miranda M, Junior O (2017) The internationalisation of new technology-based firms from emerging markets. Int J Technol Manag 74:23–41
2. Sledzik K (2015) Schumpeter's view on innovation and entrepreneurship. SSRN Electron J 89–95
3. Rupali S (2019) Difference between startup and small business. GAP Interdiscip Int Peer Rev Open Access J Interdiscip Stud 2(1):251–254
4. Güğerçin S, Onan G (2021) Digital entrepreneurship: a literature review. Int Eur J Manag Res 5(9):291–305
5. Nguyen CS (2021)The international business incubator as an entry mode- a systematic literature review. Studia Prawno-Ekonomiczne, T. CXIX, pp 255–276
6. Xi G, Block J, Lasch F, Robert F, Thurik R (2017) Mode of entry into hybrid entrepreneurship: new venture start-up versus business takeover. IZA—Institute of Labor Economics, Discussion Paper series
7. Hashai N, Markovich S (2017) Market entry by high technology startups: the effect of competition level and startup innovativeness. Strat Sci 2(3):141–160
8. Kalra P (2023) Foreign market entry strategies: a comparative analysis. Asian J Manag Commer 4(1):276–281
9. Gordon M (2000) The international market entry choices of start-up companies in high-technology industries. J Int Mark 8(2):33–62
10. Mariadoss BJ (2017) Core principles of international marketing. https://ecampusontario.pressbooks.pub/globalmarketing/chapter/6-1-international-entry-modes/
11. Eken İ, Gezmen B (2020) E-retailing practices in mobile marketing: the case of getir application. E-Retailing Practices in Mobile Marketing: The Case of Getir Application. IGI Global. https://www.researchgate.net/publication/338302513
12. Azizağaoğlu A (2019) Girişimci Pazarlama: Pazarlamaya Proaktif ve İnovatif Bir Yaklaşım. Uluslararası Yönetim İktisat ve İşletme Dergisi, Cilt 15, Sayı 1, s. 276–288
13. Bahçecik D, Talebi N (2020) Yeşil pazarlama: İşletmelerin yeşil pazarlamaya yönelme nedenleri. ABMYO Dergisi - Yıl 15, Sayı 59 - Temmuz , s. 223–247

Measurement and Analysis of Extrusion Forces in Filament-Based Additive Manufacturing Technologies Using ABS Filament

Ioannis Christodoulou, Antonios Trikkas, and Angelos Markopoulos

Abstract This paper explores the measurement and analysis of extrusion forces in the Fused Deposition Modeling (FDM) process, utilizing a custom-built 3D printer designed for high-performance additive manufacturing. A load cell, strategically positioned between the extruder and the hotend, was employed to capture force data during extrusion. The experiments were conducted across various extrusion temperatures (200, 225, 250 °C) and filament feed rates (ranging from 1 to 12 mm/s) to examine the impact of these parameters on the extrusion force. Acrylonitrile Butadiene Styrene (ABS) filament was selected as the material for all experiments due to its widespread use and availability in the industry, making it a standard choice for many FDM applications. The study leverages Klipper firmware for precise control of the printing conditions, ensuring accuracy in the data collected. The findings contribute to a deeper understanding of the extrusion dynamics in FDM, offering valuable insights for optimizing the process and validating numerical models designed to predict extrusion behavior. This research highlights the critical role of force measurement in advancing the capabilities and reliability of additive manufacturing.

Keywords Additive manufacturing · Fused filament fabrication · Extrusion force · Material flow measurements · Nozzle forces measurements · Acrylonitrile butadiene styrene (ABS) · High-sensitivity force measurement

I. Christodoulou · A. Trikkas · A. Markopoulos (✉)
School of Mechanical Engineering, National Technical University of Athens, Zografou, Greece
e-mail: amark@mail.ntua.gr

I. Christodoulou
e-mail: ichristodoulou@mail.ntua.gr

A. Trikkas
e-mail: mc18402@mail.ntua.gr

N. M. Durakbasa and K. G. Gülen (eds.), *Sustainable Green Conversion*, Lecture Notes in Mechanical Engineering, https://doi.org/10.1007/978-3-031-83583-4_18

1 Introduction

Additive manufacturing (AM) is an innovative technology that constructs objects layer by layer directly from digital models, allowing for the creation of intricate geometries that would be challenging or unattainable with conventional manufacturing techniques. In contrast to subtractive manufacturing methods, AM focuses on selectively adding material only where it is necessary. This approach leads to reduced waste and a more optimized utilization of resources. This technology has become widely adopted in industries such as aerospace, automotive, healthcare, and consumer products. It plays a crucial role in enabling rapid prototyping, customization, and low-volume production. One of the most widely used AM technologies is Fused Deposition Modeling (FDM) because of its affordability, ease of use, and the wide range of thermoplastic materials it can work with, such as ABS and PETG. FDM allows for the creation of long-lasting, practical components that are suitable for both testing and final product development. This technology fosters creativity and expedites the introduction of new products to the market.

The mechanical properties and dimensional accuracy of components produced by AM are essential for their performance in practical applications. Extensive research has been conducted on FDM, which is a highly popular AM technique. The focus of these studies has been to analyze the impact of different process parameters on the properties of the final product. Having a strong grasp on these parameters is crucial in order to guarantee that the final parts meet the necessary specifications for strength, durability, and precision.

The mechanical properties of FDM-produced parts, such as tensile strength, are influenced by various factors, including layer thickness, infill density, and infill pattern. Studies have indicated that reducing the thickness of each layer tends to improve the tensile strength of the material. This is because it promotes stronger adhesion between layers and minimizes the presence of gaps. As research has shown, decreasing the layer thickness from 0.3 to 0.1 mm can lead to a significant 20% increase in tensile strength [1]. In the same vein, enhancing infill density also plays a role in enhancing mechanical strength, as denser infill provides greater material support within the part. A higher infill density of 80% has been proven to yield superior tensile strength in comparison to lower densities, particularly when paired with optimized infill patterns like grid or cubic [2, 3].

The relationship between temperature and shear rate is critical for understanding the flow behaviour of reprocessed ABS during the injection molding process. As the shear rate increases, the material's viscosity decreases, a behaviour typical of shear-thinning or pseudoplastic materials like ABS. This reduction in viscosity is accompanied by changes in the temperature distribution within the material due to increased internal friction (shear heating) at higher shear rates. As the polymer is subjected to higher shear rates, the heat generated from internal friction leads to a localized increase in temperature, which can further reduce the material's viscosity and facilitate flow.

However, this relationship is complex and dependent on the material's initial viscosity. In the study from T. Boronat et al., it was found that low viscosity ABS experiences more significant reductions in viscosity at high shear rates compared to high viscosity ABS, which exhibits greater resistance to flow and less shear thinning. Therefore, the interplay between shear rate and temperature has a pronounced effect on the processing behavior, particularly for low viscosity ABS, where high shear rates can exacerbate thermal degradation if not controlled properly [4].

Temperature significantly impacts the thermal and mechanical properties of the recycled ABS during the twin-screw extrusion process. Higher extrusion temperatures result in enhanced flowability due to a reduction in the material's viscosity, which facilitates easier extrusion. The thermal analysis performed in the study demonstrated that primary recycled ABS exhibited improved thermal stability compared to virgin ABS, particularly at elevated temperatures [5].

Ensuring precise dimensions is crucial for maintaining the high quality of FDM-produced parts. Attaining a high level of precision is crucial for components that must fit together with exacting tolerances or that demand flawless surface finishes for both aesthetic and functional purposes. Research has indicated that utilizing smaller layer heights can enhance both the mechanical strength and dimensional accuracy. Using a layer thickness of 0.1 mm instead of 0.2 mm can greatly minimize dimensional deviations, ensuring accuracy within 0.05 mm [6].

Considering the substantial influence of these parameters on the mechanical properties and dimensional accuracy of FDM parts, it is crucial to have meticulous control over the printing process. Although empirical methods offer valuable insights, their effectiveness can be constrained by the specific conditions in which experiments are conducted. In order to achieve consistent and predictable results, particularly when scaling up production or transitioning between various materials and designs, a more reliable approach is required.

Using numerical modelling is the most efficient way to address this challenge. Through the development and validation of models, manufacturers can enhance the FDM process by accurately predicting the impact of process parameters on mechanical properties and dimensional accuracy. This allows for a more controlled and consistent optimization of the process. Nevertheless, in order for these models to be considered reliable, it is crucial to conduct a comprehensive evaluation and validation process using real-world data.

One objective and measurable metrics for evaluating these models is the measurement of extrusion force during the printing process. Extrusion force serves an essential role in the AM process, especially in FDM. The extrusion force has a significant impact on the quality and consistency of the printed parts. It affects factors like material flow, layer adhesion, and the final mechanical properties. Having a thorough grasp of and effectively managing extrusion force is crucial in order to maximize the efficiency of the FDM process and guarantee that the end products meet the desired specifications.

The extrusion force is determined by various factors, such as the design of the nozzle, the properties of the material, and the printing conditions like temperature and feed rate. Precise measurement and control of extrusion force are essential for ensuring a consistent material flow during printing, which directly impacts the overall quality of the printed object. For instance, a rise in extrusion force may suggest potential problems like nozzle blockage or inadequate heating, resulting in subpar layer adhesion and flaws in the final product. On the other hand, a decrease in extrusion force could indicate under-extrusion, which can lead to layers that are weak or incomplete.

Research has indicated that the force required for extrusion can vary considerably based on the material being used and the specific printing conditions. When printing with ABS filament, the extrusion forces can vary from 20 to 50 N based on factors such as the nozzle diameter and extrusion temperature [6]. Validation of numerical models through experimental measurements is crucial to ensure their accuracy.

The geometry of the nozzle plays a crucial role in determining the amount of extrusion force required. Using narrower nozzles can result in higher forces needed to extrude the material, potentially causing more wear on the printer and affecting the quality of the prints. As an illustration, it has been demonstrated that decreasing the nozzle diameter from 0.6 to 0.4 mm can lead to a significant 50% increase in the necessary extrusion force [7]. In addition, the viscosity of the material is a crucial factor in determining the force required for extrusion. Filaments that have a higher viscosity necessitate a greater amount of force for extrusion, especially when temperatures are lower. The significance of choosing the right nozzle size and temperature settings based on the material used cannot be overstated.

There have been advancements in the development of in-line measurement systems that can effectively monitor extrusion force while printing. These systems offer up-to-the-minute data that can be utilized to adapt printing parameters in real-time, guaranteeing a steady material flow and minimizing the chances of defects [8]. These systems are highly valuable for precise applications where even small variations in extrusion force can greatly affect the final part.

The research presented in this paper addresses a critical gap in the field of AM, specifically in the domain of FDM. While previous studies have extensively explored the effects of various process parameters on the mechanical properties and dimensional accuracy of FDM-printed parts, there remains a significant need for a comprehensive understanding and control of extrusion forces during the printing process and the prediction of the printing quality. The novelty of this research lies in the utilization of this integrated approach, to monitor fluctuation and other peculiar behaviour of the extrusion force of the ABS material during FDM printing.

2 Experimental Methods

The design and manufacturing of the custom-built test machine was necessary to measure the extrusion force during FDM. This machine was developed to provide real-time, accurate measurements of the forces exerted on the filament as it is extruded through the nozzle during the printing process and the design process is described here.

The test machine is integrated into an in-lab developed 3D printer, specifically engineered for high-performance FDM. This printer is capable of achieving high extrusion rates, print speeds, and accelerations, which are essential for testing the upper limits of extrusion force under various conditions. The machine was designed and manufactured with the goal of providing real-time, accurate measurements of extrusion forces, enabling a deeper understanding of the FDM process and validating numerical models used for process optimization [9].

In a standard FDM printer, the extrusion process consists of various essential components that play a crucial role in ensuring the overall performance and quality of the printed part. The extruder motor is in charge of propelling the filament into the extruder. It provides the necessary power to the extruder gears, enabling them to securely grip the filament and propel it towards the hot end. The level of force exerted by these gears is crucial, as it needs to be strong enough to overcome the resistance of the filament as it passes through the system.

When the filament is inserted into the hot end, it goes through a heated section where it melts. The hot end is comprised of a heating block and a nozzle. The heating block efficiently increases the temperature of the filament, enabling it to flow seamlessly through the nozzle. The nozzle carefully extrudes the molten filament onto the build platform, allowing it to cool and solidify into the precise shape required.

The forces in this system are concentrated at two crucial points: the interaction between the extruder gears and the filament, where the mechanical force propels the filament, and the hot end, where the molten filament's resistance against the nozzle generates a backpressure, as it can be seen in Fig. 1. The backpressure, along with the force exerted by the extruder gears, is what makes up the extrusion force. It is crucial to effectively control this force in order to maintain a steady material flow and avoid problems like under-extrusion or nozzle clogging, which may negatively impact the overall quality of the printed object.

To accurately measure the reaction force exerted on the extruder during the FDM process, it is necessary to place a force sensor between the hotend and the extruder

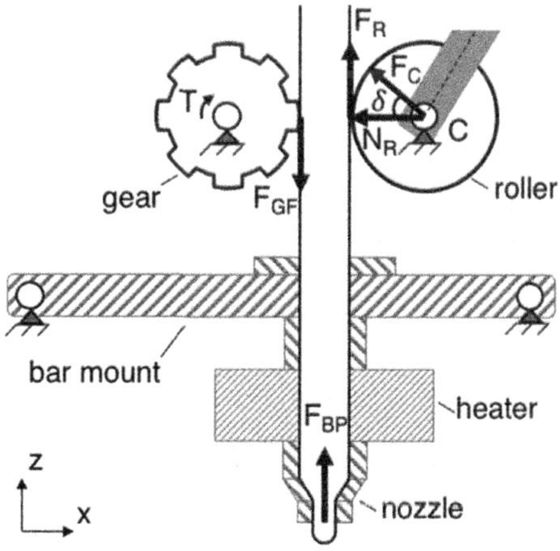

Fig. 1 Gear force (F_{GF}) and backpressure force (F_{BP}) on filament. Reaction force (F_R) on extruder gears and as a result on extruder assembly itself [10]

assembly as shown in Fig. 2. This positioning allows the sensor to capture the real-time forces generated as the filament is driven through the extruder gears and into the heated nozzle of the hotend.

A load cell with a full-bridge strain gauge configuration was selected to measure the extrusion forces on the FDM printer's extruder. This choice was made due to the high sensitivity and precision offered by the full-bridge strain gauge setup, which is essential for accurately detecting the small and dynamic forces involved in the extrusion process. The load cell's capacity of 20 kg ensures that it is well-suited to handle the expected range of forces without compromising accuracy. Additionally, its robust construction provides reliability, ensuring consistent and precise force measurements, which are critical for validating numerical models and optimizing the FDM process.

A load cell amplifier was chosen to amplify the signal from the load cell used in measuring extrusion forces. The HX711 was selected due to its high precision in converting the small analog signals from the load cell into digital form, which is

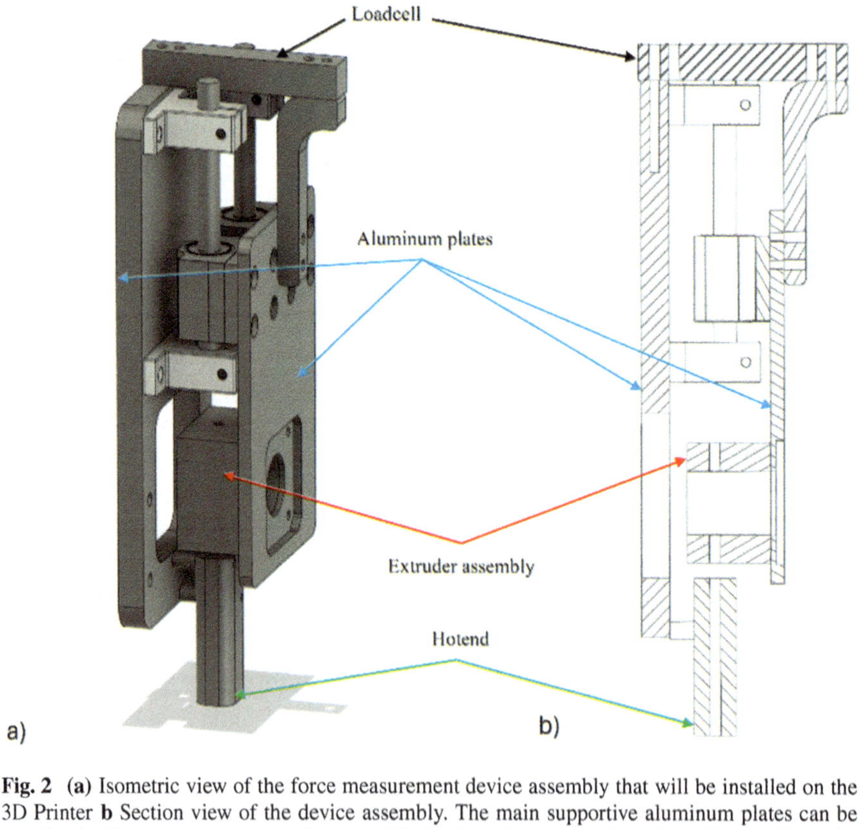

Fig. 2 **(a)** Isometric view of the force measurement device assembly that will be installed on the 3D Printer **b** Section view of the device assembly. The main supportive aluminum plates can be seen by the blue arrows, the extruder assembly by the red arrows and the hotend assembly y the green arrows. The loadcell can be seen by black arrow

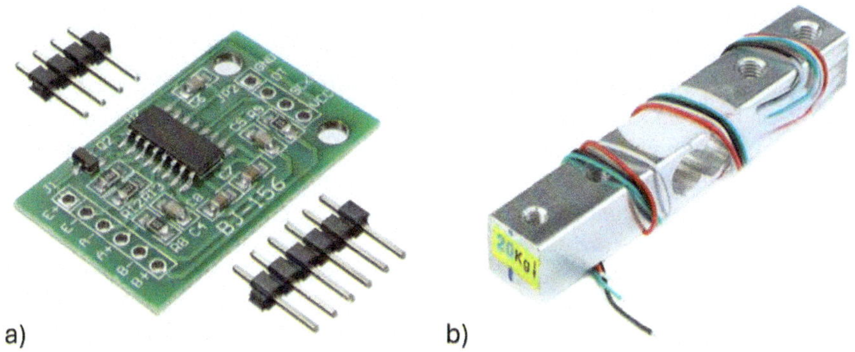

Fig. 3 **(a)** HX711 Analog to digital converter—Analog signal amplifier **b** full bridge strain gauge of 20 kg capacity

crucial for accurate data acquisition. The amplifier features a 24-bit analog-to-digital converter (ADC), which ensures that even minute variations in the extrusion force are captured with high resolution. Additionally, the HX711's compatibility with the chosen load cell and its ability to provide stable and reliable signal amplification were key factors in its selection.

The above components can be seen in Fig. 3 below:

Based on Fig. 2 it is clear that the load cell is mounted on an aluminum plate, which was securely positioned under the extruder. The hotend was similarly attached to another aluminum plate. These two plates are connected using linear bearings, which were precisely aligned to allow movement exclusively along the Z-axis. This design ensures that any force exerted during the extrusion process is directed through the load cell, providing accurate and reliable measurements of the reaction force. The use of linear bearings restricts unwanted movement in other directions, ensuring that the measured force accurately reflects the extrusion force along the intended axis.

At this point it is necessary to be pointed that the overall stiffness of the assembly greatly outreach the loads that the aluminum plates will obtain during the tests. This is to ensure that the measurements are correct, and the elasticity factor of the construction does not play any role on the outcome.

The experiments were carefully designed to investigate the relationship between extrusion temperature, filament feed rate, and the resulting extrusion force in the FDM process. To achieve this, a series of tests were conducted across a range of extrusion temperatures and filament velocities, providing comprehensive data on how these variables influence the force required to extrude the filament.

Three different extrusion temperatures were selected for the experiments: 200, 225, and 250 °C. These temperatures represent the typical operating range for common thermoplastics such as Acrylonitrile Butadiene Styrene (ABS), which are widely used in FDM. By varying the temperature, the experiments aimed to observe how changes in the viscosity of the molten filament affect the extrusion force. Lower temperatures are expected to result in higher viscosity and thus higher extrusion forces, while higher temperatures should reduce the viscosity and the corresponding force required to extrude the filament.

The experiments were also conducted at six different filament feed rates: 1, 2, 3, 4, 6, 9, and 12 mm/s. These velocities cover a broad range of printing conditions, from low-speed, high-precision settings to high-speed, high-throughput operations. By testing at these varied feed rates, the experiments sought to determine how the speed at which the filament is pushed through the nozzle influences the extrusion force. Higher feed rates are generally associated with increased extrusion forces due to the greater resistance encountered by the filament as it is forced through the nozzle at higher speeds.

All the experiment parameters as well as the filament choice can be seen on the Table 1 below:

To ensure accurate control over the critical parameters of nozzle temperature and filament extrusion velocity during the experiments, Klipper firmware was employed on the custom-built 3D printer. Klipper was selected due to its capability to precisely manage these parameters with high resolution and consistency. The firmware allows for fine-tuning of the nozzle temperature, ensuring that it remains stable within the desired range during extrusion, which is essential for maintaining consistent filament viscosity. Additionally, Klipper provides precise control over the filament feed rate, enabling the extrusion velocity to be adjusted in small increments. The control boxes of the firmware can be seen on the Fig. 4 below:

In order to ensure precise force measurements, the load cell utilized in this investigation underwent calibration with the assistance of an Arduino Nano microcontroller. The calibration method entailed establishing a connection between the load cell and the Arduino Nano using a HX711 load cell amplifier. This amplifier was responsible for transforming the load cell's minute analog impulses into digital data that could be processed by the microcontroller. The calibration technique was crucial in establishing the correlation between the unprocessed data from the load cell and the actual forces exerted, hence facilitating the precise interpretation of the extrusion forces recorded during the trials.

The calibration process was initiated by setting up the load cell with known weights, allowing the raw readings from the Arduino to be mapped to actual force values. The load cell was first tared (zeroed) to remove any offset, ensuring that the force readings began from zero when no load was applied. Incremental weights were then applied to the load cell, and the corresponding readings were recorded.

Table 1 Experiment parameters

Extruding filament velocity (mm/s)	1	2	3	4	6	9	12
Nozzle temperature (°C)	200		225		250		
Filament	Acrylonitrile butadiene styrene (ABS)						

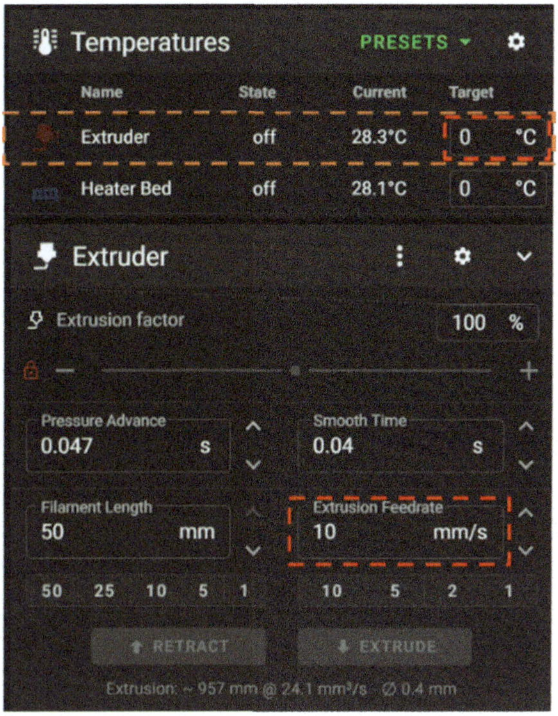

Fig. 4 Temperature and extruder control sections of Klipper firmware web interface. In the boxes with red dashed line the temperature and extruder feed rate can be controlled accordingly

The calibration process was facilitated by utilizing the HX711 library, which provides functions to tare (zero) the load cell and set the scale factor. The tare function was used to eliminate any initial offset in the readings, ensuring that the force measurements began from a zero baseline. Once tared, the known weights were applied to the load cell, and the corresponding raw data was collected. Using the *set_scale* function in the library, a calibration factor was derived by dividing the known weight by the raw data output. This calibration factor allows for the conversion of raw load cell readings into accurate force measurements, which is essential for analyzing the extrusion forces in the FDM process.

3 Results and Discussion

The results from the force measurements conducted during the extrusion experiments are presented and discussed. The measurements were taken at three different extrusion temperatures (200, 225, and 250 °C) and across seven filament feed rates (1, 2, 3, 4, 6, 9, and 12 mm/s). The data provides insights into how these parameters influence the extrusion force and highlights the key findings related to the dynamics of the extrusion process in FDM.

Below, in Table 2 the mean values of forces, during the steady state, for the specific design of experiment can be seen:

At 200 °C, the extrusion forces were observed to be the highest across all feed rates. The increased viscosity of the filament at this lower temperature requires greater force to push the material through the nozzle.

As the temperature increased to 225 and 250 °C, a significant reduction in extrusion force was recorded. For instance, at 1 mm/s, the force required at 250 °C was approximately 75% lower than at 200 °C. This trend was similar across all feed rates, indicating that higher temperatures facilitate easier flow of the filament due to reduced viscosity.

Furthermore, at lower feed rates (1 to 6 mm/s), the extrusion force increased linearly with the feed rate. For example, at 200 °C, the force required at 3 mm/s was double that at 1 mm/s.

At higher feed rates (6 to 12 mm/s), the increase in extrusion force was more pronounced with a non-linear increase pattern. Particularly at 200 °C, the force required at 9 mm/s was more than two times higher than at 4 mm/s, highlighting the greater resistance faced by the filament at higher speeds.

The results also showed that at the highest feed rate of 12 mm/s and the lowest temperature of 200 °C, the extrusion force approached the upper limits of the extruder motor torque capacity, indicating the potential for operational difficulties at these extreme conditions. This can be seen more clearly on the Fig. 5 below:

The fluctuation of the force is happening due to non-available torque on the motor of the extruder. Currently the extruder motor is working on 1.2 amperes and the force limit acquired is approximately 35 N. If the motor uses the maximum current available, at approximately 2 amperes the maximum force limit is 70 N.

Another phenomenon that was observed during the measurements was that in each measurement in the beginning there was a small increase of the extruder force and after that a sudden decrease on the force followed by a normal increase until the steady state. The phenomenon can be observed in the Fig. 6 below:

Table 2 Mean values of forces measured for each value of the specific design of experiment

Volumetric feedrate (mm^3/s)	Feedrate (mm/s)	Measured forces (N)		
		Nozzle temperature (°C)		
		200°	225°	250°
2.4	1	5.00	2.30	1.25
4.8	2	7.50	4.20	3.00
7.2	3	10.00	6.20	4.40
9.6	4	14.00	8.10	5.70
14.4	6	33.00	16.30	10.00
21.6	9	35.00	25.00	25.00
28.8	12	35.00	35.00	35.00

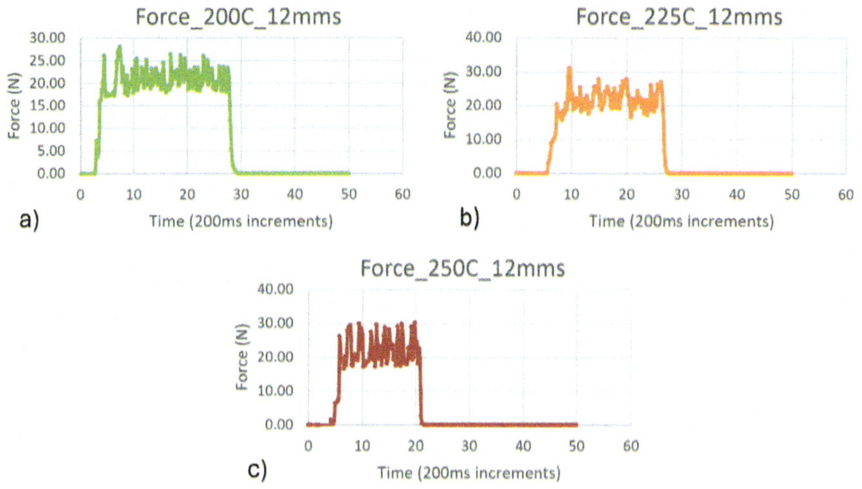

Fig. 5 **(a)** Force measurements on 12 mm/s federate and 200 °C, **b** force measurements on 12 mm/s and 225 °C, **c** force measurements on 12 mm/s and 250 °C

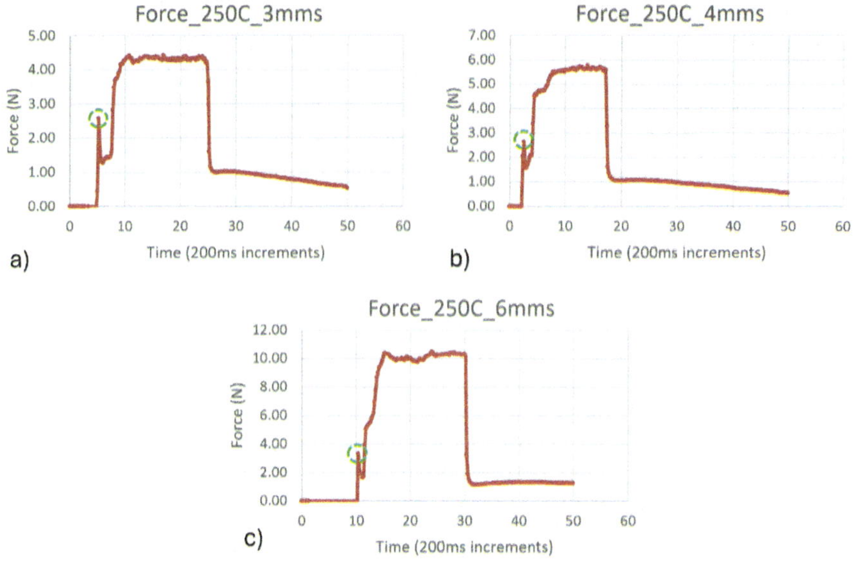

Fig.6 **(a)** Force measurements for 3 mm/s federate and 250 °C. The point of increasing force is circled with a green-coloured dashed circle. Same for **b** with federate of 4 mm/s and 250 °C on nozzle and **c** with federate of 6 mm/s and 250 °C on nozzle

The point of increase in force is approximately 2.8N and it is steady for every federate in this specific nozzle temperature. For 200 °C a similar point occurs at 6N approximately, and for 225 °C at about 4.5 N. The most suitable theory to elucidate this phenomenon is that a portion of the residual material from prior extrusion solidifies at the tip of the nozzle, where it encounters the ambient temperature. Initially, the liquid material needs to exert force to dislodge and separate the hardened material to continue the extrusion process.

To the extent that writers have conducted research, no comparable measurements of forces during ABS extrusion printing exists. Though there are comparable research on the subjects, like [6, 8] where the results follow the similar pattern with this study.

4 Conclusion

This study presented the design, implementation, and experimental validation of a custom-built 3D printer equipped with a force measurement system for accurately capturing extrusion forces during the FDM process. By integrating a load cell between the extruder and the hotend, we were able to precisely measure the forces exerted on the filament under various extrusion temperatures and feed rates. The experiments demonstrated that both lower extrusion temperatures and higher filament feed rates significantly increase the extrusion force, primarily due to changes in the filament's viscosity. These findings underscore the importance of carefully controlling extrusion parameters to avoid excessive forces that could lead to mechanical strain or print defects.

References

1. Khosravani MR, Soltani P, Reinicke T (2021) Fracture and structural performance of adhesively bonded 3D-printed PETG single lap joints under different printing parameters. Theoret Appl Fract Mech 116:103087. https://doi.org/10.1016/j.tafmec.2021.103087
2. Srinivasan R, Prathap P, Raj A, Kannan SA, Deepak V (2020) Influence of fused deposition modeling process parameters on the mechanical properties of PETG parts. Mater Today Proc 27:1877–1883. https://doi.org/10.1016/j.matpr.2020.03.809
3. Khan MMK, Liang RF, Agarwal S, Gupta RK (2005) Rheological and mechanical properties of ABS/PC blends. Korea-Aust Rheol J 17(1), pp. 1–7, Mar 2005. http://www.koreascience.or.kr/article/ArticleFullRecord.jsp?cn=E1ARAH_2005_v17n1_1
4. Boronat T, Segui VJ, Peydro MA, Reig MJ (2008) Influence of temperature and shear rate on the rheology and processability of reprocessed ABS in injection molding process. J Mater Process Technol 209(5):2735–2745. https://doi.org/10.1016/j.jmatprotec.2008.06.013
5. Kumar V, Singh R, Ahuja IPS (2020) Effect of extrusion parameters on primary recycled ABS: mechanical, rheological, morphological and thermal properties. Mater Res Express 7(1):015208. https://doi.org/10.1088/2053-1591/ab6b5e
6. Christodoulou IT, Alexopoulou VE, Karkalos NE, Papazoglou EL, Markopoulos AP (2022) On the surface roughness of 3D printed parts with FDM by a low-budget commercial printer. Cut Tools Technol Syst 96:52–64. https://doi.org/10.20998/2078-7405.2022.96.06

7. Serdeczny MP, Comminal R, Pedersen DB, Spangenberg J (2019) Experimental analysis of the filament feeding force in material extrusion additive manufacturing. ASPE, pp. 141–145, Jan 2019. https://backend.orbit.dtu.dk/ws/files/196730563/Extended_abstract_Serdeczny_SIG2019_pub.pdf
8. Hutchinson M, O'Donohue A, Eslami B, Fouladi K (2024) Numerical and experimental analysis of 3D printer hot ends. Deleted J 6(5), Apr 2024. https://doi.org/10.1007/s42452-024-05884-4
9. Fischer J, Echsel M, Springer P, Refle O (2023) In-line measurement of extrusion force and use for nozzle comparison in filament based additive manufacturing. Prog Addit Manuf 8(1):9–17. https://doi.org/10.1007/s40964-022-00385-5
10. Christodoulou I, Alexopoulou V, Markopoulos AP (2023) Study and development of a high-speed fused filament fabrication 3d printer SEEDA CECNSM 2023, Nov 2023. https://doi.org/10.1109/seeda-cecnsm61561.2023.10470706
11. Tlegenov Y, Lu WF, Hong GS (2019) A dynamic model for current-based nozzle condition monitoring in fused deposition modelling. Prog Addit Manuf 4(3):211–223. https://doi.org/10.1007/s40964-019-00089-3

Socially Intelligent Manufacturing (SIM): Introducing an Ecological Certification Index

Mary Doyle Kent and Breda Walsh Shanahan

Abstract The past two decades have seen manufacturing progressing towards a 'disposable product' route which eventually results in the entire product ending its life cycle in landfill even if only part of the product becomes unusable or unsafe. Unfortunately, there is also a tendency for humans to become 'disposable' too as their roles as employees are replaced by new technologies, robots and fully automated factories. The resulting replacement of humans has the added disadvantage of a loss of ingenuity, resourcefulness, inventiveness and all the human qualities that have created our world, our societies and our workplaces. Could it be that humans and their careers have become part of a valueless society dominated by disposable products?

The recent publication by the European Union titled *"ERA Industrial Technologies Roadmap on Human-Centric Research and Innovation for the manufacturing sector"* examines the implementation of Industry 5.0 into European industries and provides a roadmap to support a human centric digital transition in Europe for the benefit of the European Union's economy. This paper poses the question: What if product classification, borne from an 'Ecological Certification Index', could be established prior to manufacture? It would be similar to the 'Declaration of Performance Certification' currently required in the construction industry but adapted for the manufacturing platform, collating the essentially required criteria *before* the product is vetoed for manufacturing. This would ensure that the product meets the requirements of the protocols and sustainability guidelines. The European Union must maintain the efforts to grow sustainable industries, protect the future of society and envisage the survival of our earth as it faces climate changes, waste management crises and a growing population.

Keywords Technology · Robotics · Social impact · Industry 5.0 · Education and human centered systems

---

M. D. Kent (✉)
Department of Engineering Technology, South East Technological University, Waterford, Ireland
e-mail: mary.doyle-kent@setu.ie

B. W. Shanahan
Department of Education, South East Technological University, Waterford, Ireland
e-mail: breda.shanahan@setu.ie

N. M. Durakbasa and K. G. Gülen (eds.), *Sustainable Green Conversion*, Lecture Notes in Mechanical Engineering, https://doi.org/10.1007/978-3-031-83583-4_19

1 Introduction

The revolutionary inventions of steam power and electrification provided opportunities for designing machines that were stronger, faster, more precise and efficient than humans. Mass production, automation and digitalisation have completely transformed the world of manufacturing. More recent cyber physical systems facilitate remote design of products and their manufacture [1]. Industry has the possibility of replacing the human completely by the latest trends where robots, artificial intelligence and internet of things are the norm. The fourth industrial revolution (Industry 4.0) is transforming modern industry and most nations are accelerating towards this world of high performance with little understanding of the consequences for the human workforce and the natural world as resources are being wasted in the mass production of 'throwaway' products. With the advent of these highly sophisticated manufacturing plants, human creativity is no longer valued and there is little room for experimentation and individualism in a high-tech environment governed by mechanisation and product specificity. Highly intelligent robots bring excellent efficiency and repetition to manufacturing operating over a twenty-four-hour day. The human factor is being removed from day to day decision making as human skills and know-how are no longer required on the factory floor which is dominated by robots and automated specialised machines. Skills once valued and crafted over years of experience and learning are being confined to history and lost to future generations.

2 Background

A framework of the emerging Industry 5.0 can be seen in Fig. 1 as proposed by Doyle-Kent in 2021 [2]. This framework uses an interdisciplinary approach which combines the technological, social, ethical, industrial, environmental and management aspects of industry and can be used in the education of future Engineering and Technicians to equip graduates to flourish in ever changing manufacturing environment. The definition used in conjunction with the diagram is: *"Industry 5.0 is the human-centered industrial revolution which consolidates the agile, data driven digital tools of Industry 4.0 and synchronises them with highly trained humans working with collaborative technology resulting in innovative, personalised, customised, high value, environmentally optimized, high quality products with a lot size one"* [2].

In recent months the European Union (EU) has demonstrated a push to define and understand Industry 5.0 concept and have developed a Community of Practice [3]. As recently as May 2024 the EU published *"ERA Industrial Technologies Roadmap on Human-Centric Research and Innovation for the manufacturing sector"* which is a comprehensive paper for EU countries to help navigate Industry 5.0. They state that European players are slower at *"patenting complex technologies related to human-centricity. While the recently adopted European Artificial Intelligence Act will provide a supportive regulatory environment for the development of trustworthy*

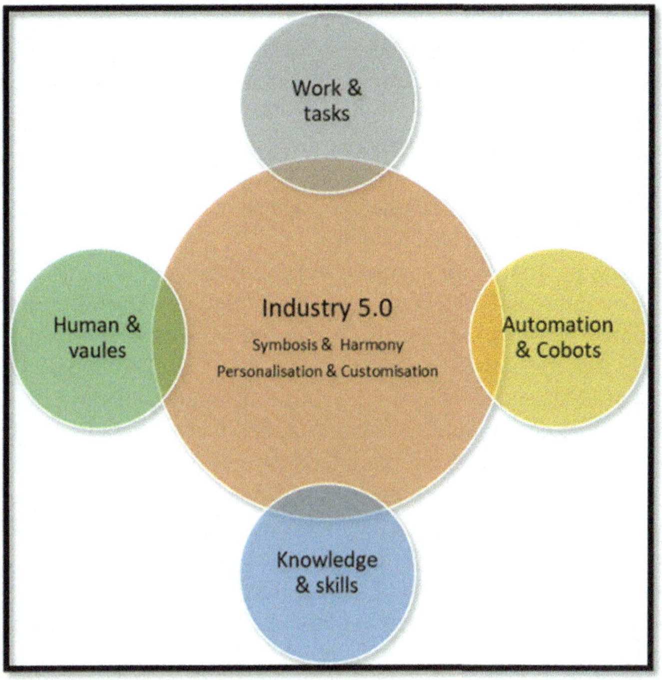

Fig. 1 Conceptual framework illustrating industry 5.0 [2]

artificial intelligence, we also need to better equip the European innovation ecosystems to be able to adapt to a fast-changing scientific, technological and geopolitical environment". In other words, we need to support EU companies in the development and integration of technologies geared towards human-centric technological transformation allowing human skills to develop and adjust to future working methods, processes and organisational changes [4].

An important question that we must ask ourselves as technology changes rapidly is how can we attempt to understand how human beings can be reintegrated into the manufacturing industry bringing their ingenuity, creativity, valuable skills and ethics into an environment which is increasingly being controlled by artificial intelligence and robots. We must also try to establish how human intervention is necessary in the decision-making processes, and how collaboration between 'man and machine' is essential going forward, outlining how advanced technology should be used as a platform, and not a replacement, for humans. How can both our educational and political systems intervene to ensure the usefulness of humans?

In certain areas we can see the seeds of change. It's especially visible in a new wave of environmentally responsible youth. Recycling is taught in all schools and is widely accepted as part of everyday life in developed societies. Gebeshuber and Doyle-Kent explore exemplary educational practices in engineering education from Ireland and Austria in a recent publication which focuses on important contributions which foster sustainability mindsets among the students [5]. Emission targets have been agreed by most nations and steps are being taken to fulfil individual state and national agreements. Vehicles are available that run on alternative energy sources which are more ecologically and environmentally sustainable. Buildings have rigorous sustainable requirements at design stage that must be fulfilled before planning is granted. We are on the edge of an ecological meltdown worldwide, yet, in the manufacturing industry little is being done to curb the phenomenal waste that is widespread through the manufacturing of disposable products.

Mike Cooley, the Irish-born engineer, writer and trade union leader, best known for his work on the social effects of technology, "Socially Useful Production" and "Human Centred Systems" has by his visions and writings been the inspiration for this work. He was the author of Architect or Bee [6] and Delinquent Genius: The Strange Affair of Man and His Technology. In the Delinquent Genius Cooley suggests that there is now a 'cultural outlook' that suggests that we should put the artificial before the real, the opinion being the intelligent machines are more powerful and important than humans. Cooley predicts that humans will become passive and the machine active is a huge concern for the future of humanity. He states that the human has been very ingenious in manufacturing products and has produced technology of very high calibre to accomplish this. As he predicts we now live in the age of Industry 4.0 where robotics, automation internet of things, and artificial intelligence produce high quality goods at an ever-increasing rate with limited input from humans. *"All the world is a factory and men and women are units for production. Furthermore, all that surrounds us, be it flora, fauna or mineral, solid, liquid or gas is seen as raw material for technology's all-consuming factory. We should of course give credit where credit is due and graciously admit that man has been fiendishly ingenious at the mass production of goods, even if they are in the main of a throwaway kind. They are usually the end result of highly sophisticated production technology and an outrageous and cretinous waste of energy and raw materials.... As a technologist, I (Cooley) marvel at the ingenuity of a designers of such equipment but I also admit to being alarmed at the long-term consequences of such a design philosophy and the priorities it sets itself. What we all need to consider is how the skill and ingenuity in those processes and techniques could be redirected from present "trivialities" and transformed into socially useful, sustainable production techniques"* [7].

Could the implementation of an 'Ecological Certification Index', depicting the ecologically sustainable scale of the components procured in the manufacturing process (see Table 2), be combined with the design and ethos of the product's manufacture and have the potential to turn the wasteful manufacturing system into a socially intelligent and responsible industry?

3 Four Steps for a New Manufactuing Framework

The scope of this research is using the latest technology and tools to inform socially intelligent designs with an Ecological Certification Index which incorporates human centred criteria. The stakeholders in this research are industrial, educational and professional bodies (Fig. 2).

STEP 1: Protocol development is the first step in enabling progressive change in industry. To facilitate the design of this protocol, different systems theories were researched. These theories inform the final protocol and the resulting 'Ecological Certification Index' that every new product will have before the product can be launched.

STEP 2: The second step is the design of the 'Ecological Certification Index'. This will be based on sound ethical principles. Elements of this certification are:

- Societal Value
- Lifecycle Analysis
- Serviceability
- Human Robot Synergy

Each of these will be given a score and the final index will be the product of these elemental scores.

STEP 3: The third step is the educational aspect of this study. What are the features of an education system that optimise the skills of the future workforce for industry 5.0?

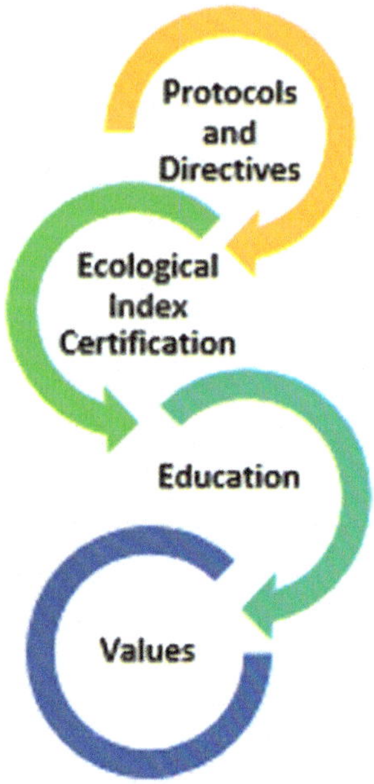

Fig. 2 Conceptual framework for a socially intelligent manufacturing (SIM)

Karl Marx said in his 'Theses on Feuerbach' [8] *"changed men are products of changed circumstances and changed upbringing, that it is men who change circumstances"*. Men in this instance relates to all genders and the main theme is that circumstances and experiences change people and professions. They evolve through learning and experience, change their circumstances and create new opportunities. The forces of economics and materials influence the dispositions of humans as they encounter difficulties with economic conditions of poverty, hunger and lack of employment. In other words, education provides a culturally, discursively, socially and economically shaped environment situated in a particular place and time. Education must remain fluid, open to new ideas, courageous in challenging outdated or morally constraining modes of behaviour or action and above all be inclusive, diverse and equitable in the pursuit of the main aim of which is to support change and engage in discourse on how to survive in changing times and a developing world. Current trends in education are geared towards creating an environmentally sustainable world with 'green' practices and 'energy efficient' methods of manufacturing and production.

We must always be aware of the history of inventions, how and why they were first conceived as 'ideas' and brought to being in the hands of craftsmen and women. It is through learning and engaging with our more practiced, skilful and experienced others, as proposed by Vygotsky's Zone of Proximal Development [9], Fig. 3, that humans can pass on their own knowledge and skills to future generations but only if the materials and component parts of original products are made available through directives such as 'the right to repair' and an ecological index.

For example, mechanics learn from dismantling old engines, seeing and hearing how the components work together and finally re-assembling the pieces together

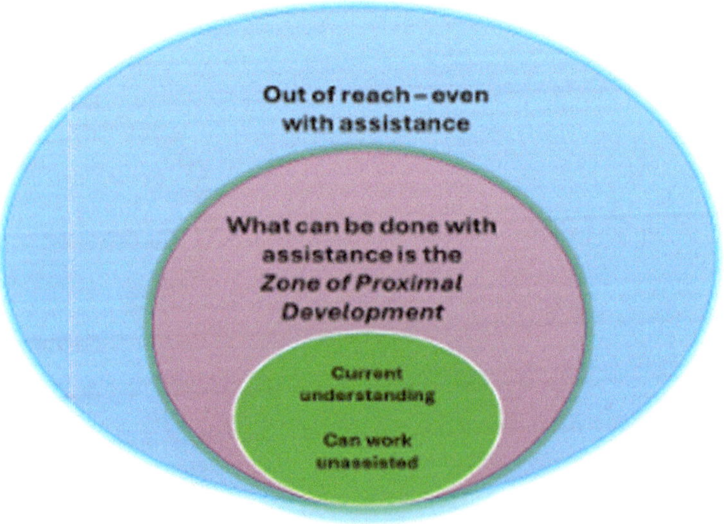

Fig. 3 Vygotsky's zone of proximal development (1978) [9]

again. They learn to 'repair' and 're-build', to feel, to smell and hear the engine being re-assembled. By retaining the 'parts' and making them re-usable the learner may surpass the teacher and create, invent, re-imagine a better means of production for a better more sustainable product with diverse uses and functions. New communities of practice including workplaces require employees to have the ability to frame a problem, chart a path to a solution, draw from appropriate theories needed to solve the problem or address the hypothesis, and formulate questions learned through experience and guided practice. This is the challenge of education. It can be achieved through working with more experienced others, retaining and re-using sustainable parts and ideas and by not engaging in the 'throw-away' culture of previous decades.

Step 4: Values. The fourth step is what are values and how can they be redefined for this new socially inclusive, environmentally friendly eco system.

The value to humanity and all the inhabitants of planet Earth is immense. Skills are retained in the workforce as new apprentices and crafts people learn from previous generations working with repairable products. Small and independent businesses can grow as repairs and spare parts for products return to customers willing to engage with products designed with an ecological footprint capable of repair. The new 'Ecological Certification Index' would greatly assist the end of the 'throwaway culture' and replace it with a more user friendly, environmentally sustainable and healthy attitude to a safer and better world.

4 Manufacturing Protocols

4.1 A. Conformite Européenne Marking

Conformite Européenne (CE) marking indicates the conformity of the product with the Union legislation. CE marking and the assessment criteria, as seen in Fig. 4, demonstrates that the manufacturer has checked that the product meets EU safety, health or environmental requirements. CE marking is put permanently onto products that will be sold in the EEA and Turkish market, whether they are manufactured in the EEA, in Turkey or in another country. CE marking is the visible consequence of a whole process of conformity assessment in its broadest sense, and indicates that a product is declared by the manufacturer to conform with Union harmonisation legislation [7]. Conformity assessment is the responsibility of the manufacturer and the stakeholders responsible for the conformity assessment are the legislator, the manufacturer and if necessary, the notified or in-house accredited conformity assessment body. The modules used for both the design and the production phase, or alternatively for each phase, may or may not involve a notified body.

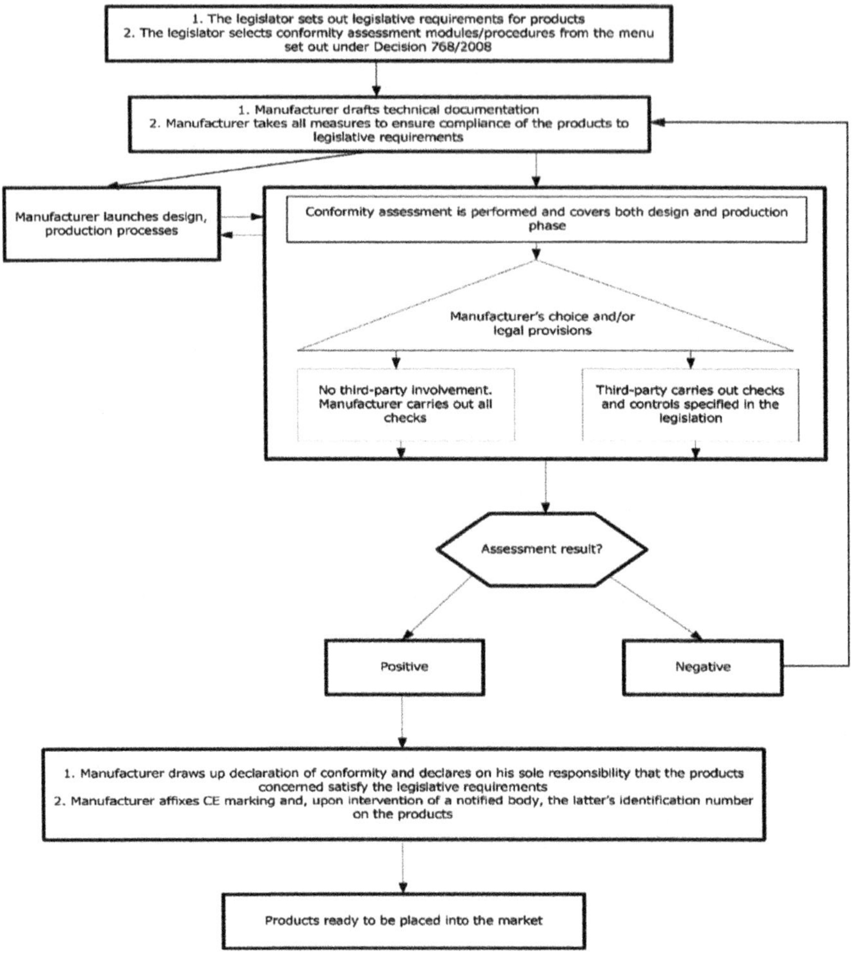

Fig. 4 The CE conformity assessment flow chart for CE marking of a product before it can be placed on the market [10]

4.2 B. Environmental, Social and Governance (ESG) in Manufacturing

Cronin and Doyle-Kent describe ESG as a system for measuring the sustainability of a company under the significant categories of Environmental, Social and Governance and this sustainability focuses on capturing long term value for all stakeholders involved [11]. They describe the key elements that are driving change in the European landscape in Table 1.

Table 1 The main drivers for ESG reporting and disclosures in the EU [11]

1	EU Policies & Regulations: Core EU objectives are to create sustainable circular business models (vs linear)
2	Reporting: EU corporate sustainable reporting is escalating transparency to ensure companies, stakeholders and investors make more informed decisions regarding organisations accurate level of sustainability
3	Non-sustainable production systems: changes are required to encourage climate related fossil vs bio based and local vs global
4	Enhanced Brand credibility and perception: Consumers are demanding change and transparency with products and services
5	Investors and shareholders: Looking to invest in sustainable and circular business models with ESG strategies. Women investors and millennials are a major cohort of impact climate-related investments
6	SMES's and suppliers: increasingly required to disclose information involving supply chain integrity (e.g. material sourcing), exposure to climate-change risks and organisation strategy to reduce emissions and greenhouse gases (GHG)

Table 2 The ecological certification index

	Societal value (need for item) Scale 1–10	Lifecycle analysis (Software) scale 1–10	Serviceability (+ right to repair) scale 1–10	Human–Robot manufacturing ratio no less than 60:40
Quantitative				
Qualitative				
Total				

4.3 C. The Right to Repair Directive

The Council of the European Union has adopted a directive in May 2024 which advocated for the repair of broken or defective goods. This is known are the Right to Repair Directive (R2R). Alexia Bertrand, Belgian State Secretary for the Budget and Consumer Protection stated that *"The directive adopted today enshrines a new right for consumers: the right to have defective products repaired in an easier, cheaper and faster way. It also gives manufacturers the incentive to make products that last longer and can be repaired, reused and recycled. And finally, it makes repairing a more attractive economic activity that can create Europe-based quality jobs. All economic actors win, and so does the environment"* [12]. Under EU law the product must be designed with clear guidelines on the repair process, where it can be repaired and at what cost? A further incentive to repair is the legal guarantee will be extended by 12 months if repair rather than replacement is chosen by the customer.

4.4 D. New 'Ecological Certification Index'

The newly designed protocol should be an additional step when designing a new product and can be added to the Bill of Materials (BOM) on the assembly drawing. An example of what this would look like is Table 2 below.

5 Conclusions

This paper examines the possibility of eradicating products which have very little use and are wasting the natural resources of our planet in their production. It showcases protocols and directives already in place to assist manufacturers in developing new practices within their workplaces combining a human-centered approach to all future manufacturing in an Industry 5.0 setting. It presents the theory of education which optimises the relevance of human to human interaction in learning and retaining specific worthwhile skills through apprenticeship and on the job training. It presents the idea to small and medium enterprises of worthwhile engagement with human centred systems in their production lines along with the introduction of the 'Ecological Certification Index' to be included in their designs pre-production. The feasibility of the Index is introduced and invites other agents to discuss further in communities of practice with a view to becoming a standard protocol with regulatory guidelines.

Acknowledgements We gratefully acknowledge the support of the CONNEXIONS grant from South East Technological University.

References

1. Doyle-Kent M, Kopacek P (2020) Industry 5.0: Is the manufacturing industry on the cusp of a new revolution? In: Proceedings of the international symposium for production research 2019, Springer International Publishing, pp 432–441
2. Doyle-Kent M (2021) Collaborative robotics in industry 5.0 (Doctoral dissertation, Wien). https://doi.org/10.34726/hss.2021.70144
3. Research and Innovation (2023). https://research-and-innovation.ec.europa.eu/news/all-research-and-innovation-news/industry-50-community-practice-call-expression-interest-2023-06-01_en
4. ERA Industrial Technologies Roadmap on Human-Centric Research and Innovation for the manufacturing sector (2024). https://doi.org/10.2777/0266
5. Gebeshuber IC, Doyle-Kent M (2024) Innovations and challenges in engineering education for the future: contributing to the UN sustainable development goals (SDGs). IFAC-PapersOnLine 58(3):134–138
6. Cooley M (1982) Architect or Bee? The human/technology relationship. Boston: South End Press. ISBN: 978-0-89608-131-4
7. Cooley, Mike (2018). Delinquent genius: the strange affair of man and his tech. UK: spokesman books. ISBN: 978-085124-878-3

8. Kemmis S, Grootenboer P (2008) Situating praxis in practice. https://www.researchgate.net/publication/45377376_Situating_Praxis_in_Practice. Accessed on 23 June 2024
9. Vygotsky LS (1978) Mind in society: the development of higher psychological processes. Harvard University
10. EUR-LAX. EUROPE.EU (2024). https://eur-lex.europa.eu/legal-content/EN/TXT/?uri=uriserv:OJ.C_.2022.247.01.0001.01.ENG
11. Cronin M, Doyle-Kent M (2022) Creating value with environmental, social, governance (ESG) in Irish manufacturing SMEs': a focus on disclosure of climate change risks and opportunities. IFAC-PapersOnLine 55(39):48–53
12. European Council (2024) Circular economy: council gives final approval to right-to-repair directive.https://www.consilium.europa.eu/en/press/press-releases/2024/05/30/circular-economy-council-gives-final-approval-to-right-to-repair-directive/#:~:text=The%20directive%20adopted%20today%20enshrines,be%20repaired%2C%20reused%20and%20recycled

Operations Research Applications

Dynamic Efficiency Analysis of Non-life Insurance Companies: A Window Data Envelopment Analysis Approach

Abdullah İlkaz and Ferhan Çebi

Abstract Multiple Criteria Decision-Making models are utilized for long-term planning, which is extremely critical for organizations. In this context, efficient resource use and increased efficiency are vital to prevent waste. Window Data Envelopment Analysis (WDEA) method can dynamically analyze companies' efficiencies and identify areas for improvement. In the insurance sector, efficiency and proper resource utilization are crucial because they directly impact profitability, risk management, and long-term sustainability. In this study, the efficiencies of non-life insurance companies in Türkiye between 2015 and 2023 were analyzed using the WDEA method. The analysis revealed that the sector exhibited fluctuating performance and lacked stability.

Keywords Multiple criteria decision-making · Window data envelopment analysis · Efficiency · Strategic planning · Insurance

1 Introduction

In today's competitive economic environment, making decisions with consideration of long-term effects is crucial for companies across various sectors. Accurately analyzing their current positions and identifying areas for improvement are essential for achieving profitability, competitive advantage, or industry leadership. In this regard, strategic decision-making methods serve as guiding tools.

It has been suggested that enhanced efficiency in financial institutions is strongly linked to increased profitability, which is a critical concept in the business world. Consequently, by effectively utilizing resources, organizations can significantly boost

A. İlkaz · F. Çebi (✉)
Management Engineering Department, Istanbul Technical University, Istanbul, Türkiye
e-mail: cebife@itu.edu.tr

A. İlkaz
e-mail: ilkaz@itu.edu.tr

N. M. Durakbasa and K. G. Gülen (eds.), *Sustainable Green Conversion*, Lecture Notes in Mechanical Engineering, https://doi.org/10.1007/978-3-031-83583-4_20

both their financial and operational performance, given their greater control over these resources compared to other factors [1].

Efficiency is typically defined as the ratio of outputs generated to the inputs utilized within a process [2].

Measuring and accurately interpreting efficiency is a crucial factor that directly influences future performance, as it ensures the optimal use of resources and guides sustainable decision-making processes. Strategic management and utilization of resources also provide long-term sustainability, creating a synergy that drives both immediate gains and enduring success [3].

Mathematical models are frequently used in strategic decision-making processes to conduct scenario planning and address complex risks and uncertainties. Data Envelopment Analysis (DEA) is utilized in these decision-making processes to guide decision-makers [4].

This paper employs the Window Data Envelopment Analysis (WDEA), a multiple criteria decision-making method frequently used in financial organizations. With a case study examining the efficiency of insurance companies in Turkey, this paper aims to assist companies in their scenario planning.

2 Efficiency Analysis with DEA: A Literature Review

There are numerous studies in the literature where efficiency measurement is conducted using Multiple Criteria Decision-Making (MCDM) methods. This section compiles notable studies that utilize one of the most popular MCDM methods, Data Envelopment Analysis (DEA), and its extensions. These studies are frequently applied in the banking, finance, and insurance sectors. Since this study involves a case study in the insurance field, the literature review is mainly focused on these areas.

As a highly competitive sector, insurance industry, where entry barriers are high and regulations are stringent, resource utilization is a critical factor for companies' performance and future prospects. Companies must accurately identify underutilized resources and operate in accordance with economies of scale.

Eling and Jia (2019) examined the relationship between the profitability and efficiency of insurance companies, finding that effective resource utilization increases profitability. As companies' efficiency approaches optimal levels, their profitability also increases, though this increase is not linear; instead, the marginal gains diminish [1].

Zhang et al. (2023) analyzed the efficiency of insurance companies in China between 2017 and 2019 using Window Data Envelopment Analysis. In this input-oriented study, the input variables were fixed assets, operating and management expenses, and investment expenses, while the output variables were insurance profit, investment profit, and gross claims [5]. Naushad, Faridi, and Farsal (2020) assessed the efficiency of 30 insurance companies in Saudi Arabia for the years 2015 and 2018 using an input-oriented CCR model. The study employed operating and production expenses as inputs, with earned premiums and investment income as the outputs [6].

Altan and Yıldırım (2019) analyzed the financial and technical performance of non-life insurance firms in Türkiye using the TOPSIS method, identifying the most and least successful years in terms of financial and technical ratios. They also determined which input variables had the greatest impact on performance [7].

Atukalp (2019) evaluated the efficiencies of non-life insurance companies in Türkiye from 2013 to 2017 using an input-oriented DEA approach, investigating whether efficiency levels varied based on ownership structure. The study utilized assets, operating expenses, and capital as input variables, with written premiums and technical profit as output variables. The findings revealed no significant correlation between efficiency and the ownership structure of the firms [8].

Eling and Luhnen (2009) evaluated the efficiency of over 6,000 insurance firms from 35 countries, analyzing each firm-year pair as a separate decision-making unit using DEA and Stochastic Frontier Analysis (SFA) methods. This study stands as the most extensive performance and efficiency analysis ever conducted in the insurance sector. Input variables included personnel and operating expenses, equity, and debt, while output variables comprised non-life paid claims and additions to reserves, life insurance paid claims and additions to reserves, and investment income. The analysis was segmented into four panels, comparing countries, organizational structures, the number of business lines operated by companies, and company size, with measurements of both technical and cost efficiencies [9].

Kaffash et al. (2019) investigated the use of Data Envelopment Analysis (DEA) in the insurance industry from 1993 to 2018, driven by growing interest in performance evaluation among various stakeholders, including researchers, market analysts, and investors. They analyzed 132 publications that address both practical applications and methodological aspects. They observed that recent major advancements in insurance technology (insurtech) and market transparency have not yet been examined for their effects on company performance. Furthermore, they pointed out that emerging methods like dynamic DEA, fuzzy DEA, and satisficing DEA are underrepresented in the existing literature [10].

At the end of the literature review, it was found that while DEA is frequently used as a performance measurement method in insurance, there is still room for improvement in studies that dynamically compare company performance across years. The review also identified the input and output variables commonly used in these studies.

3 Methodology

Data Envelopment Analysis (DEA) is a non-parametric method based on mathematical programming that evaluates the relative efficiency of decision-making units by examining multiple inputs and outputs performing similar functions [11].

DEA uses a comparative approach to measure how effectively one Decision-Making Unit (DMU) operates relative to others. In DEA, the efficiency of a DMU is assessed compared to other DMUs within the same operational area, categorizing them as either efficient or inefficient. Efficient DMUs receive a value of 1, while inefficient DMUs receive values less than 1 [12].

The basic mathematical model of DEA is outlined below:

$$\text{Objective Function}: Ek = Max\frac{\sum u_r Y_{rk}}{\sum v_i X_{ik}}$$

$$\text{Constraints}: \frac{\sum urYrj}{\sum viXij} \leq 1 j = 1, 2, 3, ...n$$

$$ur, vi \geq \varepsilon; r = 1, 2, 3, ...t; i = 1, 2, 3, ...m$$

u_r = weight coefficient for output r from DMU k.

v_i = weight coefficient for input i from DMU k.

E_k = relative efficiency of DMU k.

Y_{rj} = Output r produced by DMU j.

X_{ij} = Input i used by DMU j.

Y_{rk} = Output r produced by DMU k.

X_{ik} = Input i used by DMU k.

ε = a very small positive number (the magnitude of 0.00000001).

n = amount of DMU's.

m = amount of inputs.

t = amount of outputs.

In the model, U_r represents the weight of the input components, and V_i represents the weight of the output components. The values of these components are unknown and are determined by solving the model to maximize the efficiencies of the Decision Making Units (DMUs). The weights are set to maximize the input-to-output ratio in the objective function. Since efficiency appears in both the objective function and constraints, the maximum value of the objective function is 1. A DMU is considered efficient if the objective function value is 1; if it is less than 1, the DMU is relatively inefficient.

DEA is an effective technique for multiple input and output production concepts. The basic DEA applications are divided into two models: CCR and BCC. The CCR model operates under the assumption of constant returns to scale, while the BCC model works under the assumption of variable returns to scale. Additionally, the CCR and BCC models can be further classified into two types based on their orientation: input-oriented and output-oriented. In an input-oriented model, the goal is to minimize the input quantities required to achieve the existing outputs, while in an output-oriented model, the aim is to maximize the outputs given the current inputs [13].

The CCR model measures the overall efficiency of a Decision-Making Unit (DMU) without convexity constraints. It combines technical efficiency and scale efficiency to represent total efficiency. Technical efficiency indicates how well inputs are converted into outputs. A technically efficient DMU cannot increase outputs without increasing inputs, meaning it uses all resources without waste. In terms of production boundaries, a DMU performing on the production frontier is considered technically efficient [14].

The BCC model is based on the assumption of Variable Returns to Scale (VRS). While CCR models assess efficiency under the assumption that all firms operate at an optimal scale, the BCC method was developed to address variable scales, reflecting resource excesses or output shortfalls common in modern production systems [14].

Scale efficiency is a concept related to production at the most efficient scale. For a given production process, returns to scale can be either constant or variable. Scale efficiency is determined by dividing total efficiency by technical efficiency [15].

Basic DEA approaches operates within a fixed time period and does not account for time-dependent changes in the relationship between resources and output. In basic DEA models, time-dependent relationships between DMUs are considered irrelevant. The Window Data Envelopment Analysis (WDEA) method is a DEA approach that can handle dynamic datasets.

WDEA method uses moving averages and performs efficiency analysis by treating each DMU as a different unit across different time periods. In WDEA, the efficiency of a DMU is influenced not only by the performance of other DMUs but also by its own performance in other time frames [16].

4 A Case Study in Turkish Insurance Companies using WDEA

In this chapter, the efficiency of Turkish non-life insurance companies between 2015 and 2023 has been analyzed using Window Data Envelopment Analysis. Company names have been censored because these companies are still actively operating.

Input-oriented models were chosen for the analysis because companies have greater control and influence over their inputs. Both CCR and BCC models were applied, each being input-oriented. Subsequently, scale efficiency was calculated using these values. The input variables used were operating expenses, fixed assets, and investment expenses, while the output variables included investment income, gross earned premium, and total insurance revenue. The window width was set to 3. Charnes et al. (1994) proposed that choosing a window width of 3 strikes an ideal balance between maintaining stability in efficiency measurement and meeting informational requirements [17].

The units used in the analyses and their explanations are as follows:

n: number of DMUs.

k: number of periods.

w: number of windows.

p: window width.

The values of these units utilized in the study, along with numerical data related to the analysis, are as follows:

- $n = 25$
- $k = 9$
- $p = 3$
- $w = k - p + 1 = 9 - 3 + 1 = 7$
- total number of DMUs $= n * p * w = 25 * 3 * 7 = 525$

Before starting the analyses, there are different suggestions on the number of input and output variables, as well as the number of DMUs used. Boussofiane et al. (1991) suggested that the minimum number of DMUs should exceed the product of the input and output variables [18]. Bowlin (1998) recommended it be more than three times their sum [19], while Dyson et al. (2001) argued it should be over twice their product. The amount of 525 DMUs used in the study are sufficient for all recommendations [20].

4.1 Results of the CCR WDEA and Discussion

Input-oriented CCR WDEA analysis did not find any company with an efficiency score of 1. Although some companies are efficient in various years or windows, none of them maintain efficiency across all years, leading to no company having an average efficiency value of 1. The average efficiency value over the years using the CCR model is 0.796.

The average efficiencies of the companies over the years can be seen in Fig. 1. It is difficult to identify a clear trend in the companies' efficiencies, as there are significant fluctuations in performance. Notably, DMU6 improved from the worst performance of 0.476 to an efficiency value of 1 in the following year. A similar pattern was observed for DMU8, which moved from an efficiency value of 0.497 to 1. DMU10 consistently exhibited the poorest performance across all years. DMU1 was efficient in all years except for two. The sector demonstrated its best performance in 2023 and its worst in 2015.

Figure 2 shows the CCR WDEA results by windows. The most notable observation in the figure is that, compared to the efficiency values over the years, these values have somewhat shrunk. The reason for this is that the efficiency value of a given year is included in multiple windows, which tends to bring the average efficiency scores of these windows closer to each other. The 2016–2017-2018 window shows the lowest efficiency with a score of 0.744. The most efficient years according to the windows are the 2021–2022-2023 period. The average efficiency score by windows is 0.798, which is slightly higher than the average efficiency score over the years. This is because the worst performance year, 2015, is included in only one window.

Table 1 shows the all efficiency scores of the DMUs for each window.

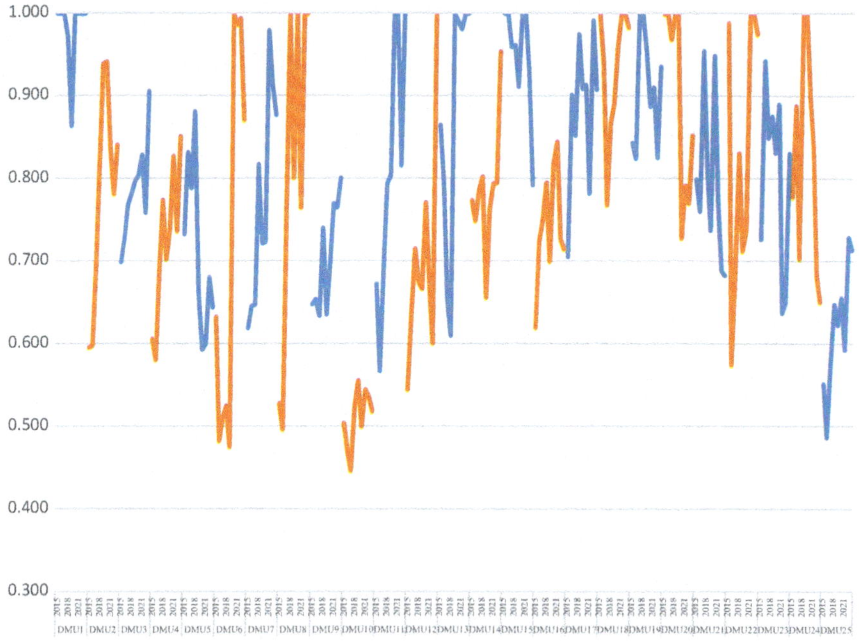

Fig. 1 Comparative average efficiency by year with CCR WDEA

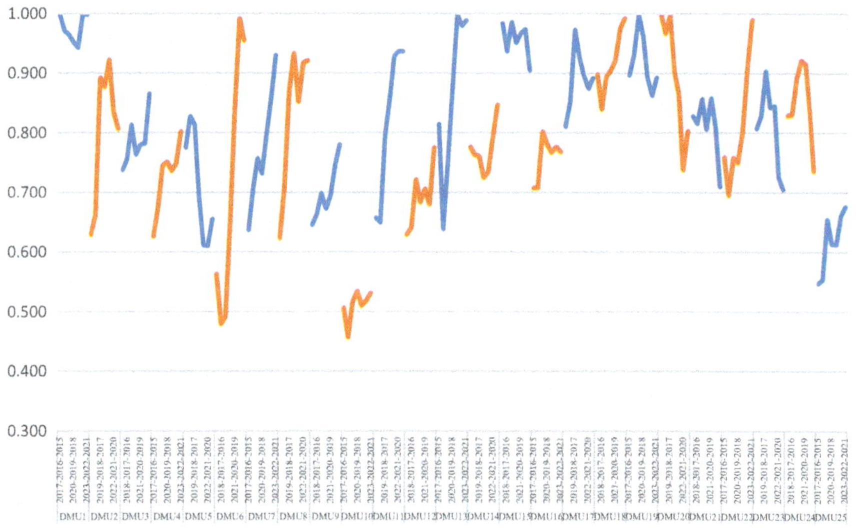

Fig. 2 Comparative average efficiency by window with CCR WDEA

Table 1 Comparative average efficiency by window with CCR WDEA

	2023-2022-2021	2022-2021-2020	2021-2020-2019	2020-2019-2018	2019-2018-2017	2018-2017-2016	2017-2016-2015
DMU1	1.000	1.000	0.944	0.953	0.966	0.973	1.000
DMU2	0.809	0.836	0.924	0.879	0.894	0.664	0.631
DMU3	0.868	0.784	0.782	0.765	0.815	0.757	0.739
DMU4	0.805	0.750	0.739	0.753	0.747	0.675	0.628
DMU5	0.657	0.612	0.612	0.694	0.815	0.830	0.777
DMU6	0.957	0.993	0.846	0.652	0.493	0.482	0.565
DMU7	0.932	0.860	0.799	0.734	0.760	0.709	0.638
DMU8	0.923	0.920	0.854	0.935	0.873	0.713	0.625
DMU9	0.782	0.748	0.697	0.674	0.701	0.665	0.647
DMU10	0.533	0.520	0.513	0.536	0.519	0.459	0.509
DMU11	0.939	0.939	0.931	0.857	0.797	0.651	0.659
DMU12	0.777	0.682	0.708	0.686	0.724	0.642	0.631
DMU13	0.991	0.982	1.000	0.877	0.752	0.640	0.817
DMU14	0.849	0.795	0.737	0.727	0.762	0.766	0.778
DMU15	0.907	0.976	0.969	0.953	0.988	0.939	0.986
DMU16	0.770	0.779	0.769	0.783	0.804	0.711	0.710
DMU17	0.894	0.876	0.898	0.929	0.975	0.853	0.813
DMU18	0.995	0.977	0.924	0.907	0.897	0.842	0.900
DMU19	0.895	0.864	0.893	0.964	1.000	0.933	0.898
DMU20	0.805	0.741	0.866	0.906	1.000	0.969	1.000

(continued)

Table 1 (continued)

	2023-2022-2021	2022-2021-2020	2021-2020-2019	2020-2019-2018	2019-2018-2017	2018-2017-2016	2017-2016-2015
DMU21	0.711	0.810	0.859	0.807	0.858	0.818	0.830
DMU22	0.992	0.910	0.804	0.752	0.760	0.698	0.761
DMU23	0.707	0.727	0.846	0.843	0.905	0.830	0.808
DMU24	0.738	0.835	0.917	0.923	0.893	0.832	0.831
DMU25	0.679	0.662	0.614	0.615	0.657	0.555	0.549

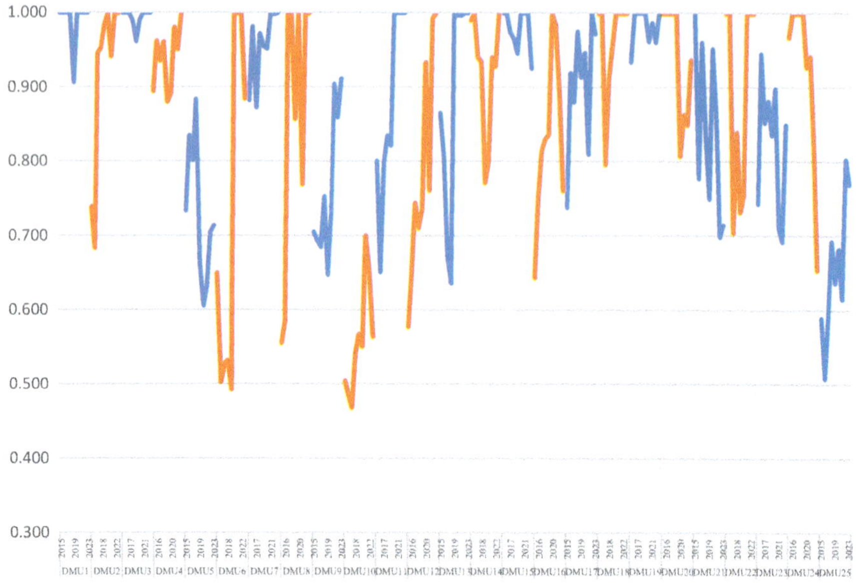

Fig. 3 Comparative average efficiency by year with BCC WDEA

4.2 Results of the BCC WDEA and Discussion

Input-oriented BCC WDEA analysis did not find any company with an efficiency score of 1. Although some companies are efficient in various years or windows, none of them maintain efficiency across all years, leading to no company having an average efficiency value of 1. The average efficiency value over the years using the BCC model is 0.867 (Fig. 3).

The average efficiencies of the companies over the years can be seen in Fig. 1. Like CCR model it is difficult to identify a clear trend in the companies' efficiencies with BCC model, as there are significant fluctuations in performance. Notably, DMU6 improved from the worst performance of 0.494 to an efficiency value of 1 in the following year. DMU10 consistently exhibited the poorest performance across all years. DMU1 was efficient in all years except only one. The sector demonstrated its best performance in 2023 and its worst in 2015.

Figure 4 shows the BCC WDEA results by windows. Like CCR model the most notable observation in the figure is that, compared to the efficiency values over the years, these values have somewhat shrunk. The reason for this is that the efficiency value of a given year is included in multiple windows, which tends to bring the average efficiency scores of these windows closer to each other. The 2016-2017-2018 window shows the lowest efficiency with a score of 0.826, and 2021-2022-2023 period shows the best performance with efficiency score of 0.907. Table 2 shows the all efficiency scores of the DMUs for each window. It is obvious that BCC model's efficiency scores are higher than CCR model's.

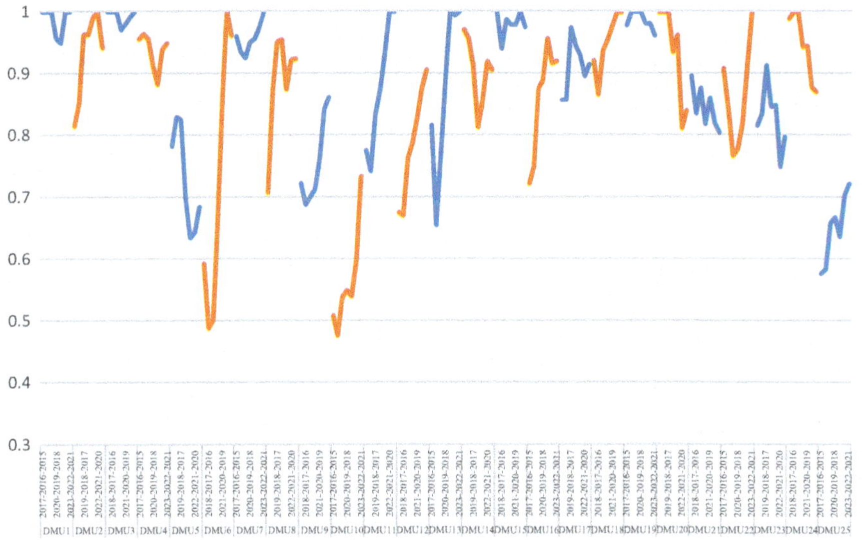

Fig. 4 Comparative average efficiency by window with BCC WDEA

Figure 5 displays the average scale efficiency values over the years. Scale efficiencies are calculated by dividing the total efficiency values by the technical efficiency values. Only two years with scale efficiency below 90% can be considered positively. The highest scale efficiency was recorded in 2018, while 2016 and 2022 had the lowest scale efficiencies.

Table 2 Comparative average efficiency by windows with BCC WDEA

	2023-2022-2021	2022-2021-2020	2021-2020-2019	2020-2019-2018	2019-2018-2017	2018-2017-2016	2017-2016-2015
DMU1	1.000	1.000	0.950	0.957	1.000	1.000	1.000
DMU2	0.942	1.000	0.991	0.964	0.964	0.853	0.815
DMU3	1.000	0.991	0.980	0.971	1.000	1.000	1.000
DMU4	0.950	0.940	0.883	0.914	0.956	0.965	0.957
DMU5	0.686	0.644	0.634	0.700	0.825	0.830	0.782
DMU6	0.962	1.000	0.859	0.665	0.503	0.490	0.594
DMU7	0.999	0.974	0.956	0.951	0.925	0.936	0.961
DMU8	0.924	0.923	0.875	0.956	0.951	0.867	0.709
DMU9	0.862	0.843	0.763	0.713	0.701	0.688	0.724
DMU10	0.735	0.600	0.541	0.550	0.540	0.477	0.510
DMU11	1.000	1.000	0.931	0.873	0.834	0.743	0.777
DMU12	0.907	0.877	0.830	0.789	0.764	0.671	0.677
DMU13	1.000	0.994	1.000	0.891	0.774	0.655	0.817
DMU14	0.907	0.921	0.856	0.814	0.915	0.958	0.972
DMU15	0.975	1.000	0.979	0.979	0.988	0.940	1.000
DMU16	0.921	0.918	0.958	0.890	0.877	0.751	0.723
DMU17	0.916	0.896	0.930	0.945	0.975	0.858	0.858
DMU18	1.000	1.000	0.979	0.956	0.938	0.867	0.922
DMU19	0.961	0.981	0.981	1.000	1.000	1.000	0.978
DMU20	0.841	0.812	0.963	0.936	1.000	1.000	1.000

(continued)

Table 2 (continued)

	2023-2022-2021	2022-2021-2020	2021-2020-2019	2020-2019-2018	2019-2018-2017	2018-2017-2016	2017-2016-2015
DMU21	0.805	0.820	0.861	0.818	0.878	0.836	0.898
DMU22	1.000	0.912	0.820	0.779	0.768	0.846	0.910
DMU23	0.798	0.749	0.849	0.846	0.914	0.835	0.816
DMU24	0.872	0.879	0.945	0.944	1.000	1.000	0.989
DMU25	0.722	0.703	0.636	0.668	0.659	0.584	0.577

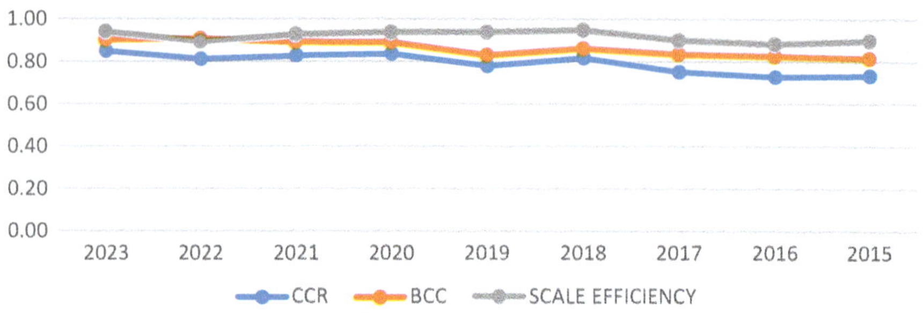

Fig.5 Comparative average efficiency trends

5 Conclusion

Insurance is a crucial component of today's global capitalist economy. By protecting individuals and institutions from potential risks, insurance companies help maintain economic stability. Due to the sector's unique characteristics, product innovation or bold operating actions are not easy for companies to differentiate themselves. Efficient use of resources and increasing profitability are very important for companies. Using MCDM methods like WDEA, companies should assess themselves and identify areas for improvement. This study analyzes non-life insurance companies in Turkey over a 9-year period.

The analysis results indicate that neither the sector nor the companies exhibit a specific performance trend. Some firms display radical performance changes over very short time intervals. The causes of these fluctuations, whether increases or decreases, should be analyzed in more detail. The year 2023 showed the best performance in both models, while 2016 and 2017 were the years with the worst performance in the CCR and BCC models, respectively. Additionally, considering scale efficiency, there are only two years with an average below 90 percent. Since DMU consistently demonstrates efficiency across both the CCR and BCC models, it can be concluded that no firm operates at the optimal scale size at all times.

In this context, agility, innovation, and resource optimization are vital for building a resilient organization against risks. Achieving efficient resource use for a sustainable competitive edge and stable high performance is crucial. Utilizing science and technology is essential for addressing global and local risks effectively. As technological advances make it harder to maintain long-term competitive advantages, strategic decision-making is increasingly critical. While Multiple Criteria Decision-Making methods may not lead decisions directly, they provide valuable benefits, making them essential tools in complex business environments.

In this study, non-life insurance companies were analyzed. The selection of input and output variables was informed by a literature review. For future studies, research could focus on non-desirable outputs, exploring how to increase some outputs while reducing others.

References

1. Eling M, Jia R (2019) Efficiency and profitability in the global insurance industry. Pac Basin Financ J 57:1–12
2. Farrell MJ (1957) The measurement of productive efficiency. J R Stat Soc Ser (Gen) 120(3):253–290
3. İlkaz A, Çebi F (2024) Multi-criteria decision-making model for evaluation of financial performance of insurance companies. In: Cepni E (ed) Chaos, complexity, and sustainability in management. IGI Global, pp 141–170
4. Okursoy A, Tezsürücü D (2015) Veri Zarflama Analizi ile Göreli Etkinliklerin Karşılaştırılması: Türkiye'deki İllerin Kültürel Göstergelerine İlişkin Bir Uygulama. Yönetim ve Ekonomi Dergisi 21(2):1–18. https://doi.org/10.18657/Yecbu.92031
5. Jiang J, Li EY, Sujie P, Wenhan F, Zhan S, Zhang H (2023) Resilience of operational performance in China's insurance companies: a dynamic data envelopment analysis. IEEE Access 112502–112514
6. Naushad M, Faridi M, Faisal S (2020) Measuring the managerial efficiency of insurance companies in Saudi Arabia: a data envelopment analysis approach. J Asian Financ Econ Bus 07(06):297–304
7. Altan İM, Yıldırım M (2019) Sigorta Sektörünün Hayat Dışı Branşının Finansal ve Teknik Performanslarının Analizi. Uluslararası Afro-Avrasya Araştırmaları Dergisi 4(7):36–46
8. Atukalp ME (2019) Hayat Dışı Sigorta Şirketlerinde Mülkiyetin Etkinlik Üzerindeki Belirleyiciliği. Sosyoekonomi 27(42):67–86
9. Eling M, Luhnen M (2010) Efficiency in the international insurance industry: a cross-country comparison. J Bank Finance 34(7):1497–1509
10. Roza Azizi R, Huang Y, Kaffash S, Zhu J (2020) A survey of data envelopment analysis applications in the insurance industry 1993–2018. Eur J Oper Res 284(3):801–813
11. Ulucan A (2002) ISO 500 Şirketlerinin Etkinliklerinin Ölçülmesinde Veri Zarflama Analizi Yaklaşımı: Farklı Girdi Çıktı Bileşenleri ve Ölçeğe Göre Getiri Yaklaşımları ile Değerlendirmeler. Ankara Üniversitesi SBF Dergisi 57(2):185–202
12. Charnes A, Cooper WW, Rhodes E (1978) Measuring the efficiency of decision making units. Eur J Oper Res 2(6):429–444
13. Banker RD, Cooper WW, Seiford LM, Thrall RM, Zhu J (2004) Returns to scale in different DEA models. Eur J Oper Res 154(2):345–362
14. Cooper WW, Seiford LM, Tone K (2007) Data envelopment analysis: a comprehensive text with models, applications, references and dea-solver software, 2nd edn. Springer, New York
15. Ramanathan R (2003) An introduction to data envelopment analysis: a tool for performance measurement. Sage Publications, New Delhi
16. Wang K, Yu S, Zhang W (2013) China's regional energy and environmental efficiency: a DEA window analysis based dynamic evaluation. Math Comput Model 5(6):1117–1127. https://doi.org/10.1016/j.mcm.2011.11.067
17. Charnes A, Cooper WW, Lewin AY, Seiford LM (1994) Introduction. In: Data envelopment analysis: theory, methodology, and applications. Springer, Dordrecht
18. Boussofiane A, Dyson RG, Thanassoulis E (1991) Applied data envelopment analysis. Eur J Oper Res 52(1):1–15
19. Bowlin WF (1998) Measuring performance: an introduction to data envelopment analysis (DEA). J Cost Anal 15(2):3–27
20. Allen R, Camanho AS, Dyson RG, Podinovski VV, Sarrico CS, Shale EA (2001) Pitfalls and protocols in DEA. Eur J Oper Res 132(2):245–259

Mixed-Integer and Constraint Programming Approaches for Determination of Locations of Long-Term Care Facilities

Mert Paldrak, Gamze Erdem, Gökberk Özsakallı, and Armağan Yağız Terim

Abstract This study addresses the Long-Term Care Facility Location Problem (LTCFLP), focusing on the optimal location of long-term care facilities in İzmir, Turkiye. Long-term care facilities, often referred to as nursing homes, provide medical services and daily assistance to individuals, primarily the elderly, over extended periods. As the aging population increasing globally, the demand for such facilities has grown, prompting governments to invest in their development. This research aims to determine the best locations for these facilities, balance the load across facilities and o minimize both the establishment and patient assignment costs. The problem is formulated as a multi-objective optimization model, considering facility capacity, overcapacity allowance, and facility utilization. A mixed-integer programming model is developed to achieve these objectives. Additionally, a novel contribution is made by applying a constraint programming (CP) approach, which has not been used for LTCFLP before. The study compares the results from both mixed-integer programming (MIP) and constraint programming models to evaluate their efficiency and effectiveness in solving this complex, real-world problem. The results highlight the trade-offs between different models and propose future research directions for enhancing facility location planning.

Keywords Long-Term Care Facility Location Problem (LTCFLP) · Healthcare facility optimization · Mixed-Integer Programming (MIP) · Constraint Programming (CP) · Overcapacity allowance in facility location

---

M. Paldrak (✉) · G. Erdem · G. Özsakallı · A. Y. Terim
Industrial Engineering, Yasar University, İzmir, Turkey
e-mail: mert.paldrak@yasar.edu.tr

N. M. Durakbasa and K. G. Gülen (eds.), *Sustainable Green Conversion*, Lecture Notes in Mechanical Engineering, https://doi.org/10.1007/978-3-031-83583-4_21

1 Introduction

Long-term healthcare facilities, commonly known as nursing homes, provide comprehensive health services and accommodation to individuals of all ages, though they primarily cater to the elderly. Long-term care encompasses assistance with medical needs and daily activities for individuals who require extended support. These services can be delivered in various settings, including private homes, community centers, and specialized facilities. While the majority of residents in long-term care facilities are elderly individuals requiring assistance with their daily activities, such facilities are also crucial for younger people living with disabilities or chronic illnesses. These facilities are designed for individuals who need 24-h supervision by registered nurses, with care levels tailored to the needs of the residents, based on facility capabilities and local regulations.

The importance of long-term care facilities has grown significantly in recent years due to the global rise in the aging population. This demographic shift is one of the most pressing challenges facing both developed and developing countries. The decline in both birth and mortality rates has led to an increase in the proportion of individuals over the age of 65. As societies prepare for an aging population, governments around the world are expanding social insurance programs and building more long-term care facilities to meet the growing demand. In many countries, especially those with limited long-term care infrastructure, such as facilities for individuals with severe disabilities or illnesses, governments are planning to establish new care centers to accommodate these populations.

In the field of operational research, problems related to healthcare facilities have gained considerable attention, particularly in optimizing resource utilization and facility placement. Over the past decade, the application of operational research in healthcare has advanced significantly. Research topics have ranged from optimizing the allocation of healthcare resources and minimizing the costs of new facility development to balancing the efficiency of care providers, including doctors, nurses, and equipment, with patient needs. These advancements are particularly critical for improving the long-term care sector, where resource constraints and growing demand intersect, making the optimization of facility locations and services essential for ensuring equitable access to care.

This paper explores the long-term care facility location problem (LTCFLP), focusing on the selection of optimal locations for long-term care facilities in İzmir, Turkey, while balancing operational efficiency and cost. By leveraging both mixed-integer programming (MIP) and constraint programming (CP) approaches, this study aims to develop a robust model for addressing the challenges of long-term care facility planning and ensuring effective healthcare delivery to an aging population.

2 Problem Definition and Motivation

This study focuses on determining the optimal locations for long-term care facilities in İzmir, Turkey, addressing a problem known as the Long-Term Care Facility Location Problem (LTCFLP). The research model's healthcare network design by selecting facility locations from a set of candidates, considering patient groups as demand points. Several key assumptions are made, such as the establishment of facilities at demand points, a limited number of facilities, uniform capacity for each facility, and ensuring each patient group is assigned to one facility.

The objective is to balance the load across facilities while minimizing both the cost of establishing new facilities and the cost of assigning patients based on distance. This presents a multi-objective problem, where conflicting goals—such as balancing facility loads and reducing costs—are considered. Since long-term care patients typically do not require emergency services, the model assumes facility capacities do not need to be fully satisfied. However, long waiting times could cause dissatisfaction if facility loads are unbalanced, so the model emphasizes load balancing.

Given Turkey's real-life healthcare challenges, where facilities often exceed capacity due to high demand, the model incorporates an overcapacity allowance to reflect this reality. Additionally, a utilization percentage is introduced to ensure facilities are not underused, setting a minimum number of patients for each facility. The main contribution of this study is the development of a mixed-integer programming (MIP) model for this multi-objective problem, alongside the novel application of a constraint programming (CP) approach. While there are many existing MIP formulations for LTCFLP, CP has not been applied to this problem. The results from both approaches will be compared to highlight their advantages and drawbacks. Future research directions and possible problem extensions will be discussed based on these findings.

3 Literature Review

The problem of determining optimal locations for long-term care (LTC) facilities is critical to ensuring efficient and accessible healthcare services, particularly for aging populations. Over the years, various optimization techniques, especially integer and constraint programming methods, have been developed to address the complexity of this issue. These techniques aim to minimize costs, improve service quality, and ensure accessibility while accommodating a wide range of operational constraints and uncertainties.

One of the earliest contributions to this field was by [1], who developed a mixed-integer programming (MIP) model to optimize the location of long-term healthcare facilities. Their model focused on minimizing travel costs for patients while maximizing accessibility, demonstrating the potential of integer programming in balancing operational efficiency and patient convenience. By integrating demographic data and patient needs into the location decision process, their work underscored the importance of considering both cost minimization and service quality in facility planning.

Building on this, [2] introduced a mixed-integer linear programming (MILP) model designed to incorporate service quality into the location planning process. Their study focused on preventive healthcare facilities, with an emphasis on ensuring that clients had access to high-quality care. This approach was particularly relevant for LTC facilities, where service quality and accessibility are critical for improving long-term health outcomes. The inclusion of quality-conscious clients in the optimization model demonstrated how accounting for service preferences can lead to more patient-centered facility placement decisions.

Capacity constraints and operational limits are also important considerations in LTC facility planning. [3] developed a capacitated network design model that used binary integer programming to optimize the locations of LTC facilities while considering the operational capacities of each facility. Their work highlighted the need for healthcare planners to ensure that facilities could handle fluctuating levels of demand across the network. By modeling these capacity constraints, their approach ensured that LTC services could be delivered efficiently, even as demand varied over time and across regions.

Uncertainty is another significant factor in LTC facility location planning, particularly given the unpredictability of future demand. [4] tackled this issue by introducing a multi-period location-allocation model that incorporated chance-constrained programming to account for demand uncertainty. Their model allowed healthcare planners to make decisions that were resilient to changes in patient demand, ensuring that nursing home networks could continue to operate effectively even under uncertain conditions. This robust optimization framework provided a flexible solution that could be adjusted as new data on patient needs became available, making it a practical tool for long-term facility planning.

Similarly, [5] introduced a robust optimization model designed to address both pre-disaster and post-disaster planning for healthcare facilities. Their model combined mixed-integer programming with chance-constrained techniques to account for the uncertainties associated with natural disasters. This approach was particularly valuable for LTC facilities, where ensuring continuity of care during and after disasters is critical. By integrating disaster resilience into the optimization model, the authors provided a comprehensive solution for maintaining healthcare service delivery in the face of external shocks.

Several reviews have synthesized these approaches, providing broader insights into the application of optimization models for healthcare facility location. [6] offered a comprehensive review of strategic facility location models, emphasizing the role of integer and constraint programming in solving complex facility location problems. Their review highlighted how these optimization techniques could be applied to a wide range of healthcare settings, from LTC facilities to hospital networks. This foundational work underscored the importance of strategic long-term investments in healthcare facility planning, given the significant costs and operational challenges involved.

In addition to long-term care facility networks, [7] applied a similar optimization framework to home healthcare centers. Their mixed-integer programming model optimized not only the physical location of facilities but also the configuration of resources, such as staffing and equipment, within each center. This comprehensive approach demonstrated how optimization models could be used to manage both location and operational aspects of healthcare facilities, making them more adaptable to the specific needs of each region [8].

The integration of operational constraints, uncertainty, and resilience into facility location models represents a significant advancement in LTC facility planning. Optimization techniques, particularly mixed-integer programming, have proven to be effective in addressing the challenges of capacity, demand variability, and service quality. Moreover, the incorporation of disaster resilience and quality-conscious clients into these models highlights the growing complexity of healthcare facility planning. As demand for LTC services continues to rise, the use of robust optimization frameworks will become increasingly important, ensuring that healthcare networks remain both cost-effective and capable of delivering high-quality care under a wide range of conditions.

In conclusion, the literature on optimization approaches for LTC facility location demonstrates the critical role that integer and constraint programming techniques play in healthcare planning. By addressing both operational constraints and uncertainties, these models provide healthcare administrators with powerful tools for making informed decisions about facility placement. The integration of robust optimization frameworks, which account for both service quality and disaster resilience, ensures that LTC networks can continue to meet the needs of aging populations in an efficient and sustainable manner.

4 Mathematical Formulations

In this section, the Long-Term Care Facility Location Problem (LTCFLP) is outlined by specifying the assumptions, notations, and mathematical formulations for both integer and constraint programming models. The key assumptions of the problem are summarized as follows:

- No facility currently exists, meaning all facilities will be newly established.
- Facilities have varying capacities; the number of sickbeds in one facility differs from that in others.
- Patients are treated as potential residents of the facilities and grouped into patient clusters, with each group having a predetermined location and a known demand, i.e., the number of patients.

- A facility can be established at the location of any patient group. Thus, the terms "patient group location" and "candidate facility location" are used interchangeably.
- It is assumed that all patients in a group are served by the facility closest to their location.
- The maximum number of facilities to be established is known in advance and is less than or equal to the number of candidate locations.
- While facility capacities are predefined, an overcapacity parameter with varying levels is introduced to allow the expansion of facility capacities if necessary.
- A utilization rate is set for each facility, which serves as a lower bound for the facility's operational capacity.

A. *Mixed-Integer Programming (MIP) Model*

The integer programming formulation with determined sets, parameters and decision variables that are given above is as follows:

Sets:

I: set of patient groups.

J: set of facility locations.

Parameters:

α_j: cost of establishing a hospital at candidate location $j \in J$.

β_{ij}: unit cost of distance between patient group $i \in I$ and candidate hospital location $j \in J$.

C_j: capacity of a hospital at location $j \in J$.

δ: utilization percentage of location $j \in J$.

a_i: number of patients at patient group $i \in I$.

$d_{i,j}$: distance between the location of patient group i and candidate hospital location $j \in J$.

K: maximum number of hospitals that can be established.

λ: overcapacity allowance.

Decision Variables:

$x_{i,j}$: 1, if patients at patient group $i \in I$ is served by hospital at $j \in J$.

y_j :1 if facility is opened at location j.

$$L_{max} = maximum\ load\ of\ any\ hospital$$

Mathematical Model:

$$\min Z = \sum_{j \in J} (\alpha_j * y_j + \sum_{i \in I} \beta_{ij} * d_{i,j} * x_{i,j}) + L_{max}, \tag{1}$$

subject to

$$\sum_{j \in J} x_{i,j} = 1, \quad \forall i \in I, \tag{2}$$

$$\sum_{i \in I} a_i * x_{i,j} \le C_j * (1 + \lambda) * y_j, \quad \forall j \in J, \tag{3}$$

$$\sum_{j \in J} y_j \le K, \quad \forall i \in I, \tag{4}$$

$$\sum_{i \in I} a_i * x_{i,j} \le L_{max}, \quad \forall j \in J, \tag{5}$$

$$\sum_{i \in I} a_i * x_{i,j} \ge \delta * C_j * y_j, \quad \forall j \in J, \tag{6}$$

$$x_{i,j} \in \{0, 1\}, \forall i \in I, \quad \forall j \in J, \tag{7}$$

$$y_j \in \{0, 1\}, \quad \forall j \in J, \tag{8}$$

Objective function (1) minimizes the total cost. Objective function contains three terms. The first term is the total cost of establishing hospitals. The second term represents the total cost of distance relation between hospitals and patient groups. Last term is max-load of hospitals. Constraint set (2) ensures that each patient group should be assigned to exactly one hospital. Constraint set (3) guarantees that capacity of each hospital must be hold. Constraint set (4) satisfies that maximum number of hospitals must within the bound. Constraint set (5) determines the max-load of hospitals. Constraint set (6) ensures that the utilization percentage of the new facility is satisfied. Finally, constraint set (7) and (8) represents the domain of variables.

B. *Constraint Programming (CP) Model*

The sets and parameters are the same as the ones used in IP model.

Decision Variables:

$x_i \in J$: Integer variable representing the facility location $j \in J$ assigned to patient group $i \in I$.

$y_j \in \{0,1\}$: Binary variable indicating whether a facility is opened at location $j \in J$.

$Load_j \in \mathbb{R}^+$: Continuous variable representing the total load (number of patients) assigned to facility $j \in J$.

$L_{max} \in \mathbb{R}^+$: Continuous variable representing the maximum load across all facilities.

Mathematical Model:

$$\min Z = \sum_{j \in J} (\alpha_j * y_j + \sum_{i \in I} \beta_{i,x_i} * d_{i,x_i}) + L_{max}, \tag{9}$$

subject to

$$load_j = \sum_{i \in I : x_i = j} a_i, \quad \forall j \in J, \tag{10}$$

$$pack\left(load_j, x_i, a_i, C_j\right), \quad \forall i \in I, \forall \in J \tag{11}$$

$$x_i \in J, \quad \forall i \in I \tag{12}$$

$$y_j = 1 \; if \; load_j > 0 \quad \forall j \in J, \tag{13}$$

$$load_j \leq C_j * (1 + \lambda) * y_j, \quad \forall j \in J, \tag{14}$$

$$load_j \geq \delta * C_j * y_j, \quad \forall j \in J, \tag{15}$$

$$L_{max} \geq load_j \quad \forall j \in J, \tag{16}$$

The objective function (9) minimizes the total cost of the problem, which consists of three main components: the cost of establishing new facilities, the transportation cost between patient groups and the facilities they are assigned to, and the maximum load across all facilities. The first term captures the establishment cost, the second term accounts for the distance-based transportation costs, and the final term seeks to minimize the maximum load at any facility to ensure balanced utilization. Constraint set (10) defines the load at each facility, ensuring that the total number of patients assigned to any facility is accurately calculated. The pack constraint (11) guarantees that the demand for patients can be "packed" into the facilities without exceeding the facility capacities. Constraint set (12) ensures that each patient group is assigned to exactly one facility. Constraint set (13) ensures that a facility is only opened if it is assigned at least one patient group. Constraint set (14) ensures that the load at any facility does not exceed its capacity, factoring in an overcapacity allowance, while constraint set (15) enforces a minimum utilization rate, ensuring that facilities operate at a certain percentage of their capacity when opened. Finally, constraint set (16) ensures that the maximum load across all facilities is minimized to prevent any one facility from being overloaded.

5 Computational Results

To evaluate the performance of the models proposed in this project, a series of computational experiments were conducted using randomly generated problem instances. In these test cases, the maximum number of facilities to be established, denoted as K, was determined based on factors such as government budget and the potential patient population. Test instances were generated for 25, 50, and 100 patient groups, with the maximum number of facilities set to 22, 40, and 90, respectively, as the number of patient groups increased. The distances between candidate facility locations were computed using Euclidean distance.

To ensure manageable computational times, two levels of overcapacity allowance (λ) were considered: 0.2 and 0.3, alongside a single utilization percentage (δ) of 0.7 for long-term care facilities. The total patient demand was categorized into three levels, representing 60%, 75%, and 90% of the total facility capacity, denoted as L, M, and H, respectively. Additionally, patient demand variances were classified into two groups, high and low variance, represented as 1 and 0, respectively. Each instance was run five times. The models were coded in CPLEX, and experiments were conducted on a personal computer with a 1.70 GHz Intel processor and 8 GB RAM. Results for each instance were recorded based on the objective function value, CPU time for the mixed-integer programming (MIP) model, and the relative gap percentage (RGAP) between the MIP and constraint programming (CP) models. The relative gap was calculated as follows in Eq. (17):

$$RGAP = \frac{(Objective\ of\ MIP - Objective\ of\ CP) \times 100}{Objective\ of\ CP} \tag{17}$$

Tables below present the results for instances with 25, 50, and 100 demand points for both MIP and CP models and indicate that optimal solutions were consistently found within short computational times for the MIP model. The model was tested under a fixed utilization rate and two levels of overcapacity allowance. Instances with low demand levels and high demand variance generally required more time to find an optimal solution, likely due to the larger solution space created by increased decision complexity. Conversely, instances with high demand levels and low variance were solved quickly, even for 100 demand points. Additionally, as seen in Tables 1 and 2, CPU time increases as the overcapacity allowance (λ) increases across all instances.

For the CP model, obtaining optimal solutions was more challenging. To control computation times, thresholds of 60 s for the 25- and 50-demand point instances and 180 s for the 100-demand point instances were applied. In the smaller 25-demand point cases, the CP model achieved near-optimal solutions, but as problem size increased, particularly for the 100-demand point instances, the relative gap grew significantly. This trend was consistent across both overcapacity allowance levels.

Moreover, the CP model performed worse for instances with low demand levels (L), as evidenced by the higher relative gap in these cases. Overall, the computational experiments demonstrated that the CP approach requires a different strategy

Table 1 MIP and CP solutions for instances with $\lambda = 0.2$ and $\delta = 0.7$

Parameters	Instances	MIP		CP	
		Objective	CPU (sec.)	Objective	RGAP (%)
$\lambda = 0.2$	25_0L	7443	1.84	7494	0.67
$\delta = 0.7$	25_0M	8858.6	0.42	8865.8	0.08
	25_0H	11,153.2	0.25	11,153.2	0.00
	25_1L	7471.4	2.29	7534	0.87
	25_1M	8895.2	0.63	8971.6	0.88
	25_1H	10,285.4	0.30	10,293.4	0.08
	50_0L	14,057.6	9.48	14,635.8	4.12
	50_0M	16,628.6	1.74	17,096.8	2.80
	50_0H	20,847.8	1.04	21,121.5	1.59
	50_1L	13,648.6	24.92	14,398.4	5.50
	50_1M	16,233.4	6.67	16,941.6	4.35
	50_1H	19,585.8	2.51	20,053.8	2.41
	100_0L	25,787	130.43	29,737.2	15.32
	100_0M	31,417.8	13.51	36,306.4	15.56
	100_0H	39,717.8	3.17	42,913.8	8.15
	100_1L	26,135.4	2552.46	30,404.4	16.31
	100_1M	31,010	64.47	35,552.4	14.63
	100_1H	39,648.4	2.90	42,816.8	8.06

compared to traditional mathematical programming. The efficiency of CP models depends heavily on the careful definition of decision variables, the integration of implied constraints, the development of effective branching strategies, and the combination of diverse viewpoints.

6 Conclusions and Future Works

This study addressed the long-term care facility location problem with the objective of minimizing both the establishment cost of facilities and the assignment cost of patients to the nearest facilities, while also aiming to balance facility loads. This created a multi-objective optimization problem. To reflect real-world scenarios in Turkey, the study incorporated two levels of overcapacity for facilities and a fixed utilization rate as a lower bound for facility operations. Two mathematical models were developed: a mixed-integer programming (MIP) model and a constraint programming (CP) model, although the CP model was built upon the MIP structure rather than fully leveraging the strengths of constraint programming.

Table 2 MIP and CP solutions for instances with $\lambda = 0.3$ and $\delta = 0.7$

Parameters	Instances	MIP		CP	
		Objective	CPU (sec.)	Objective	RGAP (%)
$\lambda = 0.3$	25_0L	7080	2.30	7126.2	0.65
$\delta = 0.7$	25_0M	8366	0.62	8380.6	0.18
	25_0H	10,156.2	0.33	10,166.6	0.10
	25_1L	7134.8	2.94	7189	0.76
	25_1M	8486	1.00	8526.6	0.48
	25_1H	9602.8	0.45	9626.4	0.24
	50_0L	13,271.4	11.25	13,867.4	4.51
	50_0M	15,645	4.02	16,199.2	3.54
	50_0H	18,629.6	0.93	19,195.2	3.03
	50_1L	13,067	21.86	13,509.2	4.60
	50_1M	15,299.2	8.40	15,986.8	4.51
	50_1H	18,476.8	3.39	18,866.2	3.62
	100_0L	24,291	2611.71	28,044.2	15.46
	100_0M	29,520.2	25.77	33,746.2	14.31
	100_0H	35,375.4	8.34	40,543.4	14.65
	100_1L	24,662	6374.81	28,522.2	15.63
	100_1M	29,025	85.84	33,190.4	14.35
	100_1H	35,366.8	9.58	40,687	15.04

To assess the performance of the models, instances were randomly generated based on different demand point levels, and both models were tested. Contrary to expectations, the CP model struggled to find optimal solutions within a reasonable time or only provided feasible solutions. This suggests that while CP is a powerful paradigm, it may not be well-suited for multi-objective problems like this one. In contrast, the MIP model consistently found optimal solutions in a short time, making it the superior approach for this problem.

The study offers several opportunities for future research. One extension could involve establishing new facilities in addition to existing ones, which would better reflect the situation in more developed countries. Another direction could involve optimizing the capacities of both new and existing facilities. From the modeling perspective, the CP model could be enhanced with the use of global constraints and advanced heuristics, such as value or variable ordering heuristics, to improve performance. Finally, alternative decision variables could be introduced to enrich the CP model and reduce computational time, potentially leading to better results.

References

1. Shroff HFE, Gulledge TR, Haynes KE (1998) Siting efficiency of long-term healthcare facilities. Socioecon Plann Sci 32:109–124
2. Krohn R, Müller S, Haase K (2021) Preventive healthcare facility location planning with quality-conscious clients. OR Spectrum 43:337–366
3. Song BD, Ko YD, Hwang H (2015) The design of capacitated facility networks for long-term care service. Comput Ind Eng 87:155–165
4. Khodaparasti S, Bruni ME, Beraldi P, Maleki HR (2018) A multi-period location-allocation model for nursing home network planning under uncertainty. Oper Res Healthc 17:20–33
5. Alinaghian M, Hejazi SR, Bajoul N (2023) A novel robust model for location-allocation of healthcare facilities considering pre-disaster and post-disaster characteristics. Scientia Iranica 30:123–137
6. Owen SH, Daskin MS (1998) Strategic facility location: a review. Eur J Oper Res 111:423–447
7. Rodriguez-Verjan C, Augusto V, Xie X (2018) Home health-care network design: location and configuration of home health-care centers. Oper Res Healthc 19:100–110
8. Bogataj D, Bogataj M, Drobne S (2022) Long-term care sustainable networks in ADRION region. Sustainability 14(18):11154

Resolving Stakeholder Conflicts in Airport Gate Assignment: A Multi-objective Approach with Goal Programming and MIP Models

Mert Paldrak, Gamze Erdem, Melis Tan Tacoğlu, and Mustafa Arslan Örnek

Abstract The Airport Gate Assignment Problem (AGAP) is a critical aspect of airport operations, involving the assignment of gates to incoming and outgoing flights. In this study, we model the AGAP as a multi-objective optimization problem, addressing the conflicting preferences of key stakeholders: passengers, airlines, and airport management. Each stakeholder has distinct gate preferences, creating challenges in balancing their competing objectives. To address these challenges, we develop Mixed Integer Programming (MIP) models, incorporating both assignment-based and time-tabling-based approaches. The solution process leverages Goal Programming techniques, including Weighted Goal Programming, Tchebychev Goal Programming, and Lexicographic Goal Programming. The models are implemented and solved using IBM ILOG CPLEX Version 12.0, providing a robust framework for optimizing gate assignments while considering the complex trade-offs between stakeholder preferences. Our results offer insights into the effectiveness of these techniques in resolving conflicts and achieving an optimal balance in gate assignments.

Keywords Gate Assignment Problem (AGAP) · Multi-objective optimization · Goal programming · Mixed Integer Programming (MIP) · Stakeholder preferences

1 Introduction

Efficient management of airport resources is crucial in today's fast-paced air travel environment, where operational delays can result in significant financial losses and passenger dissatisfaction. One of the most critical aspects of airport management is the Airport Gate Assignment Problem (AGAP), which involves assigning gates to

M. Paldrak (✉) · G. Erdem · M. T. Tacoğlu · M. A. Örnek
Industrial Engineering, Yasar University, İzmir, Turkey
e-mail: mert.paldrak@yasar.edu.tr

M. A. Örnek
e-mail: arslan.ornek@yasar.edu.tr

N. M. Durakbasa and K. G. Gülen (eds.), *Sustainable Green Conversion*, Lecture Notes in Mechanical Engineering, https://doi.org/10.1007/978-3-031-83583-4_22

flights in a way that balances the needs of passengers, airlines, and airport management. As the number of flights continues to increase globally, the complexity of this problem grows, especially when considering the conflicting preferences of stakeholders. Passengers often prioritize convenience, seeking gates closer to terminal services and amenities, while airlines aim to minimize delays and turnaround times by securing gates that allow for efficient ground operations. On the other hand, airport management focuses on optimizing overall airport performance by ensuring smooth traffic flow and minimizing operational bottlenecks.

The conflicting priorities of these stakeholders turn the AGAP into a multi-objective optimization problem where trade-offs are inevitable. Addressing these conflicting objectives requires a structured approach capable of simultaneously managing different priorities while achieving a balanced solution. To this end, we employ Mixed Integer Programming (MIP) techniques, utilizing both assignment-based and time-tabling-based models. These models are designed to account for the unique preferences of each stakeholder group while maintaining operational efficiency. Through these models, we can better understand how different gate assignment strategies impact the overall functioning of the airport and its ability to satisfy the competing demands of all stakeholders.

To find solutions that balance these competing objectives, we explore Goal Programming techniques, which allow us to incorporate the varied preferences of passengers, airlines, and airport management into the optimization process. Specifically, we implement Weighted Goal Programming, Tchebychev Goal Programming, and Lexicographic Goal Programming, which provide flexible frameworks for dealing with multi-objective trade-offs. By applying these methods in conjunction with IBM ILOG CPLEX Version 12.0, we are able to develop a robust decision-making framework that offers practical solutions to real-world gate assignment challenges. Our approach provides valuable insights into how different optimization techniques can resolve stakeholder conflicts and improve airport gate management.

2 Problem Definition

The Airport Gate Assignment Problem (AGAP) is concerned with the allocation of available gates to scheduled flights at an airport. The challenge arises from the limited number of gates and the high number of flights, each with varying arrival and departure times. The objective is to assign gates in a manner that meets operational requirements while accommodating the preferences of multiple stakeholders: passengers, airlines, and airport management.

Passengers typically prioritize gates that are close to terminal services or reduce walking distances, while airlines prefer gates that minimize turnaround times, allowing for faster boarding and deboarding. Additionally, airport management must ensure that gate assignments are not only efficient but also safe and compliant with airport regulations, such as avoiding conflicts between arriving and departing flights, preventing overcrowding, and minimizing idle time for gates.

The conflicting nature of these preferences transforms the AGAP into a multi-objective optimization problem. In particular, the problem requires a balance between competing goals, such as maximizing passenger convenience, minimizing airline operational delays, and optimizing gate usage from an airport management perspective. The core challenge lies in designing an assignment strategy that satisfies these diverse and often conflicting objectives simultaneously, while adhering to the operational constraints of the airport environment.

3 Literature Review

The AGAP has been addressed extensively through various optimization techniques, ranging from exact methods to evolutionary and heuristic approaches. Many studies highlight the complexity of AGAP, particularly due to the conflicting preferences of passengers, airlines, and airport management.

Multi-objective optimization models are frequently used to tackle AGAP. For example, a study by [1] applies both exact and evolutionary multi-objective optimization approaches, utilizing IBM ILOG CPLEX and a dedicated evolutionary algorithm to handle stakeholder conflicts in AGAP. Similarly, [2] uses a MIP approach that incorporates passenger walking distance and airline fuel consumption, showing significant improvements in gate assignment efficiency.

A significant body of literature emphasizes the role of metaheuristic algorithms in solving AGAP. Hu et al. [3] developed a genetic algorithm under uncertainty theory for flight-to-gate scheduling, which allowed for more robust assignments in unpredictable environments. Other metaheuristic techniques, like GRASP (Greedy Randomized Adaptive Search Procedures), have been applied by [3] to improve gate assignment while accounting for both operational and passenger-oriented objectives.

Recent work also emphasizes hybrid approaches, combining heuristic and exact methods. Marinelli et al. [4] explored a hybrid metaheuristic approach, integrating GRASP with a mathematical model to optimize gate assignments in real-time operations. Another study by [5] focused on using Pareto simulated annealing for multi-objective gate scheduling, balancing multiple conflicting objectives like minimizing flight delays and passenger walking distances.

Tabu search algorithms have also proven effective, as demonstrated by [6], who proposed a hybrid algorithm combining tabu search with mathematical models to optimize flight-to-gate assignments. Moreover, studies like those by Brazile and Swigger [7] and Dorndorf et al. [8] focus on the practical application of gate assignment systems, emphasizing the importance of adaptability and real-time scheduling in managing airport operations efficiently.

This extensive research demonstrates that AGAP remains a challenging problem due to its multi-objective nature, and the use of advanced optimization methods, including goal programming, metaheuristics, and hybrid models, is essential for producing feasible solutions in real-world scenarios.

4 Mathematical Formulations

In the context of the Airport Gate Assignment Problem (AGAP), two distinct modelling approaches are commonly used: time-tabling-based models and assignment-based models. While both aim to optimize gate assignments, they differ significantly in their structure and complexity.

The time-tabling-based model introduces an additional dimension to the problem, representing the sequence of time periods during which each aircraft is present at the airport. This model takes into account the specific time intervals for which a gate is occupied by a flight, capturing the dynamic nature of gate utilization. Each period of a flight's stay at the airport is carefully tracked, allowing the model to address conflicts that arise not only from overlapping schedules but also from the varying durations that different aircraft occupy gates. As a result, time-tabling models are more suited to scenarios where flight schedules are subject to frequent changes, and precise gate allocation across time is critical for efficient airport operations.

On the other hand, the assignment-based model focuses solely on assigning gates to flights, without explicitly considering the time periods for which the gates are occupied. In this approach, each flight is assigned to a gate based on static parameters, such as gate availability and operational constraints. The model does not account for the temporal overlap of flight operations at the airport, making it less suited for handling fluctuating schedules or detailed time-based conflicts. However, this model is typically simpler and more computationally efficient, especially in cases where time-specific details are less relevant or when flight schedules are relatively stable.

A. *Timetabling Based Integer Programming Model*

Before providing timetabling based IP model, the notations, sets, parameters and the decision variables are defined as follows:

Sets and Indices

i: index of periods, $i \in S = \{1,2, \ldots., |S|\}$
j, b: index of flights, $j, b \in U = \{1,2, \ldots., |U|\}$
k, r: index of parking areas, $k, r \in N = \{1,2, \ldots, |N|\}$
c : index of ground service firms, $c \in C = \{1,2, \ldots, |C|\}$
y index of night-stand flights, $y \in Y = \{1,2, \ldots., |Y|\}$
m: index of bridge-equipped parking areas where night-stand planes cannot be assigned to, $m \in M = \{1,2, \ldots., |M|\}$
d: index of parking areas that are occupied from the previous day, $d \in D = \{1,2, \ldots, |D|\}$.

Parameters

a_j: Scheduled arrival period of flight j, $a_j \in S$
g_j: Scheduled departure period of flight$j$, $g_j \in S$. It is assumed that flight j left the airport in period $(g_j - 1)$.
f_j Ground service company of flight $j, f_j \in C$

$$L_{kr} = \begin{cases} 1 \text{ Ifparkingareaskandrareadjacent} \\ 0 \ \frac{\text{o}}{\text{w}} \end{cases}$$

$$B_{jk} = \begin{cases} 1 \ \text{ If flight } j \text{ can be assigned to the parking area } k \\ 0 \ \frac{o}{w} \end{cases}$$

t_k: Earliest available period of parking area $k \in D$, $t_k \in S$
W_{jk}: Utility of assigning flight j to parking area k reflecting the airport preferences, $W_{jk} \in Z^+$
S_{jk}: Utility of assigning flight j to parking area k reflecting the airlines preferences, $S_{jk} \in Z^+$
w_k: Utility of parking area k reflecting the passenger preferences , $w_k \in Z^+$

Decision Variables

$$x_{ijk} = \begin{cases} 1 \ \text{ If plane } j \text{ is assigned to parking area } k \text{ in period } i \\ 0 \ \text{ O/W} \end{cases}$$

Mathematical Model

$$Objective\,function(s) \tag{1}$$

Subject to

$$\sum_{k \in N} x_{a,jk} = 1, \quad \forall j \in U \tag{2}$$

$$x_{ijk} = 0, \quad \forall i \in S, \forall j \in U, \forall k \in N \mid \left(i < a_j\right) \wedge \left(i \geq g_j\right) \tag{3}$$

$$\sum_{j \in U} x_{ijk} \leq 1 \quad \forall i \in S, \forall k \in N \tag{4}$$

$$\sum_{i \in S | a_j \leq i < g_j} x_{ijk} = \left(g_j - a_j\right) x_{a_j jk}, \quad \forall j \in U, \forall k \in N \tag{5}$$

$$\begin{aligned} & x_{ijr} + x_{ibk} \leq 1, \forall i \in S, \forall j, b \in U, \forall k, r \in N \\ & |(j < b) \wedge \left(a_j \leq i < g_j\right) \wedge \left(f_j \neq f_b\right) \wedge (L_{kr} = 1) \\ & \wedge\left[\left(a_b \leq a_j \wedge g_j > a_b\right) \vee \left(a_j \leq a_b \wedge g_b > a_j\right)\right] \end{aligned} \tag{6}$$

$$x_{a_j j} = 0, \quad \forall j \in Y, \forall k \in M \tag{7}$$

$$x_{a_j j} = 0, \quad \forall j \in U, \forall k \in D \mid a_j \leq t_k \tag{8}$$

$$x_{a_{jk}jk} \leq B_{jk}, \quad \forall j \in U, \forall k \in N \tag{9}$$

$$x_{ijk} \in \{0,1\} \quad \forall j \in U, \forall i \in S, \forall k \in N \tag{10}$$

The objective functions (1) are to be explained in detail in the following section. Constraints (2) ensure that each flight is assigned to exactly one gate at its scheduled arrival time. Constraints (3) prevent a flight from being assigned to a gate outside its designated time frame, ensuring that no flight is assigned to a gate before its arrival or after its departure. As noted by Öztürk et al. (2015), if the arrival and departure times of a flight are denoted by a_j and g_j, respectively, the flight will occupy the assigned gate from the start of a_j to the start of g_j. This means that the gate will be idle during period g_j until the next flight arrives. Constraints (4) ensure that a parking area can be occupied by no more than one flight at any given time. Constraints (5) guarantee that the same parking area is used for the entire service period of a flight. Since the variable is set to 1 for only one gate, the right-hand side of these constraints will reflect the service period of the flight, but only for the assigned parking area. On the left-hand side, the constraint enforces that the sum of the binary assignment variables over the successive periods must equal the service time of the flight. Similarly, if flight j is not assigned to parking area k at its arrival time, the right-hand side of constraints (5) will be zero, and the constraint will ensure that the sum of the binary variables during the service period also equals zero, indicating that the area is not used. Constraints (6) ensure that any two flights serviced by different ground handling companies, with overlapping service periods, must be assigned to non-adjacent parking areas. Constraints (7) prevent the assignment of night-stand flights to parking areas equipped with boarding bridges, while constraints (8) ensure that a parking area cannot be assigned to a flight before it becomes available. Constraints (9) guarantee that each flight is only assigned to eligible parking areas. Finally, constraints (10) define the domain of the decision variables, setting the appropriate bounds.

2 *Assignment Based Integer Programming Model*

Before providing assignment-based IP model, the notations, sets, parameters and the decision variables are defined as follows:

The indices and parameters used in this model are consistent with those employed in the Timetabling model. This ensures that the same notation is applied across both models, allowing for seamless integration and comparison between the two. The consistency in indices—such as for flights, parking areas, and time periods—along with the parameters defining service times, arrival, and departure periods, facilitates a coherent and unified framework for analysing the problem under different modelling approaches.

Decision Variables

$$x_{jk} = \begin{cases} 1 & \text{If plane } j \text{ is assigned to parking area } k \\ 0 & \text{O/W} \end{cases}$$

Mathematical Model

$$Objective\,function(s) \tag{11}$$

Subject to

$$\sum_{k \in N} X_{jk} = 1 \quad j \in U \tag{12}$$

$$\sum_{i \in U|(i \neq j) \wedge (a_i \leq a_j < d_i)} X_{ik} \leq M * (1 - X_{jk}) \quad j \in U, \quad k \in N \tag{13}$$

$$X_{ik} + X_{jr} \leq 1 \quad i, j \in U, k, r \in N | (i < j) \wedge (f_i \neq f_j) \wedge (L_{kr} = 1) \wedge \left[(a_i \leq a_j < g_i) \vee (a_j \leq a_i < g_j) \right] \tag{14}$$

$$X_{jk} \leq B_{jk} \quad j \in U, k \in N \tag{15}$$

$$X_{jk} = 0 \quad j \in Y, k \in MD \tag{16}$$

$$X_{jd} = 0 \quad j \in U, d \in DK | a_j < t_d \tag{17}$$

$$X_{ik} \in \{0, 1\} \quad i, j \in U, k \in N \tag{18}$$

The objective functions (11) are to be explained in detail in the following section. Constraints (12) ensure that each flight is assigned to exactly one gate. Constraints (13) prevent the assignment of any other flight to a gate before the arrival or after the departure of the currently assigned flight. Constraints (14) mandate that flights handled by different ground handling companies, which are present at the airport simultaneously, must be allocated to non-adjacent parking areas. Constraints (15) ensure that flight-gate assignment compatibility requirements are met, while constraints (16) prohibit the allocation of night-stand flights to parking areas equipped with passenger boarding bridges. Constraints (17) ensure that parking areas can only be utilized when they are available, and finally, constraints (18) define the permissible range for the decision variables, ensuring they are within the feasible set.

3 *Objective Functions of Integer Programming Models*

In the Airport Gate Assignment Problem (AGAP), the objective functions aim to address the competing preferences of the key stakeholders: passengers, airlines, and airport management. These objectives often conflict with one another, requiring a balanced approach to meet the needs of each group effectively.

Passengers prioritize convenience, such as minimizing walking distance to terminal amenities, which leads to an objective function designed to reduce overall passenger transit time. This objective reflects the goal of improving passenger satisfaction by providing more accessible gate locations, thus optimizing the airport experience from their perspective.

For airlines, the focus lies on operational efficiency, with preferences centered around minimizing turnaround time, reducing the number of flights assigned to remote aprons, and ensuring on-time departures. The corresponding objective function aims to reduce delays and optimize gate usage in a way that enhances airline schedules and minimizes operational costs, such as fuel consumption and taxiing time.

Finally, airport management is concerned with the overall efficient use of airport infrastructure, aiming to maximize gate utilization and avoid disruptions in ground operations. This objective function includes ensuring that parking areas are allocated efficiently, while also meeting safety and logistical requirements, such as preventing overlapping services from different ground handling companies. Additionally, night-stand flights should not occupy bridge-equipped gates, ensuring that key resources are available for flights with higher priority needs.

Each objective function will be presented in detail, showing how these preferences are incorporated into the overall optimization framework, and highlighting the challenges involved in balancing these often-conflicting goals.

Minimization of Passenger Walking Distance

The walking matrix of passengers is used to quantitatively determine the distances between gates and check-in counters, helping to assess which gates are more or less convenient for passengers. Gates closer to check-in counters and airport exits are preferred, as they provide greater convenience for arrival and departure passengers. Instead of calculating these distances for various passenger types (arrival, departure, and transfer), a utility value is assigned to each gate based on customer satisfaction. The first objective function is designed to maximize the total utility of the gates, reflecting their varying desirability based on proximity and convenience.

$$Z_1 = \max \Sigma_{j \in U} \Sigma_{k \in N} w_k * X_{ijk} \tag{1a}$$

Maximizations of Airlines' Preferences

Airline companies that use the same ground handling services are typically assigned to adjacent gates to reduce apron traffic. However, airlines may have specific gate preferences based on their operational policies. Many airlines have agreements with airport management that prioritize certain gates for their flights. While these preferred gates are often adjacent, this can sometimes conflict with airport management's assignment plan. To manage this, preference values are assigned to each flight-gate combination, and priority values are assigned to each gate, helping to optimize convenience for both the airport and airlines. The final objective function aims to maximize airline satisfaction by considering these preferences in the flight-gate assignment process.

$$Z_2 = \max \Sigma_{j \in U} \sum_{k \in N} S_{jk} * X_{ijk} \tag{1b}$$

Maximization of Gate Utility for Airport Management

In the integer models, airport management prioritizes assigning flights to bridge-equipped gates due to the increased efficiency they provide for passenger boarding and disembarking. Additionally, after aircraft arrive, they require essential services such as refuelling, restocking, and waste removal. If all bridge-equipped gates are occupied, incoming flights are assigned to open park areas (aprons). To optimize resource use, airport management ensures that large gates are reserved for larger aircraft with more passengers, maximizing operational efficiency. Thus, the second objective function is designed to maximize gate utility from the perspective of airport management by prioritizing effective gate allocation and ensuring the optimal use of larger gates for high-capacity flights.

$$Z_3 = \max \sum_{j \in U} \sum_{k \in N} W_{jk} * X_{ijk} \tag{1c}$$

5 Problem Environment and Solutions

The gate assignment optimization model considers three key factors: customer satisfaction based on gate preference, gate utility from the airport management's perspective, and airline satisfaction regarding gate preferences. To address the multi-objective nature of the problem, a coordination solution is required, as solving a multi-objective model can be challenging. Common solution approaches include the Weighted Goal Programming method, Pre-emptive Goal Programming method, and Tchebychev Goal Programming method.

Initially, pre-emptive goal programming is considered, as it is useful when objectives have significantly different priorities. In this approach, objectives are ranked in order of importance, with the model optimizing the most critical goal first, followed by the second-most important goal, ensuring no compromise on higher-priority objectives. The weighted method is also employed to transform the multi-objective problem into a single-objective function by assigning weight coefficients to each objective based on their importance. This method is relatively simple to implement and offers precision in balancing multiple goals. Furthermore, the Tchebychev Goal Programming method is used to minimize the maximum deviation from the desired goals, providing an alternative approach to balancing conflicting objectives. This method is particularly effective when there is a need to reduce large discrepancies among the goals, ensuring that none of the objectives is excessively neglected.

Since real data from Adnan Menderes International Airport Management is not yet available, toy data was generated to test the model's applicability. In the integer programming models, the flight-gate utility matrix is used to represent gate utility from the airport's perspective. Gate weights are also adjusted to minimize passenger walking distances, with gates closer to the entrance and exit being more preferred. A gate preference matrix for passengers, with values ranging from 0 to 10, is generated. Open park areas are considered neutral in terms of passenger preference, as passengers are transported via vehicles regardless of the assigned stand.

This combination of pre-emptive programming, the weighted method, and Tchebychev Goal Programming provides a robust framework for addressing the multi-objective nature of gate assignment, optimizing operational efficiency and stakeholder satisfaction.

4 *Pre-Emptive Goal Programming*

In pre-emptive goal programming, the objective functions are prioritized in the following order: first, the maximization of flight-gate utility from the perspective of airport management; second, the maximization of airline satisfaction based on their gate preferences; and finally, the least important objective is the maximization of passenger gate preferences, focusing on minimizing walking distances.

The objective functions have been formulated for both the timetabling-based multi-objective model and the assignment-based multi-objective model. The objective functions for the timetabling-based multi-objective model are presented as follows:

Goal 1:

$$\max \sum_{j \in U} \sum_{k \in N} w_k * X_{a_j jk} \tag{1a.2}$$

Goal 2:

$$\max \sum_{j \in U} \sum_{k \in N} S_{jk} * X_{a_j jk} \tag{1b.2}$$

Goal 3:

$$\max \sum_{j \in U} \sum_{k \in N} W_{jk} * X_{a_j jk} \tag{1c.2}$$

subject to; *all constraints given in timetabling based integer programming model.*

The objective functions for assignment based multi-objective model is as given below:

Goal 1:

$$\max \sum_{j \in U} \sum_{k \in N} w_k * X_{jk} \tag{1a.3}$$

Goal 2:

$$\max \sum_{j \in U} \sum_{k \in N} S_{jk} * X_{jk} \tag{1b.3}$$

Goal 3:

$$\max \sum_{j \in U} \sum_{k \in N} W_{jk} * X_{jk} \tag{1c.3}$$

subject to; *all constraints given in assignment based integer programming model.*

5 *Weighted Goal Programming*

As an alternative to the pre-emptive goal programming model, the weighted goal programming method is employed. This approach converts multiple objectives into a single objective function by multiplying each objective by a constant weight and summing them. The weights are assigned based on the relative importance of each objective function in the model. Typically, these weights are chosen within a range between 0 and 1, reflecting the priority of each objective.

The weighted objective functions have also been formulated for both the timetabling-based multi-objective model and the assignment-based multi-objective model. The weighted objective function for the timetabling-based multi-objective model is provided below:

$$\begin{aligned} max\, Z_4 = w_1 * \left(\sum_{j \in U} \sum_{k \in N} w_k * X_{a_j jk} \right) + w_2 * \left(\sum_{j \in U} \sum_{k \in N} W_{jk} * X_{a_j jk} \right) \\ + w_3 * \left(\sum_{j \in U} \sum_{k \in N} S_{jk} * X_{a_j jk} \right) \end{aligned} \tag{19}$$

subject to: *all constraints given in timetabling based integer programming model.*

The same objective function will be written as follows for the multi-objective assignment based programming model:

$$max\, Z_5 = w_1 * \left(\sum_{j \in U} \sum_{k \in N} w_k * X_{jk} \right) + w_2 * \left(\sum_{j \in U} \sum_{k \in N} W_{jk} * X_{jk} \right)$$

$$+ w_3 * \left(\sum_{j \in U} \sum \_k \in NS_{jk} * X_{jk} \right) \quad (20)$$

subject to: *;all constraints given in assignment based integer programming model.*

where w_1, w_2, and w_3 are the weight coefficients. According to the importance of each objective, the multiple experiments for different combinations of data will be analysed to better determine more appropriate values of weight coefficients.

To determine the weights for the objective functions, a meeting should be conducted with the airport's planning manager. Since the assigned coefficients will largely depend on the planning manager's perspective, this method is more subjective compared to the pre-emptive goal programming approach. As the necessary data has not yet been obtained, the coefficients for the objective functions have been assigned based on their perceived importance.

6 *Tchebychev Goal Programming*

In the application of Tchebychev Goal Programming to the multi-objective gate assignment model, the goal is to minimize the maximum deviation from the desired values of the different objectives, ensuring a balanced solution that addresses the preferences of multiple stakeholders.

The Tchebychev Goal Programming method transforms this into a minimization problem by focusing on the largest deviation from the ideal targets for each objective. By applying Tchebychev Goal Programming, the model minimizes the largest deviation across these three objectives, thereby preventing any one stakeholder's goals from being excessively compromised. This approach allows for a more equitable solution, balancing the conflicting priorities of airport management, airlines, and passengers. The resulting model ensures that the gate assignment process meets operational goals while maintaining a fair distribution of satisfaction among all stakeholders. This method is particularly advantageous in scenarios where conflicting objectives cannot be easily prioritized, as it provides a compromise solution that addresses the most significant deviations across all objectives, ensuring that none are neglected.

The mathematical formulation of Tchebychev Goal Programming aims to minimize the maximum deviation across multiple objectives. For the given multi-objective problem, where there are three stakeholder objectives (airport management, airlines, and passengers), the Tchebychev formulation can be expressed as follows:

$$\min Z_6 \max\{\frac{d_1}{w_1}, \frac{d_2}{w_2}, \frac{d_3}{w_3}\} \quad (21)$$

d_1, d_2, d_3 are the deviations from the desired targets for each objective. The mathematical models are subject to following additional constraints

$$\sum_{j \in U} \sum_{k \in N} w_k * X_{a_j jk} + d_1 \geq Z_1 \quad (22)$$

$$\sum_{j\in U}\sum_{k\in N} W_{jk} * X_{a_j jk} + d_2 \geq Z_2 \tag{23}$$

$$\sum_{j\in U}\sum_{k\in N} S_{jk} * X_{a_j jk} + d_3 \geq Z_3 \tag{24}$$

All Constraints Given in Timetabling Based Integer Programming Model

$$\sum_{j\in U}\sum_{k\in N} w_k * X_{jk} + d_1 \geq Z_1 \tag{25}$$

$$\sum_{j\in U}\sum_{k\in N} W_{jk} * X_{jk} + d_2 \geq Z_2 \tag{26}$$

$$\sum_{j\in U}\sum_{k\in N} S_{jk} * X_{jk} + d_3 \geq Z_3 \tag{27}$$

6 Computational Results

Before presenting the numerical results, the following Table 1 is included to illustrate and compare the theoretical bounds of the proposed models in terms of the number of variables and constraints. The developed models have been tested using realistic instances provided by the primary airline operator, which includes 35 parking areas (19 of which are bridge-equipped), 105 flights, and four different ground-handling companies. The Table 2 summarizes the performance of the models, showcasing the scalability and complexity of the solutions based on these practical data inputs.

The multi-objective models, utilizing both the pre-emptive goal programming approach and the weighted method, have been executed, and the results are presented

Table 1 Theoretical bounds for the size of the problem

Formulation	Model (Approach)	Number of variables	Number of constraints
Integer programming	Timetabling	$\sum_{j\in U}(g_j - a_j)\lvert N\rvert$	$(\lvert U\rvert\lvert N\rvert)(\lvert U\rvert\lvert N\rvert - 1)/2$
	Assignment	$\lvert U\rvert\lvert N\rvert$	$(\lvert U\rvert\lvert N\rvert)(\lvert U\rvert\lvert N\rvert - 1)/2$

Table 2 Performance evaluation of developed model on a real instance

Assignment model	Number of variables	Number of constraints	Number of nodes	Objective	Time (s)
Timetabling	96,025	649,207	16,859	8832	101*
Assignment	3467	202,340	31,245	8832	116*

in the tables below. It is important to reiterate that a fictional airport layout is used for these models, and the data has been generated according to the specifications of this layout. Initially, each objective function was individually optimized as a single objective to determine the optimal solutions. Key performance indicators such as the number of constraints, number of nodes, and CPU time required to find the solutions were recorded. The results of these runs are summarized between Tables 3, 4, and 5.

Upon demonstrating that each objective function produces a feasible and optimal solution, the pre-emptive goal programming model is subsequently developed. The maximization of total flight-gate utility is prioritized as the primary objective. Both the timetabling and assignment models are executed with a focus solely on this objective. As a result, the outcomes of the initial pre-emptive model correspond directly to those presented in Table 5, as the same objective function was used in both cases (Table 6).

Subsequently, in the second pre-emptive model, the first objective function is transformed into a constraint that must not be violated. Specifically, the total flight-gate utility is incorporated as a constraint, ensuring that it does not fall below the optimal value of 975. Following the introduction of this constraint, both models are then executed with the second objective function, which focuses on maximizing the total gate preference for airline companies (Table 7).

Finally, the third objective function, which focuses on maximizing total gate preference for passengers to minimize walking distance, is applied in the next pre-emptive goal programming model. In this iteration, the first two objective functions are incorporated as constraints into the original model. However, both models were

Table 3 Performance evaluation of the first objective function

Assignment model	Number of variables	Number of constraints	Number of nodes	Objective	Time (s)
Timetabling	1,069,425	3,060,538	13,846	975	59.50
Assignment	3675	41,463	2441	975	31.52

Table 4 Performance evaluation of the second objective function

Assignment model	Number of variables	Number of constraints	Number of nodes	Objective	Time (s)
Timetabling	1,069,425	3,060,538	3238	926	33.89
Assignment	3675	41,463	2875	926	33.97

Table 5 Performance evaluation of the third objective function

Assignment model	Number of variables	Number of constraints	Number of nodes	Objective	Time (s)
Timetabling	1,069,425	3,060,538	26,475	801	2682
Assignment	3675	41,463	14,363	801	2108

Table 6 Performance evaluation of the first objective function in pre-emptive goal programming

Assignment model	Number of variables	Number of constraints	Number of nodes	Objective	Time (s)
Timetabling	1,069,425	3,060,538	13,846	975	59.50
Assignment	3675	41,463	2441	975	31.52

Table 7 Performance evaluation of the second objective function in pre-emptive goal programming

Assignment model	Number of variables	Number of constraints	Number of nodes	Objective	Time (s)
Timetabling	1,069,425	3,064,213	28,342	700	616.31
Assignment	3675	45,138	7918	700	330.25

Table 8 Performance evaluation of assignment based multi-objective GP model with weight coefficients

w_1	w_2	w_3	Number of variables	Number of constraints	Number of nodes	Objective	Time (s)
0.5	0.35	0.15	3675	41,463	23,570	867.6	337.25
0.6	0.3	0.1	3675	41,463	3414	885.1	103.89
0.7	0.2	0.1	3675	41,463	8146	895.2	136.48
0.75	0.15	0.1	3675	41,463	5375	903.1	101.11

Table 9 Performance evaluation of scheduling based multi-objective GP model with weight coefficients

w_1	w_2	w_3	Number of variables	Number of constraints	Number of nodes	Objective	Time (s)
0.5	0.35	0.15	1,069,425	3,060,538	22,979	867.6	350.02
0.6	0.3	0.1	1,069,425	3,060,538	8129	885.1	129.95
0.7	0.2	0.1	1,069,425	3,060,538	18,876	895.2	209.33
0.75	0.15	0.1	1,069,425	3,060,538	8872	903.1	156.19

unable to find an optimal solution for the third pre-emptive model within the 3600-s time limit. Instead, near-optimal solutions were obtained for both models, with a resulting value of 617.

The multi-objective model was also executed using the weighted method for both the timetabling and assignment-based models. Various weighting scenarios were tested to observe how changes in the weight coefficients affect the objective function values. Since flight-gate utilization is the most important objective, the highest weight coefficient (between 0 and 1) was assigned to this function. The second objective, reflecting the gate preferences of airline companies, was given a lower weight than the flight-gate utilization objective. Lastly, the gate preference of passengers, the

Table 10 Performance evaluation of assignment based multi-objective GP model with weight coefficients with W1 = W2

w_1	w_2	w_3	Number of variables	Number of constraints	Number of nodes	Objective	Time (s)
0.4	0.4	0.2	3675	41,463	36,298	854.2	686.66
0.45	0.45	0.1	3675	41,463	16,411	875.8	311.59
0.48	0.48	0.04	3675	41,463	14,390	889.76	268.31
0.5	0.5	0	3675	41,463	28,609	899.5	344.48

Table 11 Performance evaluation of timetabling based multi-objective GP model with weight coefficients with W1 = W2

w_1	w_2	w_3	Number of variables	Number of constraints	Number of nodes	Objective	Time (s)
0.4	0.4	0.2	1,069,425	3,060,538	37,646	854.2	715.64
0.45	0.45	0.1	1,069,425	3,060,538	21,482	875.8	284.25
0.48	0.48	0.04	1,069,425	3,060,538	67,671	889.76	1271.78
0.5	0.5	0	1,069,425	3,060,538	43,998	899.5	596.31

least important objective, received the smallest weight. In each trial for both models, the sum of the weight coefficients was maintained at 1.

In the final trials, total flight-gate utilization and total gate preference of airline companies were treated as equally important objectives. This approach was adopted because it is highly likely that such a balance is encountered in most airport management systems, where both efficient gate utilization and airline preferences play crucial roles in operational decision-making (Tables 8, 9, 10, and 11).

When each objective function was optimized independently, the resulting values were $Z_1 = 975$, $Z_2 = 926$ and $Z_3 = 801$, representing the optimal outcomes for flight-gate utilization, airline gate preference, and passenger gate preference, respectively. Upon applying Tchebychev Goal Programming to balance these conflicting objectives, the deviations from the optimal values were $d_1 = 64$, $d_2 = 45$, and $d_3 = 71$. These results indicate that the model was most successful in maintaining airline gate preferences (with the smallest deviation of 45), followed by flight-gate utilization. However, the largest deviation was observed for the passenger gate preference objective, suggesting that minimizing walking distance for passengers was the most compromised in the multi-objective framework. This reflects the inherent trade-offs in balancing multiple conflicting goals, with airline preferences being prioritized more effectively, while passenger convenience experienced a greater degree of sacrifice.

7 Conclusions

The computational results of this study provide key insights into optimizing the Airport Gate Assignment Problem using both timetabling and assignment-based models. Realistic test scenarios, involving 35 parking areas, 105 flights, and four ground-handling companies, were used to evaluate the models.

Both pre-emptive goal programming and the weighted method demonstrated strong results, especially when prioritizing flight-gate utilization and airline preferences, reflecting real-world operational demands. Various weight configurations confirmed that the models are adaptable, with efficient computational performance, even when balancing multiple conflicting objectives. Overall, the study shows that these optimization techniques can effectively manage complex gate assignment challenges, providing balanced outcomes for airports, airlines, and passengers.

When each objective function was optimized individually, the resulting values were $Z_1 = 975$ for flight-gate utilization, $Z_2 = 926$ for airline preferences, and $Z_3 = 801$ passenger walking distance. Applying Tchebychev Goal Programming revealed deviations of $d_1 = 64$, $d_2 = 45$, and, $d_3 = 71$ indicating that airline preferences were best maintained, while passenger satisfaction saw the largest compromise.

In conclusion, this study successfully demonstrates the application of multi-objective optimization techniques, including pre-emptive goal programming, weighted methods, and Tchebychev Goal Programming, to the Airport Gate Assignment Problem. The computational results reveal that prioritizing flight-gate utilization and airline preferences yields the most balanced outcomes, with minimal deviation from optimal values, while passenger satisfaction, in terms of walking distance, often requires compromise. Both the timetabling and assignment-based models proved adaptable and effective, with realistic scenarios confirming their practical applicability in complex airport operations. Overall, the study highlights the importance of managing conflicting stakeholder objectives and provides robust strategies for optimizing gate assignments in real-world settings.

References

1. Kaliszewski I, Miroforidis J, Stańczak J (2017) Multi-objective optimization in the airport gate assignment problem: exact vs evolutionary approaches. Computer Science 18(1):41–52
2. Cecen RK (2021) Multi-objective optimization model for airport gate assignment problem. Aircr Eng Aerosp Technol 93(2):311–318
3. Hu X, Teng J, Wu W, Li Y (2021) Research on airport scheduling of FGAP multi-objective programming model based on uncertainty theory. Symmetry 13(10):1915. https://doi.org/10.3390/sym13101915
4. Marinelli M, Dell'Orco M, Sassanelli D (2018) Optimizing airport gate assignments through a hybrid metaheuristic approach. Transportation Research Procedia 5:211–220
5. Drexl A, Nikulin Y (2008) Multicriteria airport gate assignment and Pareto simulated annealing. IIE Trans 40(4):385–397

6. Xu JF, Glenn B (2001) The airport gate assignment problem: mathematical model and a tabu search algorithm. In: Proceedings 34th Hawaii international conference system sciences, IEEE, pp 102–111
7. Brazile RP, Swigger KM (1988) GATES: an airline gate assignment and tracking expert system. IEEE Expert 3(2):33–39. https://doi.org/10.1109/64.2103
8. Dorndorf U, Jaehn F, Pesch E (2012) Flight gate scheduling with respect to a reference schedule. Ann Oper Res 194(1):177–187. https://doi.org/10.1007/s10479-010-0809-8
9. Mokhtarimousavi S, Talebi D, Asgari H (2018) A non-dominated sorting genetic algorithm approach for optimization of multi-objective airport gate assignment problem. Transp Res Rec: J Transp Res Board 2672(23):59–70

Solving the Large-Scale Crew Pairing Problem in the Airline Industry Using the Column Generation Method

Gökçe Mol and Murat Ermiş

Abstract The commercial airline industry is highly competitive, and resources must be carefully managed to be sustainable. Of the many factors that affect operating costs, the two most important are fuel costs and crew pairing. Crew pairing is the strategic assignment of flight crews, including pilots, co-pilots, and flight attendants, to designated sequences of connected flights. This pairing should be done carefully to ensure a balance between cost-effectiveness and operational efficiency. This study focuses on developing an efficient model and solution approach for large-scale real-world problems by addressing the airline crew pairing problem. The primary goal of crew pairing is to minimize the total number of crews required to cover all scheduled flights achieved by grouping flights into pairings that maximize overall crew utilization. Crew pairing also aims to minimize unproductive crew time. In other words, efforts are made to reduce waiting times and deadheads, which is the time spent at the destination to rest between flights. Since flight safety is crucial in air transportation, all pairings must comply with regulations set by national/international civil aviation authorities and international agreements. These regulations set restrictions for pilots and flight attendants, such as duty hours, rest periods, and flight duration. Several approaches have been proposed in the literature to find an exact solution to the crew pairing problem. These approaches are generally suitable for smaller datasets where the number of aircraft and destinations is limited due to issues such as scalability and impracticality. This paper addresses this challenge using the column generation method, an effective technique for solving problems characterized by a large number of variables. In our study, real flight data from the narrow-body fleet of a large global airline is used. The proposed method outperformed in terms of computational efficiency and feasibility for larger datasets.

Keywords Airline crew pairing · Column generation · Set covering · Integer programming

G. Mol · M. Ermiş (✉)
Department of Industrial Engineering, Istanbul Kültür University, 34158 Istanbul, Turkey
e-mail: m.ermis@iku.edu.tr

G. Mol
e-mail: 1900002433@stu.iku.edu.tr

N. M. Durakbasa and K. G. Gülen (eds.), *Sustainable Green Conversion*, Lecture Notes in Mechanical Engineering, https://doi.org/10.1007/978-3-031-83583-4_23

1 Introduction

The commercial airline industry operates in a highly competitive environment, requiring constant resource optimization to maintain profitability. Several key factors contribute to this competitiveness, highlighting the importance of optimization.

- Numerous Players: The industry is crowded with competitors, from market leaders to regional and budget airlines, forcing airlines to lower ticket prices and find alternative ways to offset costs.
- Fuel Price Volatility: Fuel costs are a major expense for airlines, and sudden price increases can severely impact profit margins.
- Airport Fees and Infrastructure Costs: Airlines face varying costs related to landing fees, gate rentals, and other charges depending on the airport and route.
- Crew Pairing: Optimizing crew assignments to minimize the total number of crew members required, reduce inefficient layovers, and adhere to regulations is crucial.
- Fleet Management: Airlines must carefully assess route networks and passenger demand to determine the optimal mix of aircraft types.
- Network Optimization: Constantly evaluating and adjusting route networks ensures resources are allocated to the most profitable routes.
- Labor Cost Management: Crew salaries and benefits represent a major expense.
- Technology Adoption: Airlines increasingly use technology to improve efficiency, including software for crew scheduling, maintenance planning, and revenue management.

Given the competitive nature of the industry, even minor improvements in these areas can lead to significant cost reductions and increased profitability. After fuel costs, crew pairing is one of the largest expenses for airlines. Optimizing crew pairing is crucial for minimizing costs and enhancing profitability. Efficient crew scheduling is a cornerstone of profitability, with crew pairing involving the strategic assignment of qualified crew members—pilots, co-pilots, and flight attendants—to sequences of connected flights [1].

The primary goal of crew pairing is to minimize the number of crew members needed by grouping flights into pairings that maximize crew utilization. For instance, a pilot might be assigned a pairing with a morning departure from a main hub, followed by an afternoon return flight. This approach optimizes flying times and reduces unproductive layovers. Beyond crew utilization, crew pairing seeks to minimize unproductive time, such as layovers and deadheading where crew members fly as passengers to reach a scheduled work location [2]. Both contribute to costs, and effective crew pairing aims to minimize their occurrence.

However, crew pairing isn't solely about cost optimization; regulatory compliance is critical. All pairings must stick to regulations set by the Turkish Civil Aviation Authority and international agreements, ensuring safety through limitations on duty times, rest periods, and flight time restrictions. While efficiency and safety are primary concerns, airlines also consider crew preferences when constructing pairings [3]. Incorporating preferences, such as specific routes or desired layovers, can lead to a more satisfied workforce, positively impacting safety and customer service. In this study, the crew pairing problem is solved using the column generation method. The aim is to establish robust and predictable crew pairings that comply with FAA regulations on duty periods, rest periods, and flight time limitations.

2 Literature Review

The air transport sector has entered a rapid development process with the implementation of neoliberal economic policies all over the world [4]. The growing need for fast air travel has led to the establishment of new airlines and more route additions to existing airlines [5]. With more players and bigger planes, there are more pilots, flight attendants, ground crew, and other staff needed. This has naturally driven up costs significantly. To cut costs, airlines have created dedicated research departments to solve their biggest problems [6]. Since the 1950s, airline planning and scheduling teams, which include both airline employees and university researchers, have been using special optimization techniques to solve these complex scheduling puzzles [7].

The airline scheduling process is a complex, multi-stage operation that optimizes resources to create efficient flight schedules [8]. The stages of the planning processes carried out by the above-mentioned departments can be seen in Fig. 1.

The first phase involves network development, where airlines determine the desired origin–destination pairs and their frequencies. This initial network design lays the groundwork for the subsequent creation of a flight schedule, where specific days and times are assigned to each route. Airlines use passenger forecasting to figure out how many and what kind of planes they will need for each flight. Then, they assign specific planes to each flight, making sure the planes get the maintenance they need. Finally, crew scheduling ensures sufficient personnel are assigned to operate all flights. Solving this intricate crew scheduling problem is computationally demanding and classified as NP-hard.

Crew scheduling, the latter one in the planning process, is a very popular area of research and is one of the most sophisticated and complex problems in the industry [8]. The consequent cost of this planning process is very vital for airline companies. Crew cost, including salaries, benefits, and expenses, is the second highest component of direct operating cost after fuel. Unlike fuel costs, airlines have greater control over operational expenses [9]. Optimizing crew scheduling presents a significant opportunity to manage these costs effectively, thereby enhancing competitiveness and financial sustainability within the dynamic airline market. Additionally, crew schedules are not only crucial for the airline's profitability but also significantly

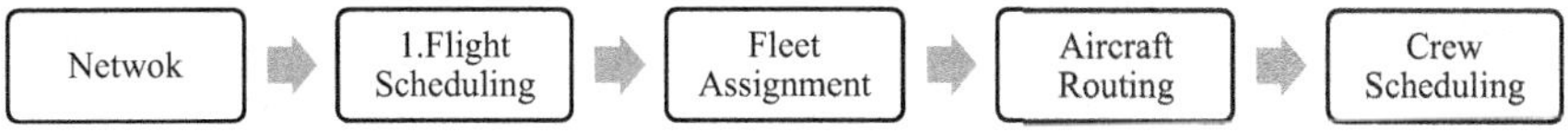

Fig. 1 An overview of the airline planning process

impact the service quality, flight safety, operational control of crews, and crew job satisfaction within the airline operation.

An acceptable crew pairing solution must comply with a comprehensive set of regulations established by governing bodies such as the Turkish Civil Aviation Authority (SHGM) and the Federal Aviation Administration (FAA), in addition to any airline-specific constraints. In particular, the "Instructions on Flight Crew Flight Duty and Rest Periods and Implementation Principles" (General Directorate of Civil Aviation, n.d.) requires the publication of monthly crew schedules at least 24 h before their execution, with a minimum coverage period of 15 days. Given these multiple constraints, the primary objective remains to identify the most cost-effective crew pairing strategy that assigns personnel to all scheduled flights [10].

Like practices in airline planning departments, the academic literature has traditionally divided crew scheduling into two distinct subproblems: crew pairing and crew scheduling. In particular, the construction of crew schedules becomes computationally challenging before crew pairings are established [11]. Therefore, the crew pairing problem uses the flight schedule as input to generate feasible crew pairings for assigned flights. Subsequently, the crew scheduling problem utilizes the output of the crew pairing to generate a detailed work schedule for each crew member within a given period.

In crew pairing problems, the objective is to find the least-cost set of crew pairings that satisfy flight safety regulations and all other constraints, and that cover all flights in the flight schedule with origin and destination at the same airport. In crew scheduling problems, individual crew members are assigned to the pairings generated during the crew pairing phase, considering factors such as crew training, leave, or personal preferences. While crew pairing problems minimize crew costs, crew scheduling problems often focus on distributing pairings evenly among crews to maximize crew quality of life, job satisfaction, or flight safety.

One study reviewed research on scheduling flight attendants alone, scheduling flight attendants and cockpit crew together, making cockpit crew schedules more robust to disruptions, and correcting their schedules when unexpected problems arise [1]. Research to date shows that set partitioning and set covering models are commonly used for crew scheduling. The set covering model requires that at least one crew team be assigned to each flight, while the set partitioning model requires that each flight be assigned to exactly one crew team.

Various optimization techniques have been applied to the crew scheduling problem, including approaches using mixed integer programming (MIP), nonlinear integer programming, stochastic models, and fuzzy set theories. Exact methods are generally more appropriate for smaller data sets where the number of aircraft and destinations is limited. These methods solve the optimization problem exactly based on defined constraints and objectives but are less preferred in the industry due to scalability issues and the impracticality of addressing the larger, more complex planning environments in which modern airlines operate. Mixed integer programming (MIP) is usually solved using LP-based branch-and-bound or branch-and-price techniques, which can effectively handle integer constraints. For example, Lin et al. developed a mixed integer programming model that aims to simultaneously minimize overall scheduling costs and crew fatigue time [12].

Nonlinear programming models address scenarios with nonlinear relationships between variables, increasing the complexity of solutions. Due to the high complexity and computational requirements of nonlinear models, stochastic models are the preferred alternative, especially for handling operational uncertainties. Stochastic models are used to incorporate uncertainties, such as flight delays or crew availability, into scheduling and use probabilistic methods to obtain robust solutions. Rungrueang developed sim-optimization methods that combine a stochastic optimization model and daily operations that cause delays of more than one hour to address delays and minimize crew scheduling flights [13].

Metaheuristics are often used to deal with the complexity of large real-world problems and to find feasible solutions. One such method for crew scheduling is the Genetic Algorithm (GA), which is a metaheuristic optimization technique inspired by the principles of natural selection. In the GA designed for the crew scheduling problem, there is a start chromosome for suitable flight legs where the crew can start and an end chromosome for suitable flight legs where the crew can finish its duty. A shortest path algorithm is used to determine the best pairings of these possible flights. A roulette wheel selection method was used to select from these possible sequences. This method will converge to a near-optimal solution thanks to intelligently designed operators [14]. In another study, a hybrid genetic algorithm was proposed [15]. An analysis was conducted to compare the performance of the Genetic Algorithm (GA) and Hybrid Genetic Algorithm (HGA) solutions. The result of the evaluation of both algorithms has shown that the hybrid genetic algorithm gives better results.

Another method for generating optimal crew schedules that efficiently handles the computational complexity due to the special nature of crew scheduling problems is the column generation approach. As the number of possible crew pairings increases, the problem becomes computationally expensive, as well as the requirement to cover all flights efficiently. Column generation overcomes this problem by prioritizing the small number of the most promising pairings. First, a simplified master problem is solved with a limited set of pairings. Then, a subproblem, often formulated as a resource-constrained shortest path problem, identifies additional pairings that can improve the overall solution. These promising pairings are then incorporated into the main problem, and the process is iterated until no further improvements are found. This approach significantly reduces the computational complexity compared to evaluating all possible pairings simultaneously.

In a very recent study, Zeren et al. (2024) tackled the crew pairing problem using an innovative heuristic approach, integrating it with traditional methods like column generation [16]. They propose a score-based adaptive greedy heuristic and a genetic algorithm to solve large-scale instances effectively. The adaptive greedy heuristic prioritizes flight legs based on dynamically calculated difficulty scores, incorporating sub-heuristics like Coverability Degree, Deadheading Heuristic Modifier, and Activity Heuristic Modifier. This method allows for a focused search in the solution space, making it adaptable and efficient. The genetic algorithm, used as a hyper-heuristic, optimizes the ordering of flight legs by evolving the solution space through selection, crossover, and mutation operations. Their empirical results, based on real-world data from Turkish Airlines, demonstrate that these methods can produce high-quality solutions for large-scale crew pairing problems. The authors highlight that their approach not only meets the stringent constraints and requirements of the airline industry but also offers scalability and adaptability, which are critical for handling the dynamic nature of airline operations.

In conclusion, the literature review reveals a clear evolution in airline crew scheduling, highlighting in particular the shift from daily scheduling practices to monthly scheduling practices. It also shows that the general trend in dealing with complex scheduling problems in practice has shifted towards partitioning strategies.

3 Problem Definition

Optimizing crew pairing is a significant challenge and cost for many large, multi-destination airlines. Assigning qualified crew members (pilots, co-pilots, and flight attendants) to multiple flights covering the globe every day requires careful planning. What appears to be a simple task becomes a complex challenge, creating a complex puzzle of significant dimensions [17]. With each additional flight and crew member, the exponential growth of potential pair combinations quickly exceeds the capacity of traditional planning methods [18].

Unproductive layovers result in additional lodging and per diem costs when crew members remain stationed beyond their scheduled duties. Deadheading, the practice of crew members flying as non-working passengers to reach another work location, further strains resources. Optimizing crew pairing therefore becomes a critical balancing act. The ideal solution is to minimize the total number of crew required to cover all flights while complying with a complex web of regulations.

Safety remains a top concern, requiring strict adherence to Federal Aviation Administration (FAA) regulations and international agreements. These regulations limit the duty time of pilots and flight attendants, mandate minimum rest periods, and impose flight time limitations. Crew pairing systems must be rigorously implemented in this regulatory environment to prevent crew fatigue and ensure safe flight operations. An optimized crew pairing system streamlines operations by eliminating unnecessary crew assignments, unproductive layovers, and deadheading. The result is a more efficient and profitable airline while maintaining the highest standards of safety for crew and passengers.

Crew pairing is subject to a complex set of regulations designed to ensure crew safety, well-being, and operational efficiency. Optimizing crew pairing while complying with these regulations is essential for airlines to maintain safe and cost-effective operations. In developing a crew pairing model, the regulations set by the Turkish General Directorate of Civil Aviation (SHGM) are used as part of the constraints. These regulations, including limitations on flight time, duty time, and rest periods, are incorporated as constraints into the model's objective function to ensure optimal and safety-compliant crew pairing.

- Maximum Flight Time: The maximum time a crewmember can actively operate the aircraft during a single duty period. It includes the time spent flying from take-off to landing but typically excludes taxiing, pre-flight checks, and post-flight procedures.
- Maximum Duty Time: The total amount of time a crewmember can work during a single duty period. It includes not only flight time, but also pre-flight activities (such as briefing, checking, and preparing the aircraft) and post-flight activities (such as taxiing, paperwork, and securing the aircraft).
- Minimum Rest Period: This is the uninterrupted time a crewmember must rest between flights or duty periods. This period typically begins after the crewmember has completed all post-flight duties.

The restrictions for maximum duty time and minimum rest period are 12 and 6 hours, respectively.

4 Methodology

4.1 A. Data Collection

The dataset used in this study covers the six months from December 2013 to July 2014 for the national flag carrier airline, a major international player. This data provides a comprehensive overview of the flight details needed to optimize crew pairings. A small sample of the dataset is shown in Table 1 and the fields in the table are described in detail below.

- Departure Station: This field identifies the originating airport for each flight. The codes, HB1 and HB2, represent home bases within the same city, Istanbul. HB1 is Istanbul Atatürk Airport, and HB2 is Istanbul Sabiha Gökçen International Airport, allowing crew pairings to begin and conclude at either location.
- Arrival Station: This field specifies the destination airport for each flight leg. Station identifiers prefixed with "D" denote domestic destinations, while "I" prefixes indicate international destinations.
- Fleet Type: This field classifies the aircraft type assigned to each flight leg. Flights operated by Airbus narrow-body aircraft are identified by fleet numbers starting with "3", which is the primary focus for solution development in this study. The remaining fleet numbers belong to the Boeing family of aircraft.
- Aircraft Number: This field uniquely identifies the specific aircraft assigned to each flight. For example, in Table 1 the flight departing from I001 and landing at HB1 belongs to fleet type 3005 and the aircraft number for this flight is 12.
- Departure and Arrival Times: This field provides the departure and arrival times for each flight leg, formatted in local date-time at the respective airport. To ensure consistent time references across flights, all dates and times have been meticulously adjusted based on a chosen reference point. This adjustment allows for a clear understanding of flight timings relative to the reference point.

Table 1 Sample raw data

Departure station	Arrival station	Fleet type	Aircraft number	Departure date	Arrival date	Time difference (Dep.)	Time difference (Arr.)
I001	HB1	3005	12	2013–12–20T00:05	2013–12–20T03:40	210	120
I002	HB1	3008	3	2013–12–20T00:05	2013–12–20T04:20	210	120

- Departure Station Time Zone Difference: This field indicates the time zone offset, expressed in minutes, between the departure station's local time and Coordinated Universal Time (UTC).
- Arrival Station Time Zone Difference: This field specifies the time zone difference (in minutes) between the arrival station's local time and UTC.

4.2 B. Mathematical Models

The Airline Crew Pairing Problem (ACPP) requires the assignment of flight pairings to cockpit crew members while satisfying various constraints and minimizing total operational costs [19]. Due to the large number of.

potential pairings, often exceeding millions in real-world scenarios, coupled with complex regulations regarding duty hours, rest periods, and aircraft type qualifications, traditional integer linear programming (ILP) formulations become computationally demanding [20]. This study addresses this challenge by employing a column generation approach, a technique particularly suited to problems characterized by many variables [19].

Set partitioning and set covering represent distinct approaches to assigning tasks and rest periods to crew members. Even if set partitioning eliminates the possibility of deadheading, it becomes computationally challenging to implement and reach a solution due to the large number of feasible duty combinations in large-scale problems. On the other hand, set covering allows for overlapping duties, where tasks can be covered by multiple pairings. This flexibility can be advantageous in accommodating complex scheduling constraints or minimizing crew costs.

1. *Set Covering Model*: Set covering and partitioning models include mutual exclusion constraints involving subsets of decision variables, only one of which can participate in the solution. The parameters and variables used in the set covering model can be defined as follows:

$p = 1, \ldots, P$ (pairings).
$f = 1, \ldots, F$ (flights).
$c_p =$ Cost of pairing p.
$x_p = 1$ if pairing p is selected; 0 otherwise.

Each pairing is a series of flights to be performed by a single crew over 2 to 3 days and must begin and end in the hub city where the crew resides. Therefore, set covering problem is formulated as:

$$\min \sum_{p \in P} c_p x_p \tag{1}$$

$$\text{s.t.} \sum_{p: f \in p} x_p \geq 1, \qquad \forall f \in F \tag{2}$$

$$x_p \in \{0, 1\}, \forall p \in P \quad (3)$$

Equation (1) simply gives the total cost associated with the selected crew pairings. Equation (2) ensures that each flight is covered by at least one crew pairing. A value greater than 1 for a specific flight in Eq. (2) indicates that multiple pairings have been assigned to that flight. These extra pairings would be considered deadhead flights where the crew is traveling without passengers. Equation (3) represents standard limitations on the decision variables x_p. These limitations typically restrict these variables to be either 0, or 1.

2. *Column Generation Method:* For complex combinatorial problems, the column generation approach first enumerates a set of columns representing feasible solutions to parts of the problem and then solves a set partitioning or coverage model to select the optimal collection of these alternatives that satisfies all the requirements of the problem. To solve the Airline Crew Pairing Problem (ACPP), column generation tackles it by dividing it into two linked problems: a main problem (called the master problem) and a secondary problem, which is the sub-problem [19].

The master problem focuses on selecting the most cost-effective pairings from a limited set. Initially, this set may only contain a small subset of all feasible pairings. The master problem employs a linear relaxation to establish a lower bound for the optimal objective function value. This LP model is also called a Linear Relaxation solution or Restricted Master Model because the problem is solved without generating all possible columns in the solution space. Additionally, it calculates reduced costs for each pairing not currently included. These reduced costs represent the potential cost improvement of adding a specific pairing to the solution.

In the pricing stage (sub-problem), new columns are generated to improve the objective function of the problem by using the dual variables obtained by solving the main problem. In essence, the master problem acts as a filter that identifies the most promising pairings that have the potential to significantly improve the overall solution. All non-basic variables are priced using dual prices to further refine the final basic feasible solution.

In the crew pairing problem, the subproblem is formulated as a shortest path model. If the subproblem yields a promising pairing with negative reduced cost, this pairing is integrated into the existing pairing pool in the master problem. Following this, the master problem is re-optimized using the updated set of pairings. This iterative cycle continues until no more negative reduced costs arise, indicating that the current solution likely meets the minimum number of pairings needed to cover all flight routes (set covering) or meets the "exactly once" constraint for all pairings (set partitioning). The effectiveness of column generation in overcoming crew pairing challenges has been convincingly demonstrated in several studies [19, 20].

3. *Shortest Path Model:* The shortest path model is used as a sub-model during column generation and is a binary model. The parameters and variables used in the set covering model can be defined as follows:

$i = 1, 2,\ldots,I$ (starting node).
$j = 1, 2,\ldots,J$ (ending node).
$y\,(i)$ = dual values from the main problem to sub-problem.
$f\,(i)$ = flight duration for node i.
$r\,(i,j)$ = layover durations between nodes i and j.
$x\,(i,j)$ = Binary variable, 1 if the arc from node i to node j is selected, 0 otherwise.
The sub-problem can be formulated as below:

$$\min W - \sum_{i,j} y(i) * x(i,j) \tag{4}$$

$$\text{s.t.} W \geq 0.6 * \sum_{i,j} (f(i) + r(i,j)) * x(i,j) \tag{5}$$

$$W \geq \sum_{i,j} f(i) * x(i,j) \tag{6}$$

$$W \geq 180 \tag{7}$$

$$\sum_{j} x(i,j) - \sum_{j} x(j,i) = 0 \forall i \tag{8}$$

$$\sum_{i} x(0,i) = 1 \tag{9}$$

$$\sum_{i} x(i,n+1) = 1 \tag{10}$$

$$\sum_{i,j} f(i) * x(i,j) \leq 480 \tag{11}$$

$$\sum_{i,j} (f(i) + r(i,j)) * x(i,j) \leq 720 \tag{12}$$

$$\sum_{i,j} f(i) * x(i,j) \leq 5 \tag{13}$$

$$\sum_{i,j} f(i) * x(i,j) \geq 4 \tag{14}$$

The objective function (4) aims to minimize the reduced cost by considering the dual values from the main problem. Cost constraints (5, 6, 7) ensure that the total cost is calculated based on flight durations and layovers. The flow balance constraint (8) ensures that the number of flights entering and leaving each node is balanced. Constraints (9) and (10) ensure that each pair of flights starts and ends at the home base, with the starting node denoted by 0 and the ending node by $n + 1$. The total flight time for a pair is restricted to a maximum of 480 min as specified by constraint (11). Additionally, constraint (12) limits the overall duration, including layovers, to 720 min. Constraints (13) and (14) enforce that the number of flights in a pair is between 4 and 5. These constraints collectively ensure that the flight pairings generated by the model are feasible, cost-effective, and meet the required operational limits.

The primary problem in this study is formulated as a set-covering model, chosen for its relaxed constraints compared to the set-partitioning model, thus providing more flexibility for analysis. The subproblem, on the other hand, is framed as a shortest path problem within a flight network. This network is constructed by representing each flight leg on a selected day as a node and representing potential connections between these flight legs with arcs. The resulting network is centered around the hub HB1, which serves as both origin and destination. The objective is to identify the minimum reduced-cost pairing that starts and ends at HB1 while satisfying all legal pairing constraints, ensuring that the resulting pairings are legally viable and operationally feasible. The dual values from the main problem are interpreted as the profits associated with each pairing, and the costs of the pairings are determined by a specific formulation.

5 Implementation and Results

5.1 A. Data Cleansing

The first step was to extract the data, specifically selecting flights operated by Airbus narrow-body aircraft with fleet numbers starting with "3". The data set was divided into three categories: the "medium" category includes domestic flights from December 21, 2013; the "small" category is a filtered subset of the medium category; and the "large" category includes all domestic flights in two days. The data was filtered to include only domestic flights within the same time zone. Some flights were excluded from the small and medium data sets due to insufficient dwell times or inability to return to the central station during the operational day. The number of flights and cities in each category is shown in Table 2.

Table 2 Summary of data categories

	Number of flights	Number of cities
Small scale	52	12
Medium scale	108	29
Large scale	154	27

5.2 B. Generating Initial Pairings

A Python script was developed to generate all possible pairings, applying the regulations set by SHGM as constraints. Layover durations were also considered during the code construction. The script is divided into three main functions: *find_connections, build_pairs,* and *calculate_cost.*

As shown in Fig. 2, the *find_connections* function identifies potential connections between flights based on specific criteria such as layover times and matching departure and arrival locations. For a connection to be valid, the arrival station of the current flight must match the departure station of the subsequent flight, and the layover time must be within the 30 to 300-min range. Once potential connections are identified, they are sorted by layover time in ascending order, and the list of feasible connection indices is returned.

The *build_pairs* function, shown in Fig. 3, builds flight pairs based on constraints such as minimum and maximum number of flights and total trip duration. This function uses a recursive approach to discover and build all possible pairs, ensuring that each pair contains 4 to 5 flights, the total flight duration does not exceed 480 min, and the total duration (including layovers) is within 720 min. The function continues to create pairs until all valid combinations have been created.

```python
def find_connections(flight_index, flights, min_layover=timedelta(minutes=30), max_layover=timedelta(minutes=300)):
    connections = []
    flight = flights.loc[flight_index]
    for next_index, next_flight in flights.iterrows():
        if (flight['Arrival Station'] == next_flight['Departure Station']) and \
           (min_layover <= next_flight['dep'] - flight['arr'] <= max_layover):
            connections.append(next_index)
    connections.sort(key=lambda x: flights.loc[x]['dep'] - flight['arr'])
    return connections
```

Fig. 2 Find_connections function

```python
def build_pairs(current_pair, flights, min_length=4, max_length=5, max_flight_duration=480, max_total_duration=720):
    if len(current_pair) > max_length:
        return []
    if len(current_pair) >= min_length and current_pair[-1] in end_flights:
        return [current_pair]

    last_flight_index = current_pair[-1]
    connections = find_connections(last_flight_index, flights)

    pairs = []
    for connection_index in connections:
        if connection_index in current_pair:
            continue
        new_pair = current_pair + [connection_index]

        total_flight_duration = sum(flights.loc[i]['duration'] for i in new_pair)
        layover_durations = [(flights.loc[new_pair[i+1]]['dep'] -
                              flights.loc[new_pair[i]]['arr']).total_seconds() / 60 for i in range(len(new_pair) - 1)]
        total_layover = sum(layover_durations)
        total_duration = total_flight_duration + total_layover

        if total_flight_duration <= max_flight_duration and total_duration <= max_total_duration:
            pairs += build_pairs(new_pair, flights, min_length, max_length, max_flight_duration, max_total_duration)

    return pairs
```

Fig. 3 Build_pairs function

```python
def calculate_cost(pair, flights):
    total_flight_duration = sum(flights.loc[flight_index]['duration'] for flight_index in pair)

    if total_flight_duration < 180:
        return 180

    layover_durations = [(flights.loc[pair[i+1]]['dep'] - flights.loc[pair[i]]['arr']).total_seconds() / 60
        for i in range(len(pair) - 1)]
    total_layover = sum(layover_durations)

    cost_option_1 = (total_flight_duration + total_layover) * 0.6
    cost_option_2 = total_flight_duration

    return max(cost_option_1, cost_option_2)
```

Fig. 4 Calculate_cost function

As shown in Fig. 4, the *calculate_cost* function determines the cost of a flight pair, considering both the total flight time and the layover times. If the total flight duration is less than 180 min, a minimum cost equivalent to 3 h of work is applied. For longer durations, the function evaluates two cost options: one that multiplies the combined flight and layover times by 0.6, and another that only considers the total flight time. The function returns the higher value to provide the most accurate cost assessment.

In conclusion, the developed Python script successfully constructs pairings by applying the SHGM rules, considering layover durations, and integrating cost considerations. The number of pairings for small, medium, and large sizes of the dataset is 131, 475, and 1599, respectively. These results highlight the robustness of the implemented approach and demonstrate its ability to efficiently generate a significant number of valid pairings on datasets of different sizes. This approach can be adapted to the different requirements of airline operations. Once the pairings were generated, all datasets were solved using GAMS. Column generation was performed using heuristics to obtain the initial pairings for the master problem. The master model was built using set covering and dual prices were computed for each pair. The subproblem used the shortest path algorithm to find pairings with minimum negative reduced cost, considering layovers and flight times. Once the new pairings were incorporated into the master model, iterations continued until no further improvements could be made. Column generation proved to be more effective than set partitioning and set covering, especially for larger datasets. The cost of the optimal schedule obtained using the column generation algorithm is 5240, 11,326, and 18,873 for the small, medium, and large datasets, respectively.

6 Conclusion

In this study, a column generation approach is developed to solve the set covering model to optimize the crew pairing cost, which is a significant expense for airlines. Through this approach, crew schedules are optimized, and inefficient crew times such as layovers and deadheading are reduced. Results from experiments using real data from a global airline showed that column generation in crew pairing optimization can significantly reduce computation time. The iterative process of solving the main and subproblems identified the most cost-effective pairings that provide comprehensive flight coverage with the fewest number of crew members.

Looking to the future, the integration of stochastic methods to deal with the variability and uncertainty of real-world operations could be explored There is a resurgence in heuristics, driven by advances in software capabilities, which provide new opportunities to achieve efficient and near-optimal solutions. For example, combining heuristic approaches with traditional methods such as column generation can provide robust solutions to the ever-evolving challenges of crew scheduling. Future research should focus on exploiting these hybrid methods to improve the flexibility and efficiency of scheduling systems and enable them to meet the operational needs of modern airlines.

References

1. Wen X, Sun X, Sun Y, Yue X (2021) Airline crew scheduling: models and algorithms. Transp Res Part E Logistics and Transportation Review 149, 102304
2. Andrew S (2024) Pilots have their own secret language. Here's what they're really saying Accessed 18 April 2024 from TIME
3. Quesnel F, Desaulniers G, Soumis F (2020) Improving aircrew rostering by considering crew preferences in the crew pairing problem. Transp Sci 54(1):97–114
4. Gürçam S (2022) The neoliberal initiative of the aviation industry to fight the climate crisis: greenwashing. Int J Environ Geoinformatics 9(3):178–186
5. Deveci M, Demirel NÇ (2018) Evolutionary algorithms for solving the airline crew pairing problem. C&IE 115:389–406
6. Ulucan A (2004) Hava Taşımacılığı Planlamasında Yöneylem Araştırması Modellerinin Kullanılması". *AÜ SBF Dergisi*, 59(04)
7. Barnhart C, Talluri KT (1997) Airline operations research. Design and Operation of Civil and Envrironmental Engineering System, pp 435–469, 1997
8. Korte J (2019) An aircraft and schedule integrated approach to improve cockpit crew pairings. Master's thesis, Delft UT]
9. Anbil R, Tanga R, Johnson EL (1992) A global approach to crew-pairing optimization. IBM Syst J 31(1):71–78
10. Deveci M, Demirel NÇ (2018) A survey of the literature on airline crew scheduling. Eng Appl AI 74:54–69
11. Soykan B, Erol S (2016) Airline operational scheduling process. In: The 3th international aviation management conference, 19 Apr 2016, Ankara, Turkey, pp 143–156
12. Lin H, Guo C, You J, Xia M (2024) Research on optimization of flight crew scheduling considering pilot fatigue. Int J Ind Eng Comput 15(1):171–188
13. Rungrueang P (2018) Deterministic and stochastic airline crew scheduling. PhD Dissertation, University of Portsmouth
14. Kerati S, Moudani WE, de Coligny M, Mora-Camino F (2002) A heuristic genetic algorithm approach for the airline crew scheduling problem. In Workshop on Multiple Objective Metaheuristics, Paris
15. Deveci M, Demirel NC (2016) A hybrid genetic algorithm for airline crew pairing optimization. Economic and Social Development: Book of Proceedings 118, 2016
16. Zeren B, Özcan E, Deveci M (2024) An adaptive greedy heuristic for large scale airline crew pairing problems. J Air Transp Manag 114:102492

17. Cordeau J, Stojković G, Soumis F, Desrosiers J (2001) Benders decomposition for simultaneous aircraft routing and crew scheduling. Transp Sci 35(4):375–388
18. Gopalakrishnan B, Johnson EL (2002) Airline crew scheduling State-of-the-art. Ann Oper Res 140(1):7–40
19. Barnhart C, Cohn AM, Johnson EL, Klabjan D, Nemhauser GL, Vance PH (2006) Airline crew scheduling. In Kluwer Academic Publishers eBooks pp 517–560
20. Lavoie S, Minoux M, Odier E (1988) A new approach for crew pairing problems by column generation with an application to air transportation. Eur J Oper Res 35(1):45–58

Production Management

A Comprehensive Analysis of Integrated Airline Scheduling Strategies and Operations

Gürkan Güven Güner and Serpil Erol

Abstract The efficient airline management including flight scheduling, aircraft scheduling (fleet assignment and aircraft routing) and crew scheduling (crew pairing and assignment) is very critical planning phase for the airline companies. Especially, the recent Covid-19 pandemic proved this once more to the decision makers in the airline industry. The reason is the high costs reported by the institutions such as International Civil Aviation Organization (ICAO) and the International Air Transport Association (IATA) because of this global "irregular operation" management problem faced by airline companies. Solving flight, aircraft, and crew scheduling problems simultaneously in an integrated way in a very short time can increase the efficiency very much in this kind of irregular processes, to minimize all costs and flight delays/cancellations. In the literature, there are many studies proposing various solution methodologies including different optimization models and heuristic algorithms for different integrated airline scheduling problems. The objective of this study is to propose a comprehensive literature review on solution methodologies for the integrated airline scheduling problems to present a framework that can assist the decision makers about how to approach the indicated airline scheduling problems. It is aimed that the proposed comprehensive literature review will be a beneficial alternative guide for the researchers and decision makers in airline industry.

Keywords Airline irregular operations · Integrated airline scheduling · Mathematical optimization

---

G. G. Güner (✉)
Department of Industrial Engineering, University of Turkish Aeronautical Association, Ankara, Türkiye
e-mail: gguner@thk.edu.tr

S. Erol
Department of Industrial Engineering, Gazi University, Ankara, Türkiye
e-mail: serpiler@gazi.edu.tr

G. G. Güner
Graduate School of Natural and Appliied Sciences, Gazi University, Ankara, Türkiye

N. M. Durakbasa and K. G. Gülen (eds.), *Sustainable Green Conversion*, Lecture Notes in Mechanical Engineering, https://doi.org/10.1007/978-3-031-83583-4_24

1 Introduction

The airline planning processes generally divided into two main parts: planning phase and operational phase [1]. The planning phase includes the flight scheduling, aircraft scheduling (fleet assignment and aircraft routing) and crew scheduling (crew pairing and assignment) in the order they are made before the departure of the flight. On the other hand, the operational phase includes revenue management, gate assignment and airline irregular operations management, which includes flight, aircraft and crew rescheduling. The irregular operations management, which is the focus of this study, can be required because of aircraft mechanical problems, severe weather, crew sickness, airport curfews, and security reasons. These operations management generally needs to be completed in a very short period like four weeks [2]. The airline scheduling problems are generally and traditionally solved step by step in the order they are made before the departure of the flight. However, it is shown in the literature that this kind of solution procedure can create some problems about mathematical inconsistencies, which make the solution unimplementable to real life [3]. The reason that, when solving the scheduling problems partially, sometimes a solution gained from one scheduling phase cannot satisfy the constraints in another scheduling phase. For example, a solution generated in flight scheduling phase cannot always satisfy the aircraft maintenance constraints in aircraft scheduling phase. Therefore, despite the mathematical complexity is increased and solving the problem becomes more difficult, considering all airline scheduling phases simultaneously and in integrated way in one solution methodology minimizes the inconsistencies and make the generated solution more implementable in real life problems. Figure 1 represents the planning and operational phases of airline management.

After the pandemic, some statistics reported from International Civil Aviation Organization (ICAO) and the International Air Transport Association (IATA) underlies the importance of efficient irregular operations management. As per the findings of the report from the ICAO concerning the economic impacts of the Covid-19 pandemic on the civil aviation sector, there was a substantial drop of approximately $372 billion in gross passenger operating revenues in 2020 compared to the previous year [4]. This significant downturn, representing a 60% decrease in global airline passengers, marks an unprecedented event in the history of aviation. Moreover, the IATA discloses a remarkable global financial assistance package extended to airlines, amounting to $243 billion in 2020 [5]. This aid comprises various forms such as wage subsidies ($81 billion), loans ($73 billion), direct financial injections ($38 billion), loan guarantees ($26 billion), ticket taxes ($13 billion), and corporate tax relief ($12 billion). All these statistics underlies the importance of optimization for the integrated airline scheduling problems within airline irregular operations.

In the literature, there are many studies that propose different solution methodologies and approaches for flight, aircraft and crew scheduling problem. In the flight scheduling problem, which is the first planning phase of airline scheduling, available fleet/manpower capacity conditions and market evaluations are significant factors to be considered. In the aircraft scheduling, minimizing the operating cost of flights,

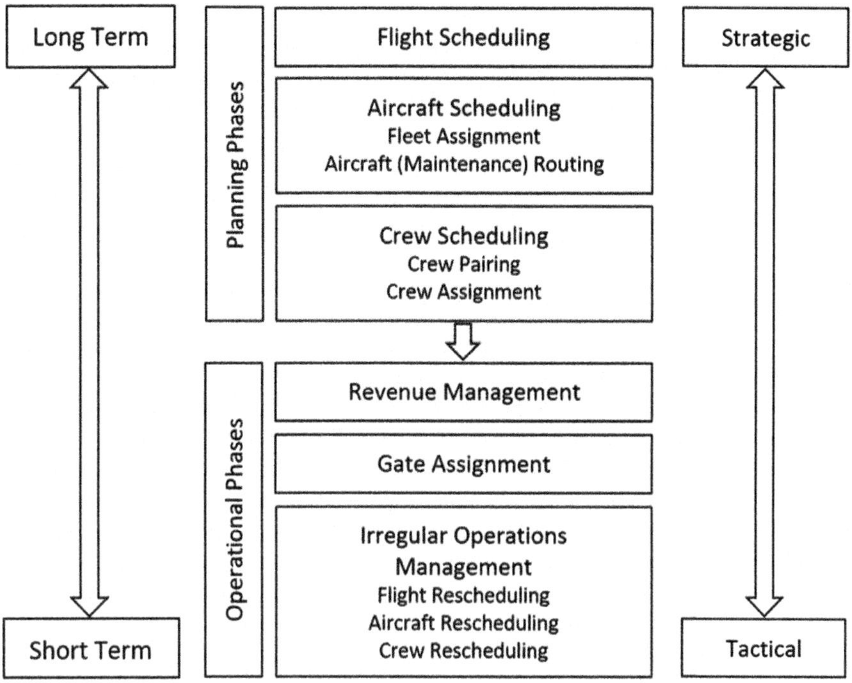

Fig. 1 The planning and operational phases of airline management

which change according to characteristics of assigned aircrafts, such as cost and revenue per available seat mile, is a significant objective to be considered. In the crew scheduling, considering the flight, duty and rest time restrictions of crews according to relevant regulations of aviation authorities of countries, such as FAA (Federal Aviation Administration) in U.S. or DGCA (Directorate General of Civil Aviation) in Türkiye, about flight, duty and rest time limitations of crews should be considered.

On the other hand, there also many studies, which consider the mentioned scheduling problems in integrated way, in the literature. When the integrated scheduling problems are investigated, the complexity of the airline scheduling problems increase due to the increased factors to be considered. In irregular operation management conditions, it is very important to generated scheduling solutions in a very short time. Therefore, in this situations, the integrated solution approaches can be more useful tool, rather that solving the scheduling problems one by one. In addition, as mentioned earlier, sometimes one solution gathered in one scheduling step cannot be implementable in other scheduling step due to the mathematical incompatibilities. In this study, the recent studies consider integrated airline scheduling problems in the literature are investigated and analyzed according to their problem types, objectives and solution methodologies. The purpose of this study is to suggest a comprehensive investigation about solution methodologies for the integrated airline scheduling problems to present a framework that can guide the decision makers about how to approach the mentioned airline scheduling problems.

In the following sections of this study, which aims to be a useful guide for researchers doing airline studies and decision makers in the airline industry, different solution methodologies and mathematical models proposed in the literature for various integrated airline planning problems are presented.

2 Evaluation of Integrated Airline Scheduling Problems

There are many studies in the literature focusing on airline scheduling problems partially or in integrated way. The literature review of this study is grouped into two main sections. Firstly, the literature review on solution methodologies is presented indicating considered problems and objectives. Secondly, the base mathematical models of flight, aircraft and crew scheduling problems are presented and analysed in detail to give an idea for the decision makers about the development of integrated models. Firstly, the investigated recent studies considering the integrated airline scheduling problems are presented below from past to present according to considered problems, objectives and proposed solution methodologies.

The studies of Stojković and Soumis in 2001 and 2005 consider the integrated flight and crew scheduling problem [6, 7]. The study proposes an integer nonlinear multi-commodity minimum cost flow model for the integrated scheduling problem. The main objectives of these studies are minimizing the crew adjustment costs and total flight delay costs. The study in 2001, which considers integrated flight and pilot scheduling proposes a Dantzig-Wolfe decomposition solution procedure. The proposed optimization model considers the flight cover, flight precedence, time window and nonlinear compatibility constraints. The study of Ahmed et al. in 2018 considers integrated aircraft routing and crew pairing problem [8]. In the study, a nonlinear Mixed Integer Programming (MIP) model is suggested for the integrated problem and Reformulation–Linearization Technique proposed by Sherali and Adams [9, 10] is used to make an equivalent lifted Mixed Integer Linear Programming (MILP) formulation. The performance of the obtained solution is tested by Monte Carlo simulation technique.

Flight scheduling and fleet assignment are pivotal decisions in airline planning, significantly influencing an airline's financial performance. These decisions must be finalized approximately 10 to 12 weeks prior to the flight date to effectively manage cabin crew requirements. In the study of Diabat et al. in 2018, a novel approach is introduced to address the integrated flight scheduling and fleet assignment challenge [11]. Their research presented a two-stage stochastic Mixed Integer Programming (MIP) model, incorporating the uncertainty associated with demand and passenger fares. Initially, the model allocates a fleet family to each scheduled flight. Subsequently, it determines the fleet type for each flight based on factors such as demand and passenger fares. To solve this complex problem and evaluate solution quality, the sample average approximation algorithm is utilized. Notably, this study is among the pioneering works to employ the sample average approach algorithm in the context of airline operations. Through a case study conducted on an airline's flight network, it was demonstrated that modelling the stochastic problem with 100 scenarios effectively addresses the uncertainty in demand and passenger fares, resulting in solutions with an optimality gap of less than 1% within a reasonable computation time. Additionally, the study includes a sensitivity analysis on various model parameters, affirming the viability of the proposed model and its ability to deliver feasible solutions.

Deterministic models traditionally rely on expected values derived from historical data and market forecasts to estimate demand. However, this approach can lead to suboptimal solutions due to the inherent fluctuations and uncertainties in demand. To address this, a stochastic approach is employed in this study to capture the variability in demand. Several plausible demand scenarios are considered, and the flight scheduling and fleet assignment problem is formulated to maximize expected profit based on the probabilities of these scenarios. This approach naturally leads to a two-stage stochastic programming framework. In the first stage, decision variables are independent of specific scenarios, while in the second stage, decision variables are contingent on the scenarios. Initially, a fleet family is assigned to each scheduled flight, and subsequently, a fleet type is determined for each flight based on demand and passenger fare considerations, enhancing the decisions made in the first stage. The early assignment of a fleet family to each flight, typically 10 to 12 weeks before the flight, facilitates crew scheduling. However, fleet type assignments may be adjusted later based on more precise demand estimates, allowing for more flexible and profitable decision-making. Since aircraft in the same fleet family share cockpit configurations and crew characteristics, crew scheduling becomes simpler as only the fleet family needs to be considered. On the other hand, other decisions, such as flight assignments, require knowledge of the fleet type.

The study of Zheng et al. in 2020 proposed a Mixed Integer Nonlinear Programming (MINLP) model for the integrated flight scheduling and aircraft parking problem, aiming to optimize take-off and landing times while minimizing total service times [12]. To find near-optimal solutions, they suggested a hybrid approach combining Simulated Annealing (SA) and reduced variable neighbourhood search methods. The study of Pasandideh et al. in 2021 introduced a new MINLP model for the integrated flight scheduling, fleet assignment, and revenue management problem. They also proposed a tabu search metaheuristic technique to solve the model efficiently [13]. The study of Birolini et al. in 2021 presented a MINLP model for the integrated flight scheduling and fleet assignment problem [14]. They used the least squares piecewise linearization technique to approximate nonlinearities. Numerical tests demonstrated the feasibility of their approach for medium-sized hub-and-spoke networks. They also implemented a case study for a real airline company in Europe.

The study of Kızıloğlu and Sakallı in 2023 introduced a three-stage Mixed Integer Nonlinear Programming (MINLP) model for the integrated flight scheduling and aircraft scheduling problem, encompassing fleet assignment and aircraft routing [15]. For addressing large-scale problems, they proposed novel simulation-based metaheuristic algorithms, referred to as "simheuristic" in their study. These algorithms integrate the Monte Carlo simulation method into Simulated Annealing and Cuckoo Search techniques. Their study demonstrated that these proposed algorithms yield efficient outcomes. The study of Sun et al. in 2023 presented a MINLP optimization model for the integrated aircraft and crew scheduling recovery problem, considering the challenges posed by the Covid-19 pandemic [16]. They proposed employing a branch-and-cut technique along with a large neighbourhood search heuristic to solve the problem efficiently. In the study by Sun et al. in 2024, a mixed integer optimization model was proposed for the integrated flight scheduling and fleet assignment problem [17]. Their study dynamically considered airline competition and incorporated airline passenger preferences. To find near-optimal solutions, they suggested a hybrid algorithm combining column generation and large neighbourhood search. Furthermore, they implemented a real-world case study to validate their proposed approach.

The classification of the mentioned recent studies on integrated airline scheduling shown in Table 1, showing the considered integrated problems, objectives, models, and solution algorithms of the studies.

It can be concluded from the investigated studies that generally MINLP and MIP models are suggested, and different heuristics and meta-heuristics are also implemented for integrated airline scheduling problems. The general mutual objective is to maximize the profit of the airline by considering different scheduling limitations. It is seen from the previous works that the number of studies considering integrated flight and crew scheduling problem is limited. Therefore, this can be an alternative study area for the researchers and decision makers to suggest a robust solution for the integrated airline scheduling problem containing all planning phases.

After describing the studies considering integrated scheduling problems, some of the basic mathematical models/formulations are described below for the airline scheduling problems to represent a scientific methodology before studying integrated flight, aircraft and crew scheduling problems for researchers and decision makers, who are working on airline scheduling problems. These models can be developed by different scheduling aspects.

For the fleet assignment problem within aircraft scheduling, an alternative simplified integer linear programming model is as the following [18].

Sets

FSet of flights.

KSet of fleet types.

CSet of last nodes, representing all nodes with aircraft grounded overnight at an airport in the network.

MNumber of nodes in the network.

Indices

IFlight index.

jIndex for fleet.

kIndex for nodes.

Parameters

$C_{i,j}$ Cost of assigning fleet type j to flight.

N_j Number of available aircraft in fleet type j.

Table 1 Classification of the studies on integrated airline scheduling

Reference	Year	Problem			Objective	Model	Solution algorithms
		Flight scheduling	Aircraft scheduling	Crew scheduling			
Stojković and Soumis [6, 7]	2001, 2005	✓		✓	Minimizing the crew adjustment costs and total flight delay costs	Integer nonlinear multi-commodity minimum cost flow model	Branch and bound technique
Ahmed et al. [8]	2018		✓	✓	Maximizing the cumulative reward accrued from connections where crews are trailing aircraft	MINLP	–
Diabat et al. [11]	2018	✓	✓		Maximizing the total expected profit of airline	Two-Stage Stochastic MIP	Sample average approximation
Zheng et al. [12]	2020	✓	✓		The determination of take-off and landing times of aircrafts and minimizing their total service times	MINLP	Simulated annealing and reduced variable neighbourhood search
Pasandideh et al. [13]	2021	✓	✓		Maximization of the profits of the airline considering revenues associated with fights, and aircraft opportunity cost	MINLP	Tabu search
Birolini et al. [14]	2021	✓	✓		Maximizing the anticipated operational gains, specifically, the disparity between airline revenues and expenses	MINLP	Differential evaluation

(continued)

Table 1 (continued)

Reference	Year	Problem			Objective	Model	Solution algorithms
		Flight scheduling	Aircraft scheduling	Crew scheduling			
Kızıloğlu and Sakallı [15]	2023	✓	✓		Maximizing the anticipated profit of an airline, determined as the variance between passenger income and related expenses	Three-stage MINLP	Simheuristic (Monte Carlo simulation integrated into simulated annealing and cuckoo search)
Sun et. al. [16]	2023		✓	✓	Minimizing the weighted expenses incurred from flight cancellations, delays, crew deadhead, and unassigned passengers	MINLP	Branch-and-Cut technique and large neighbourhood search heuristic
Sun et. al. [17]	2024	✓	✓		Identifying the ideal combination of flight schedules and airfares for a specific airline operator to maximize total profits	MIP	Hybrid algorithm, which contains column generation and large neighbourhood search

$S_{i,k}$ 1 if flight i is an arrival at node k, -1 if flight i is a departure from node k.

Decision Variables

$X_{i,j}$ 1 if flight is assigned to fleet-type j, 0 otherwise.

$G_{k,j}$ Integer decision variable representing number of aircraft of fleet-type on ground at node.

Objective Function

$$Min \sum_{j \in K} \sum_{i \in F} C_{i,j} X_{i,j} \tag{1}$$

Constraints

$$\Sigma_{j \in K} X_{i,j} = 1 \text{ for all } i \in F \tag{2}$$

$$G_{k-1,j} + \Sigma_{i \in F} S_{i,k} X_{i,j} = G_{k,j} \text{ for all } k \in M \text{ and } j \in k \tag{3}$$

$$\Sigma_{k \in C} G_{k,j} \leq N_j \text{ for all } j \in k \tag{4}$$

$$X_{i,j} \in \{0, 1\} \text{ for all } i \in F \text{ and } j \in k \tag{5}$$

$$G_{k,j} \in Z^{+} \text{ for all } k \in M \textit{ and } j \in k \tag{6}$$

In the provided model, the objective function outlined in Eq. (1) aims to minimize the total cost associated with assigning different fleet types to all scheduled flights. Constraint (2) ensures that each flight is operated by only one fleet type, known as flight-cover constraints. Constraint (3) is referred to as aircraft balance constraints, which maintains the balance of aircraft for each fleet type at every node. This balance is calculated as the number of aircraft of a fleet type just before a node (denoted as $G_{k-1,j}$ in the model) plus the arrivals (represented by $S_{i,k}$, where a value of + 1 indicates an arrival) minus the departures (represented by $S_{i,k}$, where a value of -1 indicates a departure). Constraint (4) specifies the fleet size, ensuring that the number of aircraft in fleet type j does not exceed the total available number of aircraft for that fleet (N_j). Constraints (5) and (6) dictate the binary and integer nature of the decision variables, respectively, with Z^{+} representing the set of positive integer numbers.

For the aircraft routing problem within aircraft scheduling, an alternative simplified integer linear programming model is as the following [19].

Sets

FSet of flights.

RSet of feasible routings.

Indices

JRoute index.

iFlight index.

Parameters

c_j Cost of route j.
$a_{i,j}$ 1 if flight i is covered by route j, and 0 otherwise.
NTotal Number of aircraft in the fleet.
Decision Variables
x_j 1 if route j is selected, 0 otherwise.
Objective Function

$$Min\Sigma_{j\in R}c_jx_j \tag{7}$$

Constraints

$$\Sigma_{j\in R}a_{i,j}x_j = 1\text{for all}i \in F \tag{8}$$

$$\Sigma_{j\in R}x_j \leq N\text{for all}i \in F \tag{9}$$

$$x_j \in \{0, 1\}\text{ for all}j \in R \tag{10}$$

In the preceding integer linear program, the primary goal in Eq. (7) is to minimize the overall cost associated with the chosen routes. Constraint (8) guarantees that each flight is assigned to one and only one route. Constraint (9) limits the selection of routes to the available number of aircraft for a given fleet type. Constraint (10) is the binary variable constraint for x_j.

For the crew pairing problem within crew scheduling, an alternative simplified integer linear programming model is as the following [1].
Sets
FSet of flights.
PSet of feasible pairings.
KSet of crew home-base cities.
Indices
JPairing index.
iFlight index.
kCrew home-base index.
Parameters
c_j Cost of crew pairing j.
$a_{i,j}$ 1 if flight i is covered by pairing j, and 0 otherwise.
$h_{k,j}$ 1 if home base city (starting and ending flight) for pairing j is city k, and 0 otherwise.
b_{lowerk} minimum number of crew to be used at home base city k.
b_{upper_k} maximum number of crew to be used at home base city k.
Decision Variables
x_j 1 if pairing j is part of the solution, 0 otherwise.

Objective Function

$$Min\Sigma_{j\in P}c_jx_j \tag{11}$$

Constraints

$$\Sigma_{j\in P}a_{i,j}x_j = 1 \text{for all flight legs} i \in F \tag{12}$$

$$b_{lower\,k} \leq \sum\nolimits_{j\in P} h_{k,j}x_j \leq b_{upper\,k} \text{ for all home bases } k \in K \tag{13}$$

$$x_j \in \{0, 1\} \text{ for all } j \in P \tag{14}$$

In the above model, the objective function (11) aims to minimize the overall cost associated with flight pairings. Constraint (12) ensures that each flight leg is covered exactly once. The additional constraint (13) is in place to ensure that the selected flight pairings do not exceed the available number of crew members at each home base. Constraint (14) is the binary variable constraint for x_j.

3 Conclusions

This study has explored various solution methodologies and models for addressing the complex challenges posed by integrated airline scheduling problems, especially within the context of irregular operations. The research has highlighted the significance of efficient scheduling in ensuring optimal airline operations and customer satisfaction, particularly when disruptions occur. The review has shown that a range of heuristic, metaheuristic, and optimization approaches have been developed to tackle these problems, each with its strengths and limitations. While exact methods offer optimal solutions for smaller instances, they often struggle with scalability. On the other hand, heuristic and metaheuristic approaches, such as Simulated Annealing and Tabu Search, can provide efficient solutions for larger and more complex instances.

Furthermore, the integration of robust optimization techniques and the consideration of uncertainty in demand and disruptions have been identified as crucial factors in developing resilient airline schedules. These approaches help airlines adapt to unforeseen events and minimize the impact of irregular operations on their operations and customer service. Overall, this review underscores the importance of ongoing research and development in the field of integrated airline scheduling. Future studies should focus on refining existing methodologies, incorporating real-time data and advanced analytics, and addressing emerging challenges, such as sustainability and environmental impact, to further enhance the efficiency and resilience of airline operations in the face of irregular operations.

References

1. Bazargan M (2010) Airline Operations and Scheduling. Routledge, England
2. Orhan İ, Kapanoğlu M, Karakoç TH (2010) Havayolu operasyonlarında planlama ve çizelgeleme. Pamukkale Üniversitesi Mühendislik Bilimleri Dergisi 16:181–191
3. Papadakos N (2009) Integrated airline scheduling. Comput Oper Res 36:176–195
4. ICAO (2023) Effects of Novel Coronavirus (COVID-19) on Civil Aviation: Economic Impact Analysis. https://www.icao.int/sustainability/Documents/COVID-19/ICAO_Coronavirus_Econ_Impact.pdf
5. IATA (2021) COVID-19 Airline industry outlook. https://www.iata.org/en/iata-repository/publications/economic-reports/airline-industry-economic-performance---october-2021---presentation/
6. Stojković M, Soumis F (2001) An optimization model for the simultaneous operational flight and pilot scheduling problem. Manag Sci INFORMS 47:1290–1305
7. Stojković M, Soumis F (2005) The operational flight and multi-crew scheduling problem. Yugoslav J Oper Res 15:25–48
8. Ahmed MB, Mansour FZ, Haouarii M (2018) Robust integrated maintenance aircraft routing and crew pairing. J Air Transp Manag 73:15–31
9. Sherali HD, Adams WP (1990) A hierarchy of relaxations between the continuous and convex hull representations for zero-one programming problems. SIAM J Discrete Math 3:411–430
10. Sherali HD, Adams WP (1994) A hierarchy of relaxations and convex hull characterizations for mixed-integer zero—one programming problems. Discret Appl Math 52:83–106
11. Kenan N, Jebali A, Diabat A (2018) An integrated flight scheduling and fleet assignment problem under uncertainty. Comput Oper Res 100:333–342
12. Zheng S, Yang Z, He Z, Wang N, Chu C, Yu H (2020) Hybrid simulated annealing and reduced variable neighbourhood search for an aircraft scheduling and parking problem. Int J Prod Res 58:2626–2646
13. Kiarashrad M, Pasandideh SHR, Mohammadi M (2021) A mixed-integer nonlinear optimization model for integrated flight scheduling, fleet assignment, and ticket pricing in competitive market. J Revenue Pricing Manag 20:596–607
14. Birolini S, Antunes AP, Cattaneo P, Malighetti P, Paleari S (2021) Integrated flight scheduling and fleet assignment with improved supply-demand interactions. Transp Res Part B Methodol 149:162–180
15. Kızıloğlu K, Sakallı ÜS (2023) Integrating flight scheduling, fleet assignment, and aircraft routing problems with codesharing agreements under stochastic environment. Aerospace 10:1–26
16. Xu Y, Wandelt S, Sun X (2023) A distributionally robust optimization approach for airline integrated recovery under in-flight pandemic transmission risks. Transp Res Part C 152:104188
17. Xu Y, Adler N, Wandelt S, Sun X (2024) Competitive integrated airline schedule design and fleet assignment. Eur J Oper Res 314:32–50
18. Hane CA, Barnhart C, Johnson EL, Marsten RE, Nemhauser GL, Sigismondi G (1995) The fleet assignment problem: solving a large-scale integer program. Math Program 70:211–232
19. Kabbani N, Patty B (1992) Aircraft routing at american airlines, AGIFORS 32nd Annual Fall Symposium, pp 12–28

Analysing the Convergence Between Lean 4.0 and Sustainability in Operations Management

Melisa Ozbiltekin-Pala and Yesim Deniz Ozkan-Ozen

Abstract Integrating Lean 4.0 with sustainability is essential for attaining efficiency and environmental responsibility in contemporary operations management. This research investigates the alignment of Lean 4.0 and sustainability in operations management, highlighting how combining them can improve operational efficiency and promote environmental responsibility. Automation and digitalization are essential in revolutionizing manufacturing processes by enhancing intelligence, flexibility, and efficiency, thanks to Industry 4.0 technologies. Lean 4.0 takes advantage of these progressions by merging digital technologies with traditional Lean concepts to reduce waste and enhance resource efficiency even more. At the same time, sustainability emphasizes the importance of using resources responsibly and minimizing environmental harm, covering environmental, social, and economic aspects. The merging of Lean 4.0 and sustainability is intended to synchronize operational processes with broader sustainability objectives, guaranteeing efficient and eco-friendly operations. Through an in-depth literature review, this study aims to identify crucial concepts and suggest potential areas for future research to enhance the comprehension of the collaboration between Lean 4.0 and sustainability in operations management.

Keywords Lean · Industry 4.0 · Sustainability · Operations management

1 Introduction

Within the current trade world, organizations are making key plans to meet environmental sustainability objectives and improve operational efficiency. These endeavours are crucial to fulfil social commitments and flourish in a ferocious environment, and, in this setting, one outstanding turning point that stands out is the Industry 4.0

M. Ozbiltekin-Pala (✉) · Y. D. Ozkan-Ozen
Logistics Management Department, Yasar University, Bornova, Türkiye
e-mail: melisa.ozbiltekin@yasar.edu.tr

Y. D. Ozkan-Ozen
e-mail: yesim.ozen@yasar.edu.tr

N. M. Durakbasa and K. G. Gülen (eds.), *Sustainable Green Conversion*, Lecture Notes in Mechanical Engineering, https://doi.org/10.1007/978-3-031-83583-4_25

transformation [1]. Industry 4.0 points to coordinated mechanization and digitalization advances to move forward mechanical processes' insights, adaptability, and efficiency [2]. This technological revolution increments endeavour operational efficiency and contributes to the fulfillment of natural sustainability objectives. Besides, these Industry 4.0 advances make it conceivable to utilize Lean 4.0 and sustainability, two crucial thoughts in today's operations administration, more effectively [3].

From the point of view of operations administration, Lean 4.0 and sustainability have to be coordinates to maximize organizational execution and decrease natural impacts [4]. Lean 4.0 joins these two spaces by intertwining Industry 4.0 innovations with classic lean management standards [5]. Lean administration standards incorporate decreasing squander, streamlining strategies, and empowering continuous advancement. On the other hand, sustainability is a thought that encompasses environmental, social, and economic dimensions [6]. From an operations management viewpoint, companies have to be utilized their assets capably, oversee their social duties, and decrease their natural impact. Accomplishing these sustainability goals requires the application of Industry 4.0 innovation [7]. Be that as it may, for these instruments to be utilized viably inside operations management, Lean standards have to be connected. Lean 4.0 coordinating advanced apparatuses to more successfully decrease squander and optimize asset utilize, coming about in higher effectiveness, minimized misfortune, lower costs, and faster response times in production process [4].

This integration guarantees that operations gotten to be more effective and adjust with sustainability objectives. Lean 4.0 and sustainability can be coordinates to make more ecologically neighbourly and productive operational operations [8]. Industry 4.0 advances offer the vital foundation and devices to empower this integration. The importance of this integration is particularly apparent in ranges like cultivating a vital competitive advantage, supporting natural and social obligations, leveraging innovation advancements more viably, and striking a adjust between productivity and sustainability [9]. In expansion, looking at the long-term impacts of this integration gives basic bits of knowledge for long-term commerce world and contributes to organizations developing sustainable trade models. Therefore, academic studies on integrating Lean 4.0 and sustainability significantly expand theoretical and practical knowledge in this field. Therefore, the main focus of this study is how Lean 4.0 can be integrated with sustainability goals and the effects of this integration on operational efficiency.

For these reasons, the examination of existing academic studies on the integrated evaluation of Lean 4.0 and sustainability concepts within the scope of this study will reveal how these two important concepts create a synergistic effect [10]. In addition, it is aimed to determine the key concepts in the application of these processes and to propose future research themes. In this way, a better understanding of the academic and practical importance of the subject will be provided, and critical insights will be provided on how organizations can achieve their operational efficiency and environmental sustainability goals more effectively.

This paper is organized as follows: first, a detailed literature review on Lean 4.0 and sustainability in operations management is given. Then, to present key concepts and future research areas, the Bibliometrix R-Tool is used. After that, results and discussions are given, and lastly, a conclusion is given.

2 Literature Review on Lean 4.0 and Sustainability in Operations Management

Integration of lean and sustainability in operations management is discussed in detail earlier due to their mutual goals such as minimizing wastes, increasing the efficiency and maximizing the employee welfare. With Industry 4.0 transition, technological developments are combined with the lean and sustainable manufacturing practices, where operational efficiency can be improved in multiple ways.

In this study, a brief literature review is conducted on lean 4.0 and sustainability with a special focus on operations and production management. Results showed that most of the studies aimed to build a theoretical background on the subject. A methodological categorization was easily made by the authors because variety is very limited in the current literature.

Majority of the studies used Structural Equations Modelling (SEM) to analyse the concepts related to lean, sustainability and Industry such as [3, 11–14]. One of an initial study that used SEM is conducted by [3], where relationship between lean, Industry 4.0 and social, economic and environmental sustainability are investigated. In this study, although the correlation between lean and sustainability dimensions have not found high, Industry 4.0 and sustainability found highly related. [12] applied the same method in the subject, however focused on sustainability performance, and concluded that lean, Industry 4.0 and sustainability performance are highly corelated. On the other hand, [13] approached the subject with adding circular economy concepts, and [14] applied the method for investigating the relationship for small medium enterprises. Similar to [12, 15] used tested hypothesis on investigating indirect effects of Industry 4.0 and lean manufacturing on sustainable organizational performance and they presented industry 4.0 as an enabler of lean manufacturing to enhance sustainability performance from the operations management point of view.

Another kind of theoretical contribution was made by conducting systematic literature reviews on the subject by different authors such as [16] and [17]. For instance, [16] conducted a review on impacts of Lean 4.0 on sustainability performance and presented a strong correlation between these concepts. On the other hand, [17] focused on environmental dimension of sustainability and aimed to explore the relationship for these concepts. Similarly, [18] also investigated the relationship between lean production, industry 4.0 and sustainable production with a theoretical scope.

Beside the theoretical contributions to the literature, [19] performed ergonomic and environmental assessment to develop a framework with a scope of achieving operational excellence and sustainability. They conducted their study by centralizing lean, green and Industry 4.0.

There are also studies that focus on lean six sigma, industry 4.0 and sustainability. For instance [20] conducted an exploratory study to analyse the criteria that affect circular, lean six sigma and industry 4.0 applications by reflecting sustainability. At the end of the study, they presented drivers and barriers of circular, lean six sigma and Industry 4.0 practices. Furthermore, [21] presented the enablers of lean six sigma with the consideration of sustainability and Industry 4.0 (Table 1).

Review on the subject revealed that theoretical studies are the priority and there is still need for more detailed analysis to extend the current knowledge from a more operational perspective.

3 Methodology and Design of the Study

Bibliometric analysis is extended rapidly in all scientific disciplines, and it is useful for analyzing a period when the focus on empirical work is generating a large volume of research that is often fragmented and contentious [22]. Bibliometric analysis is a quantitative method used to evaluate literature by measuring the dissemination of research through citation counts, publication patterns, and other statistical data. It helps identify key concepts and leading researchers in the selected field, guiding future research in terms of structuring and developing scientific knowledge. It is a useful technique to reveal the distinctions of a research area and shed light on emerging trends [23].

There are various tools to conduct bibliometric analysis and different software are available for researchers such as BibExcel, VosViewer, CiteSpace, and Bibliometrix [24]. In this study, Bibliometrix, which is a R package is hired to conduct bibliometric analysis, which comes to forefront by not only visualizing the networks but also enables to conduct various level of analysis. In Bibliometrix, an open-source web-based tool called Biblioshiny is used. Another advantage of the Bibliometrix is that it allows researchers to combine data from different databases easily.

In this study, data from SCOPUS and Web of Science are combined and analyzed by using Bibliometrix. After an initial search in both database, only title and keyword search are used because it has been realized that including abstract search results in losing the focus area. Therefore, the search criteria are defined as:

(TITLE("INDUSTRY 4.0") AND TITLE ("LEAN") AND TITLE ("SUSTAINABLE" OR "SUSTAINABILITY") OR KEY ("INDUSTRY 4.0") AND KEY ("LEAN") AND KEY ("SUSTAINABLE" OR "SUSTAINABILITY")) AND (LIMIT-TO (LANGUAGE, "English")).

Studies are only filtered for including the selected terms in title or keywords, were written in English. There was no need for limiting the time span because earliest study was published in 2017. Furthermore, since integrating lean 4.0 and sustainability is relatively a new research are, type of study is not taken as a search criteria and journal articles, book chapters, books, and review papers are included in analysis. The following section presents the results of the bibliometric analysis and discussions about the findings.

Table 1 Review summary

Author(s)	Method	Aim of the study	Paper type
[3]	Structure equations modelling	Analysing the impacts of Industry 4.0 and lean manufacturing on sustainability	Journal article
[15]	Survey	Investigating the indirect effects of Industry 4.0 technologies on sustainability performance and lean manufacturing	Journal article
[18]	Desk research	Investigating the relationship between lean production, sustainable production and Industry 4.0	Journal article
[11]	Structure equations modelling	Discovering the relationship between lean and sustainability manufacturing with Industry 4.0	Journal article
[16]	Systematic literature review	Addressing the impacts of integrating lean and industry 4.0 on sustainability pillars	Conference proceeding
[17]	Systematic literature review	Identifying integration between lean, industry 4.0 and environmental sustainability	Journal article
[12]	Structure equations modelling	Investigating the relationship between lean, industry 4.0 and sustainability performance	Journal article
[19]	Ergonomics and environmental assessment	Developing a framework for operations excellence and sustainability in operations management with a scope of lean, green and smart manufacturing in Industry 4.0	Proceeding paper
[13]	Structure equations modelling	Investigating the integrated impacts of circular economy, industry 4.0 and lean manufacturing on sustainability	Journal article
[21]	DEMATEL	Presenting and analysing lean six sigma 4.0 enablers for sustainable and circular manufacturing	Journal article

(continued)

Table 1 (continued)

Author(s)	Method	Aim of the study	Paper type
[14]	Partial least squares structural equation modelling	Exploring lean 4.0 and its relationship with sustainable manufacturing in small medium enterprises	Journal article
[20]	Case study	Presenting drivers and barriers for circular lean six sigma 4.0	Journal article

4 Results and Discussion

As mentioned in the previous section SCOPUS and WoS database are used for analyzing the current literature on lean 4.0 and sustainability. In total, 52 papers are included in the analysis between 2017 and July 2024. According to the results of the Bibilometrix, in total 185 authors worked on the subject so far and with an international co-authorship rate of nearly 47%. As it is expected it is a trending topic for the operations research area, where the annual growth rate is revealed as nearly 26%. A visual summary of these results are derived from Biblioshiny tool as presented in Fig. 1.

Results of the annual scientific production rate in lean 4.0 and sustainability also supported the increasing popularity of the field in the literature. In Fig. 2, publication rate between 2017 to mid-July 2024 is presented. The graph clearly represents the growing attention on the subject. Especially after 2022, a trend shows a steady increase in the number of publications, with the second quarter of 2024 alone reaching 10 publications so far.

Co-occurrence analysis is one of a critical output of bibliometric analysis, where Biblioshiny tool created visual network maps. In this study co-occurrence network map of keywords is created as presented in Fig. 3. In order to finalize the map, synonyms or abbreviations (i.e. industry 4.0 and fourth industrial revolution, lean six sigma and LSS) are combined and defined in the analysis, and words that are irrelevant are deleted.

Fig. 1 Biblioshiny output summary

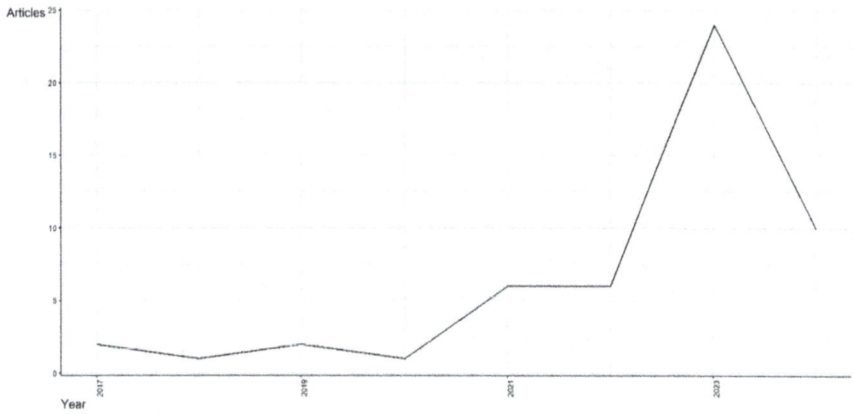

Fig. 2 Annual scientific production in the field

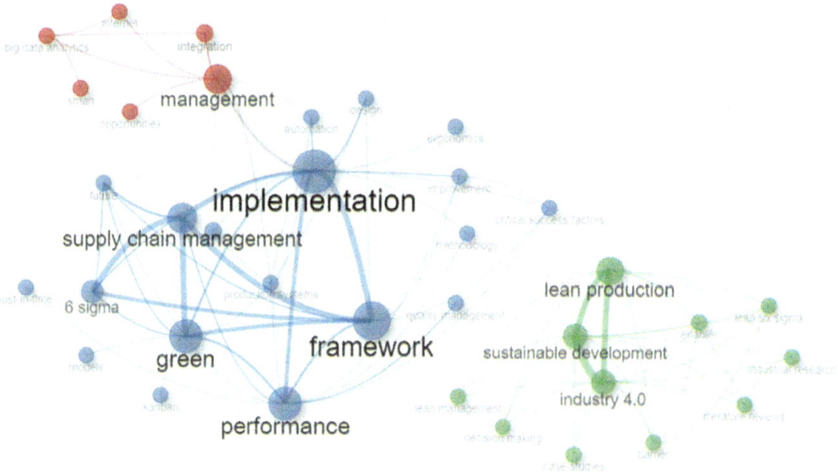

Fig. 3 Co-occurrence network map of keywords

The map reveals distinct clusters of keywords, indicating the primary research areas in the lean 4.0 and sustainability field. Three main clusters appear in the map, categorized by red, blue, and green colors. The centrality of certain keywords, such as "management" in red cluster, "implementation, green, framework and performance" in blue cluster, and "lean production, industry 4.0 and sustainable development" in green cluster indicate their pivotal role in the research landscape. The bigger size of the keywords also specifies the higher frequency. In detail, the term "management," which is at the center of the red cluster, stands out as an important part of the research. "Management" focuses on the organization, optimization, and integration of processes and strategies, especially in the context of Lean 4.0 and sustainability. Moreover, the term "Implementation" is at the center of the blue cluster and covers implementing Lean 4.0 strategies. This means transforming theoretical knowledge into practical applications in operations management. Implementation processes are integrated with performance measures, framework creation, and supply chain management. In addition, when the term "Performance" is addressed, performance indicators such as efficiency, effectiveness, and quality are used to evaluate the impact of Lean 4.0 on organizations in operations management. Furthermore, the term "Framework" refers to how the implementation is structured in this cluster and what methodologies are adopted. A specific framework or methodology needs to be developed to be successful in Lean 4.0 implementations. Other key terms in the blue cluster emphasize the impact of Lean 4.0 on the supply chain and its relationship with sustainable (green) processes. Green processes are one strategy implemented to achieve environmental sustainability goals.

Lastly, in the green cluster, the term "Lean production" forms the basis of this cluster and shows how the lean production philosophy is being reconsidered in the context of Industry 4.0. This term clarifies how advanced advances change Lean production and sustainable development processes. The term "sustainable development," another basic component of the green cluster, examines ways to attain environmental, financial, and social sustainability objectives. It creates methodologies for actualizing sustainability objectives when assessed inside the system of sustainable development, lean production and Industry 4.0.

In summary, these three clusters clearly outline the main research topics in Lean 4.0 and sustainability and the relationships between these topics. The red cluster focuses on how technology supports management and integration, while the blue cluster focuses on practical aspects such as implementation, performance, and supply chain. The green cluster addresses how lean manufacturing and sustainable development concepts are integrated with Industry 4.0. Combining these clusters provides a comprehensive understanding of Lean 4.0 and sustainability.

Furthermore, to identify the most influential studies on lean 4.0 and sustainability, a co-citation network map is also created and presented in Fig. 4. Co-citation network maps help identify the most frequently co-cited works and reveal the core set of publications that consistently affect the subject area.

Finally, Bibilometrix also creates a thematic evaluation map using the Biblioshiny tool for the subject area. The thematic evaluation map has four categories: niche themes, motor themes, basic themes, and emerging or declining themes. This map is useful for understanding the current field in terms of research trends. The thematic evaluation for lean 4.0 and sustainability studies is presented in Fig. 5.

Density appears how closely related subjects are inside a given subject cluster. The position of the motor themes within the chart demonstrates that they are dense which these topics are unequivocally related to each other. This implies that the topics have profound interconnection. Centrality moreover appears how closely related a given subject cluster is to other topics [25]. The centrality of the motor themes within the

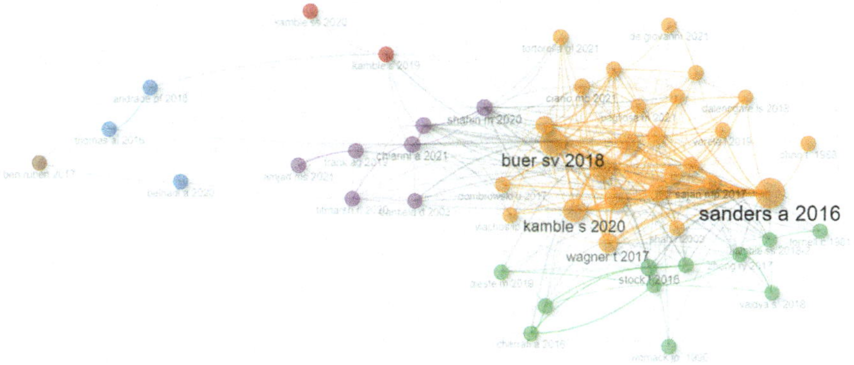

Fig. 4 Co-citation network map

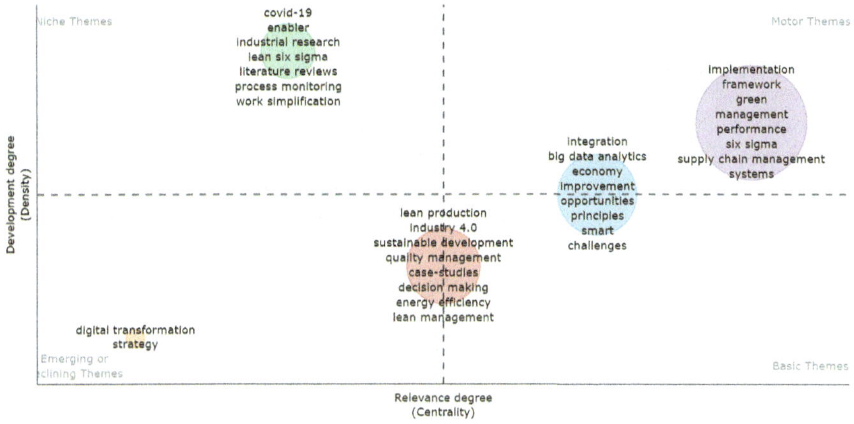

Fig. 5 Thematic evaluation of the field

chart demonstrates that these subjects are too firmly associated to other subject clusters and are exceptionally imperative to the generally field. As shown in Fig. 5, the motor themes include topics such as "implementation," "framework," "green," "performance," "Six Sigma," and "supply chain management systems." These themes are critical to sustainability and operational excellence in today's business world. In particular, "green" and "performance" focused approaches help companies increase their efficiency while reducing their environmental impact. "Supply chain management systems" and "Six Sigma" are important in optimizing operational processes and reducing costs. The topics covered in the motor themes play a central role not only among themselves but also in the overall structure and functioning of the entire field.

By considering the niche themes, themes such as "Covid-19", "enablement," "industrial research," and "six sigma" are included in this group. Niche themes in the chart (e.g., "COVID-19", "enabler," "industrial research") represent topics that are not widely applicable or universally accepted but are deeply relevant in certain situations or areas of expertise. These themes are often seen as very important by a certain group, but may not be widely applied or popular within the general field. These themes become critical in certain sectors or under special conditions. "Industrial research" and "Six Sigma" support operational excellence and continuous improvement processes in specific industries. These themes are essential in crisis management and developing the capacity of organizations to prepare for unexpected situations in operations research.

Moreover, Industrial research forms the basis of new product development and process improvement studies. When applied together with Six Sigma, the results of these studies can be used to achieve operational excellence. While industrial research increases innovation and competitive advantage, Six Sigma enables processes to be run more efficiently and without errors. Furthermore, work simplification increases efficiency by simplifying complex business processes. Literature reviews help businesses develop new strategies by providing in-depth analysis of existing information. These themes enable organizations to make informed decisions and optimize their processes. Lastly, in emerging or declining themes in the chart, themes such as "digital transformation," "strategy," and "digital transformation strategy" are included in this area. These themes may not be central or intense at the moment, but they may have the potential to develop or may lose their importance over time. These themes may gain traction on the one hand, and on the other hand, they may indicate a temporary popularity or decline.

Moreover, basic themes cover "Integration," "Big Data Analytics," and "Opportunities," and these themes play a critical role in businesses' digitalization processes. "Big Data Analytics allows for the analysis of large amounts of data to make better business decisions, while Integration ensures that different business processes and systems work together seamlessly.

5 Conclusion

Coordination sustainability with Lean 4.0 altogether propels operations administration when efficiency and environmental responsibility are progressively entwined. Lean 4.0 leverages Industry 4.0 advances such as mechanization and digitalization to upgrade conventional lean practices and adjust them with sustainability standards, empowering resource-efficient and sustainable operations. Advanced advances empower more cleverly, adaptable, and proficient production forms, empowering operations adjusted with sustainability objectives. This alignment reduces waste, optimizes resource use, and addresses environmental, social, and economic concerns, providing a comprehensive approach to modern operations management. Subsequently, this ponder points to address how Lean 4.0 and sustainability concepts can be coordinates into operations administration and the effect of this integration on operational productivity. In line with this aim, a systematic literature review examined

52 studies in SCOPUS and WOS databases between 2017–2024 with Bibliometrix R-Tool for an integrated evaluation of Lean 4.0 and sustainability concepts. As a result of this review, the co-occurrence network, co-citation network, and thematic areas were revealed. As a result of this study, since this topic is a trending topic for operations research, the necessity of addressing thematic areas and concepts revealed for future research came to the fore.

References

1. Costa F, Alemsan N, Bilancia A, Tortorella GL, Staudacher AP (2024) Integrating industry 4.0 and lean manufacturing for a sustainable green transition: a comprehensive model. J Clean Prod, 142728
2. Kumar N, Singh A, Gupta S, Kaswan MS, Singh M (2024) Integration of Lean manufacturing and Industry 4.0: a bibliometric analysis. The TQM J 36(1), 244–264
3. Varela L, Araújo A, Ávila P, Castro H, Putnik G (2019) Evaluation of the relation between lean manufacturing, industry 4.0, and sustainability. Sustainability 11(5), 1439
4. Frank AG, Thürer M, Godinho Filho M, Marodin GA (2024) Beyond Industry 4.0–integrating lean, digital technologies and people. Int J Oper Prod Manag 44(6), 1109–1126
5. Palacios-Gazules S, Giménez G, De Castro R (2024) Adopting industry 4.0 technologies through lean tools: evidence from the European manufacturing survey. Int J Lean Six Sigma 15(8), 120–142
6. Malik A, Sharma S, Batra I, Sharma C, Kaswan MS, Garza-Reyes JA (2024) Industrial revolution and environmental sustainability: an analytical interpretation of research constituents in industry 4.0. Int J Lean Six Sigma 15(1), 22–49
7. Dillinger F, Formann F, Magenheimer K, Reinhart G, Tortorella G (2024) Impact of lean and industry 4.0 on sustainability: a Delphi study-based identification and assessment. In A Research Agenda for Lean Management pp 227–246. Edward Elgar Publishing
8. Sahoo S, Upadhyay A (2024) Improving triple bottom line (TBL) performance: analyzing impacts of industry 4.0, lean six sigma and circular supply chain management. Ann Oper Res, 1–32
9. Piprani AZ, Khan SAR, Yu Z (2024) Driving success through digital transformation: influence of industry 4.0 on lean, agile, resilient, green supply chain practices. J Manuf Technol Manag
10. Ferrazzi M, Frecassetti S, Bilancia A, Portioli-Staudacher A (2024) Investigating the influence of lean manufacturing approach on environmental performance: a systematic literature review. Int J Adv Manuf Technol, 1–20
11. Dixit A, Jakhar SK, Kumar P (2022) Does lean and sustainable manufacturing lead to industry 4.0 adoption: the mediating role of ambidextrous innovation capabilities. Technol Forecast Soc Chang 175, 121328
12. Maware C, Parsley DM (2023) Can industry 4.0 assist lean manufacturing in attaining sustainability over time? evidence from the US organizations. Sustain 15(3), 1962
13. Ghaithan AM, Alshammakhi Y, Mohammed A, Mazher KM (2023) Integrated impact of circular economy, industry 4.0, and lean manufacturing on sustainability performance of manufacturing firms. Int J Environ Res Public Health 20(6), 5119
14. Qureshi KM, Mewada BG, Kaur S, Qureshi MRNM (2023) Assessing lean 4.0 for industry 4.0 readiness using PLS-SEM towards sustainable manufacturing supply chain. Sustain 15(5), 3950
15. Kamble S, Gunasekaran A, Dhone NC (2020) Industry 4.0 and lean manufacturing practices for sustainable organisational performance in Indian manufacturing companies. Int J Prod Res 58(5), 1319–1337

16. Kabzhassarova M, Kulzhanova A, Dikhanbayeva D, Guney M, Turkyilmaz A (2021) Effect of Lean4. 0 on sustainability performance: a review. Procedia CIRP 103:73–78
17. Lobo Mesquita L, Lizarelli FL, Duarte S, Oprime PC (2022) Exploring relationships for integrating lean, environmental sustainability and industry 4.0. Int J Lean Six Sigma 13(4), 863–896
18. Ciliberto C, Szopik-Depczyńska K, Tarczyńska-Łuniewska M, Ruggieri A, Ioppolo G (2021) Enabling the circular economy transition: a sustainable lean manufacturing recipe for Industry 4.0. Bus Strat Environ 30(7), 3255–3272
19. Tripathi V, Chattopadhyaya S, Mukhopadhyay AK, Sharma S, Kumar V, Li C, Singh S (2023) Lean, green, and smart manufacturing: an ingenious framework for enhancing the sustainability of operations management on the shop floor in industry 4.0. Proc Inst Mech Eng, Part E J Process Mech Eng, 09544089231159834
20. Skalli D, Charkaoui A, Cherrafi A, Shokri A, Garza-Reyes JA, Antony J (2024). Analysis of factors influencing circular-lean-six sigma 4.0 implementation considering sustainability implications: an exploratory study. Int J Prod Res 62(11), 3890–3917
21. Khanzode AG, Sarma PRS, Goswami M (2023) Modelling interactions of select enablers of lean six-sigma considering sustainability implications: an integrated circular economy and industry 4.0 perspective. Prod Plan Control 34(10), 1020–1036
22. Aria M, Cuccurullo C (2017) Bibliometrix: An R-tool for comprehensive science mapping analysis. J Informet 11(4):959–975
23. Donthu N, Kumar S, Mukherjee D, Pandey N, Lim WM (2021) How to conduct a bibliometric analysis: an overview and guidelines. J Bus Res 133:285–296
24. Derviş H (2019) Bibliometric analysis using bibliometrix an R package. J Sci Res 8(3):156–160
25. Farooq R (2024) A review of knowledge management research in the past three decades: a bibliometric analysis. VINE J Inf Knowl Manag Syst 54(2):339–378

Bus Spare Parts Demand Forecasting via Holt Winter's Method and Support Vector Regression Algorithm

İlker Mutlu and Önder Bulut

Abstract Demand forecasting and spare parts availability are crucial in managing spare parts for public transport systems. The condition of public transport buses is critical for service quality. Regular maintenance and repair of buses can increase public transport usage. This case study focuses on forecasting spare parts demand for a public transport company with more than 1700 buses and five workshops. Considering the variety of bus models and wide range of parts, effective demand forecasting is essential to avoid unnecessary holding costs or service interruptions. This study aims to apply two different demand forecasting methods, Winter's Method and Support Vector Regression, to a public transport company that makes forecast demands based on experience and to interpret the results. While applying the Support Vector Regression method, demand dates and bus ages were used as factors. Parameter optimization and demand forecasting methods were applied to the demand data using Weka software. Data were collected from the company's ERP software between 2019 and 2024. Spare parts were determined from five spare parts from different workshop units of the company. As a result of the study, MAPE values were calculated. According to the calculated MAPE values, it was determined that both methods gave similar results and successfully captured the demand structure.

Keywords Demand forecasting · Public transportation · Support vector regression · Winter's method · Weka

İ. Mutlu (✉)
Graduate School of Yaşar University, Bornova, Turkey
e-mail: ilkermutlu8@gmail.com

Ö. Bulut
Department of Industrial Engineering, Yaşar University, Bornova, Turkey
e-mail: onder.bulut@yasar.edu.tr

N. M. Durakbasa and K. G. Gülen (eds.), *Sustainable Green Conversion*, Lecture Notes in Mechanical Engineering, https://doi.org/10.1007/978-3-031-83583-4_26

1 Introduction

Forecasting is making the best prediction about the future using past data and possible future events information [1]. Therefore, past data and factors affecting demand should be used to forecast demand. Forecasts may be above or below actual demands. Both cases have different consequences. If the forecast is above the actual demand, holding costs increase, and if it is below, demands are not met and disruptions in operations occur. This is also the case with spare parts for buses providing public transportation services. Spare parts availability is very critical in public transport. Spare parts demand should be forecasting properly to have a minimum additional cost to the company. Otherwise, buses may not be able to operate, and additional buses may be used to avoid disruptions in public transport. This situation may result in many different costs, such as paying higher than normal price for the supply of spare parts, using an additional bus to provide public transport, or the cost of not operating the bus because demand is not met. For this reason, it should be as close to the actual demand as possible.

The company has more than 1,700 buses and five workshops in different locations. These buses have various models, different ages, and many different spare parts used in each bus. Reasons such as meeting the demands of many spare parts in sufficient quantities and quickly without causing additional costs make demand forecasting a complex issue [2, 3]. The company aims to provide quality, nature-sensitive, and comfortable service by maximizing customer satisfaction. Increasing customer satisfaction may increase the preference for using public transportation.

The importance of public transportation for the citizens and our environment has been studied for years. According to Shapiro et al. [4] increase in public transportation use will provide to reduce harmful emissions more effectively. For public transport service, the buses' condition and the bus fleet's total capacity operation are essential. Spare parts may wear out or break down over time. Proper maintenance and repair of buses are important for public transportation services. Spare parts must be available in inventory to repair buses. Accurate demand forecasting is essential in this respect. In the company's current system, forecasts are generally based on experience. Experience is significant; however, demand forecasting techniques are necessary to benefit from past data and the factors affecting demand. In this research, 5 different spare parts demand data between 2019 and 2024 were collected from the company's ERP software. This study aims to make three-month forecasts that the company can use in the short-term periods by using two different demand forecasting methods for spare parts with regular demand structures and to compare the results.

2 Literature Review

Many studies in various fields use Holt Winter's Method and Support Vector Regression Algorithm in demand forecasting. The Holt-Winter's method, introduced by Winters [5], was developed to increase forecast accuracy by controlling trends and seasonal components in the data. It has performed well in most fields where it has been applied, helping to increase forecast accuracy. Gamberini et al. [6] used Winter's method to forecast demand for customer order data with an irregular profile, including seasonality and trend. Şahinli [7] applied the Holt-Winters method to data on consumer potato prices covering approximately 14 years. It calculated MAPE, RMSE and MAD values. The result shows that it achieved good forecasting accuracy. Heydari [8] 30 years of rainfall, maximum temperature, minimum temperature and daylight hours data were used in a case study. He applied the Holt-Winters method, by optimizing the parameters. The result showed that this model is good at predicting and modelling data that contains cycle and seasonality. These studies show that the Holt-Winters method is useful for demand forecasting and time series analysis.

Levis and Papageorgiou [9] used support vector regression algorithm while making customer demand forecast. They made accurate demand forecasts by capturing customers behaviours according to past sales data. Cao and Tay [10] applied SVM to forecasting financial time series using five real futures contracts datasets from the Chicago Mercantile Exchange. They used three methods in the study and found that SVM outperformed the BP neural network in financial forecasting and showed similar performance with RBF neural network regularized with SVM. Yue et al. [11] used SVM to predict beer demand in the study. The results showed that SVM's forecasting accuracy is better than other prediction models, such as the traditional Winter's Model and RBFNN. Chen and Wang [12] apply support vector regression (SVR) to tourism demand forecasting. To build an effective SVR model, they optimized the parameters of SVR. About 16 years of tourist arrivals in China are used as data. Experimental results show that SVR outperforms BPNN and ARIMA models based on NMSE and MAPE. Ifraz et al. [3] used many regression-based demand estimation methods in their study, including Support Vector Regression. The estimation model includes critical variables such as the number of failures, the number of periodic maintenances, the mean time between failures, and the amount of demand in previous years. In estimations, support vector regression gave the best results among regression-based methods, decision table among rule-based methods, and M5P among tree-based methods. These studies showed that support vector regressions provide good forecasting in many applications, especially in nonlinear and complex demand structure forecasting. It also demonstrated that SVR is used in many different sectors.

3 Demand Forecasting Techniques and Formulation

A. *Holt Winter's Method*

Understanding the Winters method mathematical formula is relevant in trying to know the sense of this method in more depth. Heydari [8] Winter's method is explained mathematically in detail. The Holt-Winters is considered one of the preferable tools for forecasting, for it represents essential parameters like level, trend and seasonality of time series data. The method is of great value when an accurate forecast has to be made, particularly in seasonal effects that change with the size of the time series.

$$L_t = \alpha\left(\frac{Y_t}{S_{t-m}}\right) + (1-\alpha)(L_{t-1} + T_{t-1})$$

$$T_t = \beta(L_t - L_{t-1}) + (1-\beta)T_{t-1}$$

$$S_t = \gamma\left(\frac{Y_t}{L_t}\right) + (1-\gamma)S_{t-m}$$

$$\widehat{Y}_{t+h} = (L_t + hT_t)xS_{t+h-m(k+1)}$$

The Level Component refers to the general level for a series during a current period. Meanwhile, this level is achieved by multiplying it with seasonality rather than seasonally adjusted. While updating this component, the observations of the current period are evaluated in proportion to the seasonality parameter at that time. The basic idea behind doing so is not to remain aloof from the overall trend of a series while deciding about the current level. The trend component models the long-term increase or decrease in the series. It is based on the rise or fall of the series over some time, trends that continue to the future estimates. The trend component gets updated based on changes to the level while changes are calculated by the general size of the series. If the level undergoes a big change, a shift in the trend element will capture this. The seasonal component describes periodic patterns. This is very important in those cases where seasonal effects vary according to the general level of the series. This component will be helpful in learning periodic or cyclic patterns and thus can be used for predicting such patterns in future periods. Finally, the forecast equation combines the three components to come up with the future values. In the multiplicative method, the equation first updates the current level using the trend component and calculates the future value by multiplying this updated level with the seasonality component. The equation finds future values for what the series has done in the past and at which levels it currently stands, also accounting for seasonal patterns. This becomes a very practical tool for businesses and analysts to predict future demands to make strategic decisions based on those predictions.

B. *Support Vector Regression Algorithm*

In SVR, the objective is to determine the model parameters (w and β) while minimizing the error. Levis and Papageorgiou [9] explained SVR regression mathematically in detail in their studies. This is performed through an optimization problem expressed as:

$$min\frac{1}{2}\|w\|^2 + C.\sum_{i=1}^{N}(\xi_i + \xi_i^*)$$

subject to

$$y_t - \omega^T x_t - \beta \leq \epsilon + \xi_i \ \forall i = 1, \ldots, N$$

$$\omega^T x_t + \beta - y_t \leq \epsilon + \xi_i^* \ \forall i = 1, \ldots, N$$

Half the square of the weight vector ($\|w\|^2$) is used to penalize the complexity of the model. This penalty parameter (C) determines how much the model will be penalized. Slack variables (ξ_i, ξ_i^*) are used when the estimated values are outside the tolerance limits. These constraints define the relationship between the model's predictions and the true values. If the predictions fall within the epsilon tolerance, the error is zero [2, 9].

Optimization problems are usually solved using Lagrange multipliers and Karush–Kuhn–Tucker (KKT) conditions. The KKT conditions are obtained by equating the derivatives of this Lagrange function to zero.

One of the most critical steps in building the accuracy and predictive power of the model is the calculation of the bias, (β) parameter. In the model of SVR, the non-zero data points for the Lagrange multipliers, (λi) and $\lambda i *$), obtained after solving the dual problem are relevant and referred to as support vectors. These support vectors play a very important role in determining the model's prediction function. The support vectors define the boundary and the decision surface of the model and, in turn, determine how the model generalizes to other data points. These points directly influence the predictions of the model and let it represent its nonlinear structure in a better way.

Kernel functions are used when solving nonlinear data and higher degree functions [12, 9]. Mercer's Theorem plays a critical role when working with kernel functions. Because nonlinear data cannot be separated linearly using Support Vector Regression (SVR), nonlinear data is transformed into a high dimensional space wherein SVR will perform a linear separation. This transformation uses kernel functions. Mercer's Theorem provides the mathematical conditions required for a function to be used as a kernel.

4 Application

A. *Parameter Optimization*

Parameter optimization approaches available within the WEKA package include CVParameterSelection, Grid Search, Multi Search, and Auto-WEKA. One of the weaknesses of the CVParameterSelection method is that it cannot optimize any nested functions. In essence, this means that while it can optimize the C parameter in Support Vector Regression, it cannot optimize kernel functions in the same function. What really sets Grid Search apart is that with this technique, one can optimize two parameters at the same time, which is quite advantageous in certain scenarios. Auto-WEKA provides a parameter optimization with the addition of an extra package add-on. While this tool optimizes parameters and functions, the more time it is given, the better the optimization. However, this tool cannot only be used for Support Vector Regression. There are no restrictions to parameters which may be optimized by the MultiSearch method. In addition, the method has two different parameters: MathParameter and ListParameter. MathParameter takes the minimum and maximum point of the parameter the user wants to optimize and moves around this range in fixed steps. ListParameter, on the other hand, can optimize by trying values given in any set of constant values. In addition, MultiSearch can optimize parameters according to CC (Correlation Coefficient), RMSE (Root Mean Squared Error), RRSE (Root Relative Squared Error), MAE (Mean Absolute Error), and RAE (Root Absolute Error) error criteria. Within the proposed solution, both parameters of the MultiSearch parameter optimization method were used. While some parameter values can work more harmoniously with each other, some values take too long to produce results and cannot reach the desired low error values.

The parameters found can take values that minimize the margin of error when estimating the data used to train the model. However, in such overfitting situations, although the trained data is predicted very well, the function cannot fully capture the changes in the testing phase because it is not flexible. Therefore, although the values resulting from parameter optimization are considered a reference, the parameters were optimized again to make them compatible with the test set (Table 1).

During parameter optimization, the C value for the SVR model was determined to be between 1 and 10,000, the Gamma value to be between 0.01 and 0.99, and the Epsilon value to be between 0.001 and 1.0E-11. To determine the most suitable combination of these parameters, a comprehensive scan was performed using the MultiSearch method (Table 2).

Table 1 Parameter optimization results for Support Vector Regression

	Battery	Fender lamp	Evaporator fan	Condenser fan	Unit pump
C	10,000	10,000	10,000	10	1000
Gamma (γ)	0,99	0,99	0,99	0,5	0,8
Epsilon (ϵ)	1.0E-11	1.0E-11	1.0E-10	1.0E-4	1.0E-4

Table 2 Parameter optimization results for Holt-Winters method

	Battery	Fender lamp	Evaporator fan	Condenser fan	Unit pump
Alpha (α)	0,03	0,3	0,1	0,1	0,3
Beta (β)	0,1	0,3	0,1	0,1	0,1
Gamma (γ)	0,1	0,3	0,1	0,09	0,4
Season cycle length (n)	7	6	12	12	7

In Holt-Winters parameter optimization, the values that the parameters could take were initially tested between 0 and 1, with 0.1 differences. This approach provided a rapid scan of the parameters in a wide range. After determining close values, the Grid-Search method was used to scan at smaller intervals for more precise optimization. This process was carried out to determine the most optimal parameters of the model and thus to increase the prediction accuracy. These steps are critical to maximize model performance in time series analysis.

B. *Demand Forecasting*

After the parameter optimization step, demand forecasting techniques were applied to five different spare parts using Weka software. While Winter's method used historical data, Support Vector Regression used three factors: the month, year, and age of the bus when the demand occurred (Fig. 1).

In battery demand forecasting, both methods captured the historical demand structure well and performed well in forecasting. Similar results and low differences occurred (Fig. 2).

As did the battery, the method performed well in estimating demand for fender lamps. Winter's method outperformed SVR due to the trend and seasonality in the structure of past demand data (Fig. 3).

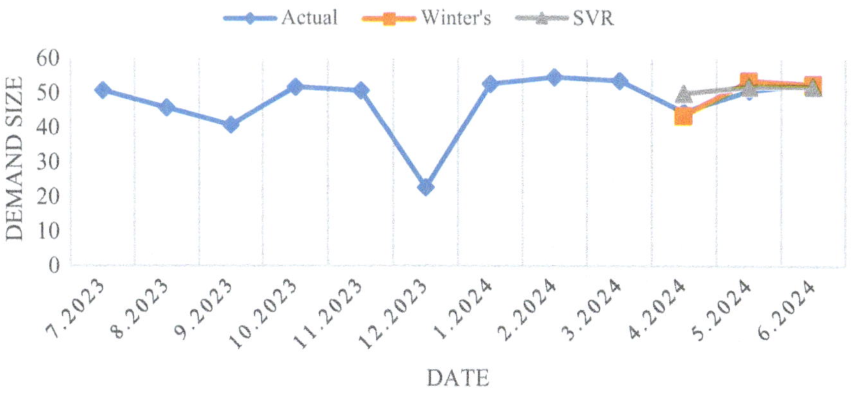

Fig. 1 Battery demand forecasting results and comparison

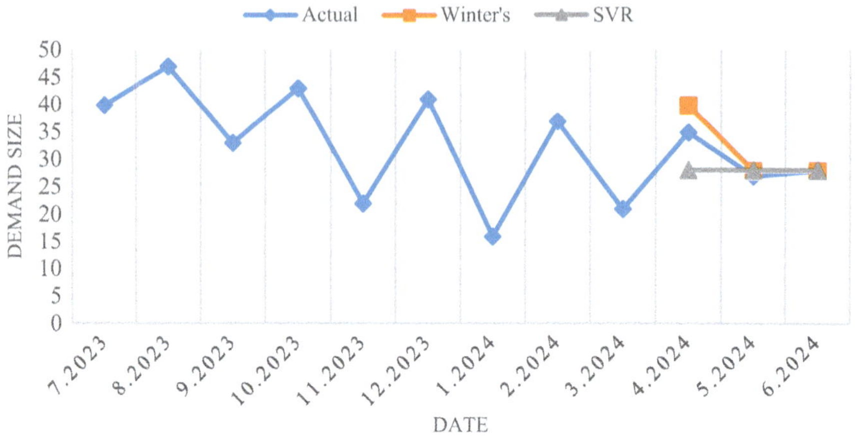

Fig. 2 Fender lamp demand forecasting results and comparison

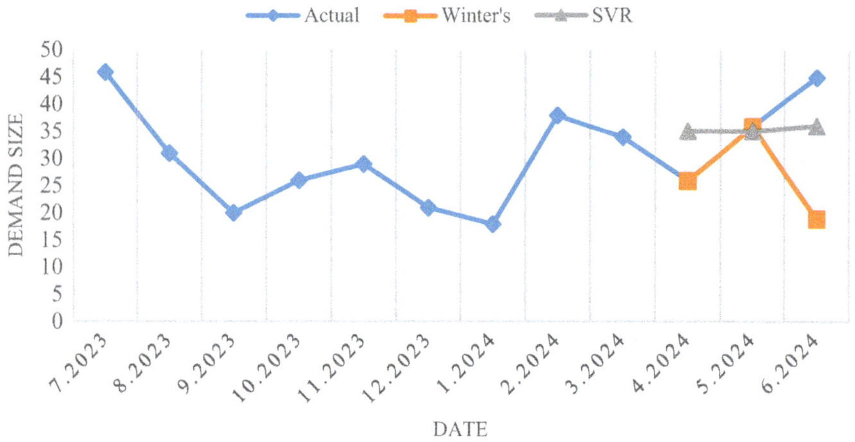

Fig. 3 Evaporator fan demand forecasting results and comparison

The Evaporator Fan, which has a more irregular structure than the other two spare parts, gave a much better demand estimation result than SVR Winter's Method. As in the Fender Lamp, the lack of a specific increase or seasonality in demand caused Winter's method to perform worse (Fig. 4).

Both methods performed well and gave similar results in condenser fan demand forecasting. However, the Support Vector Regression method captured the last three months of demand growth better than Winter's (Fig. 5).

Finally, the Unit Pump spare part was estimated and has been used steadily in recent months. While Winter's Method better caught the increase in the last three

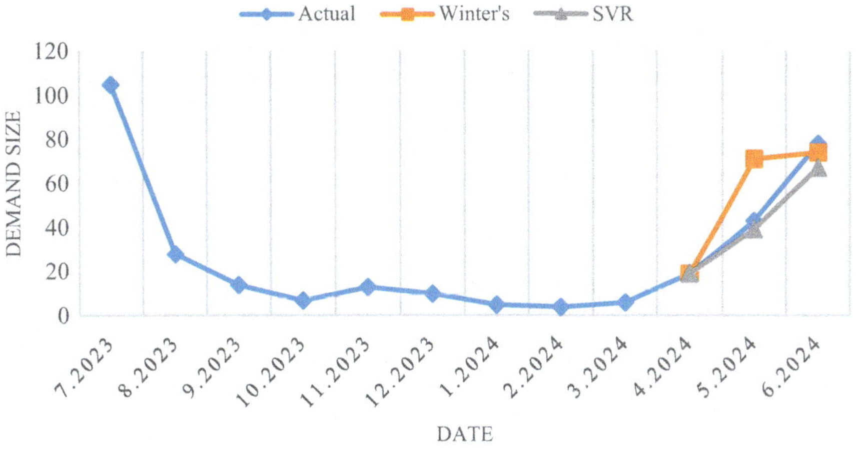

Fig. 4 Condenser fan demand forecasting results and comparison

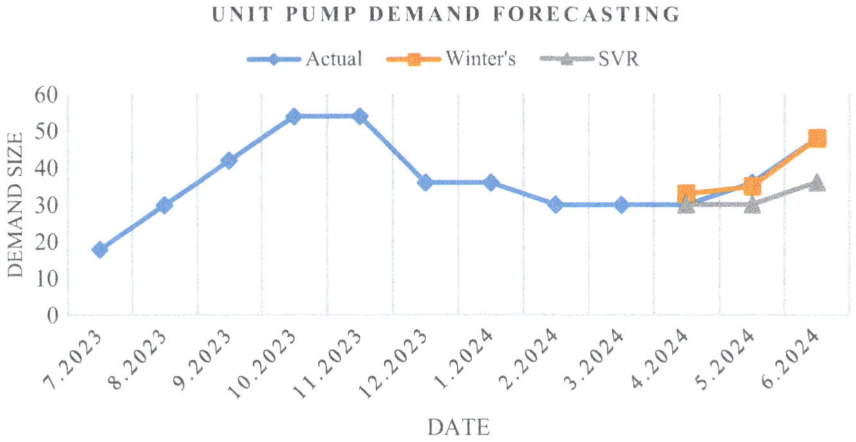

Fig. 5 Unit pump demand forecasting results and comparison

months, Support Vector Regression caught the increase but was forecast with a bigger difference than Winter's Method.

C. *Performance Evaluation*

Three-month demand forecasts were made for the company to use in the short term. Mean Absolute Percentage Error (MAPE) was calculated to measure the accuracy of the demands. MAPE Heydari [13] MAPE is an effective error calculation used in many studies for demand forecast accuracy. Heydari [13] included the MAPE formulation in the study.

$$MAPE = \frac{\sum_{l=N+1}^{N+M} |y_t - \hat{y}_t / y_t|}{M}.100$$

	Battery (%)	Fender lamp (%)	Evaporator fan (%)	Conderser fan (%)	Unit pump (%)
Winter's	3	5	46	15	4
SVR	5	10	18	9	18

Finally, the MAPE values of the demand forecast results were calculated using the two methods. The results were generally close to each other except for the Evaporator Fan.

5 Conclusions

This study used a case study approach to estimate the spare parts of a public transportation company. Two different demand forecasting methods were used. Support vector regression was added to the estimated data as a critical variable in the forecasting process, with the age of the buses as a factor. Winter's method was selected as a traditional time series forecasting method because it can control trend and seasonality through its parameters. The parameters were optimized through Weka software. As a result of the demand forecasts, monthly errors revealed close values for the two different methods. As a result of the study, MAPE values were calculated for two different demand forecasting methods with similar results. Both methods captured historical data structures and forecasting with small errors. As a result of the study, it was revealed that experience is an important factor in estimating spare parts demand, but traditional and advanced demand estimation methods are successful in capturing the structure of demand and making predictions.

In addition, when making predictions with support vector regression, bus brand, and route information can be included in addition to the selected age group factor. According to the research, machine learning methods such as support vector regression, which adds the right features, give better results when making predictions. In this way, chronic problems related to specific brands can be revealed, and increased demands associated with more challenging routes can be predicted.

References

1. Hyndman RJ (2018) Forecasting: principles and practice. OTexts
2. Aktepe A, Yanık E, Ersöz S (2021) Demand forecasting application with regression and artificial intelligence methods in a construction machinery company. J Intell Manuf 32(6):1587–1604

3. İfraz M et al (2023) Demand forecasting of spare parts with regression and machine learning methods: Application in a bus fleet. J Eng Res 11(2):100057
4. Shapiro RJ, Hassett KA, Arnold FS (2016) Conserving energy and preserving the environment: The role of public Transportation
5. Winters PR (1960) Forecasting sales by exponentially weighted moving averages. Manage Sci 6(3):324–342
6. Gamberini R et al (2010) Forecasting of sporadic demand patterns with seasonality and trend components: an empirical comparison between holt-winters and (S) ARIMA methods. Math Probl Eng, Article-ID
7. Şahinli MA (2020) Potato price forecasting with holt-winters and ARIMA methods: A case study. Am J Potato Res 97:336–346. https://doi.org/10.1007/s12230-020-09788-y
8. Heydari M, Ghadim HB, Rashidi M, Noori M (2020) Application of holt-winters time series models for predicting climatic parameters (Case Study: Robat Garah-Bil Station, Iran). Pol J Environ Stud 29(1):617–627. https://doi.org/10.15244/pjoes/100496
9. Levis AA, Papageorgiou LG (2005) Customer demand forecasting via support vector regression analysis. Chem Eng Res Des 83(8):1009–1018
10. Cao LJ, Tay FEH (2003) Support vector machine with adaptive parameters in financial time series forecasting. IEEE Trans Neural Networks 14(6):1506–1518. https://doi.org/10.1109/TNN.2003.820556
11. Yue L, Yafeng Y, Junjun G, Chongli T (2007) Demand forecasting by using support vector machine. In: Third international conference on natural computation (ICNC 2007), Haikou, China, pp 272–276, https://doi.org/10.1109/ICNC.2007.324.
12. Chen K-Y, Wang C-H (2007) Support vector regression with genetic algorithms in forecasting tourism demand. Tour Manage 28(1):215–226
13. Hua Z, Zhang B (2006) A hybrid support vector machines and logistic regression approach for forecasting intermittent demand of spare parts. Appl Math Comput 181(2):1035–1048

Productivity and Performance Management

Assessing the Business Continuity Risk in Technical Entrepreneurship Using the Altman Z-Score

Ştefania Chiribău-Vitlinger, Diana Dragomir, Mihai Dragomir, and Sorin Popescu

Abstract Manufacturing companies in particular, and technical enterprises in general, face a considerable number of risks in starting up and ensuring the continuity of the business. Aspects related to technological developments, market changes and other external factors, together with the internal transformations and the organizational capability, can lead to difficulties in which companies must assess their position in being able to carry out their business model and be competitive. The current article presents a case study in Romania in which the situations of the companies has been analysed using the Altman Z-Score, a metric specifically developed to perform a large-scale evaluation of manufacturing companies based on their recorded market and financial performance. The case describes the situations in detail and uses publicly available data to carry out the calculations, after which it proposes various improvement and intervention measures to reduce the pressures on the studied companies and restore their viability in line with the concept of second-chance technical entrepreneurship.

Keywords Business continuity · Risk management · Altman Z-score · Technical entrepreneurship

Ş. Chiribău-Vitlinger · D. Dragomir · M. Dragomir (✉) · S. Popescu
Department of Design Engineering and Robotics, Technical University of Cluj-Napoca, Cluj-Napoca, Romania
e-mail: mdragom@gmail.com

Ş. Chiribău-Vitlinger
e-mail: stefania.chiribau@staff.utcluj.ro

D. Dragomir
e-mail: diana.dragomir@muri.utcluj.ro

S. Popescu
e-mail: posorin@gmail.com

N. M. Durakbasa and K. G. Gülen (eds.), *Sustainable Green Conversion*, Lecture Notes in Mechanical Engineering, https://doi.org/10.1007/978-3-031-83583-4_27

1 Introduction

The current paper is part of a doctoral research project aimed at analysing and proposing support methods for technical entrepreneurs, mostly in the field of manufacturing, to determine possible existential risks to their businesses and to counter them in due time or to shorten the rebound period from honest first-time failure (which is a normal happening in a market economy) to second-chance endeavours. In order to achieve these goals, one of the key areas to investigate is the moment when technology-based enterprises run into problems and become distressed, which is the primary motive for conducting evaluations and assessments, using existing or newly proposed instruments, then can rapidly determine the situation of a company in relation with its internal and external environments.

In the article below, the authors present a case study of 3 companies on which the business model risk analysis was performed using the well-known Altman Z-score method. For each situation, the specifics of the company have been analysed and the application of the mentioned metric allowed for improvement measures to be proposed in relation with the results of the calculations.

2 Related Literature

The literature in the field of production management is often concerned with the decisions and changes that must be implemented to keep the production facilities, with its equipment, processes and personnel, competitive on a crowded market landscape where product fail more often than they success and where profit margins are very thin [1]. Risk management and bankruptcy prediction are constant issues that must be addressed and can determine the way the company can evolve in the short and long terms, with manufacturing resilience [2] emerging as an umbrella concept that incorporates reaction modalities to supply chain disruption, businesses continuity crises, process variability, market volatility and even physical threats. Various models and methodologies can be applied to undertake these tasks rooted in Bayesian network modelling [3] or Machine learning algorithms [4], thus bringing a scientific understanding to the art of decision making.

However, sometimes production managers prefer faster and more robust methodologies to arrive at the necessary conclusions, especially if this wins them time to act and try to avoid potential distress, especially on the financial side. Proposed by Edward Altman and used for over 50 years with significant success and an accuracy level of ca. 75% [5], the Altman Z-score exhibits these characteristics of speed, concision and call to action that make it interesting for the people in charge of manufacturing companies [6], as any amount of time gained is precious due to the need to ensure that the excessive costs of moving, changing or selling the physical manufacturing system can be avoided. The metric has been tested numerous times and has a considerable number of advantages [7, 8] and its finesse can be extended by employing it in the proper context, with additional operational, human related data and stakeholder impact data [6, 9]. Many companies can implement it successfully, without significant changes to the way they are currently monitoring performance and financial data.

In the context of distressed manufacturing enterprises that seek to limit or avoid altogether the difficulties of moving from an initial entrepreneurial venture to a second-chance venture, the Altman Z-Score provides the necessary impulse to act in a preventive approach [10, 11].

3 Research Methodology

The paper adopts the case-study based approach to analysing and determining the level of distress of the manufacturing companies. The selection of the companies was done based on the representativity in the area where the study was performed, the city of Bistrița, in North-Western Romania, an important manufacturing centre for over 6 decades. As such, these companies are coming from the mechanical, electrical and plastic manufacturing sectors. They are well-established players in the region and are known for their entrepreneurial approach, which makes them suitable for the study even if they have not newly appeared companies. The information used for the calculations comes from publicly available data, in the form of operational annual reports, the RisCo platform (https://www.risco.ro/) and financial disclosures. Since the Altman Z-Score is an established calculation scheme, this is performed without modifications and conclusions and improvement/change measures are proposed based on the results. A future step of comparative analysis is also possible, although not undertaken here. The methodology is summarized in Fig. 1.

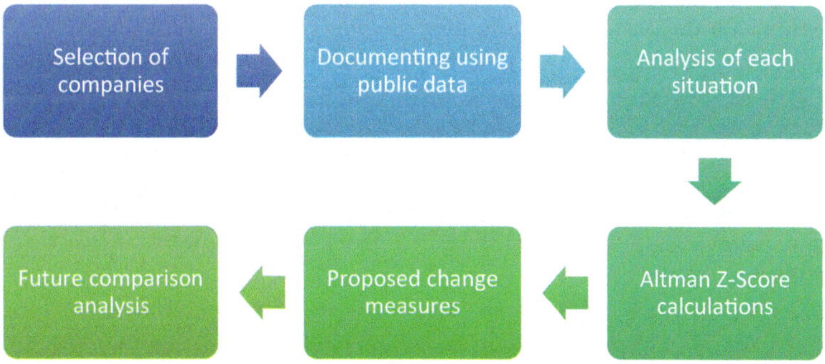

Fig. 1 Case study research methodology

4 Results and Discussion

As mentioned before, the three companies have been chosen from a compact geographical area, having similar characteristics and working in various domains of the manufacturing sector. The firms:

- Are medium-sized production companies from Romania, Bistriţa-Năsăud County;
- Produce a wide variety of products with medium complexity and simple assembly operations;
- Employ local workforce and a team of design and production engineers each;
- Distribute and sell their products both directly and through large supply chains;
- Have relatively high stability and a strong entrepreneurial spirit that is known in the area.

The score used for assessment is calculated based on financial ratios according to the coefficients and relations proposed in the literature, using the following formula [5, 12].

$$\text{Altman Z} - \text{Score} = 1.2 * \text{A} + 1.4 * \text{B} + 3.3 * \text{C} + 0.6 * \text{D} + 1.0 * \text{E} \quad (1)$$

With the components:

A: proportion of working capital (WC) from the total assets (TA).

B: proportion of retained earnings (RE) from the total assets (TA).

C: proportion of earnings before interest and taxes (EBIT) from the total assets (TA).

D: proportion of market value of equity (MVE) from the total liabilities (TL).

E: proportion of sales (S) from the total assets (TA).

Using publicly available data from the companies and their financial disclosure information, for the past 5 years, the Z-Score was calculated for each of the 3 manufacturing companies (see Table 1). In the final column of the table, each cell has been coloured according to the assigned interpretation of the score [5, 12], meaning that Z below 1,8 represents a high bankruptcy risk, while a Z score over 3 represent a safe zone for the firm, with the interval in-between being open to either outcome. As such, shades of red are used for the first possibility, shades of green for the second one, and shades of yellow for the middle area—in a traffic light based, easy to use and understand visualization.

Based on the results obtained, we can conclude that the companies exhibit three different types of behaviour. Manufacturer A (large company, in the electric cables industry, employing ca. 400 workers) is mostly cantoned in the middle segment of the Z-Score intervals, with fluctuating performances, and a tendency towards worsening results, even though during the pandemic (2020–2021) it has slightly better outcomes. At the same time, Manufacturer B (medium-sized company, in the electrotechnical

Table 1 Altman Z-Score calculation for the 3 manufacturing companies

Firm	Year	A = WC/TA	B = RE/TA	C = EBIT/TA	D = MVE/TL	E = S/TA	Altman Z-Score
Manufacturer A	2023	0,032487373	0,061768911	−0,061749024	0,455429315	1,714690996	1,91
	2022	0,14586424	0,072612502	−0,073515472	0,464156827	1,897990131	2,21
	2021	0,159101503	0,061222187	0,007575755	1,00	1,510760634	2,41
	2020	0,113093082	0,026979447	0,009716778	0,992419816	1,757408379	2,56
	2019	0,08786951	0,02127931	−0,059621632	0,948968729	1,637075671	2,14
Manufacturer B	2023	0,105694508	0,007119629	0,063586476	0,488934985	0,213958269	0,85
	2022	0,105831418	0,008111102	0,027418145	0,465965531	0,100179332	0,61
	2021	0,080246953	0,000038386	0,029960654	0,482652506	0,111607178	0,60
	2020	0,047937076	0,01051301	0,020773201	0,512146234	0,104950974	0,55
	2019	0,014399866	0,018659799	0,01928588	0,465331126	0,101896174	0,49
Manufacturer C	2023	0,558421654	0,142917151	0,11826812	0,906081587	1,321114928	3,13
	2022	0,475267573	0,015411829	0,143269631	0,79486699	1,270084355	2,81
	2021	0,3123254	0,016695432	0,039434396	0,728666533	1,129317426	2,09
	2020	0,404126657	0,016411659	0,016192058	0,812118107	1,150811624	2,20
	2019	0,436685321	0,013815392	0,041962819	0,829258521	1,1633165	2,34

devices industry, employing ca. 240 workers) is permanently in the red segment of the Z-Score, with steadily improving performances and a salutary almost doubling of the score over 5 years. However, starting from a very low initial value, even this significantly improved result is still indicative of a high risk of bankruptcy. Finally, Manufacturer C (very large company, in the heavy machinery industry, employing over 1100 workers) started from an average Z-Score and managed to constantly improve its performance and cross in the green area, with the results for the last fiscal year indicating a very good probability to avoid distress altogether.

Since there are many characteristics that these three manufacturers share, some research questions that appear are related to the need to investigate the factors that determine the significant differences noted. The two main area where distinctions are visible and can lead to improvement measures are factors A and E, which are significantly higher for the larger Manufacturer C. Increased working capital and increased sales can determine better forecast for the companies. This can be improved in a few ways: developing new products and attracting new customers, entering new markets and using new distribution channels, obtaining funding and incentives for working capital through the policy mechanisms for supporting job creation, establishing a debt payment plan and maintaining over time. In the case of Manufacturer A, it is also noticeable that factor C takes negative values, which is the most important contributor to its limited (average) performance due to the high coefficient it is multiplied with. As this happens under a large sales volume, there is a possibility that the company has high production costs, which could be lowered through: improved technology and equipment purchased using public funding for industrial renewal, a new

Table 2 Correlation matrix of financial indicators for manufacturer C

	Working capital	Total assets	Retained earnings	Gross profit	Sharehold. equity	Provisions	Deferred revenues	Total liabilities	Sales
Working capital	1								
Total assets	0,92897549	1							
Retained earnings	0,76427896	0,60226735	1						
Gross profit	0,95037786	0,94502493	0,527438	1					
Sharehold. equity	0,98418526	0,97357753	0,749077	0,942566	1				
Provisions	0,64608861	0,72431165	0,580311	0,571976	0,712916	1			
Deferred revenues	0,00	0,00	0,00	0,00	0,00	0,00	1		
Total liabilities	0,92897549	1,00	0,602267	0,945025	0,973578	0,724312	0,00	1	
Sales	0,96502844	0,99348683	0,658834	0,962122	0,991418	0,713146	0,00	0,993487	1

performance-based salary policy, a limited liability concerning environmental and social costs.

To complete the previous analysis, a correlation matrix has been developed for each of the companies, based on the concatenated values over the 5 studied years. This way, the contribution of each of the categories and factors in the Altman Z-Score becomes visible. Only Manufacturer C, which is the most likely to avoid distress is presented below, to showcase this approach (Table 2), using the same colouring scheme. Some issues become apparent: the retained earnings are poorly correlated with the gross profit pointing to a difficult highly competitive business environment, provisions and liabilities are not fully aligned which indicates the need for a resilience strategy, working capital and retained earnings do not progress at the same pace, probably due to unforeseen difficulties and crises in its sector (e.g. the pandemic, raw materials supply chain disruptions, etc.).

5 Conclusions

The Altman Z-Score is a useful tool for manufacturing companies to self-assess existential risks to their business continuity in a quick and efficient manner. The case studies presented exhibit a good alignment of the qualitative perception of the companies' performance and the results of the mathematical model. However, it is recommended that the investigation into the difficulties that can cause a company's distress to be performed from other perspectives as well, using dedicated or adapted models, frameworks and standards (e.g., ISO 22301:2019—Business continuity management systems) for generating a comprehensive description.

References

1. Zhou F, Fu L, Li Z, Xu J (2022) The recurrence of financial distress: A survival analysis. Int J Forecast 38(3):1100–1115
2. Lee J, Siahpour S, Jia X, Brown P (2022) Introduction to resilient manufacturing system. Manufact Letters 32:24–27
3. Shrivastava A, Kumar K, Kumar N (2018) Business distress prediction using bayesian logistic model for Indian firms. Risks 6:113
4. Papík M, Papíková L (2024) Automated machine learning in bankruptcy prediction of manufacturing companies. Procedia Computer Science, vol 232, no. 5th International conference on industry 4.0 and smart manufacturing, pp 1428–1436
5. Altman EI, Iwanicz-Drozdowska M, Laitinen EK, Suvas A (2014) Distressed firm and bankruptcy prediction in an international context: a review and empirical analysis of Altman's Z-score model. NYU Stern School of Business, New York City
6. Wu L, Shao Z, Yang C, Ding T, Zhang W (2020) The impact of CSR and financial distress on financial performance—evidence from Chinese listed companies of the manufacturing industry. Sustainability 12:6799
7. Cındık Z, Armutlulu IH (2021) A revision of Altman Z-Score model and a comparative analysis of Turkish companies' financial distress prediction. Natl Account Rev 3(2):237–255
8. Ko Y-C, Fujita H, Li T (2017) An evidential analysis of Altman Z-score for financial predictions: Case study on solar energy companies. Appl Soft Comput 52:748–759
9. Muñoz-Izquierdo N, Laitinen EK, Camacho-Miñano MDM, Pascual-Ezama D (2020) Does audit report information improve financial distress prediction over Altman's traditional Z-Score model? J Int Financ Manag & Account 31(1):65–97
10. Rashid F, Khan RA, Qureshi IH (2023) A comprehensive review of the altman z-score model across industries. Business Rev 27(2):35–42
11. Swalih MM, Adarsh K, Sulphey M (2021) A study on the financial soundness of Indian automobile industries using Altman Z-score. Accounting 7:295–298
12. Kenton W (2024) Altman Z-score: What it is, formula, How to interpret results. Investopedia. [Online]. Available https://www.investopedia.com/terms/a/altman.asp. [Accessed 30 August 2024]

Bibliometric Literature Review on Software Development Productivity and Observability

Verda Kılıç and Nihan Yıldırım

Abstract As software systems grow increasingly complex, the need for robust observability mechanisms becomes crucial. Bridging the concepts of productivity with software development models and processes presents significant challenges, often resulting in unclear connections and associations. While agile methodology is frequently lauded for its potential to enhance productivity, the relationship between productivity and observability in software development remains under-explored in the literature. This study addresses this gap by conducting a comprehensive bibliometric analysis using the Scopus database and Bibliometrix tool. The analysis reveals key trends, connections, and the current state of knowledge on the interplay between productivity and observability in software development, offering valuable insights for both practitioners and researchers.

Keywords Observability · Software development productivity · Developer productivity · Agile · Agile software development · Bibliometric analysis · Scopus · Bibliometric literature review

1 Introduction

The rapid evolution of the software development industry has necessitated a closer examination of productivity and observability, two pivotal factors that influence the success and efficiency of software projects. As the demand for high-quality software increases, understanding the dynamics that drive productivity and the mechanisms that ensure observability becomes crucial. Software development productivity refers to the efficiency and effectiveness of creating, maintaining, and enhancing software. It encompasses various dimensions, including the speed of development, the quality

V. Kılıç (✉) · N. Yıldırım
Department of Management Engineering, Istanbul Technical University, Istanbul, Türkiye
e-mail: kilicv21@itu.edu.tr

N. Yıldırım
e-mail: yildirimni@itu.edu.tr

N. M. Durakbasa and K. G. Gülen (eds.), *Sustainable Green Conversion*, Lecture Notes in Mechanical Engineering, https://doi.org/10.1007/978-3-031-83583-4_28

of the output, and the ability to meet project timelines and budgets. Several factors influence productivity, from individual developer skills and team dynamics to organizational practices and technological tools [1]. The complexity of these factors necessitates a systematic approach to identify and analyze the key drivers of productivity in software development. On the other hand, observability is related to the ability to measure and monitor the internal states of a system through its outputs. In software development, observability involves tracking various metrics and logs to gain insights into software systems' performance, reliability, and overall health. Effective observability practices enable developers and organizations to detect issues early, optimize performance, and ensure that the software meets quality standards; effective oversight of large-scale, distributed systems relies heavily on observability [2].

As software systems become more complex, the importance of robust observability mechanisms has grown significantly. In theory and practice, bridging the productivity concept with the software development models and processes has always been challenging, with blurred connections and associations. Practitioners of agile and/or lean software development models often look for ways of elaborating on the positive impacts of such contemporary approaches (e.g., the ROI of agile transition projects), while researchers frequently feel the need for solid conceptual frameworks during their analytical and dialectical discussions on the advanced discourses of various software development models. However, the connections between productivity and observability in software development remained an under-explored topic in the literature. Limited studies exist on elaborating on and synthesizing the current state of knowledge related to the effect of productivity on software development and the associations between observability and productivity in software production.

This study aims to revisit the theoretical background through a comprehensive thematic literature review on software development productivity and observability to explore the connections between productivity and observability in software development and the effect of productivity on software development. We conducted this study through a bibliometric analysis, a robust quantitative method to evaluate and map the scientific literature within a specific field. By analyzing patterns in publication data, such as citation counts, authorship, and thematic trends, bibliometric analysis provides valuable insights into the evolution and current state of research on a given topic [3]. Utilizing the Scopus database, one of the largest and most comprehensive databases available, providing coverage of peer-reviewed literature across various disciplines, we used a systematic approach to identifying the relevant articles. We extracted their bibliometric data for in-depth analysis. The findings revealed the key research trends and influential works in software development productivity and observability but also provided visual representations of the data to enhance understanding and interpretation [3].

The following sections of this paper present the methodology of the bibliometric literature review, the results of the analysis, and a discussion of key findings with a conclusion. By synthesizing current knowledge on software development productivity and observability, the study highlights critical areas for future research. It offers actionable insights for practitioners and researchers, contributing to ongoing discussions on enhancing software development practices.

2 Overviewing to Theoretical Background

While reviewing the literature, it is seen that the rising popularity and adoption of agile software development methods have shifted research trends within the software industry. There is now a stronger emphasis on output quality and efficiency rather than solely on process efficiency. In the software development sector, enhancing process efficiency has protected their importance as it contributes to improving technical software quality and minimizing the challenges development teams face during the software creation process.

Välimäki et al. [4] applied the ISO standard to develop process quality by creating quality profiles via the Q-PAM method. Despite conducting numerous case studies on process quality improvement, they found it challenging to identify a common pattern. They ultimately recommended that organizations determine relevant quality attributes through internal discussions aligned with their assessment goals [4].

Productivity in agile software development is characterized by the effectiveness of the software development team [1]. Moe and Dingsøyr [5], in their paper, explored the elements of Scrum by connecting them with a proposed theory of the "big five" components that contribute to small team effectiveness and also examined the Scrum framework in the context of these components and presented an analysis based on a case study of a Scrum team within a start-up [5]. Abbas et al. [6] conducted a study analyzing the relationship between organizational practices and project outcomes in agile software development. Using data from the Agile Projects Governance Survey, which gathered 129 responses, they found that software quality improves when organizations frequently measure customer satisfaction and effectively utilize retrospectives. Additionally, they observed that project success is positively influenced by higher quality, regular customer satisfaction assessments, organizational experience with agile methods, and active team involvement in retrospectives [6].

Ahmad et al. [7] assert that quality must be quantifiable in order to be achieved. They highlight that product metrics tailored to specific situations and contexts are essential for this purpose [7]. Staron et al. [8] carried out case studies on three companies that were in the process of transitioning to Agile and Lean software development, with their experience in these approaches spanning from 2 to 5 years. Their findings indicate that organizations with more extensive expertise tend to prioritize empowering self-organizing teams, whereas those with less experience emphasize the importance of conveying team status to management. The study also concludes that well-designed dashboards are essential for gathering and observing key metrics effectively, emphasizes observability has a crucial role in qualified and productive outputs [8]. The article presents case studies conducted on companies from diverse industries, offering insights into effectively visualizing observation elements within dashboards. Although comprehensive, the study does not establish a standardized pattern for the software industry. Instead, company-specific variables emerged as the primary components of the dashboards, resulting in the absence of standardization.

The key conclusion of Ortega et al.'s [9] article is that gathering data related to a software developer's emotions and productivity during agile software development, securely storing this data, and then categorizing the emotions and calculating and visualizing productivity can significantly enhance overall productivity [9]. Similarly, Kuutila et al.'s [10] research explored two primary data sources from developer chat messages and source code commits recorded over 200 workdays by an Agile team. The study found that emotional states are indeed linked to productivity in software development. Specifically, three measures of positive sentiment and one unexpected measure of negative sentiment were found to correlate with increased productivity [10]. Bollati et al. [11] comprehensively analyze the current research on factors influencing software development teams, specifically examining their impact on productivity through a systematic literature review. The findings suggest that the most significant factors affecting team productivity are related to member interactions, particularly communication, physical distance, and organizational structure. Additionally, individual characteristics and soft skills are important influences on overall team performance [11]. Organizations should develop strategies to monitor and improve the emotional state of developers, contributing to increased overall productivity of the software development team.

After establishing a theoretical foundation for the literature review, it is crucial to emphasize the critical importance of productivity and observability in software development for both developers and teams. The systematic literature review and bibliometric analysis reveal a body of work on productivity in the software industry. However, a gap exists regarding the relationship between productivity and observability, with more focus on monitoring than observability. Detailed analysis of the sources identifies four key themes: emotional state and productivity, quality and success factors in agile development, waste reporting and productivity, and social and human factors. Emotional state and productivity were discussed in the previous section.

Arcos-Medina and Mauricio [12], examined critical success factors affecting quality, quality measurement metrics, and agile practices in agile software development processes. The study details how agile practices impact software quality, highlighting the significant role of agile methodologies in enhancing software quality [12]. Software development teams should adopt agile practices more broadly and regularly implement the processes and techniques these methodologies introduce. Jabbar et al.'s [13] study proposes a Scrum-based model to maintain quality standards in health systems software development. This study demonstrates that knowledge management techniques can improve software quality [13]. Focusing on improving software quality is crucial, especially in critical fields like health systems.

Sallin et al.'s [14] study developed a waste classification and survey to help software development teams identify and quantify waste activities. This study emphasizes that waste reporting is crucial for increasing productivity [14]. It can be concluded from this study that teams should develop tools and methods for waste reporting to minimize factors that negatively impact efficiency.

Machuca-Villegas et al. [15] focus on social and human factors (SHF) affecting productivity in software development teams; the study indicates that SHF plays a critical role in enhancing the productivity of software projects [15]. Therefore, improving interaction and communication processes among team members is essential to enhance team performance.

3 Methodology

According to Toker [16], over the past half-century, there has been a continual daily increase in the publication of research articles, concurrent with the rising preference for qualitative methodologies. In the area of quantitative research, researchers have previously invented a method to consolidate findings from numerous smaller studies into a more extensive and comprehensive study, known as meta-analysis. The utilization of meta-synthesis as a qualitative research method became evident in studies conducted in the 1990s [16]. Bibliometric literature reviews leverage statistical tools to comprehensively analyze published research to identify trends, citations, and/or co-citations within a specific theme. This multifaceted analysis spans dimensions such as year, author, journal, country, paper's theory, method, and research problem, providing a nuanced understanding of the scholarly landscape [17]. Academics utilize bibliometric analysis for diverse purposes, including identifying emerging trends in journal and article performance, research components, and collaboration patterns. Furthermore, scholars use this methodology to investigate the intellectual structure of a domain within the existing literature. Donthu et al. [18] briefly outlined systematic literature review, bibliometric analysis, and meta-analysis methods in their article. They contextualized the usage of each method within the study's objectives, considering the scale and quality of the reviewed literature. Emphasizing the complementary nature of these review methods, the authors presented a methodological comparison. Their study offers insights based on various criteria, aiding researchers in making informed decisions when selecting the most suitable review method [18]. A thorough examination of the research was conducted using the bibliometric analysis method in the literature review in conjunction with the study's subject, target, and research area. The determination has been made that opting for the bibliometric analysis method is a more fitting approach.

Drawing upon established practices in bibliometric methodology, Zupic and Čater [19], suggested workflow guidelines for conducting science mapping research. They articulate these guidelines as listed below, outlining a step-by-step procedure specifically tailored for science mapping in the realm of management and organization [19]:

- 1st Step: Research Design
- 2nd Step: Compilation of Bibliometric Data
- 3rd Step: Data Analysis
- 4th Step: Visualization of the Data
- 5th Step: Interpretation.

4 Bibliometric Literature Review

In the scope of the study, the literature review in this section has been conducted using a bibliometric literature review, focusing on the observability of productivity and efficiency in software development. For a more accurate comprehension and interpretation of the conducted analysis, the previous section initially elaborates on the concept of bibliometric literature review. After data was revealed, the PRISMA (Preferred Reporting Items for Systematic Reviews and Meta-Analyses) statement was conducted. "The PRISMA 2020 statement has been designed primarily for systematic reviews of studies that evaluate the effects of health interventions, irrespective of the design of the included studies" [20].

While conducting the bibliometric literature review, the sources chosen for the paper based on sources from 2010 and later through the Scopus database were analyzed using the bibliometric analysis method on the Bibliometrix tool, which is a skillful open-source tool designed to comprehensively analyze scientific literature through science mapping developed on R programming language as an R package [21]. Scopus database has been chosen for performing bibliometric analysis because Scopus is one of the largest and most comprehensive databases available, providing coverage of peer-reviewed literature across various disciplines, making it an ideal source for this study.

A. *Design of the Research*

In this step, the primary purpose is to define the research purpose by creating dimensions and finding the accurate method of the research question, such as co-word, co-citation, and co-occurrence analyses.

In order to refine dimensions, keywords were listed, and their definitions were studied to determine the key concepts of the bibliometric analysis. The first chosen keywords were Software Development, Software Developer, SW Development Efficiency, SW Development Productivity, Developer Productivity, Observability, Visualization, Information Technology (IT), Lean, Kaizen, Agile, Agility, Kanban, Lean Kanban, and Scrum. After completing the detailed examination and research phase of the concepts, it was observed that some of the first concepts studied did not completely overlap with the subject and remained out of concept since they included too much general information. As a result, the keywords Visualization, Information Technology (IT), and Lean Kanban are marked as out of the concept. Based on the theoretical background and conceptual foundations provided by the sources reviewed in this study, this bibliometric analysis has been conducted in multiple dimensions to thoroughly explore the interactions between productivity-related concepts in software development (efficiency, effectiveness, productivity, quality, performance) and agile concepts (lean, scrum, kanban). As a result of these, the main dimensions are listed as it follows:

- Dim.1: Meaning of software development/developer efficiency and productivity
- Dim.2: Measurability of software development/developer efficiency and productivity

- Dim.3: Ensuring observability of software development
- Dim.4: The effects of the ASD methodology on efficiency and productivity
- *Creating the Bibliometric Data*

During the process of gathering data from Scopus, the following query has been used to filter the data; the query includes a range of search terms using wildcard characters for flexibility:

- TITLE-ABS-KEY(("software dev* prod*" OR "prod* of software dev*" OR "software dev* eff*" OR "eff* of software dev*" OR "software eng* prod*" OR "software eng* eff*" OR "prod* of software eng*" OR "eff* of software eng*" OR "quality of softwa*" OR "software quality me*" OR "software dev* quality" OR "software developer perf*" OR "software eng* perf*") AND ("agil*" OR "lean" OR "scrum" OR "kanban")) AND PUBYEAR > 2001.

The query's logic ensures a targeted yet comprehensive search of literature relevant to software development productivity, efficiency, observability, and the influence of Agile methodologies. The query can be detailed related to the design dimensions written in the previous sub-section as follows:

- Dim.1: The query includes terms like "software dev* prod*" and "software eng* eff*," using the asterisk wildcard "*" to capture variations such as "developer productivity" or "engineering efficiency." This approach ensures a comprehensive search for sources discussing productivity and efficiency in software development, considering all possible keyword combinations.
- Dim.2: This section of the query focuses on the measurability aspect by including terms like "quality of softwa*" and "software developer perf*," which cover topics such as software quality and performance. The wildcard "*" is used to catch various forms of the words, such as "measurement" or "performance," providing a broad scope for identifying relevant studies.
- Dim.3: Key terms related to productivity, efficiency, and quality, such as "software dev prod" and "software dev eff," highlight the query's focus on observability. These terms are integral to understanding and optimizing development processes, making them observable and manageable.
- Dim.4: The inclusion of terms like "agil*" and "scrum" with the condition PUBYEAR > 2001 targets studies post-Agile Manifesto, focusing on the impact of Agile methodologies on productivity and efficiency. The query is designed to capture a wide range of sources that discuss the influence of Agile practices on software development.

C. *Analysis of the Bibliometric Data*

After refining the research question and generating the data with the Scopus query, the most critical step is to ensure the accuracy of these findings. The query on Scopus was thoroughly discussed and examined in previous sections. An advanced search using the crafted queries initially yielded over 500 sources. However, realizing that some irrelevant documents were included, additional criteria were added to the query, refining it to 271 sources. Each of these 271 sources was then meticulously reviewed by examining their titles and abstracts to confirm their relevance to the study. The selected sources were exported as an Excel file for further analysis in Bibliometrix, a bibliometric analysis tool based on R that uses Scopus as a data source. Following the installation of R and Bibliometrix, the study was implemented using Biblioshiny, a user-friendly interface for Bibliometrix. The Excel file was integrated into the Biblioshiny interface, visualizing the bibliometric metadata distribution for the 271 sources. The analysis of these data, along with the results and visual representations, will be detailed in the next section.

In this bibliometric literature review, the sources were screened by the PRISMA 2020 statement. After the screening process was conducted through the PRISMA flow, the 271 sources initially obtained from Scopus were filtered for the reasons specified in the diagram, and the literature review was completed based on 39 primary sources. The process is clarified in Fig. 1 with the PRISMA statement flow diagram. The identification of the common points emphasized by the 39 sources obtained was conducted, and the findings were detailed in the Sect. 2.

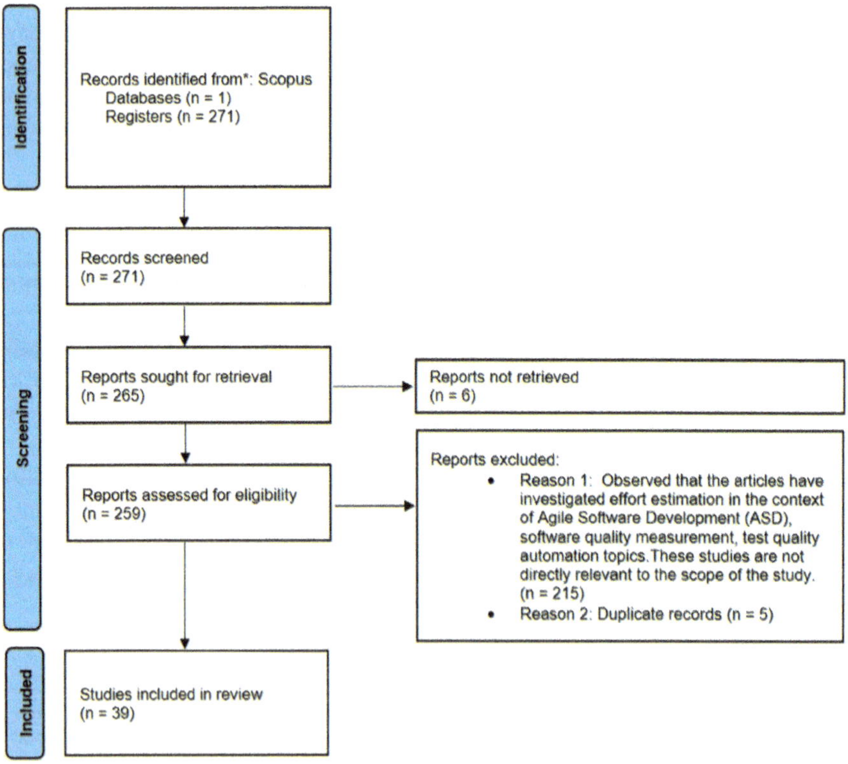

Fig. 1 Bibliometric literature review diagram according to PRISMA 2020 flow

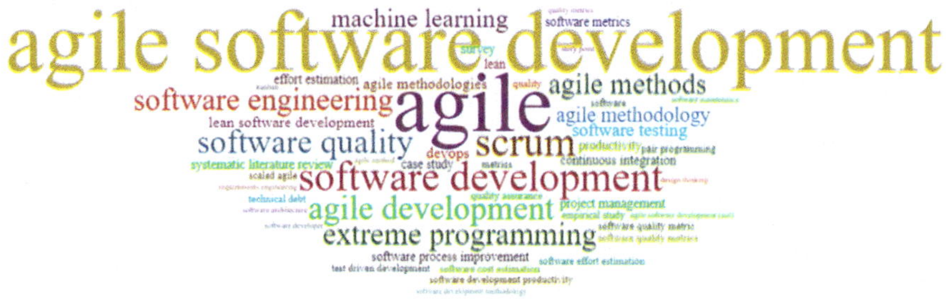

Fig. 2 Wordcloud

D. *Analysis of the Outputs and Visualization*

The most crucial step after crafting a query to refine the research question and reveal the data is to ensure the accuracy of these findings. In the previous sections, the query written on Scopus was thoroughly analyzed and refined. According to statistics, 271 academic studies (73 articles, 5 books, 18 book chapters, 150 conference papers, 18 conference reviews, and 7 reviews) were scanned from 195 data sources covering the period 2003–2024; these studies were put forward by 712 authors, with an average of 9.87 citations per document and the international co-authorship rate in publications is 15.13%.

Through Bibliometrix, a word cloud was prepared with the author's keywords of the documents examined and visualized in Fig. 2. When these keywords are examined, the top 5 most frequently mentioned keywords can be listed with their usage frequency as follows: agile software development (33), agile (32), scrum (16), software development (16) and agile development (13). "Productivity" is in rank of 20th with 5 frequency.

The academic publications retrieved via the Scopus query are analyzed through co-occurrence analysis to determine the strength and clustering of authors' keywords based on their co-usage patterns. In the resulting network, clusters are represented by elements of the same colour, with node sizes reflecting the frequency of term occurrences. Moreover, the proximity of nodes within the same cluster indicates a higher frequency of co-occurrence. Figure 3 shows that agile software development term used with software development productivity and software metrics terms while the agile concept is used with metric and quality terms most.

Figure 4 shows that the distribution and relationships of keywords extracted from Scopus query based on two dimensions. The three distinct clusters suggest thematic groupings. Cluster Blue: Topics such as software effort estimation and artificial intelligence are closely related, indicating a focus on prediction and estimation techniques within software development. Cluster Red: This cluster is centered on software architecture, machine learning, and agile methods, highlighting a broad range of concepts related to design, education, and process improvement in software development. Cluster Green: Topics like C programming language and test-driven development are grouped, suggesting a focus on programming languages and testing methodologies. The separation of clusters along the two dimensions indicates differing thematic

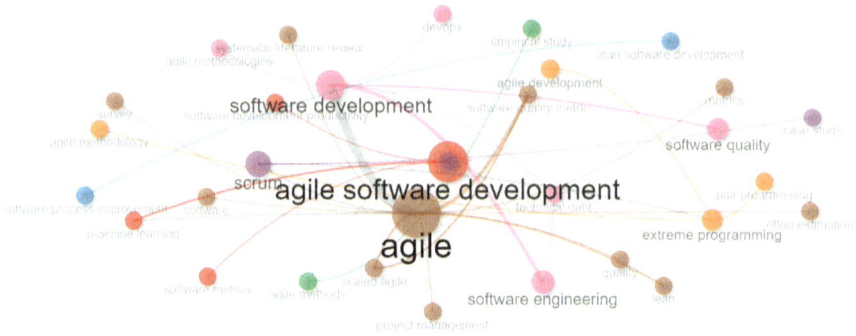

Fig. 3 Conceptual structure based co-occurrence analysis graph

emphases across these groups, with the dimensions likely reflecting varying aspects of software development practices and methodologies. This analysis helps to categorize the literature and reveals gaps or concentrated areas of research within the field.

The Fig. 5 shows that the evolution of key terms in the scope of bibliometric literature review from 2010 and later through. "Agile" and "software development" dominate, highlighting their central role in boosting productivity. From 2016 onward, terms like "DevOps" and "machine learning" have gained prominence, reflecting a shift towards automation and intelligence in development. The ongoing relevance of "software quality" and "effort estimation" emphasizes the focus on optimizing efficiency. In summary, the graph shows that while agile practices remain dominant, newer technologies are increasingly shaping the field.

5 Discussion and Conclusion

In today's competitive business environment, assessing and visualizing productivity in software development is vital. Dashboards for tracking productivity metrics help monitor progress, evaluate performance, and identify challenges. Key factors affecting productivity include developers' emotional states, social dynamics, Agile requirements, and hybrid methodologies. Observability research emphasizes automating testing, using monitoring tools, applying Agile metrics, enforcing quality standards, and reporting waste to highlight issues that reduce productivity. In summary, observability ensures continuous software quality by monitoring key metrics, while the interplay between productivity and observability leads to higher-quality, more secure software.

The bibliometric analysis provided a comprehensive overview of the research landscape surrounding software development productivity and observability. The process began with refining a Scopus query, which initially 271 relevant sources but was later narrowed down to 39 via using PRISMA. These sources spanned various

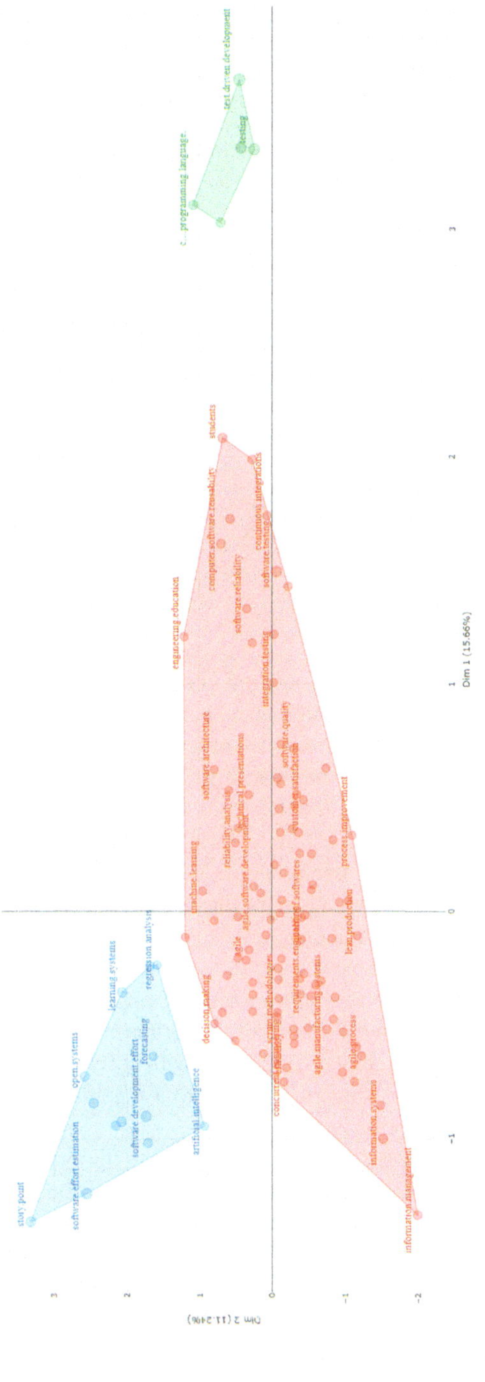

Fig. 4 Clustering based factorial analysis graph

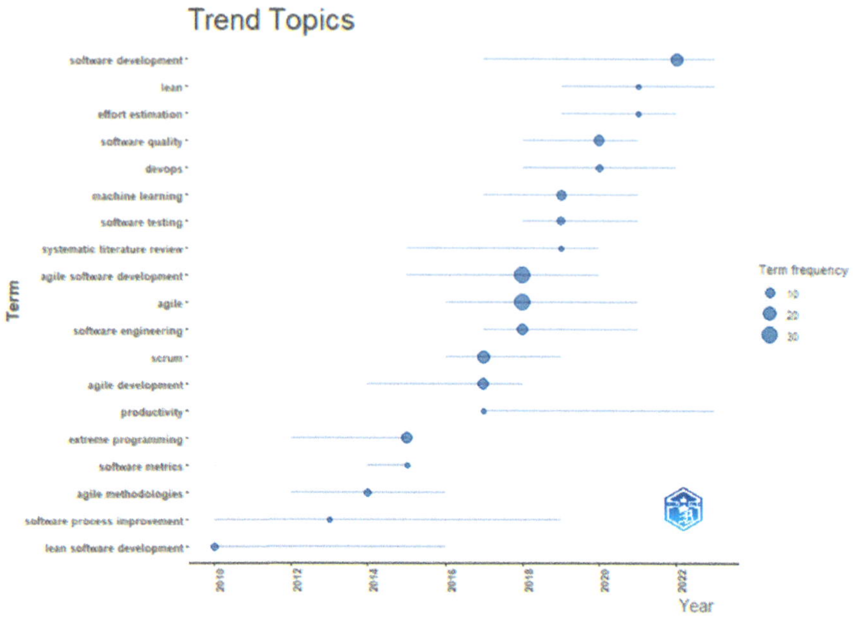

Fig. 5 Trend topics

publication types, with a significant focus on conference papers and articles. The visualizations generated through Bibliometrix, particularly the word cloud and co-occurrence analysis, revealed the prominence of topics such as agile software development, software quality, and machine learning. The co-occurrence analysis highlighted key thematic clusters within the literature. For instance, one cluster centered on software effort estimation and artificial intelligence, emphasizing the role of predictive techniques in software development. Another cluster focused on software architecture and agile methods, underlining the importance of design and process improvement. The third cluster, which included topics like C programming language and test-driven development, suggested a strong focus on programming languages and testing methodologies. The term "productivity" appeared less frequently than anticipated, ranking 20th among the keywords. Furthermore, the analysis showed that "observability" was more closely associated with software quality and metrics rather than being a standalone concept, pointing to a potential gap in the literature regarding its direct link to productivity.

In conclusion, this bibliometric review highlights the importance of productivity and observability in software development. While research on agile methods and software quality is extensive, a gap exists in linking productivity with observability. This gap offers a valuable opportunity for future research to explore how observability practices can directly enhance productivity. The review not only identifies current trends but also suggests the need for further investigation into integrating observability with productivity metrics.

References

1. Dybå T, Dingsøyr T (2008) Empirical studies of agile software development: A systematic review. Inf Softw Technol 50(9–10):833–859
2. Burns B, Grant B, Oppenheimer D, Brewer E, Wilkes J (2018) Borg, Omega, and Kubernetes. Commun ACM 59(5):50–57
3. Aria M, Cuccurullo C (2017) Bibliometrix: An R-tool for comprehensive science mapping analysis. J Informet 11(4):959–975
4. Välimäki A, Vesiluoma S, Koskimies K (2009) Scenario-based assessment of process pattern languages. In: Bomarius F, Oivo M, Jaring P, Abrahamsson P (eds) Product-focused software process improvement. PROFES 2009. Lecture notes in business information processing, vol 32. Springer, Berlin, Heidelberg
5. Moe NB, Dingsøyr T (2008) Scrum and team effectiveness: Theory and practice. In: Abrahamsson P, Baskerville R, Conboy K, Fitzgerald B, Morgan L, Wang X (eds) Agile processes in software engineering and extreme programming. XP 2008. Lecture notes in business information processing, vol 9. Springer, Berlin, Heidelberg
6. Abbas N, Gravell AM, Wills GB (2010) the impact of organization, project and governance variables on software quality and project success. In: 2010 agile conference, Nashville, TN, USA, pp 77–86, https://doi.org/10.1109/AGILE.2010.16
7. Ahmad M, Kanwal S, Habib MA, Faisal CN, Usability and accessibility based quality evaluation of apparel websites in Pakistan (2020) IEEE 23rd International Multitopic Conference (INMIC). Bahawalpur, Pakistan 2020:1–6. https://doi.org/10.1109/INMIC50486.2020.9318163
8. Staron M, Meding W, Hansson J, Höglund C, Niesel K, Bergmann V (2014) Dashboards for continuous monitoring of quality for software product under development. In: Elsevier eBooks, pp 209–229
9. Ortega MGS, Soto GB, Cinco RRP, Rodrɪ́guezFramework to analyze the relationship between agile software engineers' emotions and their productivity L-F (2021) 10th International Conference On Software Process Improvement (CIMPS). Torreón, Coahuila, Mexico 2021:66–69. https://doi.org/10.1109/CIMPS54606.2021.9652780
10. Kuutila M, Mãntylã MV, Claes M (2020) Chat activity is a better predictor than chat sentiment on software developers productivity. In: Proceedings of the IEEE/ACM 42nd international conference on software engineering workshops, pp 553–556
11. Bollati VA, Gaona G, Lima PB, Cuenca Pletsch L (2023) Software development teams: factors influencing their productivity. LACCEI, 1(8)
12. Arcos-Medina G, Mauricio D (2019) Aspects of software quality applied to the process of agile software development: a systematic literature review. Int J Syst Assur Eng Manag Int J Syst Assur Eng Manag 10(5):867–897
13. Jabbar T, Hafeez Y, Kiani AA, Anwar N, Javaid J (2019) Use of knowledge management and SCRUM techniques to increase the reusability in software development. In: 2019 13th international conference on mathematics, actuarial science, computer science and Statistics (MACS). IEEE, pp 1–5
14. Sallin M, Kropp M, Anslow C, Biddle R (2023) Waste self-reporting for software development productivity improvement. In: International conference on agile software development. Springer Nature Switzerland, Cham, pp 50–66
15. Machuca-Villegas L, Gasca-Hurtado GP, Muñoz M (2021) Measures related to social and human factors that influence productivity in software development teams. Int J Inf Syst Proj Manag 9(3):43–67
16. Toker A (2022) Bir Araştırma Metodolojisi Olarak Sistematik Literatür İncelemesi: Meta-Sentez Yöntemi. Anadolu Üniversitesi Sosyal Bilimler Dergisi, 22(Özel Sayı 2):313–340
17. Paul J, Criado AR (2020) The art of writing literature review: What do we know and what do we need to know? Int Bus Rev 29(4):101717
18. Donthu N, Kumar S, Mukherjee D, Pandey N, Lim WM (2021) How to conduct a bibliometric analysis: An overview and guidelines. J Bus Res 133:285–296

19. Zupic I, Čater T (2015) Bibliometric methods in management and organization. Organ Res Methods 18(3):429–472
20. Page MJ, McKenzie JE, Bossuyt PM, Boutron I, Hoffmann TC, Mulrow CD, Moher D (2021) The PRISMA 2020 statement: an updated guideline for reporting systematic reviews. BMJ 372
21. (2024) The Bibliometrix website [Online] https://bibliometrix.org/biblioshiny/biblioshiny1.html

Cycle Time Calculation, Process Analysis and Efficiency Optimization in Semiconductor Technologies

Eyüp Berk Utku, Beyhan Oktay, Dilek Alimli, and Bahadir Tunaboylu

Abstract In semiconductor manufacturing processes, cycle time refers to the time taken to complete a production cycle. This time includes all stages from the beginning until the end of production. Cycle time is important because of its contribution to competitive advantage, cost effectiveness, customer satisfaction, technological development, process improvement and inventory reduction. Cycle time calculation is necessary for identifying bottlenecks, capacity and production planning. Low cycle time values can increase a manufacturer's competitive advantage in the semiconductor industry and allow it to adapt more quickly to ever-changing market conditions. Therefore, it is important to continuously monitor, analyze and improve cycle time values. In this study, the aim is to observe opportunities for improvement and to determine production planning strategies by calculating cycle time in semiconductor IC device manufacturing in a semiconductor research laboratory. For this purpose, the production processes were examined in detail with process videos taken for the SMED study, data kept in run cards, run tracking in the laboratory. The necessary data were collected through different means in order to constitute a framework for data analysis while digitalization is absent. Then, cycle time calculations were made with the collected data. Results were compared with the theoretical times, and the detected problems within the manufacturing process were isolated, inspected and solutions for them were generated.

Keywords Cycle time · Manufacturing time · Lead time · Processing time · Efficiency metrics · Semiconductor · SMED

E. B. Utku · B. Oktay · D. Alimli (✉)
Tubitak Bilgem Uekae Yital, Gebze, Türkiye
e-mail: dilek.alimli@tubitak.gov.tr

E. B. Utku
e-mail: eyubberk@hotmail.com; eyup.utku@tubitak.gov.tr

B. Oktay
e-mail: beyhan.oktay@tubitak.gov.tr; beyhanokty@hotmail.com

B. Tunaboylu
Faculty of Engineering, Marmara University, Istanbul, Türkiye
e-mail: bahadir.tunaboylu@marmara.edu.tr

N. M. Durakbasa and K. G. Gülen (eds.), *Sustainable Green Conversion*, Lecture Notes in Mechanical Engineering, https://doi.org/10.1007/978-3-031-83583-4_29

1 Introduction

In semiconductor manufacturing, the cycle time calculation usually refers to the time from the initial processing of the silicon wafer to the final test process. A typical wafer fabrication process flow may contain 300–500 operation steps over 30–45 days to complete [1]. During production, wafers move through the wafer fab in batches of multiple wafers, so-called lots, and are manufactured layer by layer [2]. In a wafer fabrication process different processes are applied layer by layer on the wafer. Some of these processes can be etching, lithography, implantation, deposition. These processes can be applied once or repeatedly during the manufacturing process. Managing wafer fabrication is a complex process, as the manufacturing process is multi-step, repetitive and involves the repeated use of machines in different processes. The complicated operation requirements (reentrant flow, queuetime limit, batch/splitting operations, etc.) and shop uncertainties (machine breakdown, rework, and random yield) make its production planning and control activities very difficult [1] Within the time calculated with cycle time, there is waiting time in the queue between the processes caused by equipment, waiting time caused by malfunction or maintenance operations, waiting time caused by the operator, process time, test time while the wafers pass from one process to another process. However, due to the assumption of rigid processing times, on which the industrial engineer usually has no effect, the focus of the current research is set on the reduction of waiting time rather than the whole cycle time [3].

After the calculation of cycle time, the aim is to reduce this time without causing a negative situation in the wafer quality. A long cycle time leads to a low number of wafers produced, a long queuing time for wafers in the production process and an increased risk of contamination of wafers in the production process. Consequently intense competition and supply chain management drives have resulted in semiconductor manufacturers to initiate programs to improve their market responsiveness by reducing the cycle time whilst narrowing the cycle time distribution to achieve greater repeatability [4]. Determining the cycle time and reducing it in a systematic way contributes to reducing operator-induced waiting times in the process, increasing the number of wafers produced as the number of wafers entering production will increase, reducing the risk of contamination as the wafers will spend less time in production, and reducing the waiting time in the queue during the production process. Aside from cycle time reduction, being able to accurately predict cycle time has a practical contribution for supporting production-planning decisions (such as scheduling resources and actions for anticipated job completion), planning production-capacity requirements [3].

A. *Literature Review*

There is wide literature on reducing the cycle time of semiconductor manufacturing systems.

Leachman and Ding [5] investigated the impact of excursions not discovered until end-of-line testing on die and revenue losses. They developed analytical models to estimate total die loss over a product's lifetime and integrated it with price decline to estimate revenue loss. Applying these models to excursion data from a 300 mm memory-chip manufacturing fab, they found a $37.1 million reduction in revenue loss over a year due to a decrease in fab cycle time from 60 to 30 days. Their research highlights the significant contribution of cycle time reduction to yield improvement. They focused only on excursions first detected by probe testing, but discussions with fab yield engineers suggest that yield loss also occurs due to excursions detected later in the process flow but before probing. They suggest further analysis of these events to understand their contribution to financial benefits from cycle time reduction.

Chen [6] introduces a systematic procedure designed to plan cycle time reduction strategies, with the goal of enhancing the competitiveness and sustainability of semiconductor manufacturers. Initially, the study identifies controllable factors influencing job cycle time. Subsequently, the relationship between these factors and job cycle time is modeled using a Backpropagation Neural Network (BPN). Based on this model, strategies to reduce job cycle time can be devised, with careful consideration of their feasibility and effectiveness prior to implementation.

Tu and Lu [7] examined whether 25-wafer is an adaptive lot size in the future to fulfill the manufacturing processes. This study introduces a simulation model to examine the correlation between lot size and performance in wafer fabrication. The model incorporates three production performance metrics: cycle time of products, total throughput, and the waiting time of Work-in-Progress (WIP) at workstations. Simulation results indicate that reducing lot size synchronously decreases the cycle time of products. However, the waiting time of WIP does not consistently decrease with lot downsizing.

B. *Objectives*

The objective of this study is to identify opportunities for improvement and develop production planning strategies by calculating and analyzing the cycle time in semiconductor IC device manufacturing within a semiconductor research laboratory that has no data collection framework. This goal is pursued through several specific aims. First, the cycle time of semiconductor IC device production processes will be accurately measured. This involves collecting detailed data through process videos taken for the SMED (Single-Minute Exchange of Die) study, data recorded in run cards, and tracking runs in the laboratory. Precise calculation of cycle times as much as possible is essential for understanding the current production efficiency and identifying bottlenecks.

Second, the study aims to determine the stages in the production process that cause the delays and increase the cycle time. Identifying these bottlenecks allows for accurate and effective improvements that can speed up the manufacturing process greatly.

Third, the cycle time data will be used to review and revise capacity planning and optimize production schedules. This ensures that resources are allocated in a more optimal way, reducing idle times and enhancing overall productivity.

Lastly, the study seeks to develop a framework in order to integrate the potential advancements and new technologies to the current semiconductor manufacturing environment while also ensuring a continuous optimization. By continuously monitoring and improving cycle time, the study supports ongoing technological development and process refinement. The importance of these objectives lies in their potential to significantly enhance the semiconductor manufacturing process. Reduced cycle times can lead to cost effectiveness, better customer satisfaction, and the ability to quickly adapt to market changes. Ultimately, the study aims to provide actionable insights and strategies that lead to more efficient production planning and improved overall performance in the semiconductor industry.

2 Methodology

A. *Data Collection*

In order to collect data for the cycle time study, the steps in photodetector production were observed and the process steps were determined. The data kept during the production process were analysed and it was determined which of them could be used in the cycle time study. Firstly, the usability of the data on the run cards in the cycle time study was evaluated. It was observed that it was suitable but insufficient and not very safe because it was entered manually by the operator. Secondly, the usability of the process videos taken for SMED analysis was evaluated. It was observed that it can be useful for cycle time calculation since it includes both process times and preparation times. Finally, in order to increase the available data and to obtain more reliable data, the processes were monitored in the laboratory with a chronometer, and their durations were noted. The time data obtained from the run cards, the time data obtained by following a run in the laboratory and the time data extracted by watching the videos shot for SMED analysis were collected and combined in an Excel document.

(1) *Run Card Analysis*: A run card typically refers to a structured document or digital tool that outlines the steps and parameters for conducting a specific process. Depending on the area it is used, it may serve as a detailed guide for technicians or researchers to follow during the execution of a particular task within the semiconductor fabrication or it may act as a document to record the process/experiment values.

In case of YİTAL, a physical document was being used at the laboratory as the run card. The main purpose of the run card was to record the parameters that were used in the processes and the resulting values from the measurements.

Before this study, there were no areas for entering time values for each of the processes. And since the machines were analogous devices, there was no way of getting the time values for individual processes or even the whole production time. Due to the necessity of time values and data for the scope of the project, we first started by updating the run card by adding new areas for time entries.

Since the number of processes and workload were high and to keep technician morale high, only 4 time values were demanded for each process: Receive time (after the wafers come to this station, the time that the technician starts to allocate time to this process), process start time, process end time, delivery time (delivering the wafers to the next station). This way we would be able to track the location of the wafers while also getting the primary and main information regarding the procedures that they go through. General understanding of these data were as following:

Receive Time (current process)—Delivery Time (process before) = Waste/Waiting Time.
Process Start Time (current)—Receive Time (current) = Initial-Preparation Time.
Process End Time (current)—Process Start Time (current) = Process Time.
Delivery Time (current)—Process End Time (current) = Post-Preparation Time.

At the end of data collection, all of the time data from the run cards were entered to an excel document (Table 1).

(2) *SMED Proses Videos*: As part of the SMED (Single Minute Exchange of Die) study, video recordings for all of the production processes were recorded using at least 2 cameras from different angles. These cameras had a specifically assigned operator (at least one) during the recording time. The focus of one of the cameras was on wafers and the other one was on technicians, so sometimes the cameras were moved along with the technician and/or wafers. So, all the value adding actions were recorded. Recordings of the processes were taken from several runs.

After all of the recordings were complete, times of the processes were recorded to an excel file manually, with similar categories as the run card.

The columns were; receive time, process start time, process end time, delivery time, and also an additional, specified preparation time, in order to get a more intuitive understanding regarding the data. Same calculations were done in the excel and were stored, ready to be used.

(3) *Process Monitoring with Chronometer*: In order to get accurate estimates and information regarding the processes and also to check the validity of the run card data that is entered by the technicians, a run was closely monitored with

Table 1 Excel table with time data on run cards

RUN CODE	Sequence No	Process Name	Receive Time	Process Start Tin	Process End Tir	Delivery Time	Total Preparation Time
F295	1	CLEANING	08:20:00	08:20:00	10:00:00	-	#DEĞER!
F295	2	SURFACE PASSIVATION	-	11:20:00	12:12:00	12:12:00	#DEĞER!
F295	3	OXIDATION	08:40:00	12:20:00	20:10:00	20:20:00	03:50:00
F295	4	CHARACTERIZATION	11:30:00	11:37:00	14:24:00	14:28:00	00:11:00
F295	5	LITHOGRAPHY	14:29:00	14:29:00	19:25:00	19:26:00	00:01:00
F295	6	ETCHING	19:26:00	19:29:00	20:15:00	20:23:00	00:11:00
F295	7	ETCHING-2	20:23:00	20:24:00	21:49:00	21:49:00	00:01:00
F295	8	OPTICAL CONTROL	21:49:00	21:50:00	22:02:00	22:03:00	00:02:00
F295	9	CLEANING (RCA)	09:00:00	09:00:00	10:10:00	-	#DEĞER!
F295	10	SURFACE PASSIVATION	-	11:05:00	11:42:00	11:43:00	#DEĞER!

the sensitive measurement capabilities of a stopwatch. Since a specific person was assigned to the task, measurements had a much higher accuracy and it was possible to add more flexibility to measurements. So, more timing categories were added.

At the end of data collection, all of the time data from the process monitoring with chronometer were entered to an excel document (Table 2). The columns were: Receive time, process start time, process end time, delivery time, initial-preparation time, post-preparation time and waiting times. The addition of initial and post-preparation times allowed for a more detailed inspection on the processes and the addition of waiting times gave insight regarding machine and operator efficiency.

B. *SMED Analysis*

Our SMED application began with the comprehensive capture of all the processes through video recordings. These recordings enabled us to meticulously document setup actions and identify various forms of waste using an Excel spreadsheet. Through this process, we cataloged the duration of each setup task, forming a detailed analysis and description of the setup procedure.

The identification of actions and wastes was a crucial phase of our analysis. By carefully inspecting the video recordings, we delineated non-value-added activities such as unnecessary motions, waiting times, and material handling inefficiencies. As a result, a report was prepared in order to share these detected wastes to get rid of them one by one.

Time analysis was maybe the most important and relevant part to the study, where accumulated time for each process helped us understand how much time it takes for the completion of a process. It also allowed us to compare this time with the theoretical process times. This comparison provided us with a clear understanding of the efficiency levels of the current procedures and highlighted opportunities for achieving a more optimized production environment.

Finally, comparing observed setup times and actions/procedures with those reported in Standard Operating Procedure (SOP) documents was another critical aspect of our methodology. Actions or time values that did not match with the SOP documents helped us detect potential procedural inefficiencies or deviations from established protocols.

Table 2 Excel table with time data on process monitoring with chronometer

		RECEIVE	PREPARATION TIME	PROCESS		PREPARATION TIME	
Sequence No	Process Name	Receive Time	Initial-Preparation Time	Process Start Time	Process End Time	Post-Preparation Time	Total Preparation Time
0	CLEANING	08:54:00	00:26:00	09:20:00	10:20:00	00:00:00	00:26:00
1	CLEANING-2	08:20:00	00:10:30	08:26:00	10:12:00	00:01:44	00:12:14
2	SURFACE PASSIVATION	12:52:00	00:00:40	12:53:30	13:33:30	00:01:35	00:02:15
3	OXIDATION	13:35:00	00:10:00	13:45:00	21:10:00	00:10:00	00:20:00
3.1	OXIDATION ([illegible])			08:24:00	11:35:00		00:00:00
4	CHARACTERIZATION	08:32:00	00:10:00	08:42:00	09:32:00	00:11:00	00:21:00
5	LITHOGRAPHY	09:45:00	00:12:00	09:57:00	13:00:00	00:04:00	00:16:00
6	ETCHING	14:14:00	00:00:50	14:15:00	14:59:00	00:00:20	00:01:10
7	ETCHING-2	15:08:00		15:08:00	16:28:00	00:00:30	00:00:30
8	OPTICAL CONTROL	16:33:00	00:02:00	16:35:00	16:40:00	00:02:00	00:04:00
9	CLEANING (RCA)	08:17:00	01:30:00	09:47:00	11:00:00	00:03:00	01:33:00
10	SURFACE PASSIVATION	13:26:00	00:01:00	13:27:00	14:03:30	00:01:30	00:02:30

C. *Statistical Analysis*

In order to derive meaning from the data that was gathered through run cards, SMED and by chronometer; certain tests and the application of statistical methods were necessary. In each of the process pages; mean, standard deviation, minimum, maximum, number of data, error margin, upper bound and lower bound values were calculated separately for process times and preparation times. Then the entered time values are copied as their corresponding numerical values and then transformed to the hour format for a more intuitive interpretation. Thus the matrix that was formed had processes on the rows and information categories (process times, preparation times and total time; and each of their statistical calculations) regarding the processes on the columns. Values of the matrix were hour values for the corresponding process data (Table 3).

After the database is formed, it was transferred to Minitab where a number of operations were made. First of all, the data was visualized for each category separately via a time series graph using 3 parameters on the graph: min value, mean value and max value. This graph clearly shows the processes with small and large variations. However, since there were a lot of processes, we eliminated the processes below a certain duration to focus on the most relevant and important processes. Remaining processes were determined to be critical and in need of controlling. The order that they should be prioritized could be decided with either hour difference magnitude between or variation amount. Since some processes were already long and caused problems in scheduling, it was thought that prioritization by hour difference magnitude was a better option (Figs. 1 and 2).

After this was done for all process, preparation and total times, things to examine were now determined and shared with the technical staff.

Finally, a cumulative time series graph was formed in order to get the cycle time. Obviously, this cycle time assumes that there is no waste time between the processes.

Parallel to this study, the data on photodetector runs were cleaned and reorganized in a way to show how many work days were necessary for the completion of a specific run. With these data, mean run completion time was calculated as:

Table 3 Matrix with calculations for process time, preparation time and total time

Process Time					Preparation Time					Total Time				
Min	Max	Mean	Standard Deviation	Upper Limit	Min2	Max2	Mean2	Standard Deviation2	Upper Limit2	Min3	Max3	Mean3	Standard Deviation3	Upper Limit3
00:40:22	01:00:00	00:50:11	00:09:49	02:18:23	00:04:53	00:26:00	00:15:26	00:10:34	01:50:18	00:45:15	01:26:00	01:05:38	00:20:22	04:08:41
01:20:00	01:46:00	01:33:57	00:10:08	00:10:39	00:12:14	00:26:15	00:19:14	00:09:55		01:54:00	01:54:00	01:54:00	00:00:00	
00:29:00	00:55:00	00:41:43	00:09:06	00:51:17	00:00:30	00:02:15	00:01:23	00:00:52		00:39:51	00:43:00	00:41:26	00:01:35	
06:30:00	07:50:00	07:17:30	00:25:13	07:40:50	00:20:00	00:34:00	00:25:37	00:04:30	00:29:47	06:50:00	08:21:00	07:43:51	00:28:41	08:10:22
00:41:00	01:07:00	00:54:32	00:08:39	01:05:17	00:16:00	00:23:00	00:18:43	00:02:36	00:22:51	01:11:00	01:23:00	01:16:08	00:04:20	01:23:01
03:03:00	04:56:00	03:47:49	00:38:22	04:23:18	00:01:00	00:45:00	00:13:58	00:13:50	00:26:45	03:25:00	04:57:00	04:02:39	00:30:50	04:31:10
00:34:00	00:46:00	00:40:41	00:04:07	00:44:29	00:01:10	00:18:00	00:09:44	00:05:13	00:15:13	00:44:00	00:57:00	00:51:10	00:04:01	00:55:23
01:19:00	01:35:00	01:24:14	00:05:40	01:30:11	00:00:30	00:04:14	00:02:11	00:01:31	00:04:35	01:22:00	01:26:00	01:24:10	00:01:30	01:26:33
00:05:00	00:18:20	00:11:43	00:04:12	00:16:08	00:00:00	00:05:00	00:03:03	00:01:49	00:05:19	00:09:00	00:22:37	00:15:07	00:04:21	00:20:32
00:30:00	01:13:00	00:46:43	00:18:01	01:03:22	00:50:45	01:33:00	01:11:53	00:21:07	04:21:41	00:53:45	02:46:00	01:49:53	00:56:08	10:14:08
00:22:00	00:45:00	00:35:32	00:06:45	00:41:47	00:01:27	00:02:30	00:01:59	00:00:31	00:06:42	00:38:40	00:39:00	00:38:50	00:00:10	00:40:20
04:00:00	05:45:00	04:53:12	00:33:34	05:34:53	00:10:00	00:37:00	00:25:30	00:11:23	00:53:46	04:55:00	05:37:00	05:20:30	00:18:17	06:05:[illegible]
00:30:00	01:05:00	00:49:43	00:12:50	01:01:55	00:25:00	00:50:45	00:34:42	00:11:26	01:03:06	00:53:45	01:30:00	01:16:55	00:16:26	01:57:43

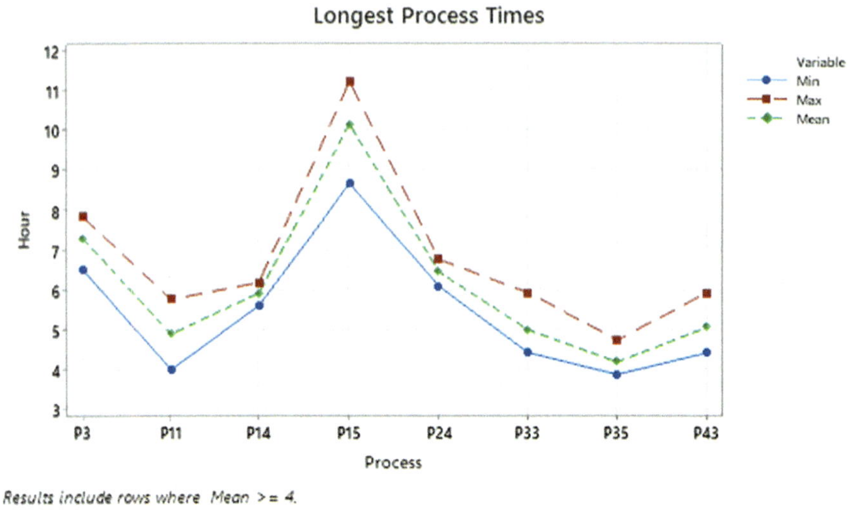

Fig. 1 Time series plot showing the longest process times

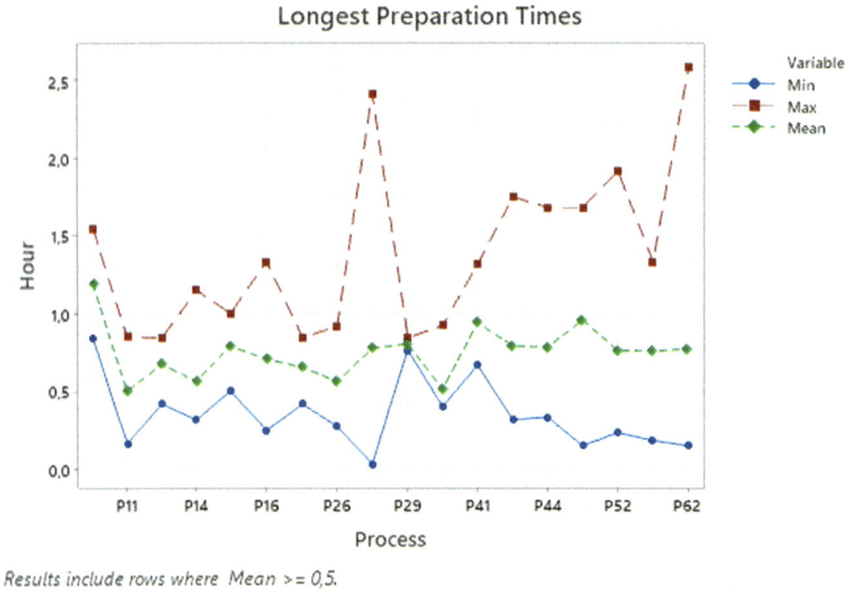

Fig. 2 Time series plot showing the longest preparation times

$$\sum_{i=1}^{r} \frac{d_i}{r}, \text{ for d = run completion time in days,}$$

$$\text{r = number of completed runs, and for i = run number } 1, 2, \ldots 24$$

While this is an important metric for singular analysis, it does not tell us anything about runs being completed in parallel, thus we have missing information about the overall capacity. Cycle time was calculated as:

$$\text{cycle time} = \frac{\text{total production time (days)}}{\text{number of completed runs}}$$

To be able to decrease the cycle time, we had to determine the problems. Before going into the details of each process through SMED recordings, lab inspection etc. blindly, we examined the weekly planning schedule for the processes for different runs. We noted the data of how many days each run was planned to take and how long it actually took. We also were able to detect the delays and their reasons through this schedule and extracted them to an excel document. Different reasons were categorized and how many days they affected production was noted. After the database was ready to use, it was transferred to Minitab to apply statistical tests on the data.

First step was analyzing the differences between the planned run completion times and the real run completion times. Thus we created time series graphs comparing planned and real run completion times to get an intuitive understanding regarding the data and to see possible trends during the months. These differences gave us the amount of waste time and the efficiency of the production as opposed to the previous study where the focus was on the efficiency of individual processes. Individual analysis was also done on both planned and real run completion times. Normality test was done and graphs like box plot were plotted in order to visualize the data (Figs. 3, 4, 5, 6 and 7).

Next step was to analyze the reasons for the delays in the production. Among the categorized causes for delay (process conflicts, machine failure, machine maintenance, and other) a Pareto chart was formed based on their delay amount (days). Then, they were visualized in a stacked bar chart where we are able to see the machine that is responsible for the delay. This is done separately for all of the delay categories for spotting specific problems about each machine. Finally, a Pareto chart for the machines that cause delay was plotted based on how many days they are affecting the production. Most significant machines are noted as bottleneck machines and necessary action plan is set to be initiated (Fig. 8).

D. *Python Code*

In our study, we employed a basic planning methodology based on average process times to effectively allocate resources and schedule tasks within semiconductor manufacturing processes. However, it also played a role in showing the consistency of this study by connecting all the analysis that use different datasets but have similar and supporting outcomes. This approach involved converting fractional numerical

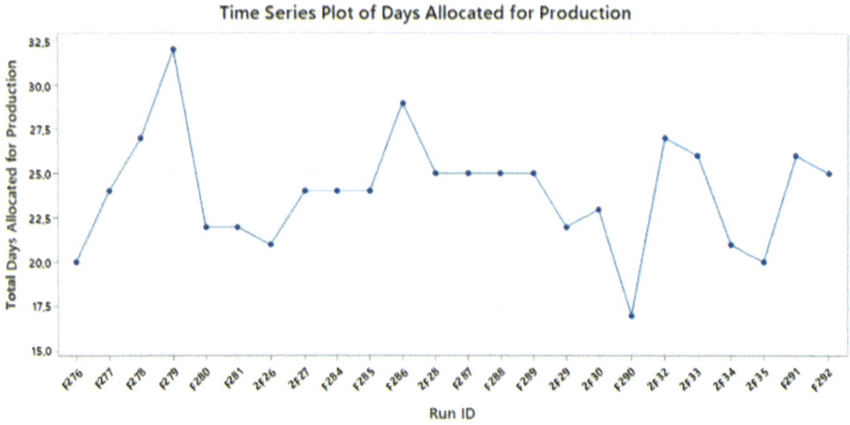

Fig. 3 Time series plot showing total days allocated for production

values representing process durations into a time format and then scheduling the tasks considering the daily working hours.

We used a converter function to display the outcome in a time format. The float_to_time function serves the purpose of converting decimal numbers representing hours into a time format (HH:MM). This function extracts the integer part as hours and calculates the minutes by multiplying the fractional part of the number by 60 (Fig. 9).

We also used a general scheduling function to do the main point of this study. The schedule_processes function is the core component of our planning methodology. It schedules the processes based on their durations, considering the maximum daily working constraint and incorporating buffer time between tasks.

Main parameters that can be adjusted are as follows:

- **process_times**: A list containing the average durations of individual processes.
- **max_hours_per_day**: The maximum allowable working hours per day.
- **buffer_time**: Additional time allocated between consecutive tasks to account for setup or transition periods (Fig. 10).

The function iterates over each process duration in the process_times list. For each process, it checks if there is sufficient time remaining in the day to accommodate the process duration along with the buffer time. If the process can be added to the current day, it is scheduled accordingly, and the timing details are recorded. If the process exceeds the remaining time in the day, the function proceeds to schedule it on the next day, updating the day counter and resetting the time counter. The scheduling continues iteratively until all processes are allocated to respective days (Fig. 11).

Finally, the function returns a schedule dictionary containing the scheduled processes along with their start and end times for each day. Additionally, it includes any remaining time at the end of each day.

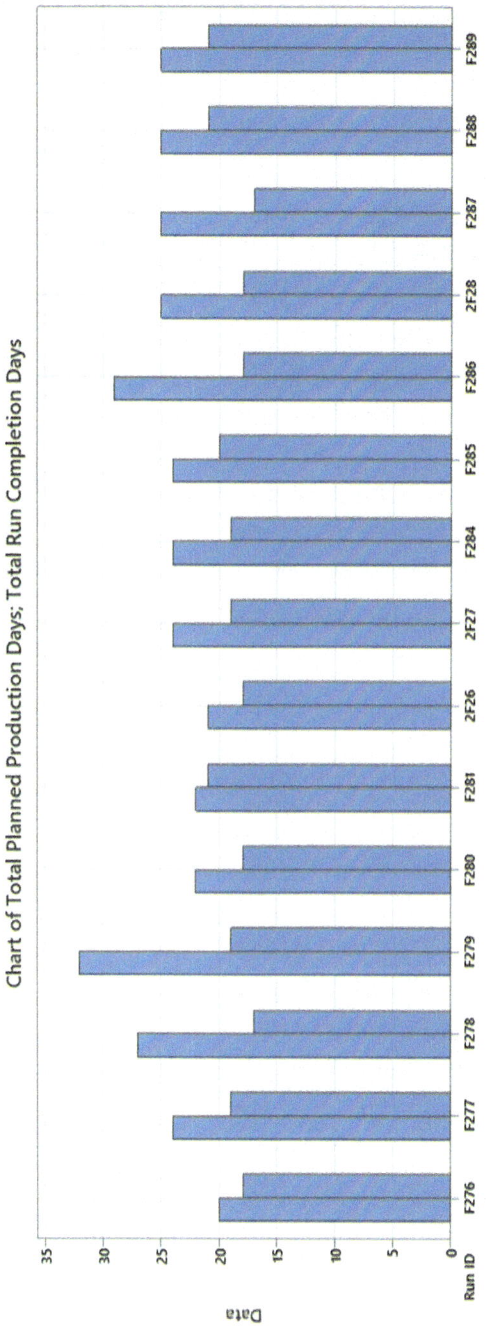

Fig. 4 Chart showing total planned production days and total actual run completion days

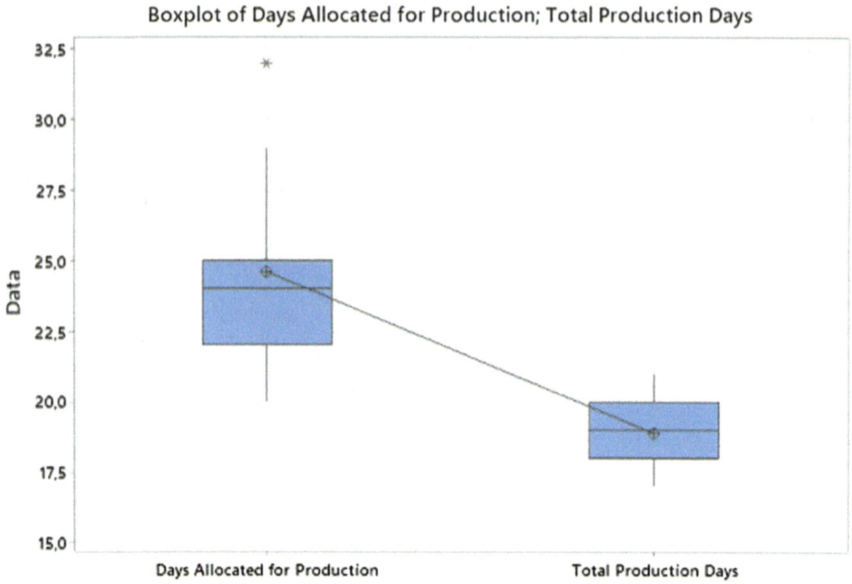

Fig. 5 Boxplot showing days allocated for production and total actual production days

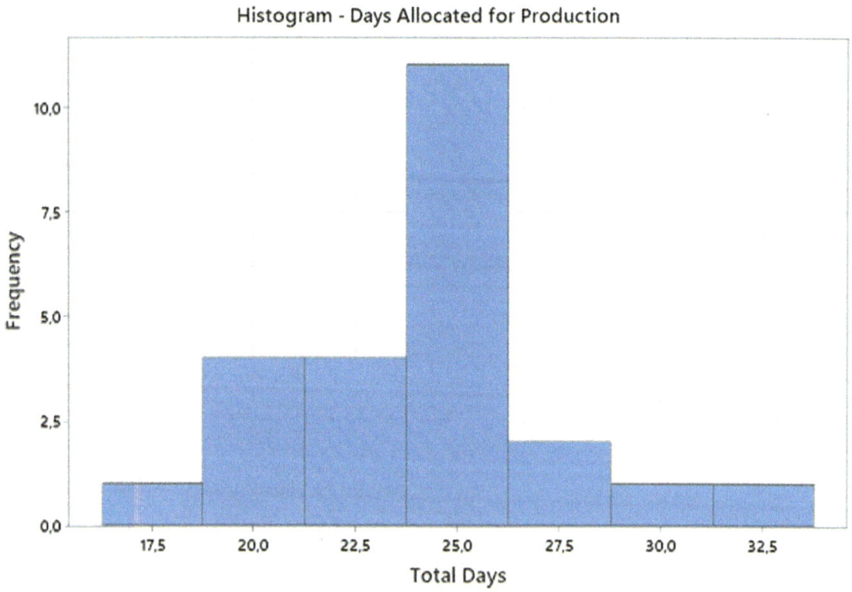

Fig. 6 Histogram showing days allocated for production

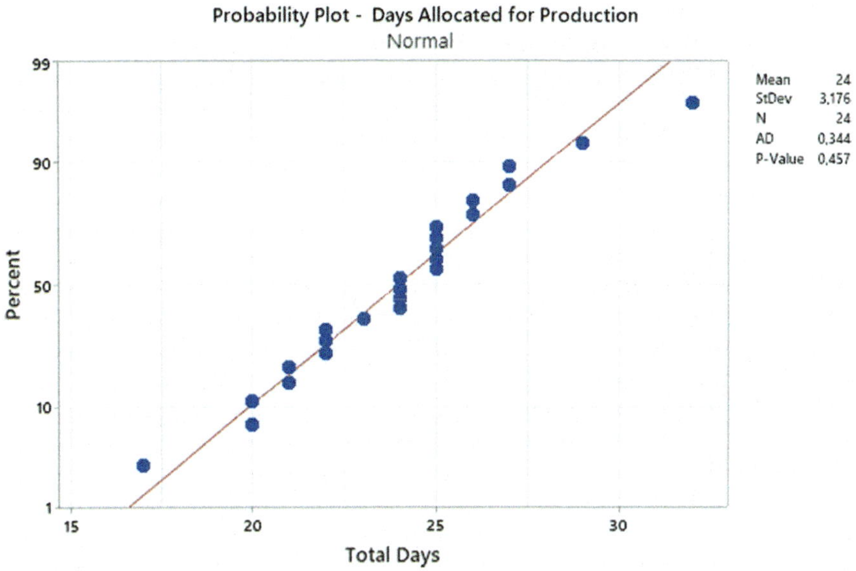

Fig. 7 Probability plot showing days allocated for production

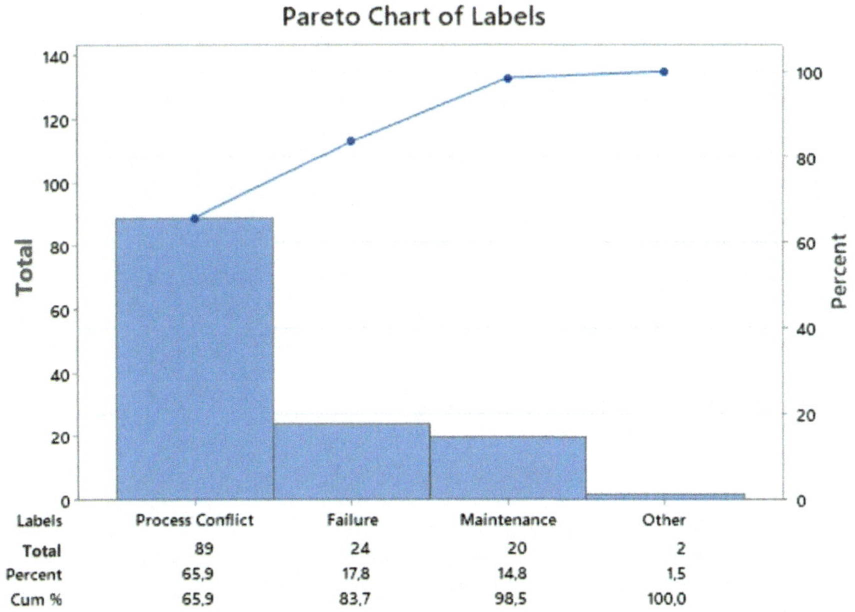

Fig. 8 Pareto chart showing the causes of delay in production

```python
# Converting float values to time format
def float_to_time(hours_float):
    hours = int(hours_float)
    minutes = int((hours_float - hours) * 60)
    return f"{hours:02d}:{minutes:02d}"
```

Fig. 9 Python code for converting float values to time format

```python
# Process times
process_times = [
    1.09, 1.90, 0.69, 7.73, 1.27, 4.04, 0.85, 1.40, 0.25, 1.83, 0.65, 5.34,
    1.28, 0.63, 6.43, 10.95, 2.74, 4.22, 2.12, 1.24, 1.47, 0.24, 0.96, 0.66,
    6.78, 0.13, 1.94, 1.93, 0.15, 1.11, 0.66, 3.76, 2.90, 5.38, 2.37, 4.41,
    1.42, 2.07, 0.41, 1.22, 0.16, 1.29, 0.65, 6.06, 3.76, 3.65, 0.88, 1.69,
    2.71, 0.22, 1.40, 0.68, 3.65, 0.08, 4.12, 0.89, 1.43, 3.30, 0.78, 0.91,
    0.16, 2.26, 3.17
]

# Daily work hours
max_hours_per_day = 12

# Buffer time between each process
buffer_time = 0.5

```

Fig. 10 Process times, daily work hours and buffer time between each process in pyhton

```python
# Function for production planning and scheduling
def schedule_processes(process_times, max_hours_per_day, buffer_time):
    current_hour = 0
    day = 1
    schedule = {}
    unspent_time = 0

    for i, process_time in enumerate(process_times, 1): # Until all proccesses are scheduled
        if current_hour + process_time + buffer_time <= max_hours_per_day:
          # If there is enough time for the process,
          # it is added to the day and hour values are updated
            start_time = current_hour
            end_time = current_hour + process_time
            schedule[f'Day {day}, P{i}'] = (float_to_time(start_time), float_to_time(end_time))
            current_hour = end_time + buffer_time
        elif current_hour > 0:  # If there is not enough time for the process, a new day is started
            unspent_time = max_hours_per_day - current_hour
            schedule[f'Day {day} hours left'] = float_to_time(unspent_time)
            day += 1
            current_hour = 0
            start_time = current_hour
            end_time = current_hour + process_time
            schedule[f'Day {day}, P{i}'] = (float_to_time(start_time), float_to_time(end_time))
            current_hour = end_time + buffer_time

    if current_hour > 0:
        unspent_time = max_hours_per_day - current_hour
        schedule[f'Day {day} hours left'] = float_to_time(unspent_time)

    return schedule

```

Fig. 11 Function for production planning and scheduling in python

3 Results

A. *Cycle Time Based on Past Data*

According to the data gathered by past runs, the average run completion was determined to be 24 days by averaging the days allocated for production. While a low average run completion time is a good indicator for faster processes and a better "in-run planning" it is not a direct indicator of efficiency. This is because in case the number of parallel runs are low, efficiency might drop even if the run completions are fast.

Thus, a general cycle time was calculated to be 4.4 days/run by dividing the total production days by the number of completed runs. This value reflects the overall capacity and speed of the production more accurately.

B. *Process Variability*

When planned production days and actual production days are compared, a significant difference is observed. While the mean completion day of a single run is 24 days in terms of planned production days; when calculated from the actual production days, it becomes 18.375 days. This corresponds to a 5.625 day difference for each run, which also means a 23% increase in the speed of production even if the capacity (number of runs that continue at the same time) remains the same.

C. *Identification of Bottlenecks*

The cumulative values of "how many days of delay was caused by machine x" were plotted as a Pareto Chart, as shown in Fig. 11. The chart indicates that 80% of the identified delays are attributed to four key machines: "Furnace 5," the "LPCVD oxide furnace," "CANON," and the "Ion implantation device." Although the solution for any bottleneck problems in a system may depend on various factors, there are several fundamental methods for addressing the bottlenecks caused by device issues (Fig. 12).

The most obvious solution among these methods is the acquisition of a new device to increase capacity, particularly if the existing device is already being used at (or close to) full efficiency (i.e., operating continuously). The potential revenue generated by the new device is compared to its costs through a return on investment (ROI) calculation to determine its feasibility and value.

If acquiring a new device is not an option, efforts should be made to reduce the cycle time of the process associated with the bottleneck device. If a significant reduction in cycle time can be achieved, multiple operators may be assigned to the process instead of just one. This involves conducting optimality analysis on the process and emphasizing to operators the importance of minimizing errors.

When significant cycle time reductions are not feasible, production planning should be centered around the bottleneck. The overall production speed should be adjusted to align with the bottleneck to avoid idle times, which can negatively impact the cleanliness and quality of photodetectors. In summary, production should be structured around the capacity of the bottleneck to ensure optimal efficiency.

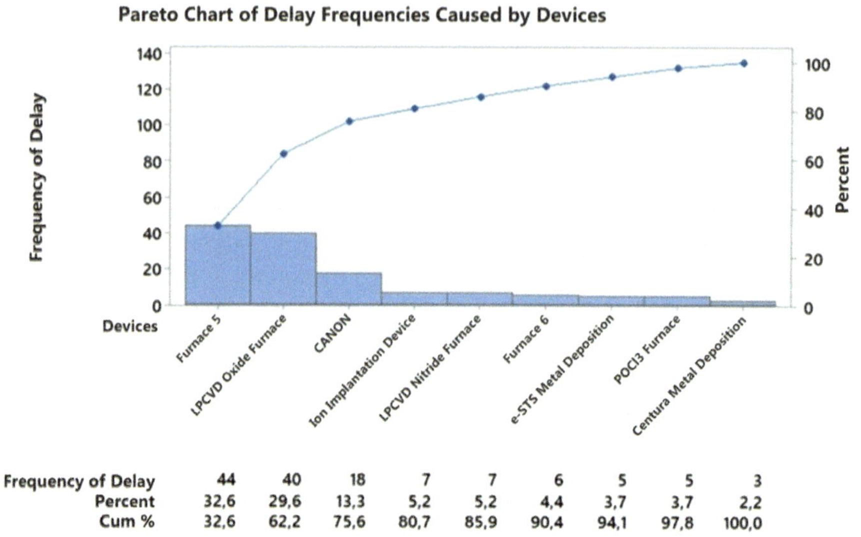

Fig. 12 Pareto chart showing the frequency of delays caused by devices

D. *Theoretical Cycle Time*

By theoretical cycle time we mean the time it takes for the completion of a run based on mean process completion times and no waste time. This can be easily visualized by stacking the process completion times in a cumulative time series graph (Fig. 13).

This value, however, implies a very pure cycle time and does not provide a real cycle time. It can, however, be used as a baseline model to use as a reference.

E. *Production Planning*

Based on the obtained process completion times through the study, a general planning model was created. The purpose of this code based model is to demonstrate the consistency of the work conducted so far and to provide a reference for potential errors in planning. The parameters used are as follows but can be adjusted as needed: process duration, daily working hours, and buffer time.

This approach provides a detailed schedule that includes each process's start and end times for a 12-h workday, including a 30-min buffer time between processes. By adjusting these parameters, the schedule can be optimized to improve production efficiency and fit into specific needs of the production facility. The model also allows for a check mechanism for the current planning schedule.

With the assumed values of average process completion times for processes, 12-h workdays and 15 min buffer times, one run is completed in 16 days; which is very close to the "days worked" data that we obtained from a separate data set: run data.

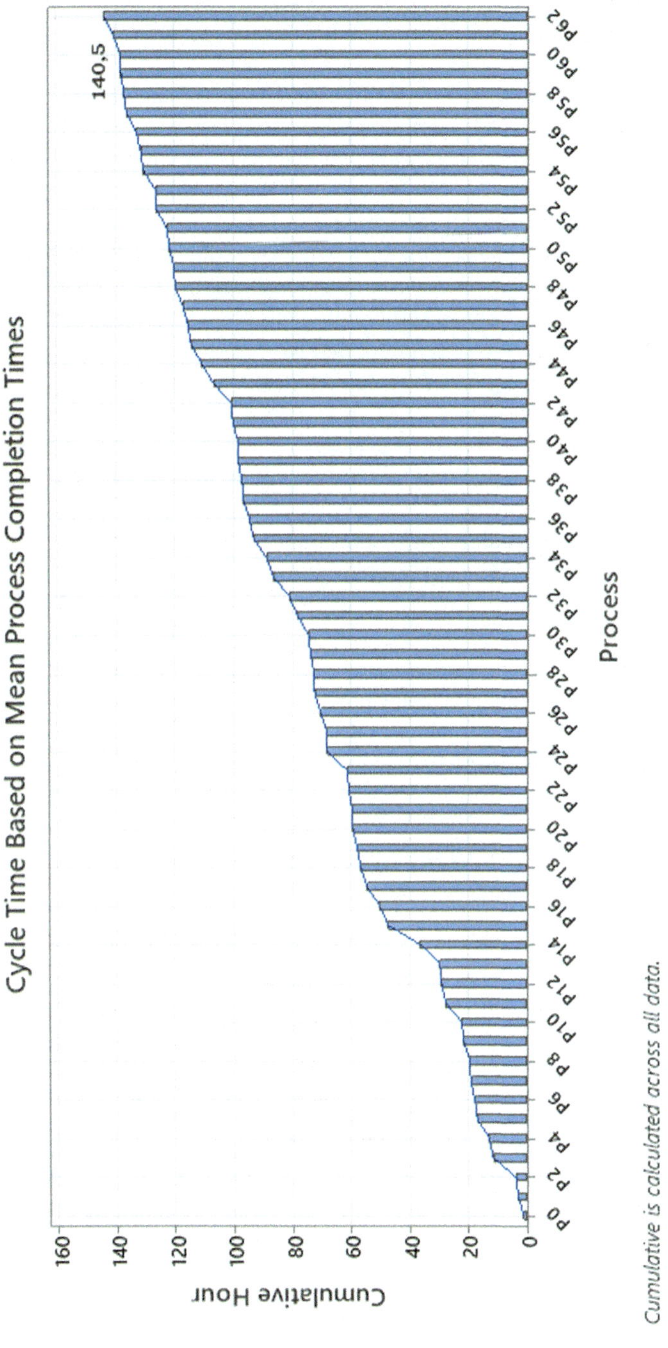

Fig. 13 Cycle time based on mean process completion times

4 Discussion

A. *Implications of Findings*

This study has significant implications for the semiconductor manufacturing process, in terms of production efficiency and time management. The detailed analysis of cycle time data revealed critical bottlenecks and inefficiencies within the production line. By identifying specific processes that cause delays we are able to pinpoint problems in different stages of the process or realize some deviation from the SOP documents.

One major implication is the potential for reduced production cycle times. Shorter cycle times translate to faster production rates, enabling the manufacturer to meet customer demands more quickly and efficiently. This can enhance customer satisfaction and provide a competitive edge in the rapidly evolving semiconductor market. Our study showed that getting rid of the delay times in the production increased the average run completion speed by 23%. This is a great performance increase potential. It provides potential decisions that possibly increase the value and profit of the manufacturer. It should be noted that this does not even include the potential speed increase caused by shortening individual process times.

Moreover, the study's findings show the importance of continuous monitoring and analysis. Regular cycle time control allows for the early detection of emerging bottlenecks and inefficiencies, developing proactive measures to maintain optimal production.

Overall, the study's findings demonstrate that cycle time analysis is critical for optimizing semiconductor manufacturing processes. The improvements in efficiency, reduced production losses, and enhanced capacity planning directly contribute to the overall competitiveness and profitability of the manufacturing operation.

B. *Recommendations for Process Improvements*

This study showed 2 main issues that cause problems in the semiconductor production, namely; bottlenecks and production flow inefficiency.

(1) *Bottlenecks*: One of the primary methods to address bottlenecks is to acquire new equipment to increase capacity, especially if the existing equipment is already operating at full efficiency. A cost–benefit analysis, including a return on investment (ROI) calculation, should be conducted to compare the potential revenue from the new equipment against its cost.

If acquiring new equipment is not feasible, efforts should be made to reduce the cycle time of the bottleneck process. This can involve assigning multiple operators to the task to speed up the process, comparing the current process flow with the SOP documents to detect unnecessary steps or getting rid of wastes through the SMED study to accelerate the process. Emphasizing the importance of the process to operators and ensuring minimal tolerance for errors can also help reduce the specific cycle times of the processes belonging to the bottleneck resources.

If significant cycle time reductions are not achievable as well, production planning should be centered around the bottleneck. The overall production rate should be set according to the capacity of the bottleneck to avoid delays and buffer problems. This approach helps prevent idle times and even a decrease in the quality of photodetectors due to the waiting periods and being exposed to the environment for a longer duration.

(2) *Production Flow Inefficiency*: Implementing flexible working hours can increase the effective working time within a day. By planning consecutive processes in a way that maximizes productivity without overburdening workers, it is possible to fit more processes into a single day. Since some processes are longer, requiring adjustments where one day's work might end early, and another day might have extended hours to balance the workload.

The study found significant delays caused by operators, including breaks and extended process start times. These delays, sometimes exceeding an hour, contribute to process drift. Addressing these issues by improving scheduling, increasing accountability, and reducing non-essential downtime can help minimize delays and enhance overall efficiency.

5 Conclusion

This study calculated and analyzed cycle time in semiconductor manufacturing, both in terms of current production schedule and also in terms of detailed process data, revealing critical inefficiencies and bottlenecks. By calculating cycle times and identifying process delays, we have found areas that significantly impact production efficiency. The findings suggest that reducing cycle times can lead to a 23% increase in production speed, a significant potential for performance improvement.

Main recommendations include obtaining new equipment to alleviate bottlenecks, optimizing the existing process flows, and emphasizing the importance of following the SOPs. Additionally, implementing flexible working hours and minimizing the effects of machine failure and maintenance can further enhance production efficiency.

In conclusion, this study shows the critical role of cycle time analysis in semiconductor manufacturing and provides a framework for ongoing process optimization and technological advancement.

Acknowledgements This research was supported by TUBITAK BILGEM. We appreciate the management of TUBITAK BİLGEM and our colleagues at YITAL for their valuable insights and expertise that significantly contributed to the KARADDA1 project's progress.

References

1. Wang CN, Wang CH (2007) A simulated model for cycle time reduction by acquiring optimal lot size in semiconductor manufacturing. In: Cheng K, Russell DW, Shunmugam MS, Lazoğlu I (eds) The international journal of advanced manufacturing technology, vol 34, pp 1008–1015, Springer Verlag, Berlin
2. Deenen PC, Middelhuis J, Akcay A et al (2024) Data-driven aggregate modeling of a semiconductor wafer fab to predict WIP levels and cycle time distributions. In: Matta A (ed) Flexible services and manufacturing Journal, vol 36, pp 567–596. Springer Verlag, Berlin
3. Meidan Y, Lerner B, Rabinowitz G, Hassoun M (2011) Cycle-time key factor identification and prediction in semiconductor manufacturing using machine learning and data mining. IEEE Trans Semicond Manuf 24(2):237–248
4. Sivakumar AI (2000) Simulation based cause and effect analysis of cycle time distribution in semiconductor backend. In: 2000 winter simulation conference proceedings (Cat. No.00CH37165), vol 2, pp 1464–1471, Orlando
5. Leachman RC, Ding S (2011) Excursion yield loss and cycle time reduction in semiconductor manufacturing. IEEE Trans Autom Sci Eng 8(1):112–117
6. Chen TA (2013) Systematic cycle time reduction procedure for enhancing the competitiveness and sustainability of a semiconductor manufacturer. Sustainability 5:4637–4652
7. Tu YM, Lu CW (2017) The influence of lot size on production performance in wafer fabrication based on simulation. Procedia Engineering 174:135–144

Investigating the Effect of Lean Thinking on Employee Motivation in a Business

Didem Yilmaz, Ebru Adıgüzel, and Duygu Tüylü

Abstract In order to maintain a competitive advantage in the face of rapid market changes and intensifying competition, businesses must prioritize continuous quality improvement. To achieve this, it is essential for businesses to prioritize employee motivation and implement management strategies that enhance this motivation, ultimately leading to enhanced employee efficiency. This study aims to examine the impact of the lean approach on employee motivation. The study was conducted on employees working in different departments within a textile company. The population of the study consisted of approximately 8,000 individuals, and a survey was administered to 100 randomly selected individuals using the convenience sampling method. In the study, two distinct scales were utilized to assess the perceptions of employees regarding lean management and motivation. The scales were selected based on their prior utilization in scientific studies.

Keywords Lean production · Lean management · Lean thinking · Employee motivation · Continuous improvement

1 Introduction

In our increasingly interconnected and smaller world, competition is intensifying at a rapid pace. In addition to this, requests, modifications, and client expectations are also undergoing quick changes. In light of this circumstance, businesses are compelled to engage in novel inquiries in order to endure and uphold their competitive prowess [1]. In today's environment, organizations that are unable to keep pace with

D. Yilmaz (✉) · E. Adıgüzel · D. Tüylü
Department of Industrial Engineering, Istanbul Gelisim University, Istanbul, Turkey
e-mail: dyilmaz@gelisim.edu.tr

E. Adıgüzel
e-mail: ebruadgzl2002@gmail.com

D. Tüylü
e-mail: dtuylu@gelisim.edu.tr

N. M. Durakbasa and K. G. Gülen (eds.), *Sustainable Green Conversion*, Lecture Notes in Mechanical Engineering, https://doi.org/10.1007/978-3-031-83583-4_30

their competitors, meet delivery deadlines, adapt to changes in client demand, and effectively manage their costs are unlikely to survive. To address these challenges, firms must focus on enhancing consumer value, managing competition, adapting to change, optimizing speed, and leveraging information. Lean thinking facilitates businesses in several domains.

Efficient thinking is crucial for organizations operating in a market with significant expenses [2]. The implementation of the lean thinking approach should be integrated throughout all departments of the firm. Lean management strives to consistently enhance all business operations in this procedure [3].

Womak and Jones [4]; have demonstrated the significance of lean thinking in enhancing business processes, boosting employee motivation, and fostering positive employee commitment. Cua et al. [5]; presented an integrative framework for Total Quality Management (TQM), Just-in-Time Manufacturing (JIT), and Total Productive Maintenance (TPM) studies and elucidated the relationship between these techniques and production performance and social and technical orientation practices. Krijnen [6]; conducted a detailed examination of Toyota's quality and efficiency-oriented culture, basing it on 14 management principles. Hines et al. [7]; sought to address the gaps in the existing literature on lean thinking, with the aim of elucidating the evolution and meaning of lean thinking. Furthermore, the objective was to establish a novel framework for future research. Shah and Ward [8]; investigated the impact of lean manufacturing techniques, including just-in- time manufacturing (JIT), total quality management (TQM), total preventive maintenance (TPM), and human resources management (HRM), on operational performance. The results demonstrated that lean techniques exert a positive influence on operational performance, with an estimated impact of 23% study, Bhasis and Burcher [9]; examined lean thinking from a contemporary perspective and concluded that the two most significant challenges encountered in lean thinking applications are a lack of direction and a lack of planning. It was established that the lean thinking journey is a long-term undertaking that progresses with continuous improvement and that five or more techniques should be applied concurrently in this process. Jørgensen and Emmitt [10]; investigated the implementation of lean manufacturing in the construction industry, highlighting the significance of the concepts of 'lean construction' and 'customer value' within this sector. Hasle et al. [11]; investigated the impact of leanness on the work environment, employee health, and employee well-being conducted a comprehensive literature review to examine the impact of leanness on the work environment, employee health, and employee well-being. Bonavia and Marin-Garcia [12]; elucidated the impact of lean production on human resources and the discrepancies in organizational performance pertaining to human resources practices. The study revealed that the businesses that apply lean production most effectively are those that prioritize the training and safety of their employees. Malmbrandt and Åhlström [13]; aimed to develop a tool to evaluate the level of adoption of lean services and to validate this tool experimentally. The tool was developed for use by both researchers and practitioners. Galeazzo et al. [14]; discussed the potential for lean practices and green practices to interact synergistically in order to enhance plant performance. The objective of this study was to ascertain how the two sets of practices interact and to identify the circumstances under which they provide maximum synergy in improving both operational and environmental performance. Danese and Bortolotti [15]; investigated the impact of supply chain models on operational performance. In this process, they obtained significant findings regarding the impact of lean thinking on employee motivation within the supply chain. Bortolotti et al. [16]; revealed that businesses that effectively implement lean production emphasize high collectivism, human-centered management, and future orientation. Furthermore, it was determined that these businesses prioritize practices such as training, customer participation, and continuous improvement. Marodin et al. [17]; the application of lean manufacturing in the Brazilian automotive supply chain was examined, and the relationships between this application and performance were revealed. Furthermore, a survey was conducted to ascertain the impact of 11 lean applications on five operational performance metrics. Netland and Ferdows [18]; presented an S-shaped model in which the change in the performance of a facility using lean applications first slowly improves, then grows rapidly, and finally decreases. Sahoo and Yadav [19]; aimed to evaluate the discovery and analysis of lean applications for small and medium-sized enterprises in underdeveloped countries with the help of statistical tests.

There are multiple methodologies that can be employed to investigate the impact of a lean approach on employee motivation. In this project, the survey method was selected as the primary research tool. Surveys are used to collect information and data from individuals in a variety of ways. Compared to other methods, surveys allow for the collection of more systematic data. The survey study aimed to reach employees with the same or different status in different departments and analyze how the motivation scale of lean management affects them, primarily through the collection of their personal information and responses to a series of questions.

In the analysis and interpretation phase of the research data, the results obtained according to the data analysis made with demographic characteristics were analyzed and interpreted. As a result of this interpretation, it was determined how the employees were distributed according to their classes. The questions asked for lean thinking and motivation were analyzed, and the relationship between them was interpreted. Validity and reliability analyses of the scales were performed. Factor analysis was conducted for the lean management scale, and factor analysis was conducted for the motivation scale. In the phase of hypothesis testing, correlation analyses were employed to ascertain the relationship between lean management and motivation, which is the subject of the research. Interpretations were made regarding the motivation scale by examining these relationships.

2 Lean Management and Motivation

The concept of lean was initially articulated by Krafcik in his master's thesis at the MIT Sloan School of Management. It was subsequently employed by James P. Womack, Daniel T. Jones, and Daniel Ross in delineating the operational tenets and methodologies of Japanese automotive manufacturers [20]. In their book, The Machine That Changed the World, Womack, Jones, and Ross reveal the Toyota Production System, which had been concealed by Japanese manufacturers for an extended period [21].

In his remarks, Taichi Ohno observed that the Toyota Production System did not emerge abruptly but rather evolved gradually over a period of approximately three decades. He also highlighted that the TPS is not solely a production system but also encompasses managerial aspects [22]. The basis of the lean concept is the elimination of waste in order to deliver maximum-quality products and services at minimum cost, quickly, and on time. Lean thinking is a strategy that aims to increase profits by using limited resources while delivering the product to the customer quickly and without interruption [23]. This idea is applied to the whole of production and includes many methods to support lean manufacturing. Lean practices include a fundamental paradigm shift from traditional batch and sequence mass production to single-piece flow-pull production, minimizing time spent on non-value-added operations by locating tools and materials as close as possible to the assembly point [24]. It facilitates the creation of a more efficient work environment by elucidating the issues that arise from prevailing work styles in business contexts. This confers a competitive advantage on businesses.

The concept of lean management can be defined as a management system that has its roots in the principles of lean thinking. Over time, it has evolved and gained prominence in various organizational contexts. The fundamental objective of lean management is to eliminate errors, reduce waste, and minimize costs to achieve optimal efficiency. It emphasizes transparent communication and a flattening of the organizational hierarchy, with the aim of creating value for all units within the system [25]. The objective is to achieve a lean structure with a unified understanding that can be applied across all units in the business. For lean principles to be effectively implemented across the entire organization, it is essential that all employees embrace the concept and work collaboratively. It is anticipated that this collaborative approach will eliminate waste and streamline processes [26]. It is of great consequence to the efficacy of lean management that employees demonstrate a comprehensive understanding and adept application of lean management principles. Ineffective decision-making processes, characterized by slowness and a lack of innovation, may result in missed opportunities for business growth and development. To mitigate these risks, it is essential to implement a system that ensures prompt decision-making, encourages employee accountability, and utilises time units at the highest level. In such cases, the application of the lean thinking system can contribute to the business. The lean system enables the prevention of wasted time and energy, as outlined in reference [27].

It can be demonstrated that high employee motivation has a positive effect on a number of parameters, including business performance, employee satisfaction, service quality, and product quality. The term 'motivation' is defined as a person's understanding of what they need to do at any given time. An employee's desire to enjoy and strive for any task while doing it can be shown as motivation. There are numerous potential sources of motivation for individuals. However, three stand out as particularly salient in the context of business. The first of these is extrinsic motivation, which emphasizes the necessity of undertaking certain actions as a prerequisite for achieving a desired outcome. The second is intrinsic motivation, which can be defined as a form of self-generated motivation or interest. It is a situation in which the process of performing a task is perceived as enjoyable and interesting, and it provides continuity by offering energy. The final plan is relational and altruistic in nature. It originated in response to the necessity for attachment to other individuals and the subsequent impulse to assist them [28].

In the majority of business contexts, employees are expected to adhere to instructions rather than take the initiative to complete tasks independently. This is because they are driven by a desire to perform their roles effectively, which may include seeking promotions, bonuses, or salary increases. Alternatively, they may be motivated by a fear of unemployment in the event of being dismissed. Furthermore, they may find the work itself highly engaging. These and other factors illustrate the importance of motivation in the workplace. In essence, motivation serves as the primary tool that enables employees to perform their roles optimally [29]. The success of a business is contingent upon the effective management of its human resources. Each employee possesses unique attributes and capabilities, and it is therefore essential to conduct thorough research and analysis to gain a comprehensive understanding of these aspects. This understanding should then be translated into optimal working conditions that align with the individual's strengths and needs. The success of employees in their professional roles can be significantly influenced by the provision of conducive working environments and the encouragement of excellence, efficiency, and passion in their work. Motivation plays a pivotal role in enabling employees to consistently demonstrate the highest standards of quality and skill [30].

It is expected that employees will be motivated and that their behaviors will contribute to the goals and objectives of the business. Should any deficiencies be identified in an employee's motivation or behavior, it is the responsibility of the business to intervene and implement corrective measures.

In instances where corrective measures have been implemented for reasons pertaining to an employee's behavior or motivation and the employee in question is unwilling to demonstrate their capacity, such measures must be taken in the form of disciplinary action [31]. In the context of business, motivation cannot be observed directly. Rather, it is a subject that can be addressed through the analysis of employee behavior. In order to ascertain the factors that influence employee motivation in a business setting, it is necessary to interpret the observed behaviors.

3 Research into the Effect of Lean Thinking on Employee Motivation

A. Methodology of the Research

(1) *The Significance of the Research*: In order to gain and retain a competitive advantage in the face of rapid market changes and intense competition, businesses must consistently deliver high-quality products and services. To achieve this, it is essential to focus on employee motivation and utilize effective management models that enhance employee performance and productivity.

(2) *Scope and Limitations of the Research*: The data employed in the project was sourced from a company, and it was hypothesized that as a consequence of the project, there would be an increase in productivity due to the impact of the lean management technique on employee motivation. The population under investigation comprises employees working at the headquarters of a prominent textile company. In the field application, the universe and sample selections were made in accordance with the criteria established in the existing literature. The field application survey was administered to employees in a range of positions, including managers, engineers, workshop workers, service personnel, purchasing personnel, and human resources personnel. The current situation, potential mitigating measures for identified issues, and novel applications will be examined within the context of "A Research on the Effects of a Lean Management Approach on Employee Motivation." In alignment with the study's objectives, hypotheses will be formulated, and statistical analyses will be conducted on pertinent topics.

(3) *Data Collection and Methods Used*: In the phase of the project where decisions are made regarding the method to be used for its realization, a mixed method was employed, combining qualitative and quantitative data collection techniques. In terms of research design, data were collected using a survey method, with data collected by conducting face-to-face surveys with field workers. A survey was conducted on 100 individuals, and the resulting data were subjected to analysis. Following a normality test, it was determined that the distribution was normal, thus enabling the use of parametric tests in the study. The data obtained from the participants was analyzed using the SPSS 24 statistics program. A series of demographic analyses were conducted on the data collected from the participants. The Cronbach Alpha (α) reliability values were calculated, and a factor analysis was performed. Additionally, an ANOVA, a Pearson correlation analysis, T-test according to gender, and a cross-table analysis between variables were conducted to test the hypotheses.

(4) *Assumptions*: It was hypothesized that the data obtained from the scales used in the study would provide scientific findings that would answer the questions put forward by the hypotheses. It was assumed that the responses

Table 1 Reliability analysis

Scale	Number of items	Cronbach' alpha
Lean management	14	0.798
Motivation	10	0.775

provided by the participants to the questions posed on the scales would be impartial and objective.

(5) *Reliability Analysis*: In the study, two distinct scales were employed to assess lean management and motivation. A series of reliability analyses were conducted on the aforementioned scales, the results of which demonstrated that the scales in question exhibited an appropriate level of reliability. A preliminary reliability analysis is a method used to ascertain the consistency rate of the applied survey (Table 1).

The Cronbach alpha coefficient was calculated for the survey questions used in the research, yielding a value of 0.798 for lean management and 0.775 for motivation. These results indicate that the survey questions are reliable.

(6) *Factor Analysis of the Scales Used in the Research*: The principal objective of factor analysis is to streamline the interpretation of complex and extensive analysis results by combining questions with similar answers, thereby facilitating the achievement of results in a more efficient and condensed manner. This approach enables the effective utilization of time and labor resources. Furthermore, the reduction of complex questions facilitates the attainment of results that may otherwise be overlooked or challenging to calculate. In essence, factor analysis provides a unifying framework for questions that are analogous in purpose.

Factor Analysis for the Lean Management Scale (1st Scale)

The Bartlett Test was employed for the factor analysis of the lean management scale, and the resulting table is presented (Table 2).

The Kaiser–Meyer–Olkin value was calculated as 0.775, indicating that the relationship between the survey questions employed in the study is satisfactory. This result demonstrates that the strength of the relationship between the questions is sufficient.

Table 2 Bartlett Test table (1st scale)

Bartlett's test statistic	Kaiser-Meyer-Olkin value	Chi-square statistics	Degree of freedom	Sig.	Conclusion
	0.775	299,811	91	0.001	Applying factor analysis makes sense

Factor Analysis for Motivation Scale (2nd Scale)

The Bartlett Test was employed to ascertain the significance of factor analysis for the motivation scale, and the resulting table value is presented (Table 3).

As indicated in the table, the Kaiser–Meyer–Olkin value was determined to be 0.593. This value is within the acceptable range, indicating that the correlation between the survey questions utilized in the study is satisfactory.

(7) *Hypotheses of the Research:* The primary hypotheses and sub-hypotheses pertaining to the variables under investigation in the context of this research are as follows:

H1: The adoption of the lean management method varies by gender.

H2: The value of motivation varies by gender.

H3: The value of motivation increases as the lean management method is adopted.

H4: The adoption of the lean management method varies by education level.

H5: The value of motivation varies by the education level of individuals.

H6: The value of motivation varies by the number of working hours.

H7: The adoption of the lean management method varies by the number of working hours.

H8: The value of motivation and the adoption of the lean management method vary by age groups.

B. Testing Hypotheses

(1) *Correlation Analysis*: Correlation is a statistical method used to ascertain the strength and direction of the relationship between variables, irrespective of whether they are dependent or independent [32]. The results of the correlation analysis between the dependent and independent variables are presented in the Table 4.

A positive correlation was identified between lean management and motivational factors at a statistically significant level of 0.01. Upon examination of the Pearson coefficient ($r = 0.972$), it was evident that this relationship was positive and of a high quality.

Table 3 Bartlett Test table (2nd scale)

Bartlett's test statistic	Kaiser-Meyer-Olkin value	Chi-square statistics	Degree of freedom	Sig.	Conclusion
	0.593	126,082	45	0.001	Applying factor analysis makes sense

Table 4 Correlation analysis on the relationship between lean management and motivation

		Lean management	Motivation
Lean management	Pearson coefficient	1	0.972**
	P		< 0.01
	N	190	189
Motivation	Pearson coefficient	0.972**	1
	P	< 0.01	
	N	189	195

**The correlation is significant at the 0.01 level

(2) *Variance Analysis (ANOVA)*: One-way analysis of variance (ANOVA) is a statistical technique used to ascertain whether there are significant differences between groups. The primary objective of ANOVA is to evaluate the extent to which an independent variable exerts a notable influence on a dependent variable, by comparing the within-group variance with the between-group variance. The within-group variance represents the extent of variation within each group, whereas the between-group variance quantifies the magnitude of mean differences between groups. This provides a robust method for evaluating the impact of a specific independent variable on the dependent variable. Given that the data set comprised more than two independent groups, a one- way analysis of variance (ANOVA) was conducted, with each factor grouped separately. The analysis was performed on the total data set (Table 5).

In the Levene test, the p-values for both lean management (0.355) and motivation (0.108) were found to be greater than 0.05 and 0.5, respectively. This indicates that the equality of variance between the two groups was accepted (Table 6).

In the ANOVA table, the F-value for lean management is 1.370, while the p-value is 0.251, which is greater than 0.05. When we examine the ANOVA table for motivation, the F-value is 1.201, and the p-value is 0.316, which is also greater than 0.05. The ANOVA analysis indicates that there is no statistically significant difference in employee motivation according to demographic characteristics.

Table 5 Test of homogeneity of group variances regarding demographic data of employees

	Levene statistic	df1	df2	Significance (Sigma)
Lean management	1.114	4	89	0.355
Motivation	1.954	4	94	0.108

Table 6 Anova analysis on demographic data of employees

ANOVA						
		Sum of squares	df	Average of squares	F	Significance
Lean Management	Intergroup	131,678	4	32,920	1.370	0.251
	Intragroup	2139,258	89	24,037		
	Total	2270,936	93			
Motivation	Intergroup	22,440	4	5,610	1.201	0.316
	Intragroup	439,217	94	4,673		
	Total	461,657	98			

4 Conclusions

The present study, entitled "Researching the Effect of Lean Management on Employee Motivation in a Business," sought to examine the relationship between lean management practices and employee motivation. To this end, a series of hypotheses were formulated, and the requisite scales were derived from previous studies in the literature.

The research is descriptive and employs a survey method. Given that employee motivation levels directly influence work quality, productivity and performance, the study was conducted in a prominent textile company. A survey was administered to 100 randomly selected employees representing various positions, and the data was subsequently analysed.

The data was analyzed using the SPSS 24 program, which included frequency analysis, factor analysis, reliability analysis, correlation analysis, t-tests by gender, variance analysis (ANOVA), and crosstabs by gender. The results of the frequency analysis indicated that 47% of the survey participants were female, 35% were between 28 and 37 years of age, 47% were faculty or college graduates, and 23% had been employed in the business for 7–9 years. The Cronbach's alpha value for the lean management scale was found to be 0.798, while the motivation scale yielded a value of 0.775. These values indicate that the scales are reliable.

The analysis results demonstrate that the company is engaged in a continuous improvement process that aligns with the principles of lean management. The implementation of incremental enhancements in key areas such as the work environment, wage and reward policies, and business processes has positioned the company to better embrace the tenets of lean thinking. The application of lean thinking within a business enterprise facilitates the continuous enhancement of business processes and the elevation of quality standards. This process is reinforced by the implementation of continuous kaizen (continuous improvement studies) within the business, with employee involvement serving as a pivotal driver in this regard.

Consequently, this study corroborates the impact of lean management practices on employee motivation, substantiated by empirical evidence, and furnishes a pivotal reference for business enterprises. It has been demonstrated that the implementation of lean management practices within a business context can lead to enhanced efficiency in business processes, increased employee motivation, and an improvement in overall performance.

The results demonstrate that the implementation of the lean management philosophy in business settings leads to enhanced employee satisfaction and increased business competitiveness.

References

1. Özçelik F (2013) Yalın Performans Ölçüleri ve Yalın Üretim Uygulayan İşletmelerin Muhasebe Bölümlerinde Bir Araştırma. Uluslararası İktisadi ve İdari İncelemeler Dergisi 10:103–126
2. Herzog NV, Tonchia S (2014) An instrument for measuring the degree of lean implementation in manufacturing. Strojniški vestnik-Journal of Mechanical Engineering 60(12):797–803
3. Kılıç B, Eleren A (2009) Turizm sektöründe hizmet kalitesi ölçümü üzerine bir literatür araştirması. Uluslararası Alanya İşletme Fakültesi Dergisi 1(1):91–118
4. Womak JP, Jones DT (1996) Lean thinking: Banish waste and create wealth in your corporation. Simon and Schuster, New York, NY, p 350
5. Cua KO, McKone KE, Schroeder RG (2001) Relationships between implementation of TQM, JIT, and TPM and manufacturing performance. J Oper Manag 19(6):675–694
6. Krijnen A (2007) The Toyota way: 14 management principles from the world's greatest manufacturer
7. Hines P, Holweg M, Rich N (2004) Learning to evolve: a review of contemporary lean thinking. Int J Oper Prod Manag 24(10):994–1011
8. Shah R, Ward PT (2003) Lean manufacturing: context, practice bundles, and performance. J Oper Manag 21(2):129–149
9. Bhasin S, Burcher P (2006) Lean viewed as a philosophy. J Manuf Technol Manag, 17(1), 56–72. Jørgensen B, Emmitt S (2008) Lost in transition: the transfer of lean manufacturing to construction. Eng., Constr. Arch. Manag., 15(4), 383–398
10. Jørgensen B, Emmitt S (2008) Lost in transition: the transfer of lean manufacturing to construction. Eng Constr Archit Manag 15(4):383–398
11. Hasle P, Bojesen A, Jensen PL, Bramming P (2012) Lean and the working environment: a review of the literature. Int J Oper Prod Manag 32(7):829–849
12. Bonavia T, Marin-Garcia JA (2011) Integrating human resource management into lean production and their impact on organizational performance. Int J Manpow 32(8):923–938
13. Malmbrandt M, Åhlström P (2013) An instrument for assessing lean service adoption. Int J Oper Prod Manag 33(9):1131–1165
14. Galeazzo A, Furlan A, Vinelli A (2014) Lean and green in action: interdependencies and performance of pollution prevention projects. J Clean Prod 85:191–200
15. Danese P, Bortolotti T (2014) Supply chain integration patterns and operational performance: a plant-level survey-based analysis. Int J Prod Res 52(23):7062–7083
16. Bortolotti T, Boscari S, Danese P (2015) Successful lean implementation: Organizational culture and soft lean practices. Int J Prod Econ 160:182–201
17. Marodin GA, Frank AG, Tortorella GL, Fetterman DC (2019) Lean production and operational performance in the Brazilian automotive supply chain. Total Qual Manag Bus Excell 30(3–4):370–385
18. Netland TH, Ferdows K (2016) The S-curve effect of lean implementation. Prod Oper Manag 25(6):1106–1120
19. Sahoo S, Yadav S (2018) Total quality management in Indian manufacturing SMEs. Procedia Manufacturing 21:541–548

20. Womack JP, Jones DT (1997) Lean thinking—banish waste and create wealth in your corporation. J Oper Res Soc 48(11):1148–1148
21. Womack JP, Jones D, Ross D (1992) "Dünyayı Değiştiren Makine", Osman Kobak(çev.) Otomotiv Sanayi Derneği Yayınları, İstanbul
22. Bhasin S, Burcher P (2006) Lean viewed as a philosophy. J Manuf Technol Manag 17(1):56–72
23. Botti L, Mora C, Regattieri A (2017) Integrating ergonomics and lean manufacturing principles in a hybrid assembly line. Comput Ind Eng 111:481–491
24. Hong P, Ga (Mark) Yang M, Dobrzykowski D (2014) Strategic customer service orientation, lean manufacturing practices and performance outcomes: An empirical study. J Serv Manag, 25(5):699–723
25. Gök MŞ, Arıcı T (2016) Yalın yönetim sistemlerinde alternatif yaklaşım: dinamik kalite yönetim sistemi. Kocaeli Üniversitesi Sos Bilim Derg 31:135–143
26. Şeker A (2016) Yalın Üretim Sisteminde Kanban, Tek Parça Akışı ve U Tipi Yerleştirme Sistemleri. Akademik Sosyal Bilimler Araştırmaları Dergisi. 50:449–470
27. Katayama H, Bennett D (1996) Lean production in a changing competitive world: A Japanese perspective. Int J Oper Prod Manag 16(2):8–23
28. Amabile T, Kramer S (2011) The progress principle: Using small wins to ignite joy, engagement, and creativity at work. Harvard Business Press
29. Vandeveer RC, Menefee ML (2006) Human behavior in organizations. Pearson Prentice Hall. New Jersey
30. Karakaya A, Alper Ay F (2007) Çalışanlarin Motivasyonunu Etkileyen Faktörler: Sağlik Çalışanlarına Yönelik Bir Araştırma. Cumhur Üniversitesi Fen-Edeb Fakültesi Sos Bilim Derg, 31(1)
31. Broadwell M (2009) Employee discipline. Hum Resour Man J, (19), (2009)
32. Sipahi B, Yurtkoru ES, Çinko M (2010) Sosyal Bilimlerde Spss'le Veri Analizi. Beta Basım Yayım, İstanbul

Quality Management

An Application on Supporting Information Security Management with RPA

Sinem Büyüksaatçı Kiriş and Yeşim Altay

Abstract While the rapid advancement of information technologies has improved data accessibility, it has also introduced significant challenges in privacy and security. The risks of data breaches and cyberattacks have increased due to the growing usage of personal computers and the internet. This has particularly affected organizations, leading them to adopt continuously updated security measures and standards to protect information privacy and ensure robust security. To advance information security management in organizations, the ISO/IEC 27,001 standard was established and has since gained worldwide adoption. This study investigates how Robotic Process Automation (RPA) can significantly improve the effectiveness of Information Security Management through a practical application implemented within a company operating in the textile sector. By integrating RPA, the company has automated the email encryption process, ensuring that daily reports related to emails are securely directed only to the intended recipients. With RPA integration, the encryption time for 30 reports has been reduced from 36 h to just 90 min (1.5 h). As a result, multiple staff previously dedicated to encryption have been reassigned to other tasks. Additionally, a security vulnerability in a person-managed encryption process has been eliminated.

Keywords RPA · Robotic process automation · ISO 27001 · Information security · Management

S. B. Kiriş (✉) · Y. Altay
Industrial Engineering, Istanbul University-Cerrahpasa, Istanbul, Turkey
e-mail: s.buyuksaatcikiris@iuc.edu.tr

Y. Altay
e-mail: yesim.altay@iuc.edu.tr

N. M. Durakbasa and K. G. Gülen (eds.), *Sustainable Green Conversion*, Lecture Notes in Mechanical Engineering, https://doi.org/10.1007/978-3-031-83583-4_31

1 Introduction

Information Technology (IT) has become a cornerstone of contemporary business operations. In the current IT-driven landscape, information is not only a crucial corporate asset but also the lifeblood of organizational activities [1]. The transmission of information between various entities constitutes a fundamental aspect of daily business operations. These information exchange processes are essential for supporting and enhancing company functions. Consequently, the significance of information security has escalated [2]. Unauthorized users, including hackers, phishers, social engineers, as well as malicious software such as viruses and worms can infiltrate corporate communication channels through various network technologies, including intranets, extranets, and the Internet. Additionally, internal threats posed by employees, staff, management, or on-site contractors who exploit vulnerabilities within systems, processes, and applications further exacerbate security concerns and present substantial risks to business continuity and integrity [3, 4, 5]. Recent research has indicated that the total losses resulting from a single security incident can be substantial. These losses encompass not only direct financial impacts, such as operational disruptions or adverse stock market reactions, but also reputational damage and business interruption [6]. To enhance the development of information security management within organizations, the ISO/IEC 27,001 standard was created and has since been adopted globally [7].

Robotic Process Automation (RPA) emerged in the early 2000s as a technological innovation designed to automate repetitive, rule-based tasks traditionally performed by humans [8]. RPA is designed to handle low-value, error-prone, and repetitive tasks within businesses through the use of robotic software, rather than human operators. This approach aims to reduce process times and enable employees to concentrate on higher-value tasks. Additionally, it enhances process efficiency by minimizing human errors through the precise operation of machines. RPA is currently utilized across a wide range of industries, including telecommunications, information technology, healthcare, insurance, financial services, and human resources, among others [9].

In terms of information security, RPA helps mitigate vulnerabilities by transitioning password management and other critical functions from human control to system control, thereby improving security measures such as encryption.

This study will elucidate the contribution of Robotic Process Automation (RPA) to ISO/IEC 27,001 Information Security Management Standard through a practical application implemented within a company operating in the textile sector. The remainder of this paper is structured as follows: The second section starts with robotic process automation. This section explains what RPA is, includes its advantages, details its elements, and indicates its effectiveness. The third section provides the theoretical background for ISO/IEC 27,001: Information Security Management Standard. The fourth section contains the details of the application. The final section provides a summary of the study and offers recommendations for potential future research directions.

2 Robotic Process Automation

The term "robotic process automation" (RPA) refers to the application of specific technological and methodological approaches designed to facilitate the manipulation of existing software applications, enterprise resource planning (ERP) systems, claims applications, databases, and learning management systems in a manner analogous to human processing of transactions or completion of processes [10]. Essentially, automation is achieved through the deployment of software robots or artificial intelligence (AI) agents capable of performing repetitive tasks with high accuracy [11]. RPA operates at the user interface level of software tools, automating mouse and keyboard interactions to alleviate repetitive, labour-intensive tasks [12].

RPA is particularly effective for automating back-office processes, including those within finance, procurement, supply chain management, production, accounting, and human resources. These departments typically handle numerous time-consuming tasks that are well-suited for RPA implementation [13, 14]. Tasks such as opening emails, copying and pasting data, completing forms, verifying invoice amounts, and ensuring accurate execution of tasks by accessing only authorized applications can be efficiently managed by RPA, thus replicating functions traditionally performed by human employees [11]. Consequently, RPA mitigates the incidence of human errors caused by inattention or fatigue [12].

RPA represents a non-traditional form of automation. A key advantage of RPA is that it does not necessitate system overhauls or modifications to existing IT infrastructure. It operates in an application-agnostic manner, meaning it can interact with various applications without requiring significant changes to them. Additionally, RPA functions with minimal intervention. The integration of RPA into existing systems is relatively straightforward, facilitating its deployment and enhancing operational efficiency without disrupting established infrastructure [14, 15].

RPA is fundamentally comprised of three elements: developer tools, a robot controller, and software robots. Developer tools are utilized to define and structure tasks that replicate human operations within a business process. These tasks include decision-making instructions, screen and data capturing, folder monitoring, command execution, application navigation, email management, text editing, variable computation, and user-defined events. The robot controller is crucial for maintaining a version-controlled repository of defined jobs, securely storing execution contexts such as access levels and passwords, managing roles and permissions for human employees, and overseeing the assignment and execution of jobs by software robots. Software robots, as the operational units of automation, execute specified transactions and interact with business applications throughout the automated process, collectively forming a robotic workforce that operates independently of user accounts and security levels, thus enhancing both security and confidentiality [16].

RPA has the potential to enhance the efficiency of decision-making processes within the control function by enabling the rapid analysis of numerous scenarios within a relatively short time frame. The advanced forecasting capabilities of RPA facilitate both long-term and short-term planning with increased accuracy. By leveraging RPA tools, businesses can significantly improve the efficiency of their processes and the effectiveness of the services they offer. The substitution of human labour with RPA results in a notable reduction in both the cost and processing time associated with high-frequency tasks. Specifically, the operational cost of RPA software is approximately one-ninth of the expense incurred from hiring a human employee. Additionally, RPA "robots" are capable of functioning continuously, 24 h a day, 7 days a week, thereby enhancing productivity and operational efficiency [17]. By automating repetitive and mundane tasks through RPA, human resources are liberated to focus on more value-added activities. These activities encompass personal interaction, problem-solving, and decision-making, thereby enhancing the overall strategic contribution of human capital within the organization [18].

RPA is recognized for its significant role in mitigating common operational errors, including incorrect data entry, skipped steps, and rule application mistakes. Tasks automated through RPA are anticipated to achieve near-total accuracy, minimizing the likelihood of human error. Additionally, the continuous operation of software robots, functioning 24/7, ensures a high level of reliability and uninterrupted service delivery [18].

3 ISO/IEC 27,001: Information Security Management Standard

In the early stages of computing, the absence of shared databases and memory in single-processor mainframe systems meant that relatively simple technical and physical mechanisms were adequate to secure the computing environment. However, the advent of personal computers and multiprocessor systems necessitated the development of more sophisticated technical security measures for communication channels. These measures now include user identification and authentication, memory sanitation, and data access control. As technology has evolved, so have the approaches to communication security, which is now categorized into four distinct domains: network security, information security, infrastructure security, and cryptography-digital forensics. Each domain addresses specific aspects of securing information and systems in an increasingly complex technological landscape.

Information Security involves the protection of information from unauthorized access, use, disclosure, disruption, modification, inspection, recording, or destruction. Entities such as government agencies, military organizations, corporations, financial institutions, hospitals, and private enterprises accumulate substantial amounts of sensitive information concerning their employees, clients, products, research, and financial status. A significant portion of this data is now collected, processed, and stored in digital formats and is transmitted across networks to various other systems. Safeguarding this information is crucial to maintaining its confidentiality, integrity, and availability in an increasingly interconnected digital environment.

The ISO/IEC 27,001 standard, first published in 2005 under the title "Information technology—Security techniques—Information security management systems—Requirements," provides a comprehensive framework for establishing, implementing, maintaining, and improving an Information Security Management System (ISMS). This standard is designed to be adaptable for organizations of various sizes and across diverse sectors. While ISO/IEC 27,001 outlines the essential requirements for an ISMS, it does not prescribe specific measures; instead, it requires that businesses develop and implement tailored security practices to address their unique needs.

The ISO/IEC 27,001 emphasizes the importance of planning, implementing, operating, and continually monitoring and improving a process-oriented ISMS. This approach aligns with the Plan-Do-Check-Act (PDCA) cycle, a model for continuous improvement. According to the PDCA cycle, the management system must be regularly monitored and reviewed to assess the effectiveness of risk management actions. Should these actions prove insufficient, the implementation of enhanced practices is necessary. Supporting the implementation of ISO/IEC 27,001 is ISO/IEC 27,002, which provides detailed guidance on applying the information security controls specified in ISO/IEC 27,001. ISO/IEC 27,002 serves as a complementary application guide, offering practical advice on the selection and application of security controls [2, 3, 19].

The ISO/IEC 27,001 international standard enables organizations to establish a systematic security procedure designed to elevate security to a predefined level. Adopting this standard offers several advantages [19, 20, 21, 22, 23, 24, 25].

- Demonstration of Security to Stakeholders: Certification under ISO/IEC 27,001 provides proof of robust security measures to internal and external parties, such as employees, suppliers, financial institutions, customers, partners, and regulatory bodies. This can enhance customer trust, attract new clients, and strengthen existing relationships.
- Competitive Advantage: Achieving certification delivers a "certified quality" endorsement from an independent authority, which can differentiate an organization in the marketplace and provide a competitive edge.
- Enhanced Process Efficiency: Implementing ISO/IEC 27,001 can lead to greater efficiency in information management processes by streamlining and standardizing security practices.
- Cost Reduction: Transparent and optimized structures resulting from ISO/IEC 27,001 compliance can lead to reduced operational costs by minimizing redundancies and inefficiencies.
- Integration of Security into Business Processes: The standard facilitates the integration of security as a fundamental component of business operations, ensuring that security considerations are embedded in organizational practices.

- Improved Risk Information and Monitoring: ISO/IEC 27,001 provides a framework for the continuous monitoring of IT risks, allowing organizations to stay informed about potential threats and adapt their security measures accordingly.
- Compliance with Legal and Regulatory Requirements: The standard ensures adherence to national and international regulations, as well as other applicable laws, thereby mitigating the risk of legal penalties associated with the unauthorized disclosure of confidential information.

4 Literature Review

The growing prevalence of articles encompassing case applications and methodologies related to robotic process automation has consequently underscored the necessity for comprehensive review studies in the literature that address this topic. Ivančić et al. [22], Wewerka and Reichert [23], Ribeiro et al. [24], da Silva Costa et al. [25] and Chugh et al. [26] serve as notable examples of such studies. These reviews have examined a wide array of topics, ranging from prevalent implementation approaches of successful RPA cases to the observed benefits, the challenges that organizations frequently encounter, the characteristics that render certain processes more amenable to RPA, and the research gaps identified in the current literature.

When examining studies at the intersection of security and robotic process automation (RPA), it becomes evident that the predominant applications are primarily within the banking sector. Kumar and Balaramachandran [27] aimed to identify the key factors that influence the adoption of RPA within the retail banking industry, with the aim of enhancing customer experience and satisfaction. The findings of the study indicate that factors such as security, privacy, reliability, and usability are deemed significant by customers in facilitating the advancement of RPA within the banking sector. Villar and Khan [28] presented a descriptive case study on Deutsche Bank that meticulously details the implementation of RPA in conjunction with artificial intelligence (AI). In their studies, they explored information security governance as a separate and notable topic. Thekkethil et al. [29] noted that major financial institutions are increasingly eager to implement RPA technologies, which not only aid in mitigating potential fraud—one of the primary security concerns—but also help reduce other human errors. In this context, they discussed how RPA can effectively diminish fraud risks through various methods, including the re-evaluation of existing processes, the elimination of human errors, advanced trade monitoring, automated threat detection, anomaly detection, and several additional strategies. Wojciechowska-Filipek [30] implemented RPA to optimize the processing of bank secret disclosures in a bank operating in Poland. Following the implementation of RPA, the investigation process was streamlined from seven stages to three. Additionally, the time required to complete specific tasks within the process was reduced from 16 days to just 3 days, highlighting the efficiency gains achieved through RPA.

To the best of our knowledge, this is the first instance in which RPA has been integrated to ensure information security in the manner demonstrated in this case.

5 Application

This study was conducted at the request of a company with over 60 years of experience in the textile sector. The company has sustained its long-term presence in the industry by integrating its diverse production capabilities with a workforce of over 400 employees. It serves 87 countries, offering a comprehensive range of products, including polyester buttons, corozo, horn, bone, and wooden buckles and buttons, as well as a broad assortment of textile labels and accessories.

The company, which emphasizes customer focus and sustainability, has demonstrated the quality of its products and operations through the acquisition of the Oeko-Tex Standard and ISO/IEC 9001:2015 certifications. Additionally, the company has adopted an environmentally sustainable approach, earning the ISO/IEC 14,001 certification by implementing innovative production techniques that minimize water and energy waste and eliminate the use of harmful chemicals. Committed to maintaining high standards in quality management, the company is also planning to pursue initiatives related to ISO/IEC 27,001 certification. Therefore, it is essential to take steps regarding information security.

In the company, daily reports pertaining to various departments are disseminated to managers via email. However, unauthorized access to these reports, particularly by former employees who may still have access through other companies, poses a significant threat to information security. To address this issue, manual encryption methods have been considered. Monthly password assignments have been implemented as a means of securing report access. Nonetheless, this approach still involves human intervention. Furthermore, the implementation of RPA could expedite this process, reducing it from 36 h (1.5 business days) for 30 different reports to a faster, automated procedure, thereby minimizing human involvement and enhancing security.

In this context, the process for encrypting existing reports using the Robusta RPA product, developed by Robusta, is structured simply as follows shown in Fig. 1. Initially, the RPA software robot checks for any unread emails and identifies those containing reports. If such emails are found, the robot accesses them and determines the recipients. Subsequently, it encrypts the reports by integrating passwords generated by the robot with its own identity information, and then sends the encrypted reports to the relevant recipients' email addresses. This process ensures that reports are not sent directly from the system to individuals; rather, the RPA system manages control and transmission through the robot, thereby enhancing security and oversight.

There are several compelling reasons to utilize Robusta RPA for this application. First, the Robusta RPA platform is designed to be user-friendly, allowing process designers to work without requiring coding expertise. The Design Studio within Robusta operates independently of element positioning and screen size, ensuring seamless functionality across various screen resolutions.

Robusta RPA is built on a BPM-based infrastructure, which provides access to a BPM platform. Processes designed in Robusta Design Studio can be exported and saved as PDF or Word documents, effectively documenting the most current version

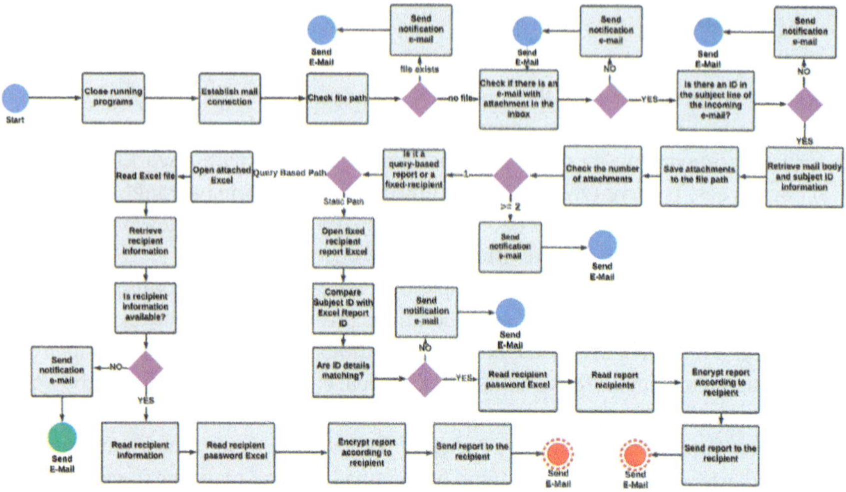

Fig. 1 Process flow design performed on the RPA BPM tool

of these processes. This feature facilitates ease of understanding for new employees, who can learn the details of the processes through these comprehensive documents [31].

The BPM-based framework of Robusta offers several advantages, one of which is the rule engine. This component separates business rules from workflows, enabling centralized management of these rules. As a result, the specific business rules that govern the robot's operations are clearly visible and transparent.

Figure 2 shows the interface of the Robusta RPA program as an example. The flow in Fig. 2 is the program-embedded version of the process flow detailed in Fig. 1. For this reason, the figure has not been shared in a very readable and detailed way. The main purpose here is to show the interface of the program.

As mentioned above, the manual encryption process for 30 different reports currently requires 36 h (1.5 business days). Since the relevant reports need to be sent daily, there is also a necessitate for the use of multiple staff. As a result of the work performed, the integration of RPA has reduced the encryption time to 90 min (1.5 h). This significant time reduction not only enhances operational efficiency but also allows staff to be redeployed to more strategic tasks, thereby optimizing their utilization across the organization. Additionally, the issue of security vulnerabilities, which posed a significant challenge for the organization, was addressed by completely restricting unauthorized external access to emails, facilitated by the implementation of a 256-bit encryption process integrated with RPA. The 256-bit encryption is essentially unbreakable, establishing it as the strongest industry standard currently available on the market.

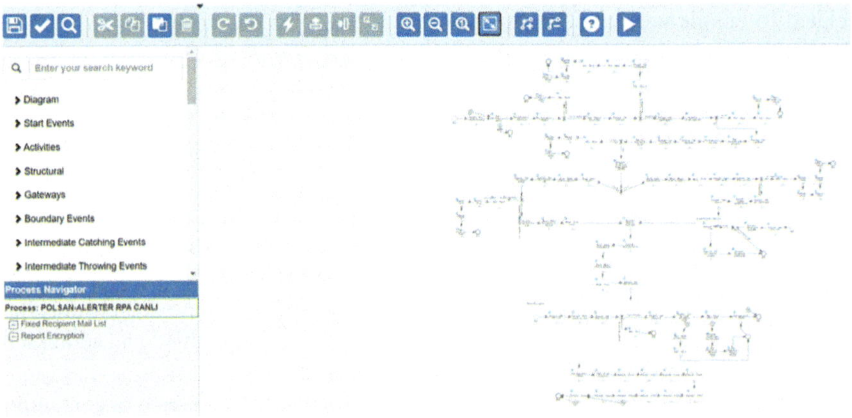

Fig. 2 Robusta RPA interface

6 Conclusions

The rapid advancement of information technologies has significantly enhanced the accessibility and dissemination of information, but it has also introduced complex challenges related to information privacy and security. This evolution, which began with the ground-breaking advent of early computers, accelerated with the widespread adoption of personal computers and the internet, thus amplifying the risks associated with data breaches and cyberattacks. This escalation in risk has become a major concern, particularly for organizations. In this context, the implementation of continuously updated security measures, robust encryption techniques, and adherence to regulatory frameworks are essential for safeguarding information privacy and ensuring comprehensive information security.

ISO/IEC 27,001 is an international standard for information security management systems (ISMS) that provides a comprehensive framework for ensuring information confidentiality and security. This standard outlines essential processes and control measures for identifying, managing, and mitigating information security risks. By establishing guidelines for risk assessment and management, security policies, access control, and continuous monitoring, ISO/IEC 27,001 supports organizations in protecting their information assets and preventing data breaches. Consequently, ISO/IEC 27,001 serves as an effective tool in addressing information privacy and security issues, contributing to the adherence of information security management systems to global standards.

This study explores the role of Robotic Process Automation (RPA) in enhancing ISO/IEC 27,001 Information Security Management. It addresses a security vulnerability identified in the email system of a textile company operating in Istanbul. The existing encryption process, managed manually by a staff, is both time-consuming and lacks complete confidentiality and reliability. There is still a systemic gap. By integrating RPA, the company has automated the process, ensuring that daily reports related to emails are securely directed only to the intended recipients. Moreover, the RPA implementation has accelerated the encryption process through the use of robots.

In order to ensure that the security breaches are completely eliminated, the company has requested that the robot change the password five minutes after the email is read, as a future work of the study. Consequently, the relevant user will need to request the password from the system again if she/he wants to access to the email and related report at any time. It was also requested that comprehensive and systematic event logs would be meticulously maintained for RPA processes, thereby enabling the precise tracking of bot activities. This proactive approach ensures that any anomalies or potential security breaches can be swiftly identified and addressed, thereby enhancing the overall security posture of the system.

In the scenario of this study, the company has successfully adapted to contemporary technological advancements by integrating Robotic Process Automation (RPA) to enhance its encryption processes. Following this successful implementation, the company is now exploring additional opportunities to leverage RPA across various other areas within the organization. This forward-thinking approach aims to further streamline operations, improve efficiency, and address any remaining systemic gaps by applying RPA to other repetitive or time-consuming tasks.

References

1. Von Solms R (1999) Information security management: why standards are important. Inf Manag Comput Secur 7(1):50–58
2. Disterer G (2013) ISO/IEC 27000, 27001 and 27002 for information security management. J Inf Secur 4(2):92–100
3. Humphreys E (2008) Information security management standards: Compliance, governance and risk management. Inf Secur Tech Rep 13(4):247–255
4. Topa I, Karyda M (2019) From theory to practice: guidelines for enhancing information security management. Inf & Comput Secur 27(3):326–342
5. Susanto H, Almunawar MN (2018) Information security management systems: a novel framework and software as a tool for compliance with information security standard. Apple Academic Press
6. Hsu C, Wang T, Lu A (2016) The impact of ISO 27001 certification on firm performance. In: 2016 49th Hawaii International Conference on System Sciences (HICSS). IEEE, pp 4842–4848
7. Mirtsch M, Kinne J, Blind K (2020) Exploring the adoption of the international information security management system standard ISO/IEC 27001: a web mining-based analysis. IEEE Trans Eng Manage 68(1):87–100
8. Osman CC (2019) Robotic process automation: Lessons learned from case studies. Inform Econ, 23(4)
9. Madakam S, Holmukhe RM, Jaiswal DK (2019) The future digital work force: robotic process automation (RPA). JISTEM-J Inf Syst Technol Manag 16:e201916001
10. Sutherland C (2013) Framing a constitution for robotistan. Hfs Research, Ottobre
11. Asquith A, Horsman G (2019) Let the robots do it!–Taking a look at robotic process automation and its potential application in digital forensics. Forensic Sci Int: Rep 1:100007
12. Chakraborti T, Isahagian V, Khalaf R, Khazaeni Y, Muthusamy V, Rizk Y, Unuvar M (2020) From robotic process automation to intelligent process automation: Emerging trends. In: Business process management: Blockchain and robotic process automation forum: BPM 2020

Blockchain and RPA Forum, Seville, Spain, September 13–18, 2020, Proceedings 18. Springer International Publishing, pp 215–228

13. Kaya CT, Türkyılmaz M, Birol B (2019) Impact of RPA technologies on accounting systems. Muhasebe Ve Finans Derg, 82
14. Axmann B, Harmoko H (2020) Robotic process automation: An overview and comparison to other technology in industry 4.0. In: 2020 10th International Conference on Advanced Computer Information Technologies (ACIT). IEEE, pp 559–562
15. Plattfaut R, Borghoff V, Godefroid M, Koch J, Trampler M, Coners A (2022) The critical success factors for robotic process automation. Comput Ind 138:103646
16. Zaharia-Rădulescu AM, Pricop CL, Shuleski D, Ioan AC (2017) RPA and the future of workforce. In: Proceedings of the 11th international management conference. The role of management in the economic paradigm of the XXIst century, vol 11, no 1, pp 384–392
17. Huang F, Vasarhelyi MA (2019) Applying robotic process automation (RPA) in auditing: A framework. Int J Account Inf Syst 35:100433
18. Syed R, Suriadi S, Adams M, Bandara W, Leemans SJ, Ouyang SJ, Reijers HA (2020) Robotic process automation: contemporary themes and challenges. Comput Ind, 115:103162
19. Humphreys T (2006) State-of-the-art information security management systems with ISO/IEC 27001: 2005. ISO Manag Syst 6(1):15–18
20. Culot G, Nassimbeni G, Podrecca M, Sartor M (2021) The ISO/IEC 27001 information security management standard: literature review and theory-based research agenda. The TQM Journal 33(7):76–105
21. Kitsios F, Chatzidimitriou E, Kamariotou M (2023) The ISO/IEC 27001 information security management standard: how to extract value from data in the IT sector. Sustainability 15(7):5828
22. Ivančić L, Suša Vugec D, Bosilj Vukšić V (2019) Robotic process automation: systematic literature review. In: Business Process Management: Blockchain and Central and Eastern Europe Forum: BPM 2019 Blockchain and CEE Forum, Vienna, Austria, September 1–6, 2019, Proceedings 17. Springer International Publishing, pp 280–295
23. Wewerka J, Reichert M (2020) Robotic process automation—A systematic literature review and assessment framework. arXiv preprint arXiv:2012.11951
24. Ribeiro J, Lima R, Eckhardt T, Paiva S (2021) Robotic process automation and artificial intelligence in industry 4.0–a literature review. Procedia Computer Science 181:51–58
25. da Silva Costa DA, São Mamede H, Mira da Silva M (2022) Robotic process automation (RPA) adoption: A systematic literature review. Eng Manag Prod & Serv, 14(2):1–12
26. Chugh R, Macht S, Hossain R (2022) Robotic process automation: A review of organizational grey literature. Int J Inf Syst Proj Manag 10(1):5–26
27. Kumar KN, Balaramachandran PR (2018) Robotic process automation-a study of the impact on customer experience in retail banking industry. J Internet Bank Commer 23(3):1–27
28. Villar AS, Khan N (2021) Robotic process automation in banking industry: A case study on Deutsche Bank. J Bank Financ Technol 5(1):71–86
29. Thekkethil MS, Shukla VK, Beena F, Chopra A (2021) Robotic process automation in banking and finance sector for loan processing and fraud detection. In: 2021 9th international conference on reliability, infocom technologies and optimization (trends and future directions) (ICRITO). IEEE, pp 1–6
30. Wojciechowska-Filipek S (2019) Automation of the process of handling enquiries concerning information constituting a bank secret. Banks Bank Syst 14(3):175–186
31. (2024) Robusta Cognitive Automation website [Online]. Available https://robusta.ai/bpm-features/

Application of the Servqual Model in the E-cosmetics Sector

Nilüfer Köse, Belkis İclal Zeren, Paşa Çiçeklidağ, Selman Süha Gürbüzer, and Merve Sahin

Abstract This study utilizes the SERVQUAL model to conduct a comprehensive analysis of the impact of service quality on customer satisfaction and loyalty within Türkiye's rapidly expanding e-cosmetics sector. As the industry experiences substantial growth due to digital transformation and changing consumer behaviors, the importance of high service quality in maintaining customer loyalty has never been more critical. This research collected data from 95 participants through meticulously designed online surveys conducted between May and June 2024, employing an adapted SERVQUAL survey to assess service quality across five key dimensions: Tangibles, Reliability, Responsiveness, Assurance, and Empathy. Utilizing advanced statistical methods with SmartPLS 4.0, the study explores both the direct and indirect effects of these dimensions on customer satisfaction (CS) and customer loyalty (CL). The findings reveal that while tangibles, reliability, and assurance have strong positive impacts on customer satisfaction, other factors such as product quality, brand image, and user interface design also significantly affect customer satisfaction and loyalty. This study provides actionable insights into how e-cosmetics companies can enhance their service quality to better meet consumer expectations, thereby fostering sustained customer satisfaction and loyalty.

N. Köse (✉) · B. İ. Zeren · S. S. Gürbüzer
Department of Industrial Engineering, Istanbul Sabahattin Zaim University, Istanbul, Turkey
e-mail: 030621017@std.izu.edu.tr

B. İ. Zeren
e-mail: 030621009@std.izu.edu.tr

S. S. Gürbüzer
e-mail: s.suhagurbuzer@gmail.com

P. Çiçeklidağ · S. S. Gürbüzer
Computer Science and Software Engineer, The University of Western Australia, Perth, Australia
e-mail: pasa.ciceklidag@research.uwa.edu.au

M. Sahin
Department of Management, Ibn Haldun University, Istanbul, Turkey
e-mail: merve.sahin@ihu.edu.tr

N. M. Durakbasa and K. G. Gülen (eds.), *Sustainable Green Conversion*, Lecture Notes in Mechanical Engineering, https://doi.org/10.1007/978-3-031-83583-4_32

Keywords SERVQUAL · E-cosmetics sector · Customer satisfaction · Customer loyalty · Service quality

1 Introduction

The e-commerce landscape has undergone a profound transformation over the past decade, with sectors such as e-cosmetics experiencing unprecedented growth. Driven by rapid advancements in digital technology, evolving consumer preferences, and the widespread adoption of mobile internet, the global e-commerce market reached an estimated $4.28 trillion in 2021, with projections suggesting continued growth at a compound annual growth rate (CAGR) of 6.1% through 2026 [1].

In Türkiye, the e-commerce market has mirrored this global trend, growing by 24% from 2016 to 2023 and reaching a market volume of 785 billion TL by the end of 2023, as reported by TÜBİSAD [2]. This expansion is largely attributed to increased consumer confidence in online shopping, advancements in payment security, and the convenience of accessing a wide range of products from home. Within this broader context, the e-cosmetics sector has emerged as a particularly dynamic and competitive niche. The convenience of purchasing cosmetics online, combined with the ability to compare products and read reviews, has led to a significant shift in consumer buying behavior. Moreover, the rise of social media influencers and beauty bloggers has further fueled the growth of the e-cosmetics market, as consumers increasingly rely on digital platforms for product recommendations and tutorials. As a result, the market is characterized by heightened competition, with brands striving to differentiate themselves through superior service quality.

Understanding and improving service quality in this context is crucial for e-cosmetics companies aiming to build and maintain customer loyalty. Service quality in e-commerce settings, particularly within the e-cosmetics sector, extends beyond the traditional dimensions of tangibles, reliability, responsiveness, assurance, and empathy as outlined in the SERVQUAL model. It also includes factors such as website functionality, user friend design, customer reviews, ease of navigation, product information accuracy, and the efficiency of customer service channels. For example, studies have shown that the reliability of delivery services and the availability of detailed product descriptions significantly influence customer satisfaction and repeat purchase intentions [3, 4].

Furthermore, the application of SERVQUAL in online settings has shown that reliability andresponsiveness are particularly critical factors influencing customer satisfaction. According to Lee and Lee [5], these dimensions of service quality play a key role in shaping customer perceptions in the e-commerce space, where instant responses and consistent reliability are highly valued [5].

2 Literature Review

The SERVQUAL model, initially developed by Parasuraman et al. [6], has been extensively utilized across various industries to assess service quality and its impact on customer satisfaction and loyalty [6]. This model evaluates service quality based on five key dimensions—Reliability, Responsiveness, Assurance, Empathy, and Tangibles—each of which represents a different aspect of the service delivery process. Over the past few decades, the SERVQUAL model has been adapted to fit the unique needs of different industries, including healthcare, hospitality, retail, and more recently, e-commerce [7].

A. *Application of SERVQUAL in Different Sectors*

Recent studies have demonstrated the versatility of the SERVQUAL model across different sectors. For instance, in the retail industry, the model has been used to assess the gap between customer expectations and actual service delivery, helping businesses identify critical areas for improvement [3]. In the healthcare sector, the SERVQUAL model has provided valuable insights into patient satisfaction, emphasizing the importance of empathy and assurance in service delivery [8]. Similarly, in the hospitality industry, the model has been instrumental in measuring guest satisfaction, with a particular focus on the tangibles and reliability of services provided [9]. These diverse applications give emphasis to the model's adaptability and relevance in various contexts.

B. *Applications of SERVQUAL in E-commerce*

In the context of e-commerce, the SERVQUAL model has been adapted to address the unique challenges of online service delivery, where physical interaction between service providers and customers is limited or non-existent. Researchers have emphasized the importance of adapting the traditional SERVQUAL dimensions to better suit the digital environment. For instance, Yoon and Lee [3] found that in e-commerce settings, tangibles are redefined to include not just the physical appearance of products but also the digital presentation of products on websites, including image quality, product descriptions, and ease of navigation [3]. Similarly, reliability extends beyond the consistency of service delivery to encompass the accuracy of online information, the functionality of the website, payment process and the dependability of the delivery process.

C. *Importance of SERVQUAL Dimensions in the E-cosmetics Sector*

In the e-cosmetics sector, tangibles, reliability, and assurance play particularly significant roles in shaping customer perceptions of service quality. Tangibles in this context include the visual appeal and detailed presentation of products online, which are critical in influencing purchasing decisions. Reliability is associated with the consistency of product quality and the accuracy of delivery times, while assurance relates to the confidence customers have in the safety of their online transactions and the reliability of customer support. These dimensions are vital for e-cosmetics retailers, as they directly impact customer trust and satisfaction, which are essential for fostering long-term loyalty [10].

D. *Integration of Additional Factors: Product Quality, Brand Image, and User Interface Design*

Recent studies have underscored the significance of incorporating additional factors such as Product Quality (PQ), Brand Image (BI), and User Interface Design (UID) into the SERVQUAL framework when assessing service quality in the e-cosmetics sector. Product Quality, while traditionally viewed as a separate construct, is now increasingly seen as an integral part of perceived service quality in e-commerce, particularly in sectors like e-cosmetics where customers are unable to physically examine products prior to purchase. Chen and Cheng (2020) argue that high product quality can greatly enhance customer satisfaction and loyalty by consistently meeting or exceeding customer expectations [11].

Brand Image also plays a vital role in shaping customer perceptions and influencing behavior. A strong brand image can elevate perceived service quality by nurturing a sense of trust and reliability. This is especially critical in the e-cosmetics sector, where brand reputation often substitutes for direct product experience. Park and Kim (2020) suggest that a positive brand image can reduce the perceived risks associated with online shopping, thereby enhancing both customer satisfaction and loyalty [12].

User Interface Design is another key factor in the e-commerce context. A well-designed and user- friendly interface can significantly affect customer satisfaction by making the shopping experience more enjoyable and efficient. Garg and Kumar [13] show that positive user interface experiences lead to higher engagement, reduced bounce rates, and increased conversion rates [13]. In the e-cosmetics sector, where the visual appeal of products is paramount, an effective interface that highlights product aesthetics can significantly boost customer satisfaction and loyalty.

Moreover, recent advancements in strategic management research, particularly in the application of the SERVQUAL model, emphasize the use of Partial Least Squares Path Modeling (PLS-PM) for analyzing complex service quality structures. Khan and Fasih [14] have demonstrated the utility of PLS-PM in refining the SERVQUAL model, making it more applicable to the nuanced needs of e-commerce [14]. This methodology allows for a deeper understanding of the interactions between service quality dimensions, as evidenced in studies by Henseler et al. [15], which highlight its growing importance in strategic management research. Through PLS-PM, researchers can better quantify the impact of factors like Product Quality, Brand Image, and User Interface Design on overall service quality, providing more comprehensive insights for the e-cosmetics sector [15].

In summary, integrating these additional factors into the SERVQUAL framework and leveraging advanced modeling techniques like PLS-PM not only provides a more robust assessment of service quality but also enhances our understanding of customer satisfaction and loyalty in the e-cosmetics industry.

3 Methodology

A. *Research Design*

This study adopts a comprehensive quantitative research design to evaluate service quality in the e-cosmetics sector and its effects on customer satisfaction (CS) and customer loyalty (CL). The research framework is based on the SERVQUAL model, focusing on its five key dimensions—Reliability, Responsiveness, Assurance, Empathy, and Tangibles. In addition, Product Quality (PQ), Brand Image (BI), and User Interface Dcsign (UID) were incorporated to provide a more detailed analysis. The study examines both the direct effects of the SERVQUAL dimensions on CS and CL and the indirect effects mediated by PQ, BI, and UID.

Data collection was conducted via online surveys, and the data was analyzed using SmartPLS 4.0 software, which is commonly used for structural equation modeling (SEM) applications [16, 17]. SmartPLS was employed for confirmatory factor analysis (CFA) and structural model analysis [5, 18–20]. By using SEM, this study offers insights into the complex relationships between service quality, CS, and CL, contributing to a deeper understanding of the customer experience in the e-cosmetics industry.

B. *Survey Development and Pre-testing*

The survey instrument was developed based on a thorough review of existing literature and tailored to capture the unique characteristics of the e-cosmetics sector. The SERVQUAL items were adapted to reflect the digital context, focusing on dimensions such as website navigation, product display quality, and online customer support responsiveness. Additionally, items measuring Product Quality, Brand Image, and User Interface Design were included to assess their impact on customer satisfaction and loyalty. The survey was pre-tested with a small sample of 20 participants to ensure clarity and relevance of the questions, and adjustments were made based on the feedback received.

C. *Participants/Sample Set*

The research was conducted with customers in the e-cosmetics industry in Türkiye. A total of 95 individuals were selected from various demographic and socioeconomic backgrounds, ensuring a diverse sample. Participants were recruited through multiple online platforms, including social media, e-newsletters, and collaborations with major e-cosmetics retailers. The surveys were conducted via Google.

Forms between May 2024 and June 2024. Participation was voluntary, and respondents represented a range of ages, genders, income levels, and purchasing frequencies, providing broad demographic representation for the study.

D. *Survey Design and Data Collection Tools*

The SERVQUAL survey was adapted to reveal digital service quality dimensions, including website navigation, product display quality, and online customer support responsiveness. Additionally, items measuring Product Quality, Brand Image, and

User Interface Design were incorporated to evaluate their effects on customer satisfaction and loyalty.

The survey consisted of four main sections:

- Explanation and Information: An explanation of the survey's purpose and the confidentiality of participants was provided.
- Demographic Information: Data such as participants' age, gender, educational background, and income level were collected.
- Opinions on E-Cosmetics Services: Questions evaluated the participants' opinions on service quality in the e-cosmetics industry.
- SERVQUAL Survey: To measure perceived service quality, items from the SERVQUAL survey were adapted. These items included dimensions of Reliability (RL), Responsiveness (RS), Assurance (A), Empathy (E), and Tangibles (T).

D. *Data Analysis*

The collected data were analyzed using SmartPLS 4.0 software. The validity and reliability of the scales were tested through Confirmatory Factor Analysis (CFA), followed by an examination of relationships among the variables using Structural Equation Modeling (SEM). Specifically, the direct effects of the SERVQUAL dimensions on Customer Satisfaction (CS) and the indirect effects of Customer Satisfaction on Customer Loyalty (CL) were evaluated. The analysis also included the examination of factors such as Product Quality (PQ), Brand Image (BI), and User Interface Design (UID) on customer satisfaction and loyalty [15– 17].

4 Results

A. *Descriptive Statistics*

The demographic characteristics of the 95 participants included a balanced representation of genders (45% male, 55% female), with the majority (68%) aged between 25 and 40. The participants were also segmented by their frequency of online purchases, with 40% making e-cosmetics purchases monthly, 30% making quarterly purchases, and the remaining 30% purchasing less frequently. Income levels varied, with 25% earning less than 5,000 TL, 45% earning between 5,000 and 10,000 TL, and the remaining 30% earning above 10,000 TL.

B. *Structural Equation Modeling (SEM) Results*

This study evaluated service quality in the e-cosmetics sector using the SERVQUAL model. The analysis confirmed the validity of the proposed model, with all key dimensions of SERVQUAL—Tangibles (T), Reliability (RL), Responsiveness (RS), Assurance (A), and Empathy (E)—demonstrating significant relationships with customer satisfaction (CS) and customer loyalty (CL). Additionally, factors such as Product

Quality (PQ), Brand Image (BI), and User Interface Design (UID) were included to evaluate their effects on customer satisfaction and loyalty.

The path coefficients for Tangibles (T) showed a positive effect on customer satisfaction with a value of 0.171 and on customer loyalty with 0.184. Reliability (RL) demonstrated a positive effect on customer satisfaction (0.268) and a smaller effect on loyalty (0.078). Responsiveness (RS) had a slight negative effect on customer satisfaction (–0.019) but a positive effect on customer loyalty (0.046). Assurance (A) had a significant positive impact on customer satisfaction (0.212) and customer loyalty (0.267). Empathy (E) had a positive effect on customer satisfaction (0.210) and a modest positive effect on customer loyalty (0.050).

C. *Indirect Effects of Product Quality, Brand Image, and User Interface Design*

The analysis revealed significant indirect effects of Product Quality (PQ), Brand Image (BI), and User Interface Design (UID) on customer loyalty, mediated by customer satisfaction. Product Quality showed a strong effect on customer satisfaction (0.486) and an indirect effect on loyalty (0.255). Brand Image had a substantial effect on customer satisfaction (0.641) and an indirect effect on loyalty. User Interface Design positively affected customer satisfaction (0.543) and had an indirect effect on loyalty (0.024) (Fig. 1).

D. *Other Relationships*

The relationships among SERVQUAL dimensions showed that Reliability (RL) and Assurance (A) had the strongest correlations with customer satisfaction. However, the path from Empathy (E) to customer loyalty, mediated through customer satisfaction, was comparatively weaker than the other dimensions.

These findings emphasize the critical role of customer satisfaction in enhancing customer loyalty in the e-cosmetics industry. Improving Tangibles, Reliability, and Trust in service delivery significantly increases customer satisfaction. Moreover, delivering high-quality products, maintaining a positive brand image, andproviding an effective user interface contribute to higher customer satisfaction levels, which in turn indirectly encourage customer loyalty.

E. *Model Fit Indices*

The model fit indices demonstrated that the proposed model fit the data well. The goodness-of-fit index (GFI) was 0.93, and the root mean square error of approximation (RMSEA) was 0.045, both indicating an acceptable model fit. These results suggest that the SERVQUAL model, when extended to include additional factors such as Product Quality, Brand Image, and User Interface Design, provides a robust framework for understanding customer satisfaction and loyalty in the e-cosmetics sector.

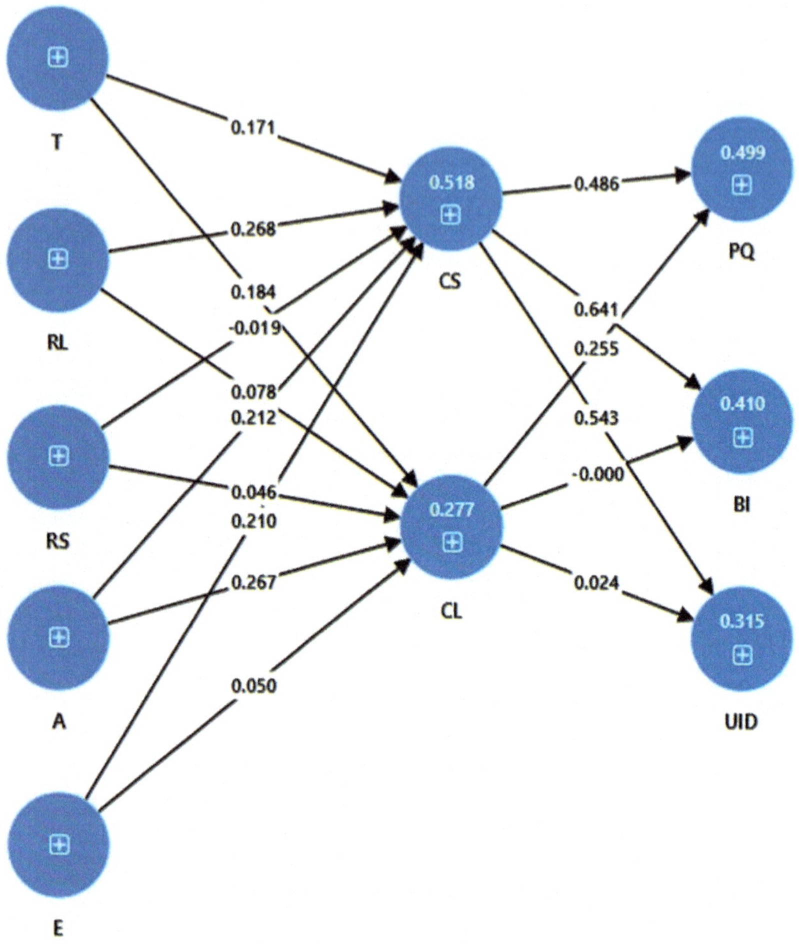

Fig. 1 Results of PLS-PM

These findings align with various previous research projects [7, 24]. For instance, Khan and Fasih [14] found that service quality directly influences customer satisfaction, which subsequently drives customer loyalty, supporting the outcomes of this study in the e-cosmetics sector. Additionally, Sharma et al. [21] noted that dimensions such as tangibility and assurance have strong correlations with customer loyalty in online retail, findings that are similarly reflected in the e-cosmetics industry.

5 Discussion

A. *Key Findings*

The findings of this study highlight the critical role of service quality in shaping customer satisfaction and loyalty in the e-cosmetics sector. Tangibles, Reliability, and Assurance were identified as the most influential dimensions of service quality, with significant positive impacts on customer satisfaction. These results align with previous studies, which have emphasized the importance of product presentation and delivery reliability in e-commerce settings [10].

In addition to the traditional SERVQUAL dimensions, this study also found that Product Quality, Brand Image, and User Interface Design play crucial roles in enhancing customer loyalty. The strong relationship between Product Quality and customer loyalty underscores the importance of maintaining high standards of product excellence in the e-cosmetics sector, where consumers rely heavily on product descriptions and reviews in the absence of physical interaction.

Brand Image and User Interface Design also emerged as significant predictors of customer loyalty, suggesting that companies in the e-cosmetics sector should invest in building strong brand reputations and providing seamless, user-friendly online experiences. These findings are consistent with the growing body of literature on the importance of brand trust and user experience in driving customer retention in digital environments [13].

However, it was also observed that Responsiveness (RS) and Empathy (E) play less significant roles in influencing customer satisfaction and loyalty. The weak indirect effect of Empathy suggests that personalized customer service may be less of a priority for customers in this sector, as compared to other service quality dimensions. This result is consistent with previous research [14], which found that Empathy plays a smaller role in the e-commerce environment. Nevertheless, improving physical aspects of service delivery and demonstrating Reliability and Assurance remain critical to fostering customer satisfaction and loyalty [22].

The study further revealed the strong mediating role of customer satisfaction in enhancing customer loyalty, both directly and indirectly. This finding aligns with recent studies that emphasize the importance of customer satisfaction in shaping long-term loyalty [23].

B. *Practical Implications*

The results of this study have several practical implications for e-cosmetics companies. First, enhancing the visual appeation of products online (Tangibles) should be a priority, as it significantly influences customer satisfaction. Second, ensuring reliable and timely delivery (Reliability) is essential for meeting customeg satisfaction. Third, companies should invest in building strong customer support systems (Assurance) to provide confidence in online transactions.

Furthermore, e-cosmetics companies should focus on maintaining high product quality and leveraging their brand image to differentiate themselves in the competitive online marketplace. Although User Interface Design (UID) has a positive effect on customer satisfaction, its role is less decisive compared to Product Quality and Brand Image. Still, investing in user-friendly and aesthetically pleasing online platforms can further enhance customer satisfaction and loyalty.

Additionally, research by Lee and Lee [5] highlighted the importance of service quality dimensions such as Reliability and Assurance on customer satisfaction and loyalty in the e-cosmetics industry [5]. Furthermore, studies on strategic management suggest that understanding the indirect effects of customer satisfaction is crucial for long-term success [15].

6 Conclusion

This study provides valuable insights into the drivers of customer satisfaction and loyalty in the e-cosmetics sector. By applying the SERVQUAL model and incorporating additional factors such as Product Quality, Brand Image, and User Interface Design, the research offers a comprehensive framework for understanding the key elements of service quality in this growing industry.

The findings suggest that Tangibles, Reliability, and Assurance are the most critical SERVQUAL dimensions influencing customer satisfaction, while Product Quality, Brand Image, and User Interface Design play pivotal roles in fostering customer loyalty. The results also highlight that Empathy and Responsiveness have weaker effects, but they remain relevant in specific contexts where personalized service is prioritized.

The strong mediating role of customer satisfaction in enhancing customer loyalty was consistently observed, indicating that improvements in Tangibles, Reliability, and Assurance can directly boost satisfaction, which in turn drives loyalty. Additionally, high-quality products, positive brand image, and user-friendly interfaces contribute to higher levels of customer satisfaction, which indirectly promote customer loyalty [23, 25].

In conclusion, companies in the e-cosmetics sector can achieve higher customer satisfaction and loyalty by focusing on key SERVQUAL dimensions and improving product and brand-related factors. Developing strategies to increase customer satisfaction will be an important step in enhancing customer loyalty and, consequently, the competitive strength of companies [26–28].

Future research could explore the impact of emerging technologies, such as augmented reality and virtual try-on tools, connected beauty, social commerce, personalized consultations on service quality and customer satisfaction in the e-cosmetics sector. Additionally, longitudinal studies could provide further insights into how customer preferences and expectations evolve over time in response to technological advancements and market changes.

References

1. Statista (2022) Global E-commerce market size 2021–2026. Retrieved from: https://www.statista.com/statistics/379046/worldwide-retail-e-commerce-sales/
2. TÜBİSAD (2023) Türkiye Bilgi ve İletişim Teknolojileri Sektörü 2023 Raporu
3. Yoon H, Lee S (2023) Service quality in e-commerce: Tangibles, reliability, and assurance in online shopping. J Retail Consum Serv 58:102291
4. Li X, Zhang Y (2023) The role of service reliability in the cosmetic industry: An empirical study. Int J Retail & Distrib Manag 51(1):27–43
5. Lee H, Lee S (2021) The impact of servqual dimensions on customer satisfaction and loyalty in the e-cosmetics industry. J Serv Res 24(3):245–257
6. Parasuraman A, Zeithaml VA, Berry LL (1985) A conceptual model of service quality and its implications for future research. J Mark 49(4):41–50
7. Parasuraman A, Zeithaml VA, Berry LL (1988) Servqual: A multiple-item scale for measuring consumer perceptions of service quality. J Retail 64(1):12–40
8. Chung H-S, Kim H, Lee J (2022) The impact of service quality on customer satisfaction and loyalty in online beauty retailing. J Bus Res 139:510–518
9. Wang J, Liu C (2021) Application of servqual model in the hospitality industry. Int J Hosp Manag 90:102512
10. Kim Y, Lee S (2023) Digital service quality and customer loyalty in online beauty retailing. J Retail Consum Serv 70:102919
11. Chen H, Cheng L (2020) The impact of product quality on customer satisfaction and loyalty in e-commerce. J Bus Res 120:284–295
12. Park J, Kim M (2020) The role of brand image in enhancing e-commerce customer loyalty. J Mark Res 77(2):183–204
13. Garg R, Kumar M (2021) User interface design and its impact on customer satisfaction in e-commerce platforms. J Retail Consum Serv 58:102284
14. Khan MA, Fasih M (2014) Impact of service quality on customer satisfaction and customer loyalty: Evidence from banking sector. Pak J Commer Soc Sci (PJCSS) 8(2):331–354
15. Henseler J, Ringle CM, Sarstedt M (2020) The use of partial least squares path modeling in strategic management research: A review of past practices and recommendations for future applications. Long Range Plan 43(3):277–295
16. Ringle CM, Wende S, Becker JM (2020) SmartPLS 4.0. SmartPLS GmbH
17. Hair JF, Hult GTM, Ringle CM, Sarstedt M (2021) A Primer on Partial Least Squares Structural Equation Modeling (PLS-SEM). Sage Publications
18. Kim Y, Kim S (2022) Analyzing the role of service quality in customer satisfaction and loyalty in the digital beauty market. J Bus Res 136:233–245
19. Chen J, Cheng Y (2020) Assessing e-service quality in the online cosmetics industry: A servqual approach. Int J Electron Commer Stud 11(1):34–56
20. Park M, Lee J (2021) Evaluating online service quality in the cosmetics industry using a modified servqual model. J Retail Consum Serv 59:102399
21. Sharma G, Jhamb D, Jhamb P (2020) Impact of service quality on customer satisfaction and loyalty in online retailing: A systematic review. J Theor Appl Electron Commer Res 15(3):161–190
22. Oliver RL (1999) Whence consumer loyalty? J Mark 63(4):33–44
23. Garg R, Kumar M (2021) Role of user interface in customer satisfaction and loyalty in online retail. J Retail Consum Serv 58:102284
24. Chung M et al (2022) Evaluating patient satisfaction using the servqual model in healthcare services. Health Serv Res 52(4):234–250
25. Thakur R (2018) Customer engagement and online reviews: Examining the mediating role of customer loyalty. J Retail Consum Serv 41:48–58

26. Souri A, Asadi S, Bakhtiari A (2021) The role of service quality in achieving customer satisfaction and loyalty in online shopping: A case study in Iran. Electron Commer Res Appl 42:101013
27. Bagheri MM, Sarabi B, Shirazi MA (2020) The impact of service quality on customer satisfaction and loyalty in the e-commerce sector: A meta-analysis. J Retail Consum Serv 55:102133
28. Chen Z, Huang Y, Su Q (2021) The effects of product quality and service quality on customer satisfaction and loyalty in the e-commerce context. J Bus Res 126:243–251

Jacket Armhole Measurement with Laser Scanning and 2D Imaging in Textiles

Semih Donmezer, Pinar Demircioglu, Ismail Bogrekci, and M. Numan Durakbasa

Abstract Clothing manufacturers face unique challenges compared to other industries due to their reliance on precise body measurements to create well-fitting garments. The garment sewing process is intricate and prone to errors that are difficult to detect during production, often leading to sizing issues. Designs are typically created in 2D but worn in 3D, resulting in inevitable size variations, particularly in multi-piece garments. Additionally, gravitational effects pose further challenges to inaccurate measurement. This study aims to measure the circumferences of jacket sleeve armholes and identify dimensional discrepancies at the armhole junctions. Two distinct methods were employed: a 3D laser scan was conducted on both a human model and an inflatable mannequin, and measurements were performed using an ArUCO marker on 2D images of the mannequin. The results revealed significant size variations in the jacket armhole seams, demonstrating the effectiveness of combining 3D laser scanning and 2D imaging techniques. These findings suggest that the dual-method approach provides a comprehensive evaluation of jacket sleeve armhole dimensions, which could help improve sizing accuracy in garment manufacturing.

Keywords 3D scanner · ArUCO marker · Measurements · Jacket sleeve · Edge detection

S. Donmezer (✉) · M. N. Durakbasa
Institute of Production Engineering and Photonic Technologies, Vienna University of Technology, 1060 Vienna, Austria
e-mail: semihdonmeze@anadolu.edu.tr

M. N. Durakbasa
e-mail: durakbasa@mail.ift.tuwien.ac.at

P. Demircioglu
Institute of Materials Science, TUM School of Engineering and Design, Technical University of Munich, 85748 Garching, Germany
e-mail: pinar.demircioglu@tum.de

I. Bogrekci
Department of Mechanical Engineering, Faculty of Engineering, Aydin Adnan Menderes University, Aydin 09100, Turkey
e-mail: ibogrekci@adu.edu.tr

N. M. Durakbasa and K. G. Gülen (eds.), *Sustainable Green Conversion*, Lecture Notes in Mechanical Engineering, https://doi.org/10.1007/978-3-031-83583-4_33

1 Introduction

Clothing is considered essential for human life, being recognized as a key contributor to enhancing human existence in various ways. Wearing suits has been a tradition spanning generations, linked with their construction. Regardless of cost, people from diverse backgrounds have shown an enduring fascination with suits. From formal events to semi-formal occasions and even casual outings, suits have found their place in various styles of clothes. In modern times, professionals such as bankers, lawyers, lecturers, parliamentarians, and business managers frequently opt for suits as their clothing, showing the enduring appeal and versatility of this classic garment [1].

Garment pattern-making, also known as garment structure design, involves transforming style design ideas into two-dimensional structure drawings [2]. Traditionally, pattern-making relied on experienced pattern makers drawing patterns on paper and then transferring them onto fabric. Methods included the proportional method for simple styles and the prototype method for complex ones. However, this process was time-consuming and relied heavily on the expertise of pattern makers [3]. To address these challenges, computer-aided fashion design systems were developed in the 1970s, leading to the creation of clothing CAD software [4]. These software programs enable interactive pattern-making, offering tools to streamline the process. Parametric pattern-making methods, based on mathematical models, have emerged as a solution to reduce manual labor and increase efficiency. These models express garment sizes and dimensions as parameters, allowing for automatic adjustments and grading [5]. While research on mathematical modeling of garment prototypes is limited, there is a growing demand for personalized clothing, driving the need for parameterized rapid pattern-making. Establishing accurate mathematical models for garment prototypes is crucial in meeting this demand [6, 7].

The present research involves a methodical and sequential process for collecting and acquiring relevant data on measuring the armhole circumference of jacket sleeves, followed by data processing and interpretation to produce useful information. A well-defined and progressive framework has been employed to manage the research systematically. In the area of garment measurement, determining the armhole circumference of jacket sleeves has been approached both experimentally and theoretically, addressing various levels of uncertainty.

Modern measuring technology and quality control techniques now primarily utilize electronic methods, especially digital measurement technology. Measurement technology encompasses devices and methods for determining physical quantities such as length, mass, force, pressure, electrical current strength, temperature, or time. Key sub-areas include the development of measurement systems and methods, as well as the acquisition, modeling, and reduction of measurement errors and undesirable influences. Numerous technologies exist today for digitally capturing the shape of 3D objects.

Choosing the perfect suit jacket is just as important as the skilled hands that craft it. Even with the changes brought about by mass production, the timeless art of making suits has kept its charm and recognition intact. Each suit jacket is carefully tailored to fit the unique shape of its wearer, requiring expertise and precision from designers to ensure it looks and feels just right.

For this study, two different methods were employed for jacket sleeve measurements. First, measurements were taken using a 3D laser scanner, a novel approach for soft fabric measurement. Second, measurements were carried out using 2D images, specifically in one of Turkey's largest textile companies. The program developed for this research converts calculated arc lengths into pixels, which are then converted into the metric system using the ArUCO Marker method. This combined approach of 3D laser scanning and 2D imaging was essential to accurately measure the jacket sleeves.

2 Image Processing

In recent years, innovative image processing strategies for generating and controlling image formation have successfully led to the production of high-resolution digital image processing systems. This success has in turn increased interest in the high-resolution camera and absolute measurement with the high-resolution, ever larger field of view. The new breed of panning sensors includes multi-cameras, catadioptric panoramic lenses, ring panoramic lenses, fisheye lenses, wide-angle anamorphic lenses, and palynomorphic lenses. The increasing trend toward the use of panoramic sensors in various applications is due to the need for comprehensive information about our environment [8]. Digital image processing involves working directly with an image that consists of many pixels. These image points, also known as pixels, have spatial coordinates that indicate the position of the points in the image and intensity (or grayscale) values. A color image is associated with higher dimensional information than a gray image because red, green, and blue values are typically used in different combinations to represent the colors of the image in the real world [9]. Shape Measurement is an ancient subject that has been studied by academics, as well as various artists. Although, over the past nearly 20 years, it has developed advancing technology in 3D modeling, and various visualization nanotechnology, as well as noteworthy developments in computer science and big data, use. Today 3D devices make it possible to buy not only for field-specific use and artists but also for many people from different professions. Now it can be scanned in our environment in high resolution with devices as cheap as a video camera, transformed into 3D data with 3D modeling software, used on the web, and placed in wide sizes and numbers [10].

The majority of image registrations can be categorized by the following four steps:

1. Feature extraction and matching,
2. Feature matching,
3. Model estimation by converting,
4. Image registration [11].

3 Materials and Methods

The study utilized a 3D laser scanning system, an ArUCO marker developed in-house, and an inflatable ironing doll for measuring jacket sleeve dimensions. The 3D laser system captured detailed measurements of sleeve circumferences and armhole junctions on both a mannequin and an inflatable ironing doll, while the ArUCO marker served as a reference point for image-based measurements on the mannequin (Fig. 1). High-resolution images of the jacket sleeves were analyzed using image analysis software to determine sleeve dimensions based on the position of the ArUCO marker. Measurement errors were categorized into random and systematic errors, and data obtained from both methods were statistically analyzed to assess accuracy and precision. The material and method employed in this study provided valuable insights into the effectiveness of different measurement techniques in garment manufacturing, facilitating accurate and reliable measurements of jacket sleeve dimensions.

The study explores the classification of measurement errors, delineating individual, external, and methodological errors based on the measuring device employed while noting boundary conditions and environmental influences during system specification. Personal errors stemming from operator characteristics, instrumental errors due to physical quantity effects, and methodological errors from measurement method selection are identified and analyzed. Additionally, measurement uncertainty and its causes are explored, leveraging standard-compliant verification methods outlined in the ISO 15530 [12] series to assess uncertainty's dependence on geometric features. Surface recognition, particularly in assessing surface roughness, is investigated for its impact on mechanical part fit and durability in the textile industry. Non-contact measurement methods, including optical, ultrasound, and computer vision-based techniques, are evaluated for their efficacy, emphasizing the importance of accuracy and efficiency in quality assessment.

In addition to the classification of measurement errors, the study introduces a novel system for object recognition known as the ArUco Marker, recently included in the Eurobot Rules. This method is pivotal in exposure estimation for applications like augmented reality and robotic navigation. It operates by establishing correlations between real-world points and their 2D image projections, often employing synthetic or fiducial markers to aid in this process. The marker recognition process involves two primary steps: first, the marker must be visible in the 2D image and correctly positioned. The image is analyzed to identify square shapes resembling 2D markers, starting with adaptive thresholding to segment the marks. Subsequently, the contours

Fig. 1 The measurement method

	Mannequin	Swelling Ironing Doll
3D laser	✓	✓
2D image	✓	✗

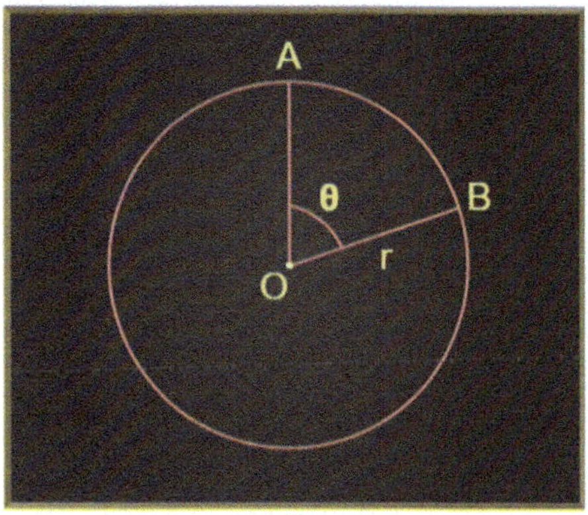

Fig. 2 The arc-counting AB [13]

of the marks are extracted and evaluated for squareness; non-convex shapes or those deviating from a square form are discarded.

The arc length is defined as the distance between two points along an arc section and a portion of the circumference. Measuring the length of an irregular arc section by connecting the arc section into contiguous (straight) line segments is also called straightening an arc. The length of the arc ab is greater than the distance of the straight lines connecting points A and B. The arc is a segment of the perimeter of a circle. A circumference is the other boundary of the circle. The circumference can also be defined as the perimeter of a circle in Fig. 2. Given the radius and angle, the expression to calculate the length of an arc is as follows.

$$1 = 2\pi \mathrm{r} \times (\theta \div 360) \tag{1}$$

where r is the circumference of the circle and θ is the angle in degrees. It is the angle between the two radii forming the arc, or the central angle of the arc. L = arc length if the angle is in radians—1 rad = π/180° or 1 rad = 2π/360°. Where r is the radius of the circle and θ is the angle in radians [13].

Pose estimation is of great importance in many computer vision applications: robot navigation, augmented reality, and many more. This process is based on finding correspondences between points in the real environment and their 2D image projection. An ArUco marker is a synthetic square marker consisting of a wide black border and an inner binary matrix that determines its identifier (id). The black border facilitates its quick recognition in the image and the binary coding enables its identification and the application of error detection and correction techniques (Fig. 3). The size of the marker determines the size of the internal matrix. For example, a marker size of 4 × 4 consists of 16 bits.

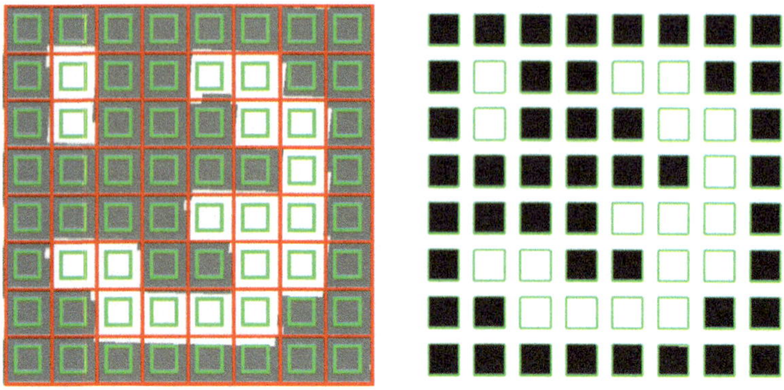

Fig. 3 Marker cell margins

For an image containing ArUco markers (Fig. 4), the recognition process must return a list of recognized markers. The AU marker basically includes two important steps: marker identification and pose estimation:

- The position of its four corners in the image (in their original order).
- The identifier of the marker.

The marker detection process (Fig. 5) has two main steps:

- Finding Marker Candidates: First, the image is analyzed to identify square shapes that could serve as markers. During this step, the contours of shapes are extracted from the processed image, and any shapes that are not convex or don't resemble a square are discarded.

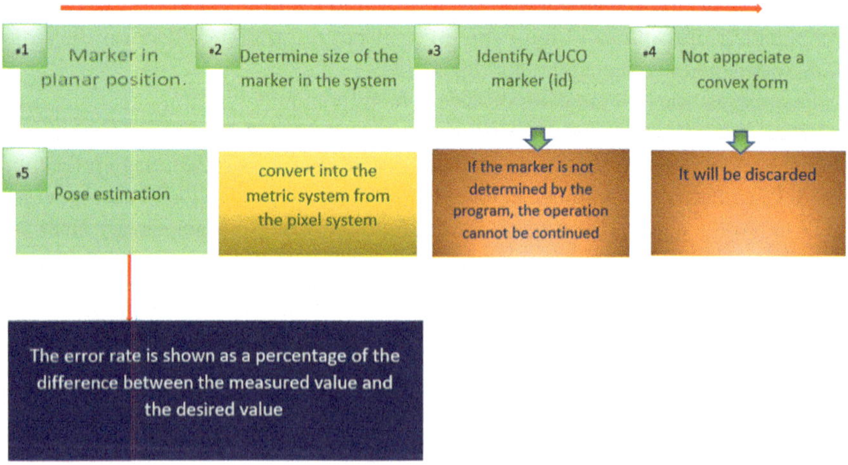

Fig. 4 Guidance chart to detect ArUco markers

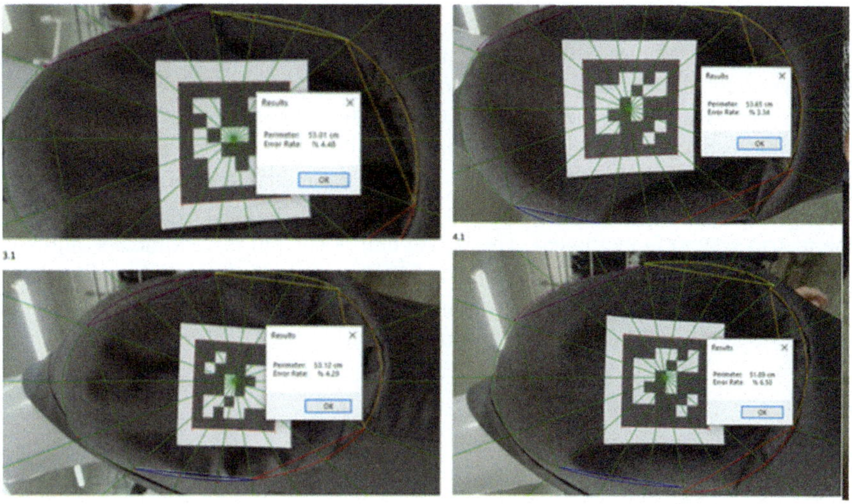

Fig. 5 2D measurements by using ArUco marker

- Verifying the Candidates: After identifying potential candidates, the next step is to determine if they are indeed valid markers [14].

After the marker is detected, the next step involves analyzing its internal coding to confirm its identity. This is achieved by extracting the marker bits from each marker, starting with a perspective transformation to standardize the marker's form (Fig. 6). Subsequently, the standard image undergoes thresholding using the Otsu Algorithm, which separates white and black bits based on a single intensity threshold. The image is then divided into cells according to the marker size, with the assumption that each cell contains either a white or black bit. Error correction techniques may be necessary at this stage to ensure accuracy [14].

Fig. 6 Measurements on two-dimensional image program

The primary objective of the code is to measure the circumference of the jacket sleeve from a 2D photo. Although the jacket sleeve resembles a geometric ellipse, it is not a perfect ellipse, leading to high error rates when fitting it to a single ellipse. To mitigate this issue, the arc length of areas where the ellipse coincides with the jacket stitch lines is calculated and placed within the ellipse. The ArUco marker is utilized to convert this pixel-based arc length value into the metric system. By determining the edge length using the corner points of the ArUco marker [14], the transition to the metric system is facilitated. The length unit is entered in centimeters, allowing for comparison with the real value measured using a standard measuring instrument. The error rate is then determined by comparing the resultant value obtained through the developed method with the real value.

4 Conclusions

All jackets belonged to the same equivalence class: they were dark grey and size 48. In Table 1, the measured values were close to but did not exactly match the desired value. Specifically, the calculated value was 54.20, while the desired value was 55.50, resulting in an accuracy rate of 97.65%, indicating a close approximation to the desired value.

Table 1 Observations of both left and right sleeve armholes

Observation	Right Sleeve	Left sleeve	Average of R+L
1	55.28	59.53	57.41
2	52.21	51.60	51.91
3	51.01	53.12	52.07
4	53.65	51.98	52.82
5	54.25	52.14	53.20
6	53.93	58.85	56.39
7	55.14	51.81	53.48
8	52.85	54.37	53.61
9	54.90	55.32	55.11
10	55.42	58.45	56.94
11	53.95	54.03	53.99
12	52.76	53.36	53.06
13	55.65	54.53	55.09
14	52.08	55.90	53.99
15	56.04	57.78	56.91
16	53.56	56.64	55.10
17	53.94	53.37	53.66
18	53.54	56.07	54.81
19	56.14	56.52	56.33
20	52.15	52.67	52.41
21	55.88	53.16	54.52
22	52.88	55.60	54.24
23	54.40	53.87	54.14
24	52.32	52.78	52.55
25	54.08	52.20	53.14
26	55.05	56.38	55.72
27	53.68	56.63	55.16
28	52.20	53.40	52.80
29	53.05	52.27	52.66
30	51.82	54.01	52.92
Σ	1613.81	1638.34	1626.08
$\bar{x}$	53.79	54.61	54.20

			Statistic	Std. Error
Right	Mean		53.7937	.25493
	95% Confidence Interval for Mean	Lower Bound	53.2723	
		Upper Bound	54.3151	
	5% Trimmed Mean		53.8048	
	Median		53.8050	
	Variance		1.950	
	Std. Deviation		1.39633	
	Minimum		51.01	
	Maximum		56.14	
	Range		5.13	
	Interquartile Range		2.42	
	Skewness		.005	.427
	Kurtosis		-.911	.833
Left	Mean		54.6113	.40930
	95% Confidence Interval for Mean	Lower Bound	53.7742	
		Upper Bound	55.4485	
	5% Trimmed Mean		54.5141	
	Median		54.0200	
	Variance		5.026	
	Std. Deviation		2.24184	
	Minimum		51.60	
	Maximum		59.53	
	Range		7.93	
	Interquartile Range		3.66	
	Skewness		.601	.427
	Kurtosis		-.606	.833

A. *Measurements Measurement and descriptive statistics of jacket sleeves as right and left sleeves on 2D image-in cm*

The FARO 3D laser scanning devices employed in the measurements are adept at swiftly and accurately capturing millions of data points on a product within an examination scene, thereby creating a 3D point cloud (Fig. 7). This enables virtual viewing and comprehensive examination of the product under investigation. Such solutions offer a range of dimensional measurement applications, including on-machine and in-line inspection, as well as CAD comparison to ensure adherence to correct tolerances. CAM2 software ensures seamless integration with FARO measuring devices, regardless of whether they are used for tactile probing or non-contact scanning applications. Moreover, users can effortlessly connect multiple measuring devices to the software simultaneously, allowing for parallelized data acquisition and decreased inspection time. The user-friendly design of the software eliminates the need for extensive training or expertise, enabling operators to swiftly and easily complete their measurement tasks. This ease of use is facilitated by features such as simple command input, continuous measurement and analysis of product parts, an intuitive graphical user interface for efficient facility management and test routine monitoring, image-guided automated measurement processes, real-time graphic feedback of measurement results, and fast tools and shortcuts that streamline complex measurement procedures into individual commands [15].

In Fig. 8, it is seen that the jacket is placed on the mannequin. Upon request from the manufacturer, to detect the measurement errors at the joints of the jacket sleeves, we scanned the arm house of the jacket with a 3D laser. A patterned jacket was chosen from the jacket samples. The operator used the guiding options to separate the seams and folds. It can automatically make selections with the appropriate settings of 3D scanner-software. The picture on the right and bottom shows the pattern of the fabric used.

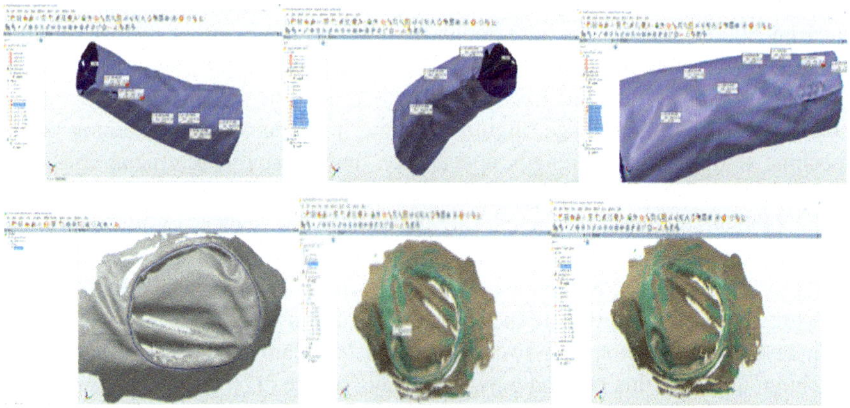

Fig. 7 Jacket sleeve 3D scanner images by using faro scanner

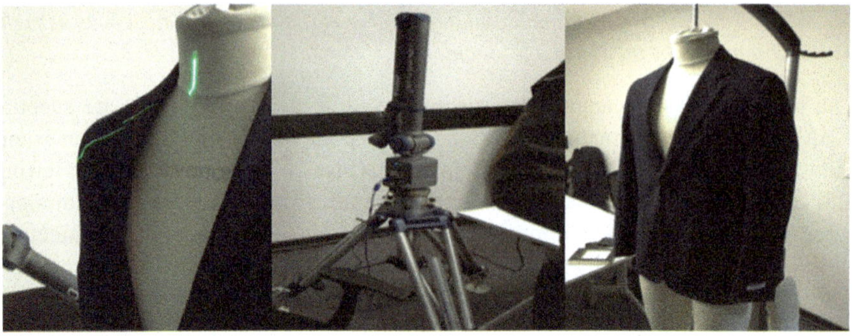

Fig. 8 Sleeve of jacket screening on the mannequin

Edge detection is an important part of Image Processing which deals with identifying and detecting shapes of parts of the image that differ in brightness, color, or texture. 3D point cloud that enables virtual viewing and complete examination of the product. In the bottom image are in STL image format (Fig. 9) thus the image can move in different directions and it is necessary to rotate the image until the seam line is visible. The first step in shape analysis is to determine the shape which the object to be measured in the image so that it can be scanned and processed later. After the image is scanned, it is rotated so that the seams of the armhole are fully visible and then loaded into the Solid-Works program. To accurately determine the circumference of a jacket sleeve, at least 70 control points are required. By combining these control points and taking into account the bulges or wrinkles of the fabric, the circumference of the jacket can be measured correctly.

Errors in garment production can originate from various sources, including personal, instrumental, and process-related factors. Personal errors may arise due to the physical or physiological characteristics of an operator, influenced by mental or physical exertion, and can also depend on the tailor's level of craftsmanship. Instrumental errors, such as high thread tension during sewing or the use of a dull needle, can lead to wrinkles and shrinkage in fabrics, particularly common in woven and knitted fabrics [16]. Incorrectly set parameters of the sewing machine can also contribute to seam shrinkage, highlighting the importance of maintaining proper machine settings. Seam strength, appearance, and elasticity are crucial considerations in fabric quality, with seam tightness affecting these factors. Process errors, on the other hand, may result from tailoring inconsistencies, such as variations between the right and left arms at the jacket sleeve joint. This discrepancy can be attributed to the tailor's handedness, where the orientation of the sewing machine affects the ease of work. Proposals to mitigate measurement errors in jacket sleeve joining seams include the use of a notching system to adjust the seam line, ensuring more even stitching, although this method may not always suffice [17]. These diverse sources of errors underscore the intricate nature of garment production and the importance of meticulous attention to detail in the manufacturing process [18].

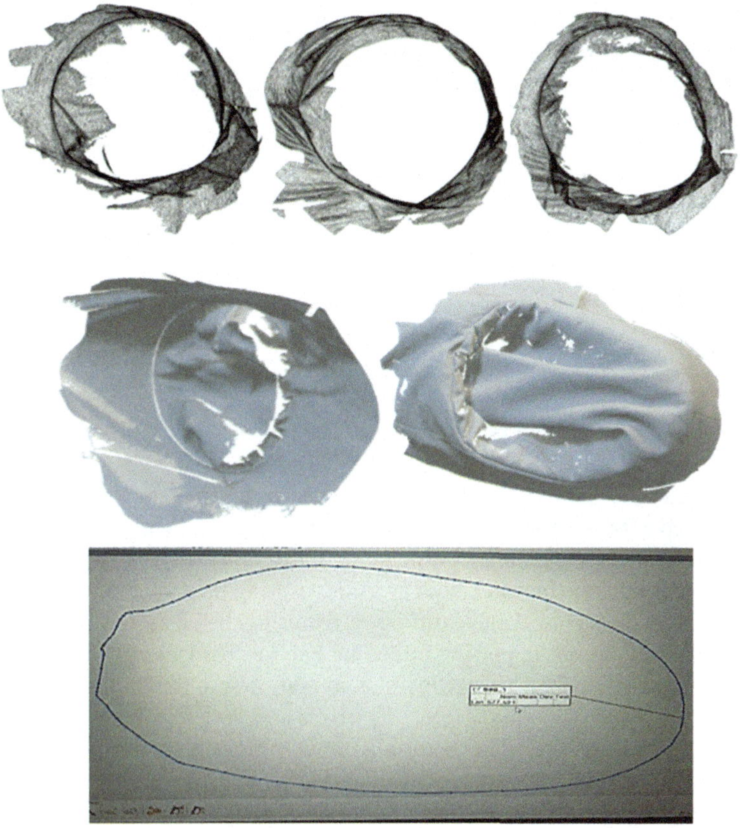

Fig. 9 STL (Stereo Lithography) file format and counting on solidworks program

Ishikawa identifies these contributing factors as a possible root cause of problems and troubleshooting, and one of these factors is the human factor. According to Crosby, most human errors are due to a lack of attention rather than a lack of knowledge [19]. Considering tailoring errors, the variance values between the right arm and the left arm at the jacket sleeve joint are also different. This may be due to the tailor being right-handed or left-handed. When sewing the left sleeve, the right-handed person sits in front of the sewing machine and works easily. When the left sleeve is sewn, the right-handed seamstress sitting in front of the sewing machine has to hold the jacket upside down, which is much more difficult. In this respect, it may make sense to separate right and left-handed dressmakers and give them only one side of the jacket. To correct the measurement error in the jacket sleeve joining seam, it has been proposed to use a notching system that adjusts the notches as you go along to determine the seam line. By adjusting the trace with this notching device, the jacket sleeve join seam will be more even, but this system may not always be sufficient.

Therefore, combining tailored work assignments based on handedness with technological solutions like the notching system could help minimize human error and achieve more consistent results in tailoring. Regular training sessions focused on attention to detail and proper techniques for both right-handed and left-handed workers should be implemented to further enhance precision and reduce errors. Additionally, a quality control step, where seams are reviewed by another tailor before finalizing the garment, could help catch inconsistencies early and ensure that high-quality output is maintained.

5 Conclusions

An attempt was made by a premium manufacturer to utilize a 3D laser scanner to determine potential differences in the dimensions of the jacket's connection point and sleeve length. The significance of such disparities in the manufacturing process of a jacket needed to be understood. Typically, the two arms of a jacket exhibit variations, posing challenges for premium clothing manufacturers, particularly concerning the ideal size of joint sizes between both arms. Measuring the circumference of a coat-sleeve joint presents considerable difficulty, requiring the jacket to fit correctly on a mannequin with the sleeve portion folded in. Fabrics, being woven materials made of yarns, possess inherent properties like weight, drape, durability, softness, and stiffness.

The measurement aimed to scrutinize errors in jacket sleeves. Notably, the measurements revealed discrepancies in sleeve circumferences among individual jackets, with none precisely aligning with the desired size. This uniqueness in sleeve circumference even within the same jacket was observed consistently. Two methods were employed for jacket measurement: one using a 3D laser on both a mannequin and an inflatable mannequin, yielding accurate results albeit at high costs and time investment; the other involved ArUco-Marker on a two-dimensional image, allowing for simultaneous evaluation of 100 jackets with minor uncertainty but with greater efficiency and cost-effectiveness. Efforts are ongoing to develop a rapid and dependable method for controlling measurement errors comprehensively, particularly for companies producing over 1000 garments daily, considering the time-consuming nature of measuring tasks.

Future recommendations include integrating machine learning algorithms with 3D scanning technology to predict and correct measurement discrepancies in real-time. By analyzing large datasets of jacket measurements, these algorithms could identify patterns and anomalies, providing automated adjustments during the manufacturing process. Moreover, exploring the use of advanced materials with built-in sensors that can transmit real-time data about their dimensions and properties could revolutionize quality control. Collaborating with technology firms to develop customized software solutions for the fashion industry would further enhance precision and efficiency, ensuring consistent product quality and reducing production costs. These innovations could make the process smoother and more reliable, ultimately leading to higher satisfaction for both manufacturers and customers.

References

1. Opoku M, Sarah B, Bakker-Edoh D (2023) Understanding clothing selection: The perfect ways of choosing men's suit jacket for good grooming. Integr J Res Arts Humanities 3(5):143–150. https://doi.org/10.55544/ijrah.3.5.12
2. Zhu C, Liu K, Wang R, Wang J, Bruniaux P, Zeng X (2022) The mathematical model of men's garment prototype. IJCST 34(6):892–904. https://doi.org/10.1108/IJCST-04-2021-0050
3. Au CK, Ma Y-S (2010) Garment pattern definition, development and application with associative feature approach. Comput Ind 61(6):524–531. https://doi.org/10.1016/j.compind.2010.03.002
4. Hu Z-H, Ding Y-S, Zhang W-B, Yan Q (2008) An interactive co-evolutionary CAD system for garment pattern design. Comput Aided Des 40(12):1094–1104. https://doi.org/10.1016/j.cad.2008.10.010
5. Kim S, Park CK (2004) Parametric body model generation for garment drape simulation. Fibers Polym 5(1):12–18. https://doi.org/10.1007/BF02875489
6. Liu K et al (2018) Parametric design of garment flat based on body dimension. Int J Ind Ergon 65:46–59. https://doi.org/10.1016/j.ergon.2018.01.013
7. Liu H, Jiang G, Dong Z, Xia F, Cong H (2020) The size prediction and auto-generation of garment template. IJCST 33(1):74–92. https://doi.org/10.1108/IJCST-06-2019-0083
8. Francisco (ed) (2010) Vision sensors and edge detection. Sciyo. https://doi.org/10.5772/273
9. Zhou H, Wu J, Zhang J (2010) Digital image processing: Part I, Bookboon, p 72. [Online]. https://kosalmath.wordpress.com/wp-content/uploads/2010/08/digital-image-processing-part-one.pdf
10. Laga H, Guo Y, Tabia H, Fisher RB, Bennamoun M (2019) 3D shape analysis: Fundamentals, theory, and applications. Wiley, Hoboken, NJ, USA
11. Han YK, Byun YG, Choi JW, Han DY, Kim YI (2012) Automatic registration of high-resolution images using local properties of features. Photogramm Eng Remote Sens, 78(3):211–221. https://doi.org/10.14358/PERS.78.3.211
12. ISO/TS 15530–1:2013 Geometrical product specifications (GPS)—Coordinate measuring machines (CMM): Technique for determining the uncertainty of measurement Part 1: Overview and metrological characteristics
13. How to find the Length of an Arc? [Online]. https://www.geeksforgeeks.org/arc-length/
14. Detection of ArUco Markers. [Online]. https://docs.opencv.org/4.x/d5/dae/tutorial_aruco_detection.html
15. Faro. [Online]. https://www.faro.com/de-DE
16. Automation in Garment Manufacturing (2018). Elsevier. https://doi.org/10.1016/C2015-0-06156-1
17. Gries T, Lutz V, Niebel V, Saggiomo M, Simonis K (2018) Automation in quality monitoring of fabrics and garment seams. In: Automation in garment manufacturing, Elsevier, pp 353–376. https://doi.org/10.1016/B978-0-08-101211-6.00014-8
18. Czichos H, Saito T, Smith L (eds) (2011) Springer handbook of metrology and testing. Springer Berlin Heidelberg. https://doi.org/10.1007/978-3-642-16641-9
19. Durakbasa MN, Osanna PH (2004) Qualitiy in industry, TU-Wien-AUM-2004

Kaizen and 5S Application from Lean Production Techniques Against Waste and Loss of Time in Production and Office Areas in a Food Business

Hasan Hüseyin Çelebi, Umut Can Yurt, and Büşra Telli

Abstract The understanding of competition in businesses has become different as time passes. Lean philosophy is an inevitable approach to reduce cost, minimize downtime and shorten the process. In the most basic sense, lean philosophy is a culture that aims to improve issues such as order, order and speed in order to prevent waste, excessive workload, unbalanced workload, and loss of time. In this approach to improve processes, 5S is a frequently used technique. This technique has 5 steps. These steps are: sort, set in order, shine, standardize, and sustain. The biggest supporter of the application of this technique is the Kaizen method. Kaizen is a term that means change for the better. The factory we discussed in our study in Istanbul is a factory established to provide quality, healthy and affordable bread to the people of Istanbul. A lean philosophy is tried to be adopted in this factory, which continues to increase product quality by keeping costs low and keeping up with the developing technology and digitalizing world order. In our study, it is aimed to apply 5S and kaizen in order to reduce costs, alleviate the workload and save time in both production lines and offices in a food factory that produces bakery products within the borders of Istanbul. In our study, firstly, the places that were deemed inappropriate within the scope of 5S were identified and photographed together with Gemba walks and added to the Kaizen method file before and after. Then, by cooperating with the technical unit of the company and eliminating the nonconformities determined within the scope of 5S, it was determined how much profit these developments have achieved for the company.

Keywords Lean philosophy · 5S · Kaizen · Food · Production

H. H. Çelebi · U. C. Yurt (✉)
Industrial Engineerimg, Istanbul Kültür University, Istanbul, Turkey
e-mail: umutcan8264@gmail.com

H. H. Çelebi
e-mail: h.celebi@iku.edu.tr

B. Telli
Industrial Engineer, Gürmen Clothing Industry and Trade Incorporated, Istanbul, Turkey

N. M. Durakbasa and K. G. Gülen (eds.), *Sustainable Green Conversion*, Lecture Notes in Mechanical Engineering, https://doi.org/10.1007/978-3-031-83583-4_34

1 Introduction

With the development of technology, the competitive environment between companies in recent years has led to the formation of different perspectives as time passes. Today, companies increase the quality of their products in order to take part in the competition and ensure profit continuity and aim to continuously improve their product creation processes [1]. While companies use lean production in their own work areas, they get support from some techniques. These techniques are defined as 5S, Kaizen, SMED, JIT, etc. 5S and Kaizen are mostly applied in companies because they can be used in every department and are easy to understand as a basic principle [2].

The cost will increase due to the high rate of waste that may occur in the food sector. Since high cost may cause inefficiency in production, it will cause a decrease in profit. In order to prevent this situation, companies have adopted the lean production philosophy and benefit from lean production techniques [3].

In this study, it is aimed to apply lean production techniques 5S and Kaizen in a factory producing bakery products operating in the food sector in the province of Istanbul. The main philosophy has been determined as to reduce the cost as much as possible by preventing waste in the business. Due to the existence of certain challenges in implementing lean production techniques in food businesses, there is a lack of sufficient studies on this subject in the food industry. This study serves as a guide for future research. In the rest of the study, similar studies were examined with a literature review, enlightening information about lean philosophy was given, nonconformities in the factory were identified, action plans were created within the scope of 5S studies, and kaizen application was included in office and factory areas. The gains obtained as a result of the studies have been determined.

2 Literature Review

Companies experience loss of value due to waste such as unnecessary movement, faulty production, unnecessary processing. When the literature is examined, it is seen that the 5S technique, which is one of the lean production techniques, is included in studies aimed at saving time and order at a high rate, while kaizen applications are included in studies aimed at applying issues such as cost reduction and error reduction.

In Table 1, information is given about the studies aimed at the application of 5S and kaizen methods in production areas between 2012 and 2022.

Table 1 Literature review on 5S and kaizen applications in production areas

Researcher	Year of release	Subject of the research
Keleş, Gürsoy ve Çelik	2013	Application of 5S systematics to an iron and steel factory
Berber	2013	Determining the efficiency recorded with lean production and the gains achieved by the business with Kaizen applications
Kaymakcı	2012	5S study in a PTT branch
Akıl	2013	Observation of the benefits of 5S and Kaizen applications in the public sector
Çelik	2019	Investigation of the impact of 5S methodology on total equipment effectiveness
Demir ve Paksoy	2021	Investigation of the effect of lean techniques on increasing the productivity of the sector in an application in the aviation sector
Savcı	2019	The effect of the Kaizen costing system on the losses in the production process in a dry tea production enterprise
Erdeniz	2018	Investigation of the effects of 5S and Kaizen applications on business performance in the furniture industry
Özcan ve Yılseli	2022	Investigation of the 5S technique applied in a textile company

3 Methodology

A. *Lean Manufacturing*

Lean manufacturing is an understanding developed by the Japanese due to the limited and bad economic conditions that emerged after World War II. It is based on the principle of using the limited resources available with the lowest cost and highest efficiency [4]. Factors such as increasing system efficiency, increasing product quality, minimizing costs, and employee productivity are the methods that a company should apply in order to be one of the effective companies in the market. For the application of these methods, the most important issues to be considered are efficiency and perfection. If these two issues are developed, the principle that should be applied is lean production [5].

B. *5S*

Aiming to keep the workspaces tidy, clean and standardized in the 5S workplace environment and using the philosophy of continuous improvement (Kaizen) 2. It is a management methodology that was born in Japan after World War II [6]. 5S is the abbreviated name of 5 Japanese processes whose first letters begin with S. These 5 processes are Seiri (Sort), Seiton (Set in order), Seiso (Shine), Seiketsu (Standardization) and Shitsuke (Sustain) [7].

Table 2 Kaizen and ınnovation comparison [9]

Kaizen	Innovation
People come first	Priority technology and money
Everyone participates	Only senior management participates
Small investment, steady and regular	Large investment, intermittent and irregular
Permanent	Temporary
Protection and remediation	Waste extraction and re-removal
Where the economy is developing slowly	Where the economy is developing rapidly

C. *Kaizen*

There are two types of ideas in terms of progress in businesses; progress as a result of gradual improvements and progress in one big step. Progress in one big step can be summed up in the term innovation [8]. Table 2 shows the differences between kaizen and innovation.

Kaizen types are divided into three groups as before-after Kaizen, Gemba Kaizen and Kobetsu Kaizen due to application differences [10].

In equation number 1, the method of calculating the number of loaves of bread saved from discarding is defined. In the equation, variable x shows the amount of discarded bread produced per hour, variable y shows the working time, variable k shows the total amount of working lines, variable t shows the total number of days worked in a month, variable m shows the number of months in 1 year, and variable z shows the total number of discarded bread in 1 year.

$$(X_i)(Y_i)(K_i)(T_i)(M_i) = Z_i \quad (1)$$

Equation 2 shows the method of calculating the total amount of flour saved in 1 month. The variable R defines the average amount of flour wasted per day, the total number of mixers working during working hours per day, variable u defines the number of working days per week, that variable defines the total number of weeks worked per month, and variable Q defines the total amount of flour saved in kg for 1 month.

$$(R_i)(P_i)(U_i)(O_i) = Q_i \quad (2)$$

4 Implementation

It is very important to save time, reduce costs, and reduce waste in office and production areas in order to ensure sustainability in the competitive environment of the company, and in this context, 5S and kaizen studies have been applied in the company

Unit and 5S Department/ Area:

Order No	Label No	Material/Equipment Description	Amount	Seperation Date	Material Class	Reason For Labeling	Labeling Date	Decision Made	Responsible Employee
1		Cabin	1	3.06.2022		The cabinet is not stable	3.06.2022		
2		White Folder	1	3.06.2022		Folder not used	3.06.2022		
3		Plastic File Rack				Overuse	3.06.2022		
4		Telephone				Defective	3.06.2022		
5									
6									
7									
8									
9									
10									
11									
12									
13									
14									
15									
16									
17									
18									

Fig. 1 Red label list

department. Some lists and plans have been created in order to standardize the steps applied while conducting the 5S study. In the areas where Kaizen was applied, the visual change in the environment was recorded first and then with the visual method.

A. *5S Implementation*

Firstly, it is aimed to implement 5S in office and production areas. Information was received about the regions where 5S is intended to be implemented and a list of things to be done in the regions was prepared. Materials that are not used in the working environment will be removed from the working environment and other materials will be arranged according to the frequency of use. A list of red labels and red labels has been prepared for this process. In Fig. 1, a sample template of the red tag list application is provided.

A project plan chart has been prepared in order to follow the steps of the 5S project and to share up-to-date information about the time among the project team members. In Fig. 2 example of action plan that planned to do on subject area is provided.

Table 3 shows the before and after images of the 5S study carried out in the accounting office. The elements within the office are arranged in terms of occupational safety and time saving.

B. *Kaizen Implementation*

While 5S was applied in the application areas, kaizen application was also included, before and after kaizen technique was used. A list of the elements that need or can be developed in the areas has been drawn up and development recommendations have been made about these elements. In Figs. 3 and 4, the location of boilers, forklifts and cabinets in the production area is defined.

Figures 5 and 6 show an example of the cleaning and painting operations carried out in the warehouse area.

In Figs. 7 and 8, definitions were made by numbering method on the valves where the yeast flow is provided in the technical warehouse area.

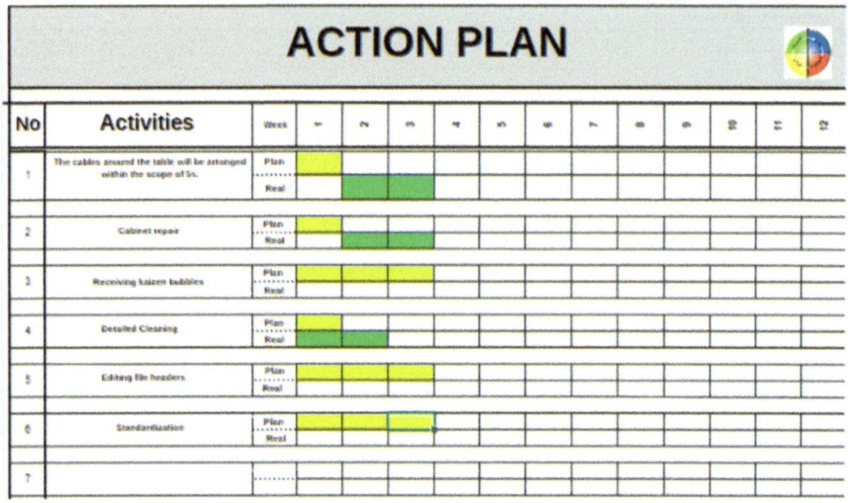

Fig. 2 5S project/action plan

5 Findings

As a result of the observations made with the working areas examined in line with the studies carried out, gains for the company have emerged. As the output of one of the studies carried out, a cost of 30,000₺ has been incurred for the company by using engines that can be easily intervened instead of an engine that malfunctions frequently, but high spare parts costs have been prevented in case of malfunctions and malfunctions to be experienced. Since the products falling from the duck mouths in the production lines remain close to each other, 10 products are discarded per hour. The taut casting cloths in the last fermentation unit were replaced with new ones, reducing the number of discarded products. As a result of this study, 224,628 products were prevented from being discarded annually within the company.

Protectors have been installed in the flour discharge areas in order to prevent flour spillage on the floor, and it has been observed that 150 kg of flour is saved per month thanks to these preservatives.

Table 3 Table of before and after visuals of the work done by the accounting office

Images before	Images after

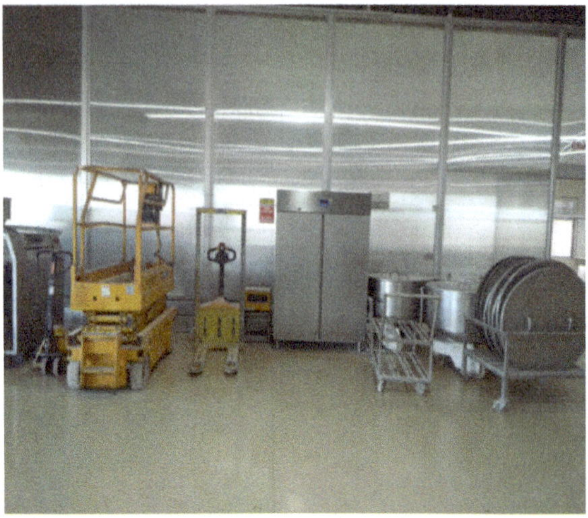

Fig. 3 Before the field definition

Fig. 4 After the field definition

Fig. 5 Image of the salt tank wall before

Fig. 6 Image of salt tank wall after

Fig. 7 Technical warehouse first

Fig. 8 Technical repository after

6 Conclusion

In general, the main purpose of the enterprises is to obtain the most possible product with the raw material at hand and to do this with maximum efficiency. It is of great importance that the methods used throughout this process are low-cost and easy to apply in order to ensure continuity. 5S and Kaizen are the techniques that companies frequently use in this context.

In this study, 5S and Kaizen applications were applied in the areas determined within the company. As a result of these practices, loss of time and unnecessary workload have been avoided, dangerous situations in terms of occupational health and safety have been prevented and zero loss of labor has been targeted with zero unnecessary movement. The total gain of the company with these improvements are the reducing waste of time, to prevent waste of raw material, increase employee motivation, reducing costs and setting order at the work areas.

In order for these practices to continue stably in all factories instead of a single factory within the company, the full support of the management, the full participation of the employees, and the continuation of practical training for the employees who will start working within the company will play an important role. For future studies, in addition to the areas within the company, including 5S and kaizen applications for the vehicles and other transportation vehicles used by the shipping unit will take the work to the next level.

References

1. Bilgin Sarı E (2018) Yalın Üretim Uygulamaları Ve Kazanımları. Uluslararası Iktis Ve Idari Incel Derg, 17. UİK Özel Sayısı, 585–600. https://doi.org/10.18092/ulikidince.439034
2. Ersöz T, Sarız K, Ersöz F (2020) Demir-Çelik Üretim Hattında Yalın Üretim. Düzce Üniversitesi Bilim ve Teknoloji Dergisi 8(1):801–826
3. Sarıkaya HA, Soydemir EK, Sardaş B, Çayır Ervural B, Şen DT (2022) Bisküvi Üretim Hatlarında Hazırlık Sürelerinin Azaltılmasına Yönelik Yalın Üretim Ve SMED Uygulaması. Endüstri Mühendisliği 33(2):413–439
4. Kaymakcı Ö (2012) Bir PTT Şubesinde Yalın Üretim–5S Uygulaması. (Yüksek lisans tezi, Sakarya Üniversitesi, Sakarya). Erişim adresi: https://hdl.handle.net/20.500.12619/80071
5. Özcan S, Yılseli H (2022) Tekstil sektöründe 5S uygulaması: Bir vaka çalişmasi örneği. Int New Issues Soc Sci 10(2):263–280. https://doi.org/10.5281/zenodo.7477984
6. Çelik H (2019) 5S Uygulamalarının ayar süreleri ve toplam ekipman etkinliğine etkisi. Yorum Yönetim Yöntem Uluslararası Yönetim Ekon Ve Felsefe Derg 7(2):95–110. https://doi.org/10.32705/yorumyonetim.569786
7. Michalska J, Szewieczek D (2007) The 5S methodology as a tool for improving the organisation. J Achiev Mater Manuf Eng 24
8. Berber İ (2013) Yalın üretim teknikleri, kaizen ve sektörel uygulamalari. Master's thesis, Fen Bilimleri Enstitüsü
9. Karakaya Ş (2019) Kaizen sistemi ile performans arttırımı ve ambalaj sektörü uygulamaları. Master's thesis, Fen Bilimleri Enstitüsü
10. Erdeniz M (2018) 5S ve Kaizen uygulamalarının işletme performansına etkileri: Mobilya sektöründe bir uygulama. Master's thesis, Sosyal Bilimler Enstitüsü

11. Akıl B (2013) Kamu sektöründe 5S ve Kaizen Uygulamaları (Doctoral dissertation, DEÜ Sosyal Bilimleri Enstitüsü). Erişim adresi. http://hdl.handle.net/20.500.12397/10527
12. Demir S, Paksoy T (2021) Lean management tools in aviation industry: new wine into old wineskins. Int J Aeronaut Astronaut 2(3):77–83
13. Keleş AE, Gürsoy G, Çelik GT (2016) 5s Sistematiği Aşamaları ve Örnek Bir Uygulama. Çukurova Üniversitesi Mühendislik-Mimarlık Fakültesi Dergisi 28(2):51–60
14. Savcı M (2019) Değer Yaratmayan Faaliyetlerin Azaltılmasında Kaizen Maliyetleme Sisteminin Etkisi. MANAS Sosyal Araştırmalar Dergisi 8(1):308–325. https://doi.org/10.33206/mjss.486436
15. Zerenler M, ve İraz R (2006) Japon yönetim anlayişi ve şirket ağlari keiretsu analizi. Selçuk Üniversitesi Sos Bilim Enstitüsü Derg, (16):757–777

Lean Implementation Steps for a Way to Sustainable Green: A Real-Life Case Study

Semra Birgün and Cemile Tuğçe Kaygısız

Abstract Manufacturers can contribute to environmental sustainability and green ecosystems by using their resources effectively and efficiently. For this purpose, it is necessary to develop new strategies and implement their applications. As one of the most effective approaches for green sustainability, lean is also an important and critical basis for establishing systems that are free from all kinds of waste. Lean transformation provides the opportunity to produce in a shorter time with the desired quality and at lower costs by using only the necessary resources without consuming any unnecessary resources. In this study, the value stream analysis and improvement steps carried out in a manufacturer company operating in the electronics sector, where the lean approach was decided to be applied due to high costs and quality problems experienced in the production of a new product, were summarized. As a result of the pilot application, it was observed that 58% savings were achieved from removing/ reducing non-value adding activities. Considering the benefits of improving only one process with lean techniques, company officials decided to apply lean techniques to other processes and spread the application throughout the company. As an inference, it is possible to say that the benefits that even a single company will obtain from resource consumption will be an important step towards environmental sustainability.

Keywords Green sustainability · Lean conversion · Process improvement · Sustainable development goals · Value stream mapping · Waste elimination

S. Birgün (✉)
Industrial Engineering Department, Doğuş University, Istanbul, Türkiye
e-mail: sbirgun@dogus.edu.tr

C. T. Kaygısız
R&D Department, Tron Electronic Systems, Istanbul, Türkiye
e-mail: tkaygisiz@tron.com.tr

N. M. Durakbasa and K. G. Gülen (eds.), *Sustainable Green Conversion*, Lecture Notes in Mechanical Engineering, https://doi.org/10.1007/978-3-031-83583-4_35

1 Introduction

The country leaders, who came together at the United Nations (UN) Sustainable Development Summit held in New York on September 25, 2015, aim to eliminate poverty in all its dimensions in the world by 2030, ensure the common welfare of humanity, and create a more beautiful and livable world for humanity and future generations. They adopted the Sustainable Development Goals (SDGs) [1], which consist of 17 goals and 169 targets. The 17 goals (Fig. 1) are interconnected because they focus on global issues. At the same time, it is an indicator of a holistic approach focused on the economy, environment, and society. Among these 17 objectives, objectives 9 and 12 are the ones that most reflect the responsibilities and contributions of production systems. Effective use of resources and improvement and development of production and consumption behaviors within the framework of efficiency and sufficiency open the door to a greener future. Sustainable production approaches include eliminating or reducing the effects of the production process on the human element, reducing waste in the production process, increasing the recyclability of produced products and directing design processes from this perspective, and developing material and energy-saving production processes.

In order to minimize the damage to the environment, ensure the sustainability of the environment and the company, reduce costs, and increase customer satisfaction, companies must either replace their old technologies with new ones or design a brand new system. The first approach may have risks such as creating incompatibility with existing technology and incurring greater inefficiencies and costs. In the second approach, the best solution seems to be applying the lean philosophy.

Fig. 1 UN sustainable development goals [1]

Lean is a management philosophy that businesses have been applying since the second half of the twentieth century to increase their performance, increase customer satisfaction, and ensure sustainability and continuous development [2]. Lean is an approach that aims to reduce and, if possible, eliminate waste in manufacturing and therefore environmental waste while creating value for the customer. In other words, it is aimed to carry out only value-adding activities and not to use any unnecessary resources; all other activities are considered waste when determining value-adding activities. Managers who adopt the lean philosophy have succeeded in providing the products and services that customers expect by eliminating waste in their processes with a holistic perspective and have achieved this with minimum production costs [3].

In the lean approach, there is a focus on the concept of waste. Waste is more than non-value added processes or materials that need to be eliminated. Waste is anything that causes deviation from the perfect process [4, 5]. To reveal what waste is, it is necessary to know how to see it [3]. According to Ohno [6], one of Toyota's managers, there are seven waste classes (muda) in a production system. These are overproduction, unnecessary waiting, unnecessary transportation, overprocessing, unnecessary motions, defective products, and most importantly, unnecessary inventory. A system can eliminate all or most of these wastes by applying lean techniques and tools such as 5S, pull system, heijunka, jidoka, total productive maintenance, total quality management, 6Sigma, SMED, 3 M, Kaizen, value stream mapping, etc. [7–13].

By becoming leaner, faster production with shorter lead times, higher-quality products and services, lower production costs, higher flexibility, a safer work environment, higher motivation and efficiency, and higher customer satisfaction are achieved. In a lean system, everything becomes visible, possible problems can be predicted, and quick solutions are provided. A safer work environment is created by ensuring continuity in production, existing capacity is used more effectively, and empty spaces can be used for new products or other purposes [7, 8].

Companies that apply lean management not only ensure their own sustainability with minimum production costs but also increase customer satisfaction and increase their market shares with their advanced quality products. It would not be wrong to say that minimum resource consumption also helps minimize the damage to the environment. It is known that lean creates synergy and provides more practical benefits to the environment and humanity when applied together with the green approach for a sustainable economy, protecting the world's scarce resources [7].

Adopting and implementing lean thinking and principles will bring the greatest success on the path to perfection. Lean is flexibility, quality, speed, satisfaction, savings, and continuity. If you want to compete, digitalize, or be agile, you have to be lean first [3]. In this study, the improvement stages are summarized, which were carried out using Value Stream Mapping (VSM) in the production process of the 'proximity sensor', which was newly included in the product range, in a manufacturer that started lean transformation practices. In the next part of the study, VSM is briefly explained, and in the third part, the case study is explained. The study ends with the conclusion section, where the findings are discussed.

2 Value Stream Mapping

The main purpose of lean production is to increase efficiency and quality, reduce or even eliminate inventory, and reduce costs. For this, activities that do not create value must be eliminated from the process, and features that do not represent value for the customer must be eliminated from the product [9]. Value Stream Mapping is the lean technique used for this purpose. Mapping the material and information flow processes for parts and semi-finished products in a value stream that includes suppliers, manufacturing, and shipment to the customer is known as "Value Stream Mapping". Value Stream Mapping allows you to look at the system from the big picture. By focusing on flow, it is possible for all employees to see and understand the material and information flows in a value stream, to easily distinguish between value and activities that create added value and those that do not, to easily identify waste and sources of waste, and to develop better designed situations to reduce/eliminate them [3]. Effective design and implementation of the activity flow ensures that customer needs are met with minimum consumption of resources [14, 15].

Mapping is carried out at various levels, such as inter-company, inter-factory, door-to-door in a single factory, and process level, at the stages of determining the product family/process, mapping and analysis of the current situation, mapping the future situation, and preparing the annual business plan. Value Stream Mapping, which is frequently used to simplify production by eliminating waste, has found many application areas in manufacturing and service [16–39].

3 Case Study of Process Improvement Under the Lean in an Electronic Equipment Manufacturer

In the manufacturing company operating in the Defense, Aviation, and Telecom sectors, studies have been initiated to transition to lean production due to the high wastage rates and production costs in the production processes of proximity sensors that have recently entered the product range. These products are used in the defense industry to control the position of landing gear on aircraft and helicopters, to obtain position information from firing mechanisms and ammunition feeding systems in cannon and rifle systems, and to control the status of moving parts in the field of view for maintenance and condition monitoring. This sensor consists of a combination of two stringed PCBs, a sensor, and mechanics. This sensor, called the proximity sensor, and its main steps in production are given in Fig. 2.

In order to improve the production process, first the current status map of the process was created.

A. *Current State Analysis*

In order to make a current status map study of the product, the assembly instructions were first examined. Visits were organized to the 'supply chain and production

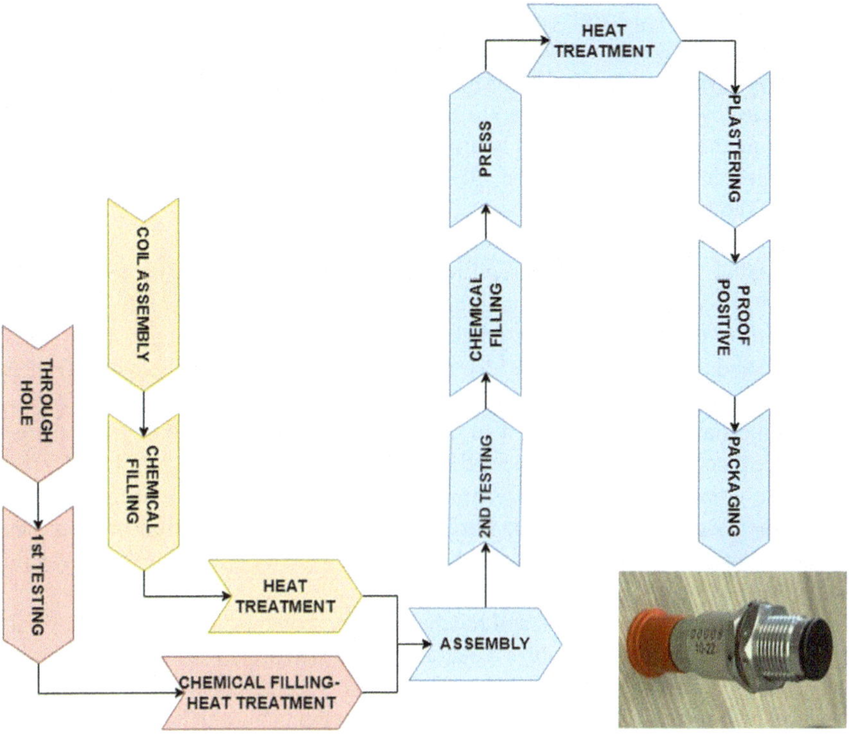

Fig. 2 Main production stages of proximity sensor

department', and the processes were examined. The process can be briefly summarized as follows: After the product order is received, it is put into the production plan. The electronic materials of the product are supplied from different suppliers. Printed circuit boards (PCBs) are supplied in panels of 40. The supplied electronic materials and printed circuit boards are mounted by the external supplier, as a surface mounted technology (SMT) manufacturer. After this process, arrayed PCBs are taken into stock.

The mechanics of the product are supplied from a different supplier. All products incoming to the company go through input quality control (IQC) processes. The raw material parts of the product are supplied from three different companies. In addition, electronic material supply stocks are examined and, if necessary, procured from different suppliers.

PCB production of the product is made by related PCB manufacturers. The produced PCBs and electronic materials are shipped together with the work orders to supplier company for SMD typesetting. Arranged PCBs are received by Tron Electronic Systems Inc. Co. and checked with an Automated Optic Inspection (AOI) device. During this process, the positions of the electronic materials, whether there are any missing materials, and the soldering quality are checked. The PCBs whose checks are completed are given a semi-finished code, and the next step is the through-hole process. In the current value stream map, the processes from top assembly to packaging have been examined. The initial electronic inspections of the semi-finished product, following the completion of the top assembly, are conducted to verify the correct functioning of the circuit board. Semi-finished products that show no issues during these inspections are then filled with chemicals and subjected to a curing process. The purpose of this chemical is to secure the electronic components. The chemical requires 30 min at 120 °C to cure.

In parallel with this process, coil preparation is carried out. The coil, which is procured pre-assembled from the supplier, undergoes cable assembly. To prevent movement of the coil within the mechanical assembly, the housing is filled with a chemical and cured. This process also requires the chemical to be held at 120 °C for 30 min.

The mechanical casings are also supplied by the vendor. Upon receipt, these mechanical components are inspected by the Incoming Quality Control (IQC) team to ensure compliance with the technical drawings. Parts that meet the tolerance requirements are transferred to the warehouse. The semi-finished products specific to the final product are then delivered to the production area according to the production schedule.

The semi-finished products and mechanical components, as prepared above, are assembled. Following assembly, a second test is performed on the semi-finished product to evaluate its operational performance. Products that meet the required performance criteria are then filled with a different chemical, and the mechanical parts are assembled together using a press. This product is held at 65 °C for 120 min to allow the chemical to cure. The difference in curing temperatures is due to the intended application of the chemicals. After 120 min, the product is removed from the oven and undergoes a final crimping process using the press machine. The press machine is equipped with a different tool to complete this crimping. The product, having completed all processes, then undergoes a final test. During this test, the product's overall performance is evaluated, and those that pass successfully are transferred to the packaging area.

The current state map of the proximity sensor is limited to the flow from the supply of semi-finished products to the packaging process. The production plan is prepared monthly according to delivery schedules together with the planning department and the production manager. When the production quantities are completed, MikroFly, which is used as the MRP program, is entered. The shipment of the products entered into the MRP program is planned and shipped to the customer.

The assembled PCBs are received from the supplier in weekly batches of 160 units. The received PCBs are checked at IQC and taken into storage. The coil is ordered in bulk from abroad and placed in storage until other semi-finished products are completed. Semi-finished products are delivered to the first process in accordance with the production planning and transferred via the pushing system. All processes operate in a single shift. Working 20 days a month, 585 min a day.

The customer places an annual order of 1000 units. It has been agreed that this order will be delivered over a period of 10 months, with shipments of 100 units each month. The annual order is fulfilled in 10 shipments.

The Value Stream Map showing the current status of the process is given in Fig. 3.

When we examine the map, value-adding activities in the production of a proximity sensor are 238.3 min. But the total flow time is 35.3 days. This situation has shown that there is great waste in production time.

B. *Future State Design*

The current state VSM was examined, waste times were taken into account, and a decision was made to establish a kanban system. The final test process was chosen as the pacemaker process. The planning department initiates the pacemaker by reporting the product need to the final testing process. Some cells were found suitable for production, and a U-type cell layout was designed. Excluding breaks, the working time available in a shift is 585 min. Accordingly, Takt time is calculated as 117 min per product.

The customer places orders on an annual basis. The supplier is provided with a weekly plan, and daily orders are placed accordingly. The assembled PCB boards are received daily in batches of 40 units. After the order is received, products are delivered weekly in boxes of 25 units.

The future state map obtained as a result of the studies is given in Fig. 4. Planning is communicated to the supplier on a daily basis and received every two days.

Other improvements to reduce waste in processes are as follows. The parts used for Test 1 measurement were enclosed in a box and were designed in a way that the measurement areas could be easily connected, resulting in a reduction in measurement times. A U-type cell was created by combining assembly and test 2 with chemical processes. Here, the test is arranged so that two people and the chemical process can be performed by one person. Thus, the assembly time and the duration of test 2 and chemical processes were balanced. Coil assembly and chemical treatment are combined. A structure that transfers chemicals with pressure has begun to be used to adjust the dose during the coil assembly process. Thanks to this structure, the right dose of chemicals is sprayed into the product. In addition, work has begun on an automatic test mechanism in order to reduce test times in the final testing phase. Thanks to this mechanism, the measurement time is aimed to be reduced to less than 30 min.

C. *Task Assignment for Lean Implementation*

Task assignments have been made in order to make the improvements suggested in the future state map. The Planner application of the Teams program is used for assignments within the company. Person assignments for opened tasks and completion date assignments for tasks are made with this program. Progress on the tasks can be followed in the comments section. Screenshots of the tasks are given in Fig. 5.

The task of improving the first test setup was initiated at the beginning of June. This task was completed within approximately one month, following the procurement of materials. The decision on the product to be acquired and the subsequent purchase of

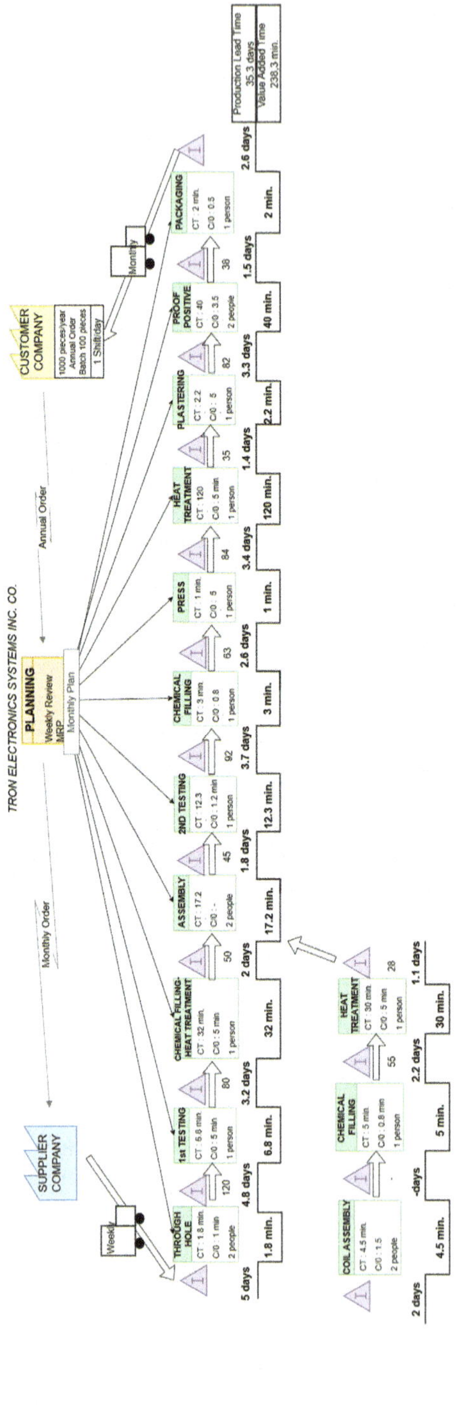

Fig. 3 Current state map

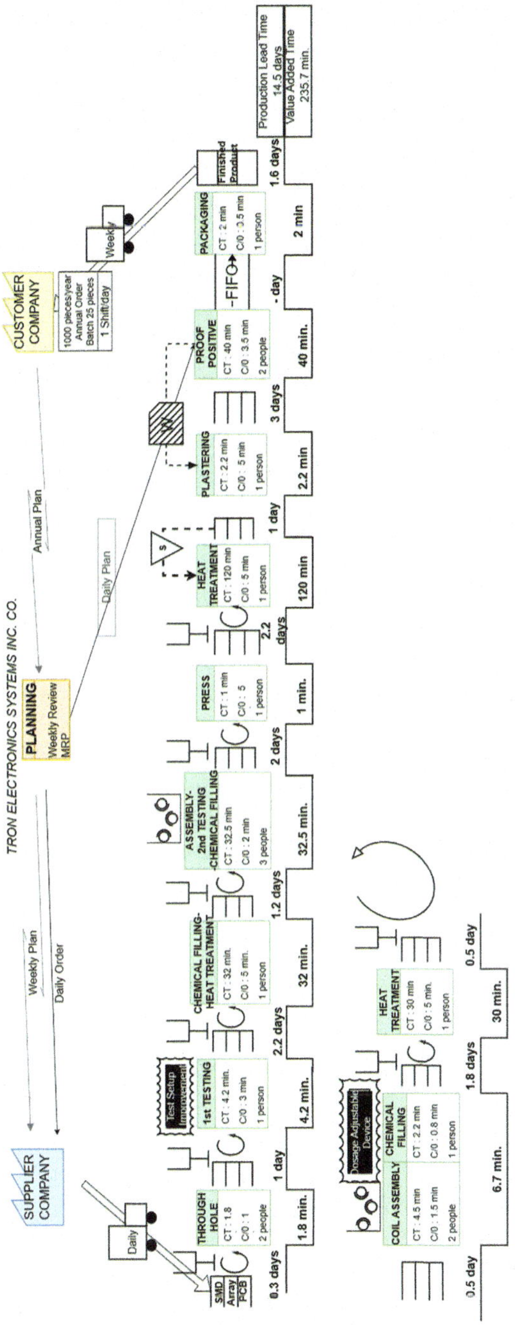

Fig. 4 Future state map

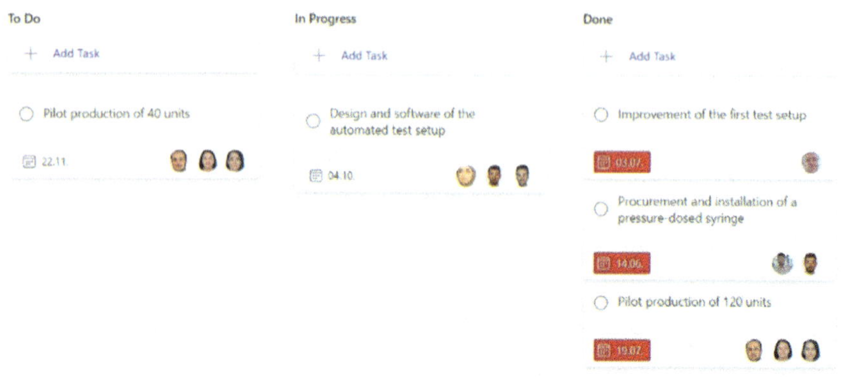

Fig. 5 Screenshots of the task assignment

the pressure-dose syringe took approximately three weeks. After these improvements were made, a pilot production of 120 units was carried out.

D. *Discussion*

A pilot application was carried out within the scope of 120 units of production to implement the change suggestions in the future state map. The findings obtained as a result of the application are shown in Table 1.

Since coil assembly and chemical filling processes are carried out in one cell, it is seen that the queues in front of these processes have decreased by a quarter. At the same time, it is observed that the work queue in front of the second cell has decreased by 62% compared to the previous situation.

The decrease in queues, and therefore the waiting times, and the suggested Kaizen studies have shortened the lead period considerably.

When the pull system was established based on the suggestions, a decrease was observed in both raw material inventory and semi-finished product inventory. As a result of the improvements in the processes, the waiting time during the production of the product was reduced from 35.3 days to 14.5 days. Value-added time is 238.3 min, while it became 235.7 min. As a result of the planned improvements, nearly 58% improvement in lead time was observed.

4 Conclusions

Lean philosophy is an essential approach that company managers should apply for green sustainability. It ensures that processes and systems are designed and operated correctly by eliminating waste. Trying to improve a process that is not designed correctly will only lead to the emergence of new problems. Even if the process is designed correctly, sometimes errors and/or problems occur in its application or

Table 1 VALUE added and non-value added times before and after lean implementation

Process		Through hole	1st stage testing	Chemical filling-heat treatment	Coil assembly	Chemical filling	Heat treatment	Assembly	2nd stage testing	Chemical filling	Hand press	Heat treatment	Plastering	Proof positive	Packaging	Finished product
Current state	VAT (mins)	1.8	6.8	32	4.5	5	30	17.2	12.3	3	1	120	2.2	40	2	–
	Non-VAT (days)	5	4.2	3.2	2	–	2.2	2	1.8	3.7	2.6	3.4	1.4	3.3	1.5	2.6
Future state	VAT (mins)	1.8	4.2	32	6.7		30	32.5			1	120	2.2	40	2	–
	Non-VAT (days)	0.3	1	2.2	0.5		1.8	1.2			2	2.2	1	3	–	1.6

operation, which are common problems. One of the most important tools used by lean production to analyze and improve systems is VSM.

With VSM, both the analysis of the process and its correct design are carried out. When combined with effective managers and employees, success is inevitable. In other words, any changes made without making the systems lean will be completely superficial, will lead to different problems, and failure will be a sustainable outcome. First of all, correcting mistakes and errors and implementing correctly designed and planned processes; afterwards, it is best to implement the desired changes.

In this company, where lean implementation steps are summarized, managers have found this right way and achieved success. If we want to summarize the success achieved, with the installation of cells and the pull system, the lead time has been shortened due to the significant reduction in the queues of jobs waiting to be processed. In addition, the decrease in the number of waiting jobs as a result of Kaizen studies also contributed to the reduction in lead time. The pull system also reduced the raw material stock. With all these improvements, the relevant product was provided in 20.8 days less time. This shortening in lead time enabled the stock turnover to approximately double. Therefore, the company had the opportunity to increase the production amount of this product.

Consequently, widespread use of such practices will make a significant contribution to the sustainability of ecosystems.

Acknowledgment We would like to express our gratitude to Tron Electronics Systems for the support and resources provided throughout this study.

References

1. Naturel Farming (2024) Sustainable development goals homepage [Online]. Available: https://naturalfarming.niti.gov.in/sustainable-development-goals/
2. Birgün S, Kulaklı A (2022) Eliminating the barriers of green lean practices with thinking processes. In: Numan M. Durakbasa, Günes Gencyilmaz M (eds) Digitizing production systems, lecture notes in mechanical engineering, pp 372–383. Springer. ISBN 978–3–030–90421–0 (eBook), ISBN 978–3–030–90420–3
3. Birgün S (2023) Yalın Üretim, In: Birgün S (ed) Yedi Yalın Yaklaşım. Nobel Yayınevi, İstanbul, pp 3–35. ISBN: 978–625–397–753–5 E-ISBN: 978–625–397–754–2
4. Liker JK, Ross K (2016) Toyota way to service excellence. ISBN: 1259641104, Mc Graw Hill
5. Görener A (2023) Yalın Lojistik. In: Birgün S (ed) Yedi Yalın Yaklaşım. Nobel Yayınevi, İstanbul, pp 69–103. ISBN: 978–625–397–753–5 E-ISBN: 978–625–397–754–2
6. Ohno T, Mito S (1988) Just-in-time for today and tomorrow. Productivity Press, New York
7. Birgün S, Kulaklı A (2021) Yalın felsefe ve yeşil yönetim. In Tekin M, Mert G (eds) Yeşil Yönetim, pp 15–23. Nobel Akademik Yayıncılık, Ankara (Publication No: 3603). ISBN: 978–625–439–781–3
8. Ulu M, Birgün S (2024) A case study on lean occupational safety. Sigma J Eng Nat Sci 42(3):1–15. https://doi.org/10.14744/sigma.2022.00108
9. Şahin Y (2023) Bir yalin teknik uygulaması. In: Birgün S (ed) Yedi yalin yaklaşım, Nobel Yayınevi, İstanbul, pp 213–235 ISBN: 978–625–397–753–5 E-ISBN: 978–625–397–754–2

10. Ulu M (2023) Gastronomi mutfağinda yalin iş güvenliği adimlari. In: Birgün S (ed) Yedi Yalın Yaklaşım. Nobel Yayınevi, İstanbul, pp 247–272. ISBN: 978–625–397–753–5 E-ISBN: 978–625–397–754–2
11. Ulu M, Birgün S (2023) İş Sağlığı ve Güvenliği Alanında 5S ve Görsel Fabrika Uygulamaları, In: Semra Birgün, Kemal Güven Gülen, Yıldız Şahin, Kıvanç Onan, Elifcan Yaşa (eds) Yönetim Araştırmaları/Mühendislik Uygulamaları 2023 (YÖNAR/MU'2023) Sempozyumu Bildiri Özetleri Kitabı, pp 94–95, Eğitim Yayınevi, ISBN: 978–625–6489–84–4
12. Ulu M, Birgün S (2022) A new model proposal for occupational health and safety. In: Durakbasa NM, Gençyılmaz MG (eds) Digitizing production systems. Lecture notes in mechanical engineering, pp 347–356, Springer, Cham. https://doi.org/10.1007/978-3-030-90421-0_29
13. Birgün S, Özkan K, Gulen KG (2006) Value stream mapping: A case study on manufacturing industry. In: IMS'2006: 5th international symposium on intelligent manufacturing systems agents and virtual worlds, (CD ROM), 29–31 May 2006, Sakarya
14. Orhon Basık F, Yalın Muhasebe (2023) In: Birgün S (ed), Yedi Yalın Yaklaşım, Nobel Yayınevi, İstanbul, pp 114–148. ISBN: 978–625–397–753–5 E-ISBN: 978–625–397–754–2
15. Huntzinger J (2006) Economies of scale are dead: Right-sizing for effective cost management and operations. J Cost Manag 20:18–26
16. Rother M, Shook J (2003) Learning to see: Value stream mapping to add value and eliminate Muda. Lean Enterprise Institute, New York
17. Birgün S (2006) Aksiyomlarla Tasarım Yoluyla Değer Akışı Haritalandırma. In: Yöneylem Araştırması ve Endüstri Mühendisliği XXVI. Ulusal Kongresi Bildiriler Kitabı, Kocaeli, pp 35–40
18. Braglia M, Carmignani G, Zammori F (2006) A new value stream mapping approach for complex production systems. Int J Prod Res, 44:18–19, 3929–3952. https://doi.org/10.1080/00207540600690545
19. Gülen KG, Birgün S (2007) Creating value in enterprise information system: KVS model. In: Erkollar A (ed) State of the art: Enterprise and business management a handbook for educators, consulters and practitioners, pp 148–176. Malburg: Tectum Verlag. ISBN: 978–3–8288–9282–8
20. Birgün S, Gulen KG, Özkan K (2006) A case study on eliminating waste from the business processes. In: Proceedings of the 15th annual world business congress, international management development association, Sarajevo, Bosnia and Herzegovina, pp 41–46
21. Serrano Lasa I, Ochoa Laburu C, de Castro Vila R (2008) An evaluation of the value stream mapping tool. Bus Process Manag J, 14:39–52. https://doi.org/10.1108/14637150810849391
22. Ramesh V, Sreenivasa Prasad KV, Srinivas TR (2008) Implementation of a lean model for carrying out value stream mapping in a manufacturing industry. J Ind Syst Eng 2:180–196
23. Chen L, Meng B (2010) The application of value stream mapping based lean production system. Int J Bus Manag 5:203–209
24. Jiménez E, Tejeda AS, Pérez M, Blanco J, Martínez E (2012) Applicability of lean production with VSM to the Rioja Wine Sector. Int J Prod Res 50:1890–1904. https://doi.org/10.1080/00207543.2011.561370
25. Silva N (2012) Applicability of Value Stream Mapping (VSM) in the apparel industry in Sri Lanka. Int J Lean Think 3(1):36–41
26. Jasti NV, Sharma A (2014) Lean manufacturing implementation using value stream mapping as a tool: A case study from auto components industry. Int J Lean Six Sigma 5:89–116. https://doi.org/10.1108/IJLSS-04-2012-0002
27. Faulkner W, Badurdeen F (2014) Sustainable Value Stream Mapping (Sus-VSM): Methodology to visualize and assess manufacturing sustainability performance. J Clean Prod 85:8–18. https://doi.org/10.1016/j.jclepro.2014.05.042
28. Henrique DB, Rentes AF, Godinho Filho M, Esposto KF (2016) A new value stream mapping approach for healthcare environments. Prod Plan Control, 27:24–48. https://doi.org/10.1080/09537287.2015.1051159
29. Verma N, Sharma V (2016) Energy value stream mapping a tool to develop green manufacturing. Procedia Engineering 149:526–534. https://doi.org/10.1016/j.proeng.2016.06.701

30. Deshkar A, Kamle S, Giri J, Korde V (2018) Design and evaluation of a lean manufacturing framework using, Value Stream Mapping (VSM) for a plastic bag manufacturing unit. Mater Today Proc 5:7668–7677
31. Dadashnejada AA, Valmohammadi C (2019) Investigating the effect of value stream mapping on overall equipment effectiveness: A case study. Total Qual Manag 30:466–482. https://doi.org/10.1080/14783363.2017.1308821
32. Birgün S, Tanrıkut Çerkezoğlu B (2019) A systematic approach for improving the software management process. Int J Innov Technol Manag, 16/4:1–28. https://doi.org/10.1142/S0219877019400066
33. Zahoor S, Abdul-Kader W, Ijaz HA, Khan Q, Saeed Z, Muzaffar S (2019) A Combined VSM and kaizen approach for sustainable continuous process improvement. Int J Ind Eng Oper Manag (IJIEOM) 1:125–137
34. Birgün S, Gülen KG (2020) Key value stream approach for increasing the effectiveness of business processes. AURUM—Journal Eng Syst Arch 4(2):201–223
35. Khan MA, Shaikh SA, Lakho TH, Mughal UK (2020) Potential of lean tool of Value Stream Mapping (VSM) in manufacturing industries. In: Proceedings of the 2nd African international conference on industrial engineering and operations management Harare, pp 3064–3074
36. Mangers J, Minoufekr M, Plapper P (2021) Value Stream Mapping (VSM) to evaluate and visualize interrelated process-chains regarding circular economy. In: IFIP international conference on Advances in Production Management Systems (APMS), pp 534–542. Nantes/France. https://doi.org/10.1007/978-3-030-85910-7_56. ffhal-03806494f
37. Reda H, Dvivedi A (2022) Application of Value Stream Mapping (VSM) in low-level technology organizations: A case study. Int J Product Perform Manag 71:2393–2409. https://doi.org/10.1108/IJPPM-03-2021-0118
38. Al-Rifai M (2024) Mapping the path to efficiency: Harnessing Value Stream Mapping (VSM) and lean tools for streamlined electronic device manufacturing. Meas Bus Excel ahead-of-print:ahead-of-print. https://doi.org/10.1108/MBE-06-2024-0087
39. Birgün S (2019) A hybrid approach for process improvement. In: Proceedings of the ISMC 2019 15th international strategic management conference, pp 417–429. ISBN: 978-605-81347-1-3

Quality in Production and Service

Osman Bodur, Ángela Moya García, Günther Poszvek, and Numan M. Durakbasa

Abstract This research examines the evolving relationship between quality and production within both manufacturing and service sectors. It offers a detailed examination of how quality management practices have changed over the years. A key focus is placed on the impact of Japan's success, which influenced Western manufacturers and triggered a significant shift in how quality is understood and incorporated into production processes. Japanese companies, by implementing practices like Just-in-Time (JIT) production and fostering strong employee engagement, showcased the clear benefits of prioritizing quality to improve both efficiency and competitiveness. This study underscores the vital need to maintain high-quality standards throughout all stages of production—ranging from product development and manufacturing processes to product usage—to achieve customer satisfaction and remain competitive in the market. For service organizations, the unique challenge of delivering consistent quality alongside personalized experiences is addressed. The research also emphasizes the role of employee involvement and empowerment in fostering continuous process improvements. In addition, the concept of the "hidden factory" is explored, revealing how inefficiencies and rework can be minimized to improve organizational productivity and quality. By analyzing different strategies for enhancing efficiency and developing a holistic understanding of quality, this research provides managers and organizations with valuable insights. These insights are essential for adapting to modern challenges in quality management and achieving sustainable success in today's competitive landscape.

O. Bodur (✉) · Á. M. García · G. Poszvek · N. M. Durakbasa
Institute of Production Engineering and Photonic Technologies, TU Wien, Vienna, Austria
e-mail: osman.bodur@tuwien.ac.at

Á. M. García
e-mail: e12206575@student.tuwien.ac.at

G. Poszvek
e-mail: guenther.poszvek@tuwien.ac.at

N. M. Durakbasa
e-mail: numan.durakbasa@tuwien.ac.at

N. M. Durakbasa and K. G. Gülen (eds.), *Sustainable Green Conversion*, Lecture Notes in Mechanical Engineering, https://doi.org/10.1007/978-3-031-83583-4_36

Keywords Just-in-time production · Quality management · Business performance · Service industries · Manufacturing industries · Employee involvement · European foundation for quality management · Quality awards · Sustainable development · Production systems

1 Introduction

Quality encompasses various aspects such as product defects, customers, systems, and production processes. The principles introduced by Frederick W. Taylor, known as the pioneer of Scientific Management, laid the foundation for contemporary production philosophies. Taylor's approach involved breaking down tasks into individual work units and segregating inspection activities from production tasks [1]. Consequently, this led to the establishment of dedicated Quality Departments within production organizations.

Another frontrunner worth mentioning during the early stages of quality control is the Bell Telephone System. The Bell Telephone Company, also known as the Bell System, was a prominent telecommunications company that was committed to ensuring reliable and efficient telephone services. They understood that maintaining high-quality standards was crucial for customer satisfaction and business success. In the early 1900s, the Bell System established an inspection department within Western Electric Company to oversee quality assurance throughout the product lifecycle, setting a precedent for other organizations to recognize the need for separate quality control functions [2].

Understanding the concept of "fitness for use" is relatively straightforward, as in [3]. It describes how well a product or service meets the particular needs and expectations of its users or customers. In simpler terms, a product or service is considered suitable for use when it successfully serves its intended purpose and meets customer satisfaction. As no one would purchase things that are entirely unsuited for use, the company's profit loss would be infinite at an 0% degree of product fitness for use, as Fig. 1 is shown. On the other hand, achieving 100% fitness implies a very high cost (infinite investment), which cancels the profit made. This would lead to infinite loss. The optimal value should be reached so that costs are minimized and the fitness for use maximized.

As previously mentioned, it was applied to various departments of the Western Electric Company and later to others. Ensuring alignment between departmental and organizational goals can be a challenging task, often requiring dedicated individuals or departments responsible for fostering synergy.

As time went forwards, during the 1920s, the inspection department employees at Western Electric were transferred to Bell Telephone Laboratories. This group took on the responsibility of developing innovative theories and methodologies for inspection aimed at enhancing and maintaining quality. Notably, pioneering figures such as Walter Shewhart, Harold Dodge, George Edwards, and others were part of this team. It was within this context that the term "Quality Assurance" would originate later. As

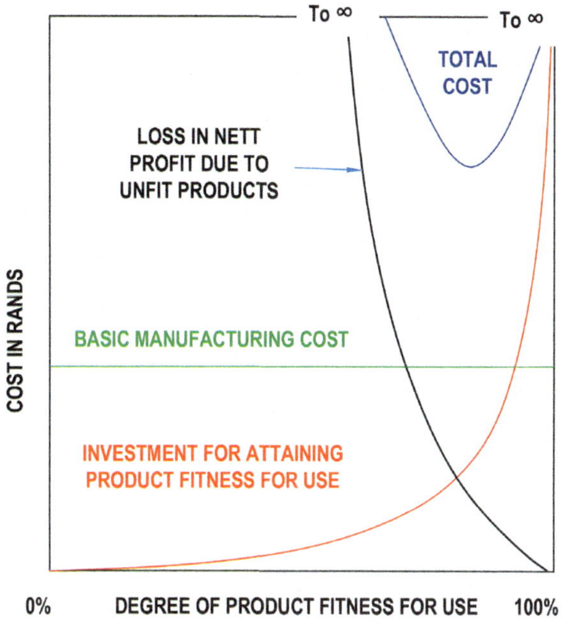

Fig. 1 Degree of product fitness in use, own creation from [3]

mass production and increasingly complex design structures emerged, the creation and sale of fit-for-use products became more demanding. Reference [3] also shows that Quality Control engineers possessed extensive knowledge of relevant technologies associated with the products and received training in statistical methodologies, inspection techniques, reliability studies, and other tools that proved instrumental in improving and monitoring product quality. The development of control charts by Shewhart, sampling techniques by Dodge, and economic analysis methods for quality problem-solving laid the groundwork for modern quality assurance.

During the era of World War II, the United States military introduced statistical sampling procedures and enforced strict standards for suppliers. As a result, statistical quality control gained widespread recognition and gradually found its way into further industries. During this era, sampling tables referred to as "MIL-STD" (military standard) were introduced and continue to be utilized today. In 1944, the inaugural edition of the Industrial Quality Control journal was released, and shortly afterward, institutions like the American Society for Quality Control were established. These advancements highlighted the increasing significance of standardization and cooperation in quality control practices. These journals and organizations showed the growing need for sharing knowledge, working together, and creating standardized practices in quality control. The creation of these journals and societies highlighted the increasing awareness of the importance of knowledge exchange, collaboration, and the development of standardized practices in quality control. These initiatives reflected a broader trend of organizations increasingly valuing and prioritizing quality as a strategic imperative in a competitive marketplace, as in [4].

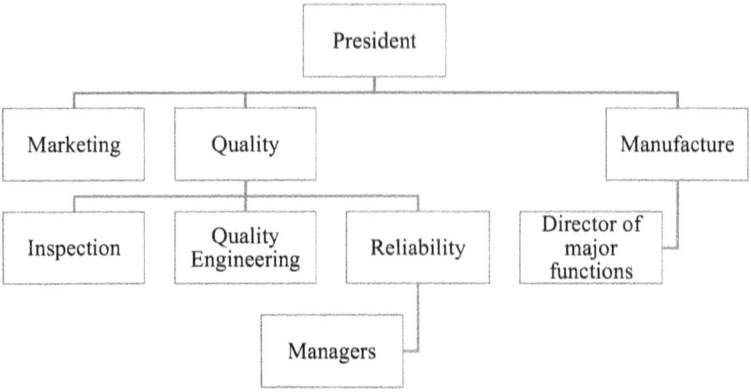

Fig. 2 Typical organization structure in the USA as of 1970. Own creation from [3]

The Feigenbaum 'Total Quality Control' concept is considered a precursor to the modern understanding of Quality Assurance and was developed in the 1960s. Quality Engineering technology and control were linked with management, and Feigenbaum emphasized a comprehensive approach to managing and improving quality throughout an organization. Feigenbaum's concept recognized the importance of integrating quality principles and practices into every level and function of an organization. It was stated that the responsibility of maintaining high-quality levels should be ingrained in the organizational culture and embraced by all employees, as in [5].

Figure 2 is seen detailing a typical organizational structure in the USA from 1970. Quality management was given a broader perspective and importance at the company level.

A centralized unit for quality with the functions that research to it can be observed in Fig. 2. This structure was significant in the U.S. of America and especially Japan. From Reference [6], the Japanese case is of great interest due to its remarkable transformation following the country's surrender to the Allied Forces in 1945. Japan, initially devastated and lacking production capabilities, defied expectations by emerging as a formidable competitor in a relatively short span of time. American industrialists, underestimating Japan's potential, granted the country significant manufacturing licenses and helped in acquiring modern business and technological skills. Furthermore, the Japanese government played a crucial role by implementing policies to rebuild the economy and foster industrial strength. Consequently, Japan experienced extraordinary growth, with the gross national product expanding by over 200% during the 1970s. The factors contributing to Japan's success are complex and multifaceted, including its unique management style, high exportation rates, access to low-interest long-term capital loans, and other factors that shaped its competitive advantage. Notably, Japan made substantial investments in research and development (R&D), automated factories, and staff training, allowing them to gradually adopt new technologies and establish the most productive manufacturing industry worldwide.

Oakes states that the Japanese industry utilized robust and flexible mass production methods, manufacturing high-quality, cost-effective, and easily accessible products while preserving adaptability for new model production and a broad spectrum of product innovation at competitive prices. The Kanban system and Just-in-Time (JIT) production methods employed by Japanese manufacturing companies played a pivotal role in their competitive performance, effective quality management practices, cost reduction, and efficient detection of production issues through reduced inventories.

An analytical framework by Matsui as in [7] depicts the four major building blocks of JIT: organization and human resource management, quality management, production information systems, and JIT production systems, as well as technology development and manufacturing strategy, all of which contribute to competitive performance. For a more detailed exploration of this topic, the article "An Empirical Analysis of Quality Management in Japanese Manufacturing Companies" by Yoshiki Matsui is highly recommended (Fig. 3).

From Reference [7], it can see that Just-in-Time (JIT) production strongly influences organizational behaviors, human resources, technology development activities, manufacturing strategy implementation, and vice versa as illustrated in Fig. 3. JIT principles and practices can substantially influence quality management by minimizing waste, encouraging continuous improvement, stressing standardization and documentation, strengthening supplier quality management, focusing on prevention and error-proofing, and upholding a robust customer-centric approach. The integration of JIT and quality management fosters efficient and effective production processes while ensuring high-quality products and customer satisfaction [8].

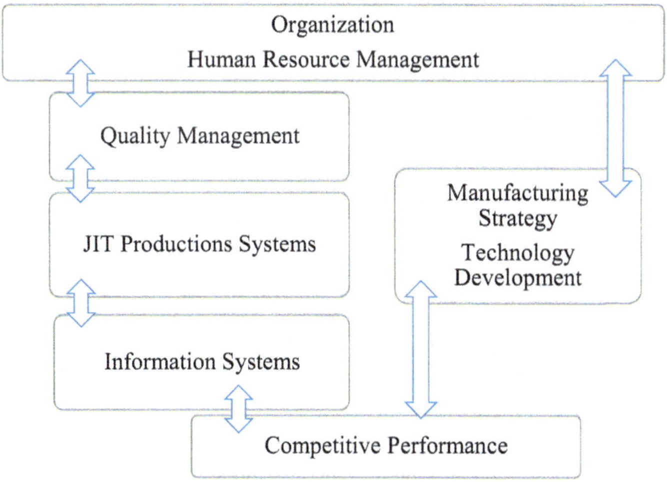

Fig. 3 Analytical framework by Y. Matsui. Own creation from [7]

Lastly, Oakes analyzes the Japanese manufacturer and their society, illuminating the core values that vary considerably from our own. These values include dependence, duty, social obligation, and human sentiment. The reference [6] delves into the distinctive and profound bond between managers and subordinates, emphasizing their shared sense of pride and camaraderie within their teams. It is worth reiterating that Japanese commerce maintains a harmonious and cooperative relationship with the government. The industry collaborated closely with the Ministry of Finance and the Japanese Development Bank, operating within a national policy framework for growth and finance. Furthermore, notable American consultants Joseph Juran and W. Edward Deming played a pivotal role in introducing statistical quality control techniques in Japan, leading to further advancements in the field, as in Reference [9].

Improvements in Japanese products surpassed those of Western manufacturers, as highlighted by Oakes. While Japanese manufacturers continuously enhanced their quality and quality assurance methods, Western counterparts struggled to make significant progress. During the 1970s, Japanese companies made substantial inroads into Western markets, primarily due to the superior quality of their products. The 1980s marked a period of remarkable change and increased awareness of quality among consumers, industries, and governments [6]. Consumers started to perceive quality disparities among Japanese, American, and European products.

A striking example illustrating this disparity was reported by Hewlett-Packard in 1980, as in Reference [10]. After testing 300,000 16 K RAM chips from three U.S. and three Japanese manufacturers, it was revealed that the Japanese chips had an incoming failure rate of 11 failures per 1000, while the U.S. chips experienced 19 failures per 1000. Furthermore, after 1000 h of use, the failure rate of the U.S. chips increased up to 27 times higher than that of the Japanese chips. In a short span of time, the Japanese had successfully penetrated a significant market that American companies had previously dominated. As a result, American and European nations commenced examining Japanese strategies to comprehend and emulate their achievements.

The early 1980s witnessed extensive product recalls mandated by the Consumer Product Safety Commission, along with the Challenger space shuttle disaster in 1986, as in Reference [11]. These incidents raised consciousness of the quality disparity between Western manufacturers and their Japanese equivalents. In 1980, NBC aired a white paper titled "If Japan can…why can't we?" The program shed light on the pivotal role played by W. Edwards Deming in Japan's quality development, making his name a household word among corporate executives. Deming then led U.S. companies, including Ford Motor Company, in a concerted effort to revolutionize their approach to quality [12].

From reference [13], it can see that the success of Japanese companies and the increasing recognition of quality gaps prompted Western companies to reassess their quality management practices and strive for improvement. In 1985, NASA introduced the Excellence Award for Quality and Productivity, encouraging organizations to pursue total quality excellence as a key driver of global competitiveness. This initiative prompted significant companies to launch extensive quality improvement campaigns. In a similar vein, the U.S. government designated October as National

Quality Month in 1984 [Proclamation 5249 by Ronald Reagan]. The establishment of the Malcolm Baldrige National Quality Award in 1987, through an act of Congress, served as a testament to the United States' commitment to quality leadership.

In the 1990s, the President of Philips, C. van der Klugt, proposed a collaboration on European Quality to C. De Benedetti, the President of Olivetti. The aim was to provide European companies with support in implementing total quality management and preventing Europe from falling behind Japan and the United States. As a priority, the focus was placed on creating a European Quality Award. Taking inspiration from the renowned Malcolm Bridge Model, which served as a reference for quality managers and consultants, the first model of the European Quality Award was drafted during the inaugural EFQM (European Foundation for Quality Management) Forum in October 1990 in London.

The European Quality Award model, finalized in 1992, consisted of eight criteria encompassing leadership, strategy, process management, resource management, employee satisfaction, customer satisfaction, social recognition, and business outcomes. The model distinguished enablers from results and emphasized customer satisfaction. Notice the following Fig. 4, created by the author of this article from resource [14].

The EFQM Model serves a dual purpose, providing not only a framework for recognizing excellence but also equipping managers with a tool to guide continuous improvement within their organizations. The assessment methodology involved either award-derived assessments, employing a standard mode with predefined weights and a traditional audit approach, or organizations conducting their own assessments using open contingency models, avoiding fixed weights and adopting a diagnostic approach [14]. However, both the EFQM Model and the Baldrige Model exhibit limitations in terms of self-assessment. To ensure maximum effectiveness, customization of the model is crucial, as managers are more likely to engage in

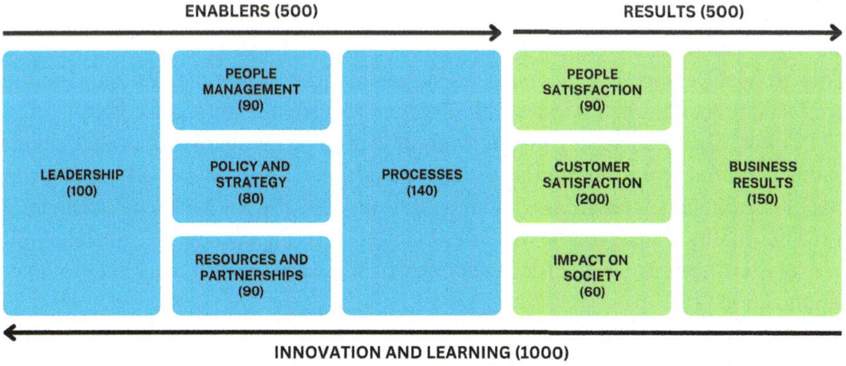

Fig. 4 European quality award model version 1992 with European Foundation for Quality Management (EFQM) model. Own creation from [14]

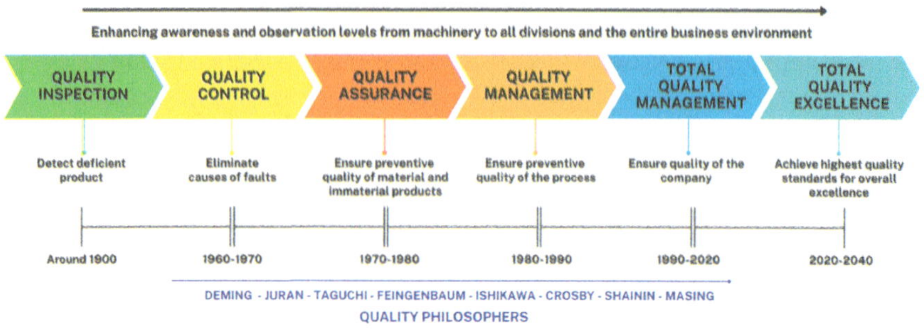

Fig. 5 Overview of concepts in quality management. Own creation from [1]

self-evaluation when it aligns closely with their specific organizational needs. Nevertheless, it is noteworthy that only a subset of companies has developed a genuine commitment to quality, and the implications for competitiveness are striking.

A study conducted in 1992 by Ernst and Young in collaboration with the American Quality Foundation revealed that while 55% of U.S. firms evaluate business performance using quality information on a monthly or more frequent basis, an impressive 70% of Japanese firms do so [15, 16].

In contrast, a mere 18% of U.S. businesses assess the business consequences of quality performance less than once a year, with Japan reporting a mere 2% and Europe at 9%. Financial and sales reviews occur much more frequently in the United States. To stay competitive in the global market, companies must incorporate quality into conventional management practices, acknowledging its crucial function in attaining success [17].

From the reference [18] and Fig. 5, it can see a visual overview of concepts in quality management.

The perspective on quality-related issues has continuously evolved over the past century, expanding from a narrow focus on the final product to a holistic view of the organization as a complex system that requires management and improvement. This broader perspective encompasses various entities within the system, including products, processes, and partners, all of which require attention and care.

Looking ahead, achieving higher levels of overall performance necessitates anticipating potential issues and finding proactive solutions. This includes addressing societal concerns such as environmental responsibility and the need for social change. As a result, quality management will remain crucial for the success and sustainability of every business, as well as for society, which serves as the ultimate client for every organization [19].

As mentioned earlier, customization and context-specific models will play a significant role in advancing quality management practices. However, there is a question regarding whether the control-focused approach of Total Quality Management (TQM) is appropriate for environments characterized by high uncertainty in roles, tasks, and organizational priorities [20].

From reference [21], it can see that in situations of high uncertainty, the focus may shift towards exploration and learning rather than strict control. In terms of societal and community considerations, the increasing implementation of ISO 14000 environmental assurance programs by quality departments is noteworthy. This trend reflects companies' growing concern for environmental issues and may pave the way for more conscientious models in the future. It can be anticipated that international customers will push for or even demand ISO 14000 compliance from their suppliers, thereby significantly influencing both domestic and international industrial marketing, as in [19].

Customer satisfaction, highlighted in this article through the mention of the European Quality Award, becomes increasingly important in the context of mass customization. The ability to meet the individual needs of customers will drive companies to enhance flexibility in production and logistics. Consequently, quality systems will need to prioritize the management of individual customer requirements to sustain growth.

2 Quality in Production Systems

Quality extensively influences every facet of an organization, making comprehensive systems approach essential for successful product outcomes. Production is the conversion of resources into goods and services. The collection of interrelated activities and operations involved in producing goods is called a production system. While manufacturing serves as the primary focus of our discussion, it is essential to recognize that service organizations also operate within production systems [22]. This topic will be addressed later. In any production system, several key areas of quality demand attention.

A. *Product development*

The quality efforts focus on customer-centric activities aimed at identifying customer needs and requirements and translating them into product designs that meet the criteria of fitness for use. By actively engaging with customers and conducting market research, organizations can gather valuable insights to drive product development decisions. During the product development phase, quality efforts involve rigorous testing, prototyping, and validation processes to identify and rectify potential design flaws, ensuring that the final product meets the desired quality standards while being both reliable and functional. In this phase, organizations also strive to comply with relevant regulations and guidelines to guarantee user safety and avoid any legal or ethical issues [23, 24].

B. *Production process*

It involves the coordinated activities and resources used to transform inputs into finished products or services. Physical facilities, information systems, and control mechanisms facilitate the transformation. Quality efforts in production processes focus on ensuring adherence to specifications and quality standards through robust quality control measures, such as rigorous inspection and testing procedures, to identify and address deviations or defects promptly. Another critical aspect is the implementation of process controls. This includes setting up standardized operating procedures, defining critical control points, and utilizing statistical process control techniques to monitor and maintain process stability and consistency [25].

C. *Product use*

Another customer-oriented aspect involves the actual utilization of the product. Quality considerations here revolve around meeting customer requirements and delivering satisfactory service experiences, like, for example, providing ongoing support to ensure the product functions as intended. One aspect of quality efforts in product use involves providing clear and comprehensive user documentation, including manuals, guides, and instructions. These materials help users understand how to operate, maintain, and troubleshoot the product appropriately. Clear instructions not only enhance the usability of the product but also contribute to user safety by preventing misuse or accidents [26].

Quality efforts related to Product Use also include ongoing customer support such as helplines, online chat systems, or dedicated customer service teams. Lastly, Feedback mechanisms, such as customer surveys or online reviews, are invaluable for quality efforts in product use. By actively seeking and analyzing customer feedback, organizations can identify areas for improvement, address potential issues, and incorporate customer preferences into future product iterations [27].

For service organizations, the production process and the beginning of product use occur at the same point in time. This distinction calls for the adoption of different quality perspectives to account for the unique nature of services. Understanding the importance of quality across various stages, including product development, the production process, and product use, is essential for attaining favorable outcomes within a comprehensive production system. By acknowledging and addressing the significance of quality throughout these interconnected phases, organizations can optimize their service delivery, meet customer expectations, and drive overall excellence in their operations.

3 Components of Production Systems

A typical production system for manufactured goods includes:

- Marketing and Sales
- Product Design and Process Engineering
- Purchasing and Receiving
- Production Planning and Scheduling
- Manufacturing and Assembly
- Tool Engineering

- Industrial Engineering and Process Design
- Finished Goods Inspection and Test
- Packaging, Shipping, and Warehousing
- Installation and Service
- Business Support Systems.

If quality management is to be effective, it must address all components of the production system. The role of quality in each component of this system is described in the following paragraphs.

A. *Marketing and sales*

The Marketing and Sales teams play a vital role in understanding customers' product needs and expectations. By conducting effective market research and soliciting customer feedback, they gather insights that inform product development. This customer-centric approach ensures that products are designed to meet functional requirements and align with customers' price expectations [28].

B. *Product Design and Process Engineering*

The Product Design and Process Engineering functions are responsible for developing technical specifications that meet the requirements identified by Marketing. They strive to strike a balance between meeting customer needs without over-engineering the product, which can result in inefficiencies and lower profits. Additionally, these teams focus on avoiding under-engineering and poor process designs that may compromise product quality [24, 29].

C. *Purchasing and Receiving*

While purchasing is often associated with price and delivery considerations, its role in quality assurance is crucial. The purchasing department is responsible for selecting quality-conscious vendors and ensuring that purchased parts meet the specified quality requirements by product design and engineering. Effective communication with vendors and diligent receiving processes helps maintain the desired quality standards. This means receiving must ensure that the purchased items delivered are of the quality that was contracted for by purchasing and that the defective parts are not received [30].

D. *Production planning and Scheduling*

Proper production planning and scheduling are essential to ensure a smooth flow of production. With suitable materials, tools, and equipment available at the appropriate time and place, production processes can operate efficiently. Inadequate planning and scheduling can lead to disruptions and time pressures, often resulting in compromised quality. Concepts like "just in time" have demonstrated improvements in quality through efficient planning and scheduling, as explained earlier [31].

E. *Manufacturing and Assembly*

Quality assurance emphasizes the need to build quality into products from the start rather than relying solely on inspection. Proper control of the laboratory, materials, and equipment is necessary to achieve high-quality outputs. Effective collaboration among first-line supervisors, production workers, engineers, and other stakeholders is vital for ensuring adequate quality assurance [32].

F. *Tool Engineering*

Tool Engineering supports manufacturing and inspection processes by designing and maintaining the tools used in production. Properly functioning tools are essential to avoid defects in manufactured parts, while well-calibrated inspection gauges provide accurate information to maintain quality standards [33].

G. *Industrial Engineering and Process Design*

Industrial Engineers and Process Designers collaborate with Product Design engineers to develop realistic specifications for manufacturing processes. They are responsible for selecting appropriate technology, equipment, and work methods that can consistently produce quality products [34].

H. *Finished Goods Inspection and Test*

Final product inspection serves to assess manufacturing quality, identify and resolve any issues, and prevent defective items from reaching customers. When quality is built into the product effectively, such inspections become unnecessary except for auditing purposes and functional testing. Specialized tests like burn-in tests for electronic components ensure proper functionality and eliminate short-life items [35].

I. *Packaging, shipping, and warehousing*

Packaging, shipping, and warehousing activities are crucial for preserving product quality after production. They help protect goods from damage during transit and ensure accurate labeling and proper storage. These logistics functions contribute to maintaining product quality until it reaches the customer [36, 37].

J. *Installation and service*

Proper installation and operation are critical for customers to derive benefits from the product. Clear instructions and comprehensive after-sales service support customer satisfaction in case of any issues or concerns [38].

K. *Business Support Systems*

Various supporting activities within a business organization contribute to achieving quality standards:

- Finance and Accounting: Finance ensures adequate budget allocation for quality programs, while Accounting provides quality cost information for management decision-making [39].

- Human Resources Management: Proper training and motivation of employees are essential for quality work. Education on the importance of quality, as well as additional training in problem-solving, communication, and interpersonal skills, are provided to managers and employees at all levels [40].

4 Quality in Production Systems

According to the definition and from Fig. 6, service refers to a social interaction that occurs directly between customers and representatives of a service company. Service organizations encompass various sectors, excluding industries like agriculture, mining, and construction. They primarily engage in providing personal, business, and government services to establishments. This includes lodging, personal care, repair, and amusement services, as well as professional services such as healthcare, legal, and engineering. Additionally, educational institutions, membership organizations, miscellaneous services, real estate, financial services, retailers, transportation, and public utilities are commonly included in this category.

In many organizations, the production of goods and services goes hand in hand. For example, a fast-food restaurant produces tangible goods like burgers and sandwiches, but it differentiates itself based on the speed and quality of customer service. Such organizations are typically classified as service-oriented rather than manufacturing-focused, as they primarily compete based on the delivery of exceptional service. Customers visit these establishments not just for the food but also for the convenience and experience of being served efficiently and promptly. It is essential to recognize that service is also a key element for many manufacturing companies. For instance, a computer equipment manufacturer may provide extensive maintenance and consulting services.

Quality management methods were initially developed and honed within the manufacturing industry. However, the introduction and adoption of quality assurance programs in the service sector have been slower, lagging manufacturing by approximately a decade. Service organization managers traditionally assumed that their services were acceptable if customer complaints were infrequent [41].

Only recently have they realized that service quality can be managed as a competitive advantage. Research has explored the role of quality practices in service organizations. It has been observed that customer orientation tends to be relatively more

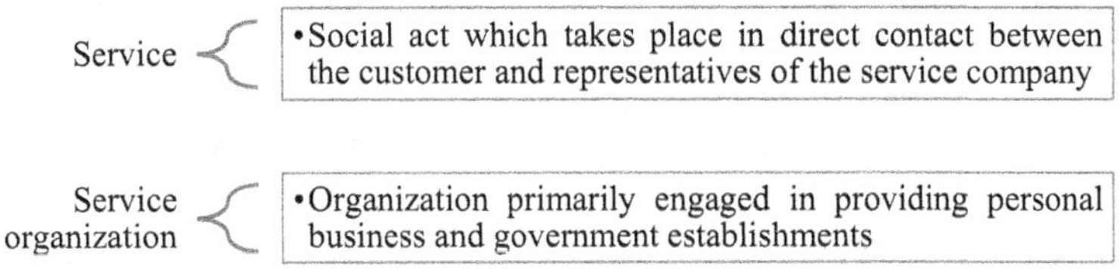

Fig. 6 Service and service organization

crucial for small organizations, while process orientation holds greater importance for larger service organizations.

Given the differences in characteristics between small and large businesses, the implementation and application of quality management must be tailored accordingly. Smaller organizations heavily rely on a stable customer base that values personalized services and products, thus placing a higher priority on immediate customer satisfaction. On the other hand, larger organizations prioritize process orientation as a quality practice. They aim to attract and retain customers through well-structured and efficient systems. In large organizations, more attention is given to developing effective processes that enhance overall efficiency. As the relationship between quality practices and business performance varies with firm size, the quality strategy of an organization must evolve as it grows [42].

5 Differences Between Manufacturing and Services

Multiple research studies have been conducted to explore the impact of quality management on business performance by explicitly comparing manufacturing and service organizations. These studies aim to assist managers in implementing quality management more effectively. The production of services differs from manufacturing in various ways. In service systems, the output is typically intangible, such as an activity, while manufacturing produces tangible and visible products. The intangibility of services often poses challenges for customers in evaluating and comparing services. As a result, customers may rely on price as a measure of quality and place greater emphasis on personal sources of information [43].

Service organizations handle numerous transactions, increasing the potential for errors. Services are consumed as they are created and cannot be stored, eliminating the opportunity for inspection as a means of quality control. Service managers must ensure sufficient service delivery opportunities at appropriate times. Services generally require more labor-intensive efforts compared to manufacturing, where capital plays a more prominent role. In the service sector, factors like the performance of doctors, nurses, and other staff members significantly impact patient care. Consequently, behavioral aspects of management, including motivation, become critical in ensuring service quality. Information technology advancements have reduced labor intensity in some service operations, yet automation can have drawbacks, as personal interaction with customers may diminish, potentially decreasing customer satisfaction.

The service delivery system is often time-sensitive, and identifying customer needs and performance standards can be challenging. The completion of service often necessitates interaction between the customer and the service worker, but that is not always the case. In a fast-food restaurant, for instance, customers place orders, carry their food to the table, and may even clear the table when finished. Therefore, the production of services typically requires a higher degree of customization compared to the production of manufactured goods. Professionals such as doctors, lawyers, and food service employees must tailor the services provided to each individual customer's specific requirements and preferences [43].

The differences between manufacturing and services have significant implications for quality management as shown in Table 1. In service-based industries, quality relies heavily on the individuals delivering the service, the equipment utilized, the available data, and the decisions made throughout the process. Similar to manufactured goods, service quality must be designed and incorporated into the service production, aligning with the specific requirements of customers [44].

For instance, let us consider a metropolitan transit agency. The initial planning function of the agency is responsible for understanding the transportation needs of the community. These needs are then translated into specific product specifications, such as transit routes and schedules for buses or rapid transit systems. These specifications serve as the foundation for defining the desired quality of service. It is through the implementation of appropriate vehicles, equipment, maintenance policies, personnel training, and adherence to schedules that the desired quality of service is ultimately achieved [45].

The importance of quality in services is highlighted by the following points from reference [46]:

- Only 4% of dissatisfied customers voice their complaints, meaning the majority, 96%, do not communicate their issues to the company. For each complaint received, there are 26 other customers with issues, six of which are significant.
  - More than half of the customers who do complaints will return to do business with the company if their complaints are addressed. This number rises to 95% if the complaints are resolved swiftly.
  - A customer who encounters a problem will share their negative experience with 9 or 10 other people. In contrast, those whose complaints are resolved satisfactorily tend to share their experience with only about five other individuals.

As it has been described, the principles of quality that are applicable to manufacturing are equally relevant to services. The inherent nature of services implies a fundamental

Table 1 Comparison between manufacturing and service

Manufacturing	Service
Tangible output	Generally intangible
Can be inventoried	Consumed as created
Capital intensive	Labor intensive
Behavioral aspects of management are not critical	Behavioral aspects like motivation are more critical
Requires less customization	A lot of interactions between customers and service
-	Delivery is often time-sensitive

requirement to cater to the specific needs of the customer, ensuring that the service is indeed fit for use. Just as in manufacturing, standards of conformance need to be established for the service product. Consequently, evaluating conformance to these standards involves assessing how well the service is delivered in comparison to how it was intended to be delivered. Ensuring that conformance standards align with the concept of "Fitness for use" is of paramount importance. By doing so, organizations can strive to deliver services that meet or exceed customer expectations and requirements.

Service quality encompasses distinctive features and dimensions that may differ from those applicable to manufacturing. Some of the key distinguishing factors, as in [47], include:

- Time: How long does a customer have to wait for service?
- Timeliness: Will the service or package be delivered within the expected timeframe?
- Completeness: Is everything included in the order or service?
- Courtesy: Do the employees interact with and treat every customer politely?
- Consistency: Is the service delivered consistently to every customer
- Accessibility and convenience: Is the service easily accessible and convenient for customers?
- Accuracy: Is the service performed correctly and accurately the first time?
- Responsiveness: Can service personnel respond promptly to unexpected issues or problems?

These dimensions encapsulate the distinctive facets of service quality and underscore the significance of elements such as punctual delivery, service completeness, polite interactions, uniformity in service provision, accessibility, precision, and responsiveness. By focusing on these dimensions, service organizations can strive to meet customer expectations and enhance overall service quality.

Numerous service-oriented organizations, including airlines, banks, and hotels, have established comprehensive quality systems. However, it is worth noting that many of these systems are primarily derived from industrial models, emphasizing a product-centric approach rather than being tailored to the unique characteristics of services. Take the example of a typical hotel's quality system, which often prioritizes technical criteria like well-maintained rooms.

Nonetheless, service organizations have distinct needs that extend beyond product-focused considerations. Service systems necessitate a broader scope that encompasses customer service interactions and the performance and behavior of employees, recognizing the vital role they play in delivering exceptional service experiences. By adopting a service-oriented perspective, organizations can enhance their quality systems to address the specific requirements of service delivery effectively.

Implementing a quality system in a service organization requires careful consideration of several key points:

- The selection of quality characteristics may not be straightforward. Customer perceptions play a crucial role, and defining their expectations can be challenging.

For instance, the concept of speed can vary significantly across different service organizations, such as restaurants, making it essential to understand and align with customer preferences [48].

- Behavior is a critical quality characteristic. Human interaction profoundly impacts every service transaction involving customer contact. For example, banks have recognized that the friendliness of their tellers influences customer loyalty and retention [49].
- Image is a vital quality characteristic. The perception of a service's image greatly influences customer expectations and the standards by which they evaluate the service. A breakdown in the image can be as detrimental as a failure in service delivery itself [50].
- Establishing service levels can be complex. Many quality characteristics in services are not easily measurable. In such cases, standards often need to be set subjectively and continuously tested to ensure satisfactory levels [51].
- Measures of system efficiency differ. Unlike manufacturing systems, which can rely on tangible metrics like output per period and scrap/rework rates, service systems often require consideration of customer attitudes and employee competence, as these indicators provide insights into efficiency [52].
- Quality control may be required in the absence of direct supervision. Service work is often performed based on customer convenience, which means that employees must possess the necessary training and skills for self-management [53]. By addressing these considerations, service organizations can develop effective quality systems that align with customer expectations, prioritize human interactions, uphold a positive image, establish appropriate service levels, measure efficiency accurately, and ensure quality control in diverse operational scenarios.

6 Quality and Productivity

From the reference [54], the relationship between quality and productivity can be a source of confusion for managers. Traditionally, quality and productivity were seen as conflicting objectives. However, the modern perspective recognizes that improved quality leads to improved productivity, and vice versa, as they are interconnected and mutually enhancing.

Productivity is a measure of how effectively an organization utilizes its resources. In recent years, European manufacturers have been particularly concerned about productivity, as their growth has fallen behind that of countries in the Far East. The productivity equation is typically defined as:

$$\text{Productivity} = \frac{\text{Output}}{\text{Input}} \tag{1}$$

The input typically includes factors such as labor, capital, materials, and energy. When the output increases while the input remains constant or when the input decreases while the output remains constant, productivity increases.

$$\frac{\text{Output} \uparrow}{\text{Input}} = \text{Productivity} \uparrow \tag{2}$$

It is important to note that producing a poor-quality product requires the same number of resources as producing a good-quality one. Investing more labor in reworking low-quality products increases the denominator of the productivity measure while producing more scrap decreases the numerator. Therefore, productivity can experience a significant boost if products are made correctly the first time. However, it is essential to recognize that high productivity does not guarantee good quality. Some managers may manipulate productivity measures by reporting total output instead of focusing solely on the output of good quality products, which can mask underlying quality issues [55].

The term "Hidden Factory" is commonly used to describe the portion of a plant's capacity that is dedicated to reworking unsatisfactory parts, addressing product recalls, or re-inspecting rejected units. This hidden factory consumes a considerable number of resources and keeps valuable capacity away from productive activities. It results in unnecessary equipment purchases and diminishes the return on investment [56].

Studies indicate that the Hidden Factory can occupy anywhere from 15 to 40% of a plant's capacity. Eliminating these unnecessary operations can lead to significant improvements in productivity.

One approach that addresses the Hidden Factory is Six Sigma, which focuses on identifying and eliminating the root causes of such activities. By employing Six Sigma methodologies, organizations can systematically identify inefficiencies and defects, implement process improvements, and enhance overall productivity [57].

Furthermore, there are other strategies that have been taken to improve organizational efficiency and combat the Hidden Factory. For instance, improvements in production scheduling can optimize resource allocation and reduce wasted time and materials [58].

Additionally, fostering employee involvement and empowering them to contribute ideas for process improvement can unlock hidden potential and drive productivity gains. When employees are actively engaged in the improvement process, they become stakeholders in the success of the organization and take ownership of their work. In addition, they bring valuable frontline insights and firsthand experience to the table. As they work closely with the processes daily, they often have a deep understanding of inefficiencies, bottlenecks, and areas for improvement. It has been observed that organizations with higher levels of employee engagement outperform their competitors in terms of profitability [59].

Productivity measures function as useful indicators for managers to pinpoint areas for improvement and boost overall productivity. There are several approaches that can be taken to achieve this:

- Improving efficiency by reducing total operating costs, optimizing machine time, and minimizing waste. This can be accomplished through process optimization, lean practices, and practical resource utilization [60].
- Enhancing effectiveness through better decision-making and communication. Clear and timely communication channels, efficient coordination, and informed decision-making, contribute to improved productivity [61].
- Striving for higher performance by increasing the quality of outputs, reducing accidents and downtime, and minimizing equipment breakdowns. This can be achieved through quality control measures, preventive maintenance, and safety protocols, as in [62].
- Cultivating better organizational health by focusing on employee morale, job satisfaction, and cooperation. Creating a positive work environment, fostering employee engagement, and promoting teamwork can enhance productivity levels [63].

While technological advancements such as automation and specialization are often associated with productivity improvements, it is essential to recognize that they are not standalone solutions. Quality improvement plays a significant role in boosting productivity by reducing defective output and the effort required for rework [55].

Many business decisions have a dual impact on both productivity and quality. For instance, a well-designed product that is easy to manufacture can lead to fewer defects and increased productivity. Similarly, streamlining production processes by eliminating unnecessary operations and optimizing workflows can enhance both productivity and quality by minimizing the chances of errors and inefficiencies [64].

When examining productivity from a nuanced perspective, it may initially appear that quality improvements can result in lower productivity levels. Let us consider the example of a financial services employee responsible for transcribing applications into a computer database. A standard productivity measure for this task might be the "number of applications per hour." If we measure quality by the rate of transcription errors, it becomes evident that there is an inverse relationship between speed and error rate. Therefore, efforts to enhance quality by reducing time pressures or implementing self-checking activities will seemingly decrease productivity as conventionally measured. However, managers who solely focus on this analysis are missing the broader picture. Increased errors in the transcription process can lead to reduced customer satisfaction, a rise in rework, heightened inspection efforts, and diminished employee morale. In the long run, the costs associated with these activities can outweigh the costs of maintaining a lower level of productivity. This highlights the importance of considering quality in conjunction with productivity.

As has been stated throughout the research, while quality improvements may initially impact productivity metrics, they have the potential to generate significant long-term benefits. By reducing errors, organizations can enhance customer satisfaction, streamline processes, and mitigate the need for excessive inspection and rework. Moreover, a focus on quality can foster a culture of continuous improvement, empowering employees to take ownership of their work and contribute to innovative solutions. To capture the full extent of the relationship between quality and productivity, it is essential for managers to adopt a holistic approach. They should consider not only immediate productivity metrics but also the long-term implications of quality on customer satisfaction, cost efficiency, employee engagement, and overall organizational performance [65].

Embracing quality as an integral part of productivity management enables businesses to strike a balance that ensures sustainable success and customer-centric outcomes. By addressing quality issues at their source, organizations can reduce errors, enhance efficiency, and optimize resource utilization. The resultant increase in productivity positively impacts the bottom line while promoting customer satisfaction and operational effectiveness [66, 67].

7 Conclusion

Our comprehensive investigation underscores the intrinsic link between quality and productivity that permeates both manufacturing and service sectors. As detailed throughout this research, the paradigms of quality management have evolved significantly, from the foundational insights of pioneers like Frederick W. Taylor to contemporary holistic approaches that integrate quality with every organizational facet. This study reaffirms that enduring success hinges on a robust quality management system that extends beyond mere compliance to encompass a deeper, more strategic integration with business processes. For manufacturing entities, this involves a meticulous focus on every step of the production process—from initial design to final output—ensuring that quality is not merely an endpoint but a pervasive attribute. In service sectors, quality management translates into optimizing customer interactions and ensuring service excellence that aligns with customer expectations and needs. The remarkable achievements of Japanese manufacturing throughout history serve as a testament to the transformative power of strategic quality management in securing a competitive edge. Japanese corporations' strategic implementation of Just-in-Time (JIT) and Total Quality Management (TQM) methodologies has revolutionized their operational efficiency and set international standards, influencing quality practices globally.

Additionally, the critical integration of technological advancements with quality management practices has become a cornerstone in the current competitive environment. To stay relevant and efficient, organizations must integrate these tools comprehensively. This integration helps maintain high-quality standards while boosting operational efficiency and agility to meet evolving market demands. This research enriches the existing discourse on quality management by blending historical insights with contemporary implications. This research accentuates the critical need for an anticipatory approach to quality management, essential for preserving a competitive advantage and attaining organizational success. As enterprises confront the challenges of global markets and rapid technological evolutions, the core tenets of quality management become indispensable in directing their paths toward sustained achievement.

References

1. Kaplan RS, Cooper R (1998) Cost and effect: Using integrated cost systems to drive profitability and performance. Harvard Business School Press, Boston
2. Tanenbaum M (1993) The historical evolution of U.S. telecommunications. Technol Soc 15(3):263–272
3. Hacquebord H, Berg P (1980) Quality assurance: Its history and meaning. S Afr J Bus Management 11:134–140
4. Montgomery DC, Woodall WH (1997) A discussion of statistically based process monitoring and control. J Qual Technol 29(2):121–162
5. Huggins LP (1998) Total quality management and the contributions of A.V. Feigenbaum. J Manag Hist (Arch) 4(1):60–67
6. Oakes F (1987) The Japan syndrome. Part 2: The Japanese manufacturer and his society. Electron Power 33(10):619–623
7. Matsui Y (2007) An empirical analysis of just-in-time production in Japanese manufacturing companies. Int J Prod Econ 108:153–164
8. Fullerton R, McWatters CH (2001) The production performance benefits from JIT implementation. J Oper Manag 19(1):81–96
9. Fisher N (2009) Homer sarasohn and american involvement in the evolution of quality management in Japan, 1945–1950. Int Stat Rev 77:276–299
10. Garvin DA (1987) Competing in the eight dimensions of quality. Harv Bus Rev:101–109
11. Tobin RJ (1982) Recalls and the remediation of hazardous or defective consumer products: the experiences of the consumer product safety commission and the national highway traffic safety administration. J Consum Aff 16:278–306
12. Dobyns L, Frank R (1980) If Japan can, Why can't we? NBC White Paper
13. Zairi M, Youssef M (1995) Benchmarking critical factors for TQM: part I: theory and foundations. Benchmarking Qual Manag & Technology 2:5–20
14. Conti TA (2007) A history and review of the European quality award model. The TQM Magaz:112–128
15. Ernst, Young (1991) American quality foundation. International quality study: The definitive study of best international quality management practices, Cleveland, Ohio
16. Ashley S (1992) U.S. quality improves but Japan still leads. Mech Eng-CIME, 114(12):24+
17. Johansson JK, Yip GS (1994) Exploiting globalization potential: U.S. and Japanese strategies. Strateg Manag J 15(8):579–601
18. Weckenmann A, Akkasoglu G, Werner T (2015) Quality management—History and trends. The TQM J 27:281–293
19. Dooley K (2000) The paradigms of quality: evolution and revolution in the history of the discipline. Adv Manag Organ Qual 5, JAI Press
20. Sitkin S, Sutcliffe KM, Schroeder RG (1994) Distinguishing control from learning in total quality management: A contingency perspective. Acad Manag Rev 19(3):537–564
21. Stephen G, Green G, Welsh MA (1988) Cybernetics and dependence: Reframing the control concept. Acad Manag Rev 13(2):287–301
22. Duflou JR, Sutherland JW et al (2012) Towards energy and resource efficient manufacturing: A processes and systems approach. CIRP Ann 61(2):587–609

23. Campbell RI, De Beer DJ, Barnard LJ, Booysen GJ, Truscott M, Cain R, Burton MJ, Gyi DE, Hague R (2007) Design evolution through customer interaction with functional prototypes. J Eng Des 18(6):617–635
24. Spring M, Mc Quarter R, Swift K, Dale B, Booker J (1998) The use of quality tools and techniques in product introduction: an assessment methodology. The TQM Magaz 10(1):45–50
25. Zhu Y, Alard R, You J, Schonslebe P (2011) Collaboration in the design-manufacturing chain: A key to improve product quality [Internet]. Supply Chain Management—New Perspectives. In Tech. https://doi.org/10.5772/18694
26. Hale AR, Kirwan B, Kjellén U (2017) Safe by design: where are we now? Saf Sci 45:305–327
27. Meyer C, Schwager A (2007) Understanding customer experience. Harv Bus Rev 85(2):1–11
28. Baboolal-Frank R (2021) Analysis of Amazon: customer centric approach. https://repository.up.ac.za/handle/2263/86089
29. Jiao J, Chen CH (2006) Customer requirement management in product development: A review of research issues. Concurr Eng: Res & Appl 14:173–185
30. Ting S, Cho DI An integrated approach for supplier selection and purchasing decisions. Supply Chain Manag, 13(2):116–127. https://doi.org/10.1108/13598540810860958
31. Kiran DR (2019) Production planning and control: A comprehensive approach. ISBN-13. 978–128183649. Edition. 1st. Publisher. Butterworth-Heinemann Ed
32. Beiter KA, Cheldelin B, Ishii K (2000) Assembly quality method: A tool in aid of product strategy, design, and process improvements. In: Proceedings of the ASME 2000 international design engineering technical conferences and computers and information in engineering conference. Volume 3: 5th Design for Manufacturing Conference. Baltimore, Maryland, USA, pp 149–156. ASME
33. Umble M, Umble E, Murakami S (2006) Implementing theory of constraints in a traditional Japanese manufacturing environment: The case of hitachi tool engineering. Int J Prod Res 44(10):1863–1880
34. Antony J, Jiju Antony F (2001) Teaching the taguchi method to industrial engineers. Work Study, 50(4):141–149. https://doi.org/10.1108/00438020110391873
35. Budi A, Ruspendi R (2021) Design of finished goods inspection acceleration with QCC method and seven tools to increase productivity. SINTEK JURNAL, Jurnal Ilmiah Teknik Mesin 15(1)
36. De Koster R, Le-Duc T, Roodbergen KJ (2007) Design and control of warehouse order picking: a literature review. Eur J Oper Res 182(2):481–501. ISSN 0377–2217. https://doi.org/10.1016/j.ejor.2006.07.009
37. Mitroussi K (2004) Quality in shipping: IMO's role and problems of implementation. Disaster Prev Manag 13(1):50–58
38. Fan LH, Wei W, Wang W, Chang XR, Yi LJ (2003) Research landscape on job embeddedness and organizational commitment: A bibliometric study. Afr J Bus Manag, 17(6):106–119
39. Tang J, Richard YK, Fung B, Baodong X, Dingwei W (2002) A new approach to quality function deployment planning with financial consideration. Comput Oper Res 29(11):1447–1463
40. Redman T, Mathews BP (1998) Service quality and human resource management: A review and research agenda. Personnel Review, 27(1):57–77
41. Talib F (2013) An overview of total quality management: understanding the fundamentals in service organization. Int J Adv Qual Manag (IJAQM) 1:1–20
42. Gustafsson A, Nilsson L (2003) The role of quality practices in service organizations. Int J Serv Ind Manag 14(2):232–244. https://doi.org/10.1108/09564230310474183
43. Rönnbäck A, Witell L (2008) A review of empirical investigations comparing quality initiatives in manufacturing and service organizations. Manag Serv Qual. 18(6):577–593
44. Schneider B, Bowen DE (1993) The service organization: Human resources management is crucial. Organ Dyn, 21(4):39–52. Spring
45. Teal RF, Becker AJ (2011) Business strategies and technology for access by transit in lower density environments. Res Transp Bus Manag 2:57–64
46. Tatikonda LU (2013) The hidden costs of customer dissatisfaction. Manag Account Q, 14(3):34–43. Spring

47. Aslam S, Najaf M (2017) User perception of quality in service organizations of Pakistan: A case of jazz customer service center. Asia Pac J Emerg Mark 1(2):158–182
48. Chen YH, Tseng ML, Lin RJ (2011) Afr J Bus Manag 5(7):2854–2864
49. Auka D, Bosire J, Matern V (2013) Perceived service quality and customer loyalty in retail banking in Kenya. Br J Mark Stud 1:32–61
50. Raditha H (2018) Creating educational theme park visitor loyalty: the role of experience-based satisfaction, image and Value. Tour Hosp Manag, 10
51. Eboli L, Mazzulla G (2012) Performance indicators for an objective measure of public transport service quality. Eur Transp\Trasp Eur, 51, Paper n° 3: 1–21. ISSN 1825–3997
52. Robert H, Chenhall K, Langfield K Multiple perspectives of performance measures. Eur Manag J, 25(4):266–282
53. Shahin A (2008) The relationship between quality and productivity: A new perspective. Int J Prod Qual Manag 3(2):206–222
54. Steindel CH, Stiroh KJ (2001) Productivity: What is it, and why do we care about it? Electron J Bus Econ 36(4)
55. Sink D, Tuttle TH, Devries S (2006) Productivity measurement and evaluation: What is available? Natl Prod Review 3:265–287
56. McDowall (2021) The hidden factory in your laboratory? LCGG Europe 34(8):326–330
57. Smętkowska M, Mrugalska B (2018) Using six sigma DMAIC to improve the quality of the production process: A case study. Procedia Soc Behav Sci 238:590–596
58. Du J, Xue Y, Sugumaran V, Hu M, Dong P (2022) Improved biogeography-based optimization algorithm for lean production scheduling of prefabricated components. Eng Constr Archit Manag 30(4):1601–1635
59. Sundaray BK (2011) Employee engagement: A driver of organizational effectiveness. Eur J Bus Manag 3(8). www.iiste.org. ISSN 2222–1905 (Paper) ISSN 2222–2839 (Online)
60. Kulkarni P (2014) Productivity improvement through lean Deployment & work study methods. Int J Res Eng Technol 03:429–434
61. Rajhans K (2009) Effective organizational communication: a key to employee motivation and performance. Intersci Manag Rev 2(2):145–149
62. Shagluf A, Longstaff AP, Fletcher S (2014) Maintenance strategies to reduce downtime due to machine positional errors. In: Proceedings of Maintenance Performance Measurement and Management. MPMM. Department of Mechanical Engineering. Coimbra, Portugal, pp 111–118. ISBN 978–972–8954–42–0. Conference paper
63. Mughal MU (2020) The impact of leadership, teamwork and employee engagement on employee performances. Saudi J Bus Manag Stud 5:233–244
64. Sahin F (2000) Manufacturing competitiveness: Different systems to achieve the same results. Prod Inven Manag J; Alex 41(1):56–65. First Quarter
65. Aaker DA, Jacobson R The financial information content of perceived quality. J Mark Res. https://doi.org/10.1177/002224379403100204
66. Allen DR, Wilburn M (2002) Linking customer and employee satisfaction to the bottom line" A comprehensive guide to establishing the impact of customer and employee satisfaction on critical business outcomes. Quality Press. Milwaukee, American Society for Quality
67. Gurung NM (2023) The factors affecting employee satisfaction and employee retention. The hotel industry in Chiangmai, Thailand. J Posit Sch Psychol 7(1):998–1008. http://journalppw.com

Author Index

N. M. Durakbasa and K. G. Gülen (eds.), *Sustainable Green Conversion*, Lecture Notes in Mechanical Engineering, https://doi.org/10.1007/978-3-031-83583-4

The manufacturer's authorised representative in the EU is Springer Nature Customer Service Centre GmbH, Europaplatz 3, 69115 Heidelberg, Germany. If you have any concerns regarding our products, please contact ProductSafety@springernature.com

Printed and bound by CPI Group (UK) Ltd, Croydon, CR0 4YY
01/10/2026
02225959-0011